D0900324

Springer Finance

Editorial Board
M. Avellaneda
G. Barone-Adesi
M. Broadie
M.H.A. Davis
E. Derman
C. Klüppelberg
E. Kopp
W. Schachermayer

Springer Finance

Springer Finance is a programme of books addressing students, academics and practitioners working on increasingly technical approaches to the analysis of financial markets. It aims to cover a variety of topics, not only mathematical finance but foreign exchanges, term structure, risk management, portfolio theory, equity derivatives, and financial economics.

Ammann M., Credit Risk Valuation: Methods, Models, and Application (2001)
Back K., A Course in Derivative Securities: Introduction to Theory and Computation (2005)
Barucci E., Financial Markets Theory. Equilibrium, Efficiency and Information (2003)
Bielecki T.R. and Rutkowski M., Credit Risk: Modeling, Valuation and Hedging (2002)
Bingham N.H. and Kiesel R., Risk-Neutral Valuation: Pricing and Hedging of Financial Derivatives (1998, 2nd ed. 2004)
Brigo D. and Mercurio F., Interest Rate Models: Theory and Practice (2001, 2nd ed. 2006)
Buff R., Uncertain Volatility Models –Theory and Application (2002)
Carmona R.A. and Tehranchi M.R., Interest Rate Models: An Infinite Dimensional Stochastic Analysis Perspective (2006)
Dana R.-A. and Jeanblanc M., Financial Markets in Continuous Time (2003)
Deboeck G. and Kohonen T. (Editors), Visual Explorations in Finance with Self-Organizing Maps (1998)
Delbaen F. and Schachermayer W., The Mathematics of Arbitrage (2005)
Elliott R.J. and Kopp P.E., Mathematics of Financial Markets (1999, 2nd ed. 2005)
Fengler M.R., Semiparametric Modeling of Implied Volatility (2005)
Filipovic D., Term-Structure Models (2008 forthcoming)
Fusai G. and Roncoroni A., Implementing Models in Quantitative Finance (2008)
Geman H., Madan D., Pliska S.R. and Vorst T. (Editors), Mathematical Finance – Bachelier Congress 2000 (2001)
Gundlach M., Lehrbass F. (Editors), CreditRisk$^+$ in the Banking Industry (2004)
Jondeau E., Financial Modeling Under Non-Gaussian Distributions (2007)
Kabanov, Y.A. and Safarian M., Markets with Transaction Costs (2008 forthcoming)
Kellerhals B.P., Asset Pricing (2004)
Külpmann M., Irrational Exuberance Reconsidered (2004)
Kwok Y.-K., Mathematical Models of Financial Derivatives (1998, 2nd ed. 2008)
Malliavin P. and Thalmaier A., Stochastic Calculus of Variations in Mathematical Finance (2005)
Meucci A., Risk and Asset Allocation (2005)
Pelsser A., Efficient Methods for Valuing Interest Rate Derivatives (2000)
Prigent J.-L., Weak Convergence of Financial Markets (2003)
Schmid B., Credit Risk Pricing Models (2004)
Shreve S.E., Stochastic Calculus for Finance I (2004)
Shreve S.E., Stochastic Calculus for Finance II (2004)
Yor M., Exponential Functionals of Brownian Motion and Related Processes (2001)
Zagst R., Interest-Rate Management (2002)
Zhu Y.-L., Wu X. and Chern I.-L., Derivative Securities and Difference Methods (2004)
Ziegler A., Incomplete Information and Heterogeneous Beliefs in Continuous-time Finance (2003)
Ziegler A., A Game Theory Analysis of Options (2004)

Yue-Kuen Kwok

Mathematical Models of Financial Derivatives

Second Edition

Yue-Kuen Kwok
Department of Mathematics
Hong Kong University of Science & Technology
Clear Water Bay
Kowloon
Hong Kong/PR China
maykwok@ust.hk

ISBN 978-3-540-42288-4 e-ISBN 978-3-540-68688-0

Springer Finance ISSN 1616-0533

Library of Congress Control Number: 2008924369

Mathematics Subject Classification (2000): 60Hxx, 62P05, 90A09, 91B28, 91B50
JEL Classification: G12, G13

© Springer Berlin Heidelberg 2008
Revised and enlarged 2nd edition to the 1st edition originally published by Springer Singapore 1998 (ISBN 981-3083-25-5).

This work is subject to copyright. All rights are reserved, whether the whole or part of the material is concerned, specifically the rights of translation, reprinting, reuse of illustrations, recitation, broadcasting, reproduction on microfilm or in any other way, and storage in data banks. Duplication of this publication or parts thereof is permitted only under the provisions of the German Copyright Law of September 9, 1965, in its current version, and permission for use must always be obtained from Springer. Violations are liable to prosecution under the German Copyright Law.

The use of general descriptive names, registered names, trademarks, etc. in this publication does not imply, even in the absence of a specific statement, that such names are exempt from the relevant protective laws and regulations and therefore free for general use.

Cover design: WMX Design GmbH, Heidelberg
The cover design is based on a photograph by Stefanie Zöller

Printed on acid-free paper

9 8 7 6 5 4 3 2 1

springer.com

To my wife Oi Chun,
our two daughters Grace and Joyce

Preface

Objectives and Audience

In the past three decades, we have witnessed the phenomenal growth in the trading of financial derivatives and structured products in the financial markets around the globe and the surge in research on derivative pricing theory. Leading financial institutions are hiring graduates with a science background who can use advanced analytical and numerical techniques to price financial derivatives and manage portfolio risks, a phenomenon coined as *Rocket Science on Wall Street*. There are now more than a hundred Master level degreed programs in Financial Engineering/Quantitative Finance/Computational Finance in different continents. This book is written as an introductory textbook on derivative pricing theory for students enrolled in these degree programs. Another audience of the book may include practitioners in quantitative teams in financial institutions who would like to acquire the knowledge of option pricing techniques and explore the new development in pricing models of exotic structured derivatives. The level of mathematics in this book is tailored to readers with preparation at the advanced undergraduate level of science and engineering majors, in particular, basic proficiencies in probability and statistics, differential equations, numerical methods, and mathematical analysis. Advance knowledge in stochastic processes that are relevant to the martingale pricing theory, like stochastic differential calculus and theory of martingale, are introduced in this book.

The cornerstones of derivative pricing theory are the Black–Scholes–Merton pricing model and the martingale pricing theory of financial derivatives. The renowned risk neutral valuation principle states that the price of a derivative is given by the expectation of the discounted terminal payoff under the risk neutral measure, in accordance with the property that discounted security prices are martingales under this measure in the financial world of absence of arbitrage opportunities. This second edition presents a substantial revision of the first edition. The new edition presents the theory behind modeling derivatives, with a strong focus on the martingale pricing principle. The continuous time martingale pricing theory is motivated through the analysis of the underlying financial economics principles within a discrete time framework. A wide range of financial derivatives commonly traded in the equity and fixed income markets are analyzed, emphasizing on the aspects of pricing, hedging, and their risk management. Starting from the Black–Scholes–Merton formulation of the option pricing model, readers are guided through the book on the new advances in the state-of-the-art derivative pricing models and interest rate models. Both analytic techniques and numerical methods for solving various types of derivative pricing models are emphasized. A large collection of closed form price formulas of various exotic path dependent equity options (like barrier options, lookback options, Asian options, and American options) and fixed income derivatives are documented.

Guide to the Chapters

This book contains eight chapters, with each chapter being ended with a comprehensive set of well thought out exercises. These problems not only provide the stimulus for refreshing the concepts and knowledge acquired from the text, they also help lead the readers to new research results and concepts found scattered in recent journal articles on the pricing theory of financial derivatives.

The first chapter serves as an introduction to the basic derivative instruments, like the forward contracts, options, and swaps. Various definitions of terms in financial economics, say, self-financing strategy, arbitrage, hedging strategy are presented. We illustrate how to deduce the rational boundaries on option values without any distribution assumptions on the dynamics of the price of the underlying asset.

In Chap. 2, the theory of financial economics is used to show that the absence of arbitrage is equivalent to the existence of an equivalent martingale measure under the discrete securities models. This important result is coined as the Fundamental Theorem of Asset Pricing. This leads to the risk neutral valuation principle, which states that the price of an attainable contingent claim is given by the expectation of the discounted value of the claim under a risk neutral measure. The concepts of attainable contingent claims, absence of arbitrage and risk neutrality form the cornerstones of the modern option pricing theory. Brownian processes and basic analytic tools in stochastic calculus are introduced. In particular, we discuss the Feynman–Kac representation, Radon–Nikodym derivative between two probability measures and the Girsanov theorem that effects the change of measure on an Ito process.

Some of the highlights of the book appear in Chap. 3, where the Black–Scholes–Merton formulation of the option pricing model and the martingale pricing approach of financial derivatives are introduced. We illustrate how to apply the pricing theory to obtain the price formulas of different types of European options. Various extensions of the Black–Scholes–Merton framework are discussed, including the transaction costs model, jump-diffusion model, and stochastic volatility model.

Path dependent options are options with payoff structures that are related to the path history of the asset price process during the option's life. The common examples are the barrier options with the knock-out feature, the Asian options with the averaging feature, and the lookback options whose payoff depends on the realized extremum value of the asset price process. In Chap. 4, we derive the price formulas of the various types of European path dependent options under the Geometric Brownian process assumption of the underlying asset price.

Chapter 5 is concerned with the pricing of American options. We present the characterization of the optimal exercise boundary associated with the American option models. In particular, we examine the behavior of the exercise boundary before and after a discrete dividend payment, and immediately prior to expiry. The two common pricing formulations of the American options, the linear complementarity formulation and the optimal stopping formulation, are discussed. We show how to express the early exercise premium in terms of the exercise boundary in the form of an integral representation. Since analytic price formulas are in general not available for American options, we present several analytic approximation methods for pricing American options. We also consider the pricing models for the American barrier options, the Russian option and the reset-strike options.

Since option models which have closed price formulas are rare, it is common to resort to numerical methods for valuation of option prices. The usual numerical approaches in option valuation are the lattice tree methods, finite difference algorithms, and Monte Carlo simulation. The primary essence of the lattice tree methods is the simulation of the continuous asset price process by a discrete random walk model. The finite difference approach seeks the discretization of the differential operators in the Black–Scholes equation. The Monte Carlo simulation method provides a probabilistic solution to the option pricing problems by simulating the random process of the asset price. An account of option pricing algorithms using these approaches is presented in Chap. 6.

Chapter 7 deals with the characterization of the various interest rate models and pricing of bonds. We start our discussion with the class of one-factor short rate models, and extend to multi-factor models. The Heath–Jarrow–Morton (HJM) approach of modeling the stochastic movement of the forward rates is discussed. The HJM methodologies provide a uniform approach to modeling the instantaneous interest rates. We also present the formulation of the forward LIBOR (London-Inter-Bank-Offered-Rate) process under the Gaussian HJM framework.

The last chapter provides an exposition on the pricing models of several commonly traded interest rate derivatives, like the bond options, range notes, interest rate caps, and swaptions. To facilitate the pricing of equity derivatives under stochastic interest rates, the technique of the forward measure is introduced. Under the forward measure, the bond price is used as the numeraire. In the pricing of the class of LIBOR derivative products, it is more effective to use the LIBORs as the underlying state variables in the pricing models. To each forward LIBOR process, the Lognormal LIBOR model assigns a forward measure defined with respect to the settlement date of the forward rate. Unlike the HJM approach which is based on the non-observable instantaneous forward rates, the Lognormal LIBOR models are based on the observable market interest rates. Similarly, the pricing of a swaption can be effectively performed under the Lognormal Swap Rate model, where an annuity (sum of bond prices) is used as the numeraire in the appropriate swap measure. Lastly, we consider the hedging and pricing of cross-currency interest rate swaps under an appropriate two-currency LIBOR model.

Acknowledgement

This book benefits greatly from the advice and comments from colleagues and students through the various dialogues in research seminars and classroom discussions. Some of materials used in the book are outgrowths from the new results in research publications that I have coauthored with colleagues and former Ph.D. students. Special thanks go to Lixin Wu, Min Dai, Hong Yu, Hoi Ying Wong, Ka Wo Lau, Seng Yuen Leung, Chi Chiu Chu, Kwai Sun Leung, and Jin Kong for their continuous research interaction and constructive comments on the book manuscript. Also, I would like to thank Ms. Odissa Wong for her careful typing and editing of the manuscript, and her patience in entertaining the seemingly endless changes in the process. Last but not least, sincere thanks go to my wife, Oi Chun and our two daughters, Grace and Joyce, for their forbearance while this book was being written. Their love and care have always been my source of support in everyday life and work.

Final Words on the Book Cover Design

One can find the Bank of China Tower in Hong Kong and the Hong Kong Legislative Council Building in the background underneath the usual yellow and blue colors on the book cover of this Springer text. The design serves as a compliment on the recent acute growth of the financial markets in Hong Kong, which benefits from the phenomenal economic development in China and the rule of law under the Hong Kong system.

Contents

1

Introduction to Derivative Instruments

The past few decades have witnessed a revolution in the trading of *derivative securities* in world financial markets. A financial derivative may be defined as a security whose value depends on the values of more basic underlying variables, like the prices of other traded securities, interest rates, commodity prices or stock indices. The three most basic derivative securities are forwards, options and swaps. A *forward contract* (called a *futures contract* if traded on an exchange) is an agreement between two parties that one party will purchase an asset from the counterparty on a certain date in the future for a predetermined price. An *option* gives the holder the *right* (but not the obligation) to buy or sell an asset by a certain date for a predetermined price. A *swap* is a financial contract between two parties to exchange cash flows in the future according to some prearranged format. There has been a great proliferation in the variety of derivative securities traded and new derivative products are being invented continually over the years. The development of pricing methodologies of new derivative securities has been a major challenge in the field of financial engineering. The theoretical studies on the use and risk management of financial derivatives have become commonly known as the *Rocket Science* on Wall Street.

In this book, we concentrate on the study of pricing models for financial derivatives. Derivatives trading is an integrated part in portfolio management in financial firms. Also, many financial strategies and decisions can be analyzed from the perspective of options. Throughout the book, we explore the characteristics of various types of financial derivatives and discuss the theoretical framework within which the fair prices of derivative instruments can be determined.

In Sect. 1.1, we discuss the payoff structures of forward contracts and options and present various definitions of terms commonly used in financial economics theory, such as self-financing strategy, arbitrage, hedging, etc. Also, we discuss various trading strategies associated with the use of options and their combinations. In Sect. 1.2, we deduce the rational boundaries on option values without any assumptions on the stochastic behavior of the prices of the underlying assets. We discuss how option values are affected if an early exercise feature is embedded in the option contract and dividend payments are paid by the underlying asset. In Sect. 1.3, we consider the pricing of forward contracts and analyze the relation between forward price and futures price under a constant interest rate. The product nature and uses of interest rate swaps and currency swaps are discussed in Sect. 1.4.

1.1 Financial Options and Their Trading Strategies

First, let us define the different terms in option trading. An option is classified either as a call option or a put option. A *call* (or *put*) option is a contract which gives its holder the *right* to buy (or sell) a prescribed asset, known as the *underlying asset*, by a certain date (*expiration date*) for a predetermined price (commonly called the *strike price* or *exercise price*). Since the holder is given the right but not the obligation to buy or sell the asset, he or she will make the decision depending on whether the deal is favorable to him or not. The option is said to be *exercised* when the holder chooses to buy or sell the asset. If the option can only be exercised on the expiration date, then the option is called a *European* option. Otherwise, if the exercise is allowed at any time prior to the expiration date, then the option is called an *American* option (these terms have nothing to do with their continental origins). The simple call and put options with no special features are commonly called *plain vanilla options*. Also, we have options coined with names like *Asian option, lookback option, barrier option*, etc. The precise definitions of these exotic types of options will be given in Chap. 4.

The counterparty to the holder of the option contract is called the *option writer*. The holder and writer are said to be, respectively, in the *long* and *short* positions of the option contract. Unlike the holder, the writer does have an obligation with regard to the option contract. For example, the writer of a call option must sell the asset if the holder chooses in his or her favor to buy the asset. This is a zero-sum game as the holder gains from the loss of the writer or vice versa.

An option is said to be *in-the-money* (*out-of-the-money*) if a positive (negative) payoff would result from exercising the option immediately. For example, a call option is in-the-money (out-of-the-money) when the current asset price is above (below) the strike price of the call. An *at-the-money* option refers to the situation where the payoff is zero when the option is exercised immediately, that is, the current asset price is exactly equal to the option's strike price.

Terminal Payoffs of Forwards and Options

The holder of a forward contract is obligated to buy the underlying asset at the forward price (also called delivery price) K on the expiration date of the contract. Let S_T denote the asset price at expiry T. Since the holder pays K dollars to buy an asset worth S_T, the terminal payoff to the holder (long position) is seen to be $S_T - K$. The seller (short position) of the forward faces the terminal payoff $K - S_T$, which is negative to that of the holder (by the zero-sum nature of the forward contract).

Next, we consider a European call option with strike price X. If $S_T > X$, then the holder of the call option will choose to exercise at expiry T since the holder can buy the asset, which is worth S_T dollars, at the cost of X dollars. The gain to the holder from the call option is then $S_T - X$. However, if $S_T \leq X$, then the holder will forfeit the right to exercise the option since he or she can buy the asset in the market at a cost less than or equal to the predetermined strike price X. The terminal payoff from the long position (holder's position) of a European call is then given by

$$\max(S_T - X, 0).$$

Similarly, the terminal payoff from the long position in a European put can be shown to be

$$\max(X - S_T, 0),$$

since the put will be exercised at expiry T only if $S_T < X$. The asset worth S_T can be sold by the put's holder at a higher price of X under the put option contract. In both call and put options, the terminal payoffs are guaranteed to be nonnegative. These properties reflect the very nature of options: they will not be exercised if a negative payoff results.

Option Premium

Since the writer of an option is exposed to potential liabilities in the future, he must be compensated with an up-front premium paid by the holder when they together enter into the option contract. An alternative viewpoint is that since the holder is guaranteed a nonnegative terminal payoff, he must pay a premium get into the option game. The natural question is: What should be the fair option premium (called the option price) so that the game is fair to both the writer and holder? Another but deeper question: What should be the optimal strategy to exercise prior to expiration date for an American option? At least, the option price is easily seen to depend on the strike price, time to expiry and current asset price. The less obvious factors involved in the pricing models are the prevailing *interest rate* and the degree of randomness of the asset price (characterized by the *volatility* of the stochastic asset price process).

Self-Financing Strategy

Suppose an investor holds a portfolio of securities, such as a combination of options, stocks and bonds. As time passes, the value of the portfolio changes because the prices of the securities change. Besides, the trading strategy of the investor affects the portfolio value by changing the proportions of the securities held in the portfolio, say, and adding or withdrawing funds from the portfolio. An investment strategy is said to be *self-financing* if no extra funds are added or withdrawn from the initial investment. The cost of acquiring more units of one security in the portfolio is completely financed by the sale of some units of other securities within the same portfolio.

Short Selling

Investors buy a stock when they expect the stock price to rise. How can an investor profit from a fall of stock price? This can be achieved by short selling the stock. Short selling refers to the trading practice of borrowing a stock and selling it immediately, buying the stock later and returning it to the borrower. The short seller hopes to profit from a price decline by selling the asset before the decline and buying it back afterwards. Usually, there are rules in stock exchanges that restrict the timing of the short selling and the use of the short sale proceeds. For example, an exchange may impose the rule that short selling of a security is allowed only when the most recent movement in the security price is an uptick. When the stock pays dividends, the short seller has to compensate the lender of the stock with the same amount of dividends.

No Arbitrage Principle

One of the fundamental concepts in the theory of option pricing is the absence of arbitrage opportunities, is called the *no arbitrage principle*. As an illustrative example of an arbitrage opportunity, suppose the prices of a given stock in Exchanges A and B are listed at \$99 and \$101, respectively. Assuming there is no transaction cost, one can lock in a riskless profit of \$2 per share by buying at \$99 in Exchange A and selling at \$101 in Exchange B. The trader who engages in such a transaction is called an *arbitrageur*. If the financial market functions properly, such an arbitrage opportunity cannot occur since traders are well aware of the differential in stock prices and they immediately compete away the opportunity. However, when there is transaction cost, which is a common form of market friction, the small difference in prices may persist. For example, if the transaction costs for buying and selling per share in Exchanges A and B are both \$1.50, then the total transaction costs of \$3 per share will discourage arbitrageurs.

More precisely, an *arbitrage opportunity* can be defined as a self-financing trading strategy requiring no initial investment, having zero probability of negative value at expiration, and yet having some possibility of a positive terminal payoff. More detailed discussions on the "no arbitrage principle" are given in Sect. 2.1.

No Arbitrage Price of a Forward

Here we discuss how the no arbitrage principle can be used to price a forward contract on an underlying asset that provides the asset holder no income in the form of dividends. The forward price is the price the holder of the forward pays to acquire the underlying asset on the expiration date. In the absence of arbitrage opportunities, the forward price F on a nondividend paying asset with spot price S is given by

$$F = Se^{r\tau}, \tag{1.1.1}$$

where r is the constant riskless interest rate and τ is the time to expiry of the forward contract. Here, $e^{r\tau}$ is the growth factor of cash deposit that earns continuously compounded interest over the period τ.

It can be shown that when either $F > Se^{r\tau}$ or $F < Se^{r\tau}$, an arbitrageur can lock in a risk-free profit. First, suppose $F > Se^{r\tau}$, the arbitrage strategy is to borrow S dollars from a bank and use the borrowed cash to buy the asset, and also take up a short position in the forward contract. The loan with loan period τ will grow to $Se^{r\tau}$. At expiry, the arbitrageur will receive F dollars by selling the asset under the forward contract. After paying back the loan amount of $Se^{r\tau}$, the riskless profit is then $F - Se^{r\tau} > 0$. Otherwise, suppose $F < Se^{r\tau}$, the above arbitrage strategy is reversed, that is, short selling the asset and depositing the proceeds into a bank, and taking up a long position in the forward contract. At expiry, the arbitrageur acquires the asset by paying F dollars under the forward contract and closing out the short selling position by returning the asset. The riskless profit now becomes $Se^{r\tau} - F > 0$. Both cases represent arbitrage opportunities. By virtue of the no arbitrage principle, the forward price formula (1.1.1) follows.

One may expect that the forward price should be set equal to the expectation of the terminal asset price S_T at expiry T. However, this expectation approach does not enforce the forward price since the expectation value depends on the forward holder's view on the stochastic movement of the underlying asset's price. The above no arbitrage argument shows that the forward price can be enforced by adopting a certain trading strategy. If the forward price deviates from this no arbitrage price, then arbitrage opportunities arise and the market soon adjusts to trade at the "no arbitrage price".

Volatile Nature of Options

Option prices are known to respond in an exaggerated scale to changes in the underlying asset price. To illustrate this claim, we consider a call option that is near the time of expiration and the strike price is \$100. Suppose the current asset price is \$98, then the call price is close to zero since it is quite unlikely for the asset price to increase beyond \$100 within a short period of time. However, when the asset price is \$102, then the call price near expiry is about \$2. Though the asset price differs by a small amount, between \$98 to \$102, the relative change in the option price can be very significant. Hence, the option price is seen to be more volatile than the underlying asset price. In other words, the trading of options leads to more price action per dollar of investment than the trading of the underlying asset. A precise analysis of the elasticity of the option price relative to the asset price requires detailed knowledge of the relevant pricing model for the option (see Sect. 3.3).

Hedging

If the writer of a call does not simultaneously own a certain amount of the underlying asset, then he or she is said to be in a *naked position* since he or she has no protection if the asset price rises sharply. However, if the call writer owns some units of the underlying asset, the loss in the short position of the call when the asset's price rises can be compensated by the gain in the long position of the underlying asset. This strategy is called *hedging*, where the risk in a portfolio is monitored by taking opposite directions in two securities which are highly negatively correlated. In a *perfect hedge* situation, the *hedger* combines a risky option and the corresponding underlying asset in an appropriate proportion to form a riskless portfolio. In Sect. 3.1, we examine how the *riskless hedging principle* is employed to formulate the option pricing theory.

1.1.1 Trading Strategies Involving Options

We have seen in the above simple hedging example how the combined use of an option and the underlying asset can monitor risk exposure. Now, we would like to examine the various strategies of portfolio management using options and the underlying asset as the basic financial instruments. Here, we confine our discussion of portfolio strategies to the use of European vanilla call and put options. We also assume that the underlying asset does not pay dividends within the investment time horizon.

The simplest way to analyze a portfolio strategy is to construct a corresponding *terminal profit diagram*. This shows the profit on the expiration date from holding the options and the underlying asset as a function of the terminal asset price. This simplified analysis is applicable only to a portfolio that contains options all with the same date of expiration and on the same underlying asset.

Covered Calls and Protective Puts

Consider a portfolio that consists of a short position (writer) in one call option plus a long holding of one unit of the underlying asset. This investment strategy is known as *writing a covered call*. Let c denote the premium received by the writer when selling the call and S_0 denote the asset price at initiation of the option contract [note that $S_0 > c$, see (1.2.12)]. The initial value of the portfolio is then $S_0 - c$. Recall that the terminal payoff for the call is $\max(S_T - X, 0)$, where S_T is the asset price at expiry and X is the strike price. Assuming the underlying asset to be nondividend paying, the portfolio value at expiry is $S_T - \max(S_T - X, 0)$, so the profit of a covered call at expiry is given by

$$\begin{aligned} & S_T - \max(S_T - X, 0) - (S_0 - c) \\ = & \begin{cases} (c - S_0) + X & \text{when } S_T \geq X \\ (c - S_0) + S_T & \text{when } S_T < X. \end{cases} \end{aligned} \tag{1.1.2}$$

Observe that when $S_T \geq X$, the profit is capped at the constant value $(c - S_0) + X$, and when $S_T < X$, the profit grows linearly with S_T. The corresponding terminal profit diagram for a covered call is illustrated in Fig. 1.1. Readers may wonder why $c - S_0 + X > 0$? For hints, see (1.2.3a).

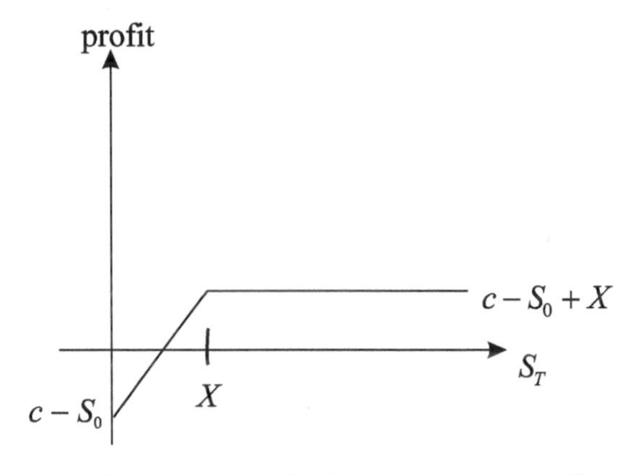

Fig. 1.1. Terminal profit diagram of a covered call.

The investment portfolio that involves a long position in one put option and one unit of the underlying asset is called a *protective put*. Let p denote the premium paid for the acquisition of the put. It can be shown similarly that the profit of the protective put at expiry is given by

$$\begin{aligned} & S_T + \max(X - S_T, 0) - (p + S_0) \\ &= \begin{cases} -(p + S_0) + S_T & \text{when } S_T \geq X \\ -(p + S_0) + X & \text{when } S_T < X. \end{cases} \end{aligned} \tag{1.1.3}$$

Do we always have $X - (p + S_0) < 0$?

Is it meaningful to create a portfolio that involves the long holding of a put and short selling of the asset? This portfolio strategy will have no hedging effect because both positions in the put option and the underlying asset are in the same direction in risk exposure—both positions lose when the asset price increases.

Spreads

A spread strategy refers to a portfolio which consists of options of the same type (that is, two or more calls, or two or more puts) with some options in the long position and others in the short position in order to achieve a certain level of hedging effect. The two most basic spread strategies are the price spread and the calendar spread. In a *price spread*, one option is bought while another is sold, both on the same underlying asset and the same date of expiration but with different strike prices. A *calendar spread* is similar to a price spread except that the strike prices of the options are the same but the dates of expiration are different.

Price Spreads

Price spreads can be classified as either bullish or bearish. The term *bullish* (*bearish*) means the holder of the spread benefits from an increase (decrease) in the asset price. A bullish price spread can be created by forming a portfolio which consists of a call option in the long position and another call option with a higher strike price in the short position. Since the call price is a decreasing function of the strike price [see (1.2.6a)], the portfolio requires an up-front premium for its creation. Let X_1 and X_2 ($X_2 > X_1$) be the strike prices of the calls and c_1 and c_2 ($c_2 < c_1$) be their respective premiums. The sum of terminal payoffs from the two calls is shown to be

$$\begin{aligned} & \max(S_T - X_1, 0) - \max(S_T - X_2, 0) \\ &= \begin{cases} 0 & S_T < X_1 \\ S_T - X_1 & X_1 \leq S_T \leq X_2 \\ X_2 - X_1 & S_T > X_2. \end{cases} \end{aligned} \tag{1.1.4}$$

The terminal payoff stays at the zero value until S_T reaches X_1, it then grows linearly with S_T when $X_1 \leq S_T \leq X_2$ and it is capped at the constant value $X_2 - X_1$ when $S_T > X_2$. The bullish price spread has its maximum gain at expiry when both calls expire in-the-money. When both calls expire out-of-the-money, corresponding to $S_T < X_1$, the overall loss would be the initial set up cost for the bullish spread.

Suppose we form a new portfolio with two calls, where the call bought has a higher strike price than the call sold, both with the same date of expiration, then a bearish price spread is created. Unlike its bullish counterpart, the bearish price spread leads to an up front positive cash flow to the investor. The terminal profit of a bearish price spread using two calls of different strike prices is exactly negative to that of its bullish counterpart. Note that the bullish and bearish price spreads can also be created by portfolios of puts.

Butterfly Spreads

Consider a portfolio created by buying a call option at strike price X_1 and another call option at strike price X_3 (say, $X_3 > X_1$) and selling two call options at strike price $X_2 = \frac{X_1+X_3}{2}$. This is called a *butterfly spread*, which can be considered as the combination of one bullish price spread and one bearish price spread. The creation of the butterfly spread requires the set up premium of $c_1 + c_3 - 2c_2$, where c_i denotes the price of the call option with strike price X_i, $i = 1, 2, 3$. Since the call price is a convex function of the strike price [see (1.2.13a)], we have $2c_2 < c_1 + c_3$. Hence, the butterfly spread requires a positive set-up cost. The sum of payoffs from the four call options at expiry is found to be

$$\max(S_T - X_1, 0) + \max(S_T - X_3, 0) - 2\max(S_T - X_2, 0)$$
$$= \begin{cases} 0 & S_T \le X_1 \\ S_T - X_1 & X_1 < S_T \le X_2 \\ X_3 - S_T & X_2 < S_T \le X_3 \\ 0 & S_T > X_3 \end{cases}. \tag{1.1.5}$$

The terminal payoff attains the maximum value at $S_T = X_2$ and declines linearly on both sides of X_2 until it reaches the zero value at $S_T = X_1$ or $S_T = X_3$. Beyond the interval (X_1, X_3), the payoff of the butterfly spread becomes zero. By subtracting the initial set-up cost of $c_1 + c_3 - 2c_2$ from the terminal payoff, we get the terminal profit diagram of the butterfly spread shown in Fig. 1.2.

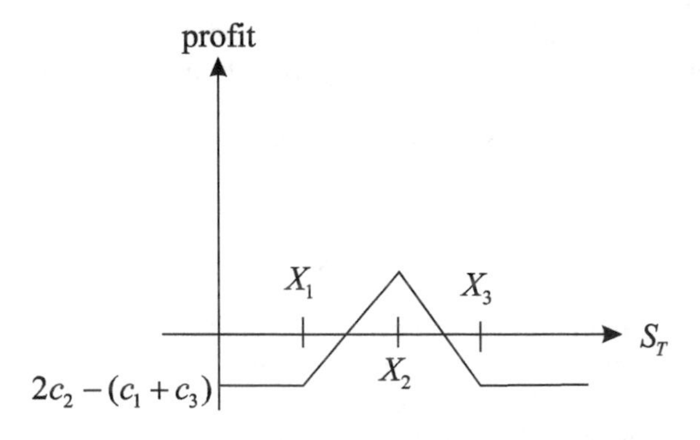

Fig. 1.2. Terminal profit diagram of a butterfly spread with four calls.

The butterfly spread is an appropriate strategy for an investor who believes that large asset price movements during the life of the spread are unlikely. Note that the terminal payoff of a butterfly spread with a wider interval (X_1, X_3) dominates that of the counterpart with a narrower interval. Using the no arbitrage argument, one deduces that the initial set-up cost of the butterfly spread increases with the width of the interval (X_1, X_3). If otherwise, an arbitrageur can lock in riskless profit by buying the presumably cheaper butterfly spread with the wider interval and selling the more expensive butterfly spread with the narrower interval. The strategy guarantees a non-negative terminal payoff while having the possibility of a positive terminal payoff.

Calendar Spreads

Consider a calendar spread that consists of two calls with the same strike price but different dates of expiration T_1 and T_2 $(T_2 > T_1)$, where the shorter-lived and longer-lived options are in the short and long positions, respectively. Since the longer-lived call is normally more expensive,[1] an up-front set-up cost for the calendar spread is required. In our subsequent discussion, we consider the usual situation where the longer-lived call is more expensive. The two calls with different expiration dates decrease in value at different rates, with the shorter-lived call decreasing in value at a faster rate. Also, the rate of decrease is higher when the asset price is closer to the strike price (see Sect. 3.3). The gain from holding the calendar spread comes from the difference between the rates of decrease in value of the shorter-lived call and longer-lived call. When the asset price at T_1 (expiry date of the shorter-lived call) comes closer to the common strike price of the two calls, a higher gain of the calendar spread at T_1 is realized because the rates of decrease in call value are higher when the call options come closer to being at-the-money. The profit at T_1 is given by this gain minus the initial set-up cost. In other words, the profit of the calendar spread at T_1 becomes higher when the asset price at T_1 comes closer to the common strike price.

Combinations

Combinations are portfolios that contain options of different types but on the same underlying asset. A popular example is a *bottom straddle*, which involves buying a call and a put with the same strike price X and expiration time T. The payoff at expiry from the bottom straddle is given by

$$\max(S_T - X, 0) + \max(X - S_T, 0) = \begin{cases} X - S_T & \text{when } S_T \le X \\ S_T - X & \text{when } S_T > X. \end{cases} \tag{1.1.6}$$

Since both options are in the long position, an up-front premium of $c + p$ is required for the creation of the bottom straddle, where c and p are the option premium of the European call and put. As revealed from the terminal payoff as stated in (1.1.6), the

[1] Longer-lived European call may become less expensive than the shorter-lived counterpart only when the underlying asset is paying dividend and the call option is sufficiently deep-in-the-money (see Sect. 3.3).

terminal profit diagram of the bottom straddle resembles the letter "V". The terminal profit achieves its lowest value of $-(c+p)$ at $S_T = X$ (negative profit value actually means loss). The bottom straddle holder loses when S_T stays close to X at expiry, but receives substantial gain when S_T moves further away from X in either direction.

The other popular examples of combinations include *strip*, *strap*, *strangle*, *box spread*, etc. Readers are invited to explore the characteristics of their terminal profits through Problems 1.1–1.4.

There are many other possibilities to create spread positions and combinations that approximate a desired pattern of payoff at expiry. Indeed, this is one of the major advantages of trading options rather than the underlying asset alone. In particular, the terminal payoff of a butterfly spread resembles a triangular "spike" so one can approximate the payoff according to an investor's preference by forming an appropriate combination of these spikes. As a reminder, the terminal profit diagrams presented above show the profits of these portfolio strategies when the positions of the options are held to expiration. Prior to expiration, the profit diagrams are more complicated and relevant option valuation models are required to find the value of the portfolio at a particular instant.

1.2 Rational Boundaries for Option Values

In this section, we establish some rational boundaries for the values of options with respect to the price of the underlying asset. At this point, we do not specify the probability distribution of the asset price process so we cannot derive the *fair* option value. Rather, we attempt to deduce reasonable limits between which any acceptable equilibrium price falls. The basic assumptions are that investors prefer more wealth to less and there are no arbitrage opportunities.

First, we present the rational boundaries for the values of both European and American options on an underlying asset that pays no dividend. We derive mathematical properties of the option values as functions of the strike price X, asset price S and time to expiry τ. Next, we study the impact of dividends on these rational boundaries for the option values. The optimal early exercise policies of American options on a non-dividend paying asset can be inferred from the analysis of these bounds on option values. The relations between put and call prices (called the *put-call parity relations*) are also deduced. As an illustrative and important example, we extend the analysis of rational boundaries and put-call parity relations to foreign currency options.

Here, we introduce the concept of time value of cash. It is common sense that \$1 at present is worth more than \$1 at a later instant since the cash can earn positive interest, or conversely, an amount less than \$1 will eventually grow to \$1 after a sufficiently long interest-earning period. In the simplest form of a bond with zero coupon, the bond contract promises to pay the par value at maturity to the bondholder, provided that the bond issuer does not default prior to maturity. Let $B(\tau)$ be the current price of a zero coupon default-free bond with the par value of \$1 at maturity, where τ is the time to maturity (we commonly use "maturity" for bonds and "expiry" for options). When the riskless interest rate r is taken to be constant and interest is compounded continuously, the bond value $B(\tau)$ is given by $e^{-r\tau}$. When r is nonconstant but a deterministic function of τ, $B(\tau)$ is found to be $e^{-\int_0^\tau r(u)\,du}$. The formula for $B(\tau)$ becomes more complicated when the interest rate is assumed to be stochastic (see Sect. 7.2). The bond price $B(\tau)$ can be interpreted as the discount factor over the τ-period.

Throughout this book, we adopt the notation where capitalized letters C and P denote American call and put values, respectively, and small letters c and p for their European counterparts.

Nonnegativity of Option Prices
All option prices are nonnegative, that is,

$$C \geq 0, \quad P \geq 0, \quad c \geq 0, \quad p \geq 0. \tag{1.2.1}$$

These relations are derived from the nonnegativity of the payoff structure of option contracts. If the price of an option were negative, this would mean an option buyer receives cash up front while being guaranteed a nonnegative terminal payoff. In this way, he can always lock in a riskless profit.

Intrinsic Values
Let $C(S, \tau; X)$ denote the price function of an American call option with current asset price S, time to expiry τ and strike price X; similar notation will be used for other American option price functions. At expiry time $\tau = 0$, the terminal payoffs are

$$C(S, 0; X) = c(S, 0; X) = \max(S - X, 0) \tag{1.2.2a}$$
$$P(S, 0; X) = p(S, 0; X) = \max(X - S, 0). \tag{1.2.2b}$$

The quantities $\max(S - X, 0)$ and $\max(X - S, 0)$ are commonly called the *intrinsic value* of a call and a put, respectively. One argues that since American options can be exercised at any time before expiration, their values must be worth at least their intrinsic values, that is,

$$C(S, \tau; X) \geq \max(S - X, 0) \tag{1.2.3a}$$
$$P(S, \tau; X) \geq \max(X - S, 0). \tag{1.2.3b}$$

Since $C \geq 0$, it suffices to consider the case $S > X$, where the American call is in-the-money. Suppose C is less than $S - X$ when $S > X$, then an arbitrageur can lock in a riskless profit by borrowing $C + X$ dollars to purchase the American call and exercise it immediately to receive the asset worth S. The riskless profit would be $S - X - C > 0$. The same no arbitrage argument can be used to show condition (1.2.3b).

However, as there is no early exercise privilege for European options, conditions (1.2.3a,b) do not necessarily hold for European calls and puts, respectively. Indeed, the European put value can be below the intrinsic value $X - S$ at sufficiently low asset value and the value of a European call on a dividend paying asset can be below the intrinsic value $S - X$ at sufficiently high asset value.

American Options Are Worth at Least Their European Counterparts

An American option confers all the rights of its European counterpart plus the privilege of early exercise. Obviously, the additional privilege cannot have negative value. Therefore, American options must be worth at least their European counterparts, that is,

$$C(S, \tau; X) \geq c(S, \tau; X) \tag{1.2.4a}$$

$$P(S, \tau; X) \geq p(S, \tau; X). \tag{1.2.4b}$$

Values of Options with Different Dates of Expiration

Consider two American options with different times to expiry τ_2 and τ_1 ($\tau_2 > \tau_1$), the one with the longer time to expiry must be worth at least that of the shorter-lived counterpart since the longer-lived option has the additional right to exercise between the two expiration dates. This additional right should have a positive value; so we have

$$C(S, \tau_2; X) > C(S, \tau_1; X), \quad \tau_2 > \tau_1, \tag{1.2.5a}$$

$$P(S, \tau_2; X) > P(S, \tau_1; X), \quad \tau_2 > \tau_1. \tag{1.2.5b}$$

The above argument cannot be applied to European options because the early exercise privilege is absent.

Values of Options with Different Strike Prices

Consider two call options, either European or American, the one with the higher strike price has a lower expected profit than the one with the lower strike. This is because the call option with the higher strike has strictly less opportunity to exercise a positive payoff, and even when exercised, it induces a smaller cash inflow. Hence, the call option price functions are decreasing functions of their strike prices, that is,

$$c(S, \tau; X_2) < c(S, \tau; X_1), \quad X_1 < X_2, \tag{1.2.6a}$$

$$C(S, \tau; X_2) < C(S, \tau; X_1), \quad X_1 < X_2. \tag{1.2.6b}$$

By reversing the above argument, the European and American put price functions are increasing functions of their strike prices, that is,

$$p(S, \tau; X_2) > p(S, \tau; X_1), \quad X_1 < X_2, \tag{1.2.7a}$$

$$P(S, \tau; X_2) > P(S, \tau; X_1), \quad X_1 < X_2. \tag{1.2.7b}$$

Values of Options at Different Asset Price Levels

For a call (put) option, either European or American, when the current asset price is higher, it has a strictly higher (lower) chance to be exercised and when exercised it induces higher (lower) cash inflow. Therefore, the call (put) option price functions are increasing (decreasing) functions of the asset price, that is,

$$c(S_2, \tau; X) > c(S_1, \tau; X), \quad S_2 > S_1, \tag{1.2.8a}$$

$$C(S_2, \tau; X) > C(S_1, \tau; X), \quad S_2 > S_1; \tag{1.2.8b}$$

and

$$p(S_2, \tau; X) < p(S_1, \tau; X), \quad S_2 > S_1, \tag{1.2.9a}$$

$$P(S_2, \tau; X) < P(S_1, \tau; X), \quad S_2 > S_1. \tag{1.2.9b}$$

Upper Bounds on Call and Put Values

A call option is said to be a *perpetual call* if its date of expiration is infinitely far away. The asset itself can be considered an American perpetual call with zero strike price plus additional privileges such as voting rights and receipt of dividends, so we deduce that $S \geq C(S, \infty; 0)$. By applying conditions (1.2.4a) and (1.2.5a), we can establish

$$S \geq C(S, \infty; 0) \geq C(S, \tau; X) \geq c(S, \tau; X). \tag{1.2.10}$$

Hence, American and European call values are bounded above by the asset value. Furthermore, by setting $S = 0$ in condition (1.2.10) and applying the nonnegativity property of option prices, we obtain

$$0 = C(0, \tau; X) = c(0, \tau; X),$$

that is, call values become zero at zero asset value.

The price of an American put equals its strike price when the asset value is zero; otherwise, it is bounded above by the strike price. Together with condition (1.2.4b), we have

$$X \geq P(S, \tau; X) \geq p(S, \tau; X). \tag{1.2.11}$$

Lower Bounds on Values of Call Options on a Nondividend Paying Asset

A lower bound on the value of a European call on a nondividend paying asset is found to be at least equal to or above the underlying asset value minus the present value of the strike price. To illustrate the claim, we compare the values of two portfolios, A and B. Portfolio A consists of a European call on a nondividend paying asset plus a discount bond with a par value of X whose date of maturity coincides with the expiration date of the call. Portfolio B contains one unit of the underlying asset. Table 1.1 lists the payoffs at expiry of the two portfolios under the two scenarios $S_T < X$ and $S_T \geq X$, where S_T is the asset price at expiry.

At expiry, the value of Portfolio A, denoted by V_A, is either greater than or at least equal to the value of Portfolio B, denoted by V_B. Portfolio A is said to be dominant

Table 1.1. Payoffs at expiry of Portfolios A and B

Asset value at expiry	$S_T < X$	$S_T \geq X$
Portfolio A	X	$(S_T - X) + X = S_T$
Portfolio B	S_T	S_T
Result of comparison	$V_A > V_B$	$V_A = V_B$

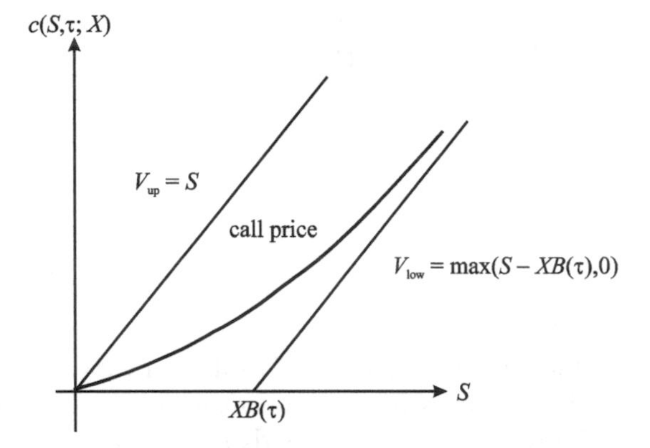

Fig. 1.3. The upper and lower bounds of the option value of a European call on a nondividend paying asset are $V_{up} = S$ and $V_{low} = \max(S - XB(\tau), 0)$, respectively.

over Portfolio B. The present value of Portfolio A (dominant portfolio) must be equal to or greater than that of Portfolio B (dominated portfolio). If otherwise, arbitrage opportunity can be secured by buying Portfolio A and selling Portfolio B. The above result can be represented by

$$c(S, \tau; X) + XB(\tau) \geq S.$$

Together with the nonnegativity property of option value, the lower bound on the value of the European call is found to be

$$c(S, \tau; X) \geq \max(S - XB(\tau), 0).$$

Combining with condition (1.2.10), the upper and lower bounds of the value of a European call on a nondividend paying asset are given by (see Fig. 1.3)

$$S \geq c(S, \tau; X) \geq \max(S - XB(\tau), 0). \tag{1.2.12}$$

Furthermore, as deduced from condition (1.2.10) again, the above lower and upper bounds are also valid for the value of an American call on a nondividend paying asset. The above results on the rational boundaries of European option values have to be modified when the underlying asset pays dividends [see (1.2.14), (1.2.23)].

Early Exercise Polices of American Options

First, we consider an American call on a nondividend paying asset. An American call is exercised only if it is in-the-money, where $S > X$. At any moment when an American call is exercised, its exercise payoff becomes $S - X$, which ought to be positive. However, the exercise value is less than $\max(S - XB(\tau), 0)$, the lower bound of the call value given that the call remains alive. Thus the act of exercising prior to expiry causes a decline in value of the American call. To the benefit of the holder, an American call on a nondividend paying asset will not be exercised prior to expiry. Since the early exercise privilege is forfeited, the American and European call values should be the same.

When the underlying asset pays dividends, the early exercise of an American call prior to expiry may become optimal when the asset value is very high and the dividends are sizable. Under these circumstances, it then becomes more attractive for the investor to acquire the asset through early exercise rather than holding the option. When the American call is deep-in-the-money, $S \gg X$, the chance of regret of early exercise (loss of insurance protection against downside move of the asset price) is low. On the other hand, the earlier acquisition of the underlying asset allows receipt of the dividends paid by the asset. For American puts, irrespective whether the asset is paying dividends or not, it can be shown [see (1.2.16)] that it is always optimal to exercise prior to expiry when the asset value is low enough. More details on the effects of dividends on the early exercise policies of American options will be discussed later in this section.

Convexity Properties of the Option Price Functions

The call prices are convex functions of the strike price. Write $X_2 = \lambda X_3 + (1-\lambda)X_1$ where $0 \leq \lambda \leq 1$, $X_1 \leq X_2 \leq X_3$. Mathematically, the convexity properties are depicted by the following inequalities:

$$c(S, \tau; X_2) \leq \lambda c(S, \tau; X_3) + (1 - \lambda)c(S, \tau; X_1) \qquad (1.2.13a)$$

$$C(S, \tau; X_2) \leq \lambda C(S, \tau; X_3) + (1 - \lambda)C(S, \tau; X_1). \qquad (1.2.13b)$$

Figure 1.4 gives a graphical representation of the above inequalities.

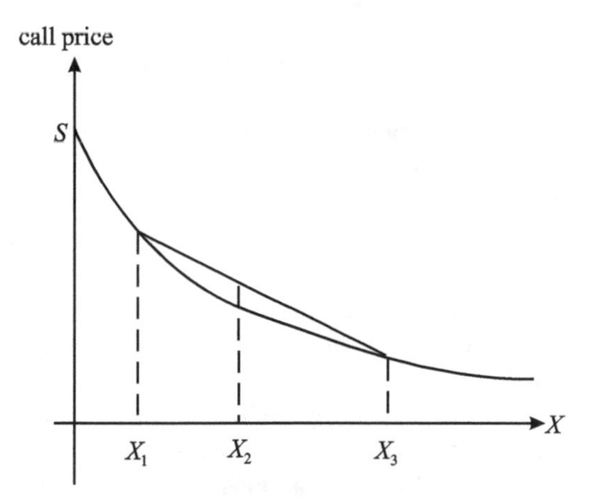

Fig. 1.4. The call price is a convex function of the strike price X. The call price equals S when $X = 0$ and tends to zero at large value of X.

Table 1.2. Payoff at expiry of Portfolios C and D

Asset value at expiry	$S_T \leq X_1$	$X_1 \leq S_T \leq X_2$	$X_2 \leq S_T \leq X_3$	$X_3 \leq S_T$
Portfolio C	0	$(1-\lambda)(S_T - X_1)$	$(1-\lambda)(S_T - X_1)$	$\lambda(S_T - X_3)+$ $(1-\lambda)(S_T - X_1)$
Portfolio D	0	0	$S_T - X_2$	$S_T - X_2$
Result of comparison	$V_C = V_D$	$V_C \geq V_D$	$V_C \geq V_D$	$V_C = V_D$

To show that inequality (1.2.13a) holds for European calls, we consider the payoffs of the following two portfolios at expiry. Portfolio C contains λ units of call with strike price X_3 and $(1 - \lambda)$ units of call with strike price X_1, and Portfolio D contains one call with strike price X_2. In Table 1.2, we list the payoffs of the two portfolios at expiry for all possible values of S_T.

Since $V_C \geq V_D$ for all possible values of S_T, Portfolio C is dominant over Portfolio D. Therefore, the present value of Portfolio C must be equal to or greater than that of Portfolio D; so this leads to inequality (1.2.13a). In the above argument, there is no factor involving τ, so the result also holds even when the calls in the two portfolios are allowed to be exercised prematurely. Hence, the convexity property also holds for American calls. By changing the call options in the above two portfolios to the corresponding put options, it can be shown by a similar argument that European and American put prices are also convex functions of the strike price.

Furthermore, by using the linear homogeneity property of the call and put option functions with respect to the asset price and strike price, one can show that the call and put prices (both European and American) are convex functions of the asset price (see Problem 1.7).

1.2.1 Effects of Dividend Payments

Now we examine the effects of dividends on the rational boundaries for option values. In the forthcoming discussion, we assume the size and payment date of the dividends to be known. One important result is that the early exercise of an American call option may become optimal if dividends are paid during the life of the option.

First, we consider the impact of dividends on the asset price. When an asset pays a certain amount of dividend, no arbitrage argument dictates that the asset price is expected to fall by the same amount (assuming there exist no other factors affecting the income proceeds, like taxation and transaction costs). Suppose the asset price falls by an amount less than the dividend, an arbitrageur can lock in a riskless profit by borrowing money to buy the asset right before the dividend date, selling the asset right after the dividend payment and returning the loan. The net gain to the arbitrageur is the amount that the dividend income exceeds the loss caused by the difference in the asset price in the buying and selling transactions. If the asset price falls by an amount greater than the dividend, then the above strategical transactions are reversed in order to catch the arbitrage profit.

Let $D_1, D_2, \cdots, D_n$ be the dividend amount paid at $\tau_1, \tau_2, \cdots, \tau_n$ periods from the current time. Let D denote the present value of all known discrete dividends paid between now and the expiration date. Assuming constant interest rate, we then have

$$D = D_1 e^{-r\tau_1} + D_2 e^{-r\tau_2} + \cdots + D_n e^{-r\tau_n},$$

where r is the riskless interest rate and $e^{-r\tau_1}, e^{-r\tau_2}, \cdots, e^{-r\tau_n}$ are the respective discount factors. We examine the impact of dividends on the lower bound on the European call value and the early exercise feature of an American call option, with dependence on the lumped dividend D. Similar to the two portfolios shown in Table 1.1, we modify Portfolio B to contain one unit of the underlying asset and a loan of D dollars (in the form of a portfolio of bonds with par value D_i and time to expiry $\tau_i, i = 1, 2, \cdots, n$). At expiry, the value of Portfolio B will always become S_T since the loan of D will be paid back during the life of the option using the dividends received. One observes again $V_A \geq V_B$ at expiry so that the present value of Portfolio A must be at least as much as that of Portfolio B. Together with the nonnegativity property of option values, we obtain

$$c(S, \tau; X, D) \geq \max(S - XB(\tau) - D, 0). \tag{1.2.14}$$

This gives us the new lower bound on the price of a European call option on a dividend paying asset. Since the call price becomes lower due to the dividends of the underlying asset, it may be possible that the call price falls below the intrinsic value $S - X$ when the lumped dividend D is deep enough. Accordingly, the condition on D such that $c(S, \tau; X, D)$ may fall below the intrinsic value $S - X$ is given by

$$S - X > S - XB(\tau) - D \text{ or } D > X[1 - B(\tau)]. \tag{1.2.15}$$

If D does not satisfy the above condition, it is never optimal to exercise the American call prematurely. In addition to the necessary condition (1.2.15) on the size of D, the American call must be sufficiently deep in-the-money so that the chance of regret on early exercise is low (see Sect. 5.1). Since there will be an expected decline in asset price right after a discrete dividend payment, the optimal strategy is to exercise right before the dividend payment so as to capture the dividend paid by the asset. The behavior of the American call price right before and after the dividend dates are examined in detail in Sect. 5.1.

Unlike holding a call, the holder of a put option gains when the asset price drops after a discrete dividend is paid because put value is a decreasing function of the asset price. Using an argument similar to that above (considering two portfolios), the bounds for American and European puts can be shown as

$$P(S, \tau; X, D) \geq p(S, \tau; X, D) \geq \max(XB(\tau) + D - S, 0). \tag{1.2.16}$$

Even without dividend ($D = 0$), the lower bound $XB(\tau) - S$ may become less than the intrinsic value $X - S$ when the put is sufficiently deep in-the-money (corresponding to a low value for S). Since the holder of an American put option would not tolerate the value falling below the intrinsic value, the American put should be exercised prematurely. The presence of dividends makes the early exercise of an American put option less likely since the holder loses the future dividends when the asset is sold upon exercising the put. Using an argument similar to that used in (1.2.15), one can show that when $D \geq X[1 - B(\tau)]$, the American put should never be exercised prematurely. The effects of dividends on the early exercise policies of American puts are in general more complicated than those for American calls (see Sect. 5.1).

The underlying asset may incur a *cost of carry* for the holder, like the storage and spoilage costs for holding a physical commodity. The effect of the cost of carry on the early exercise policies of American options appears to be opposite to that of dividends received through holding the asset.

1.2.2 Put-Call Parity Relations

Put-call parity states the relation between the prices of a pair of call and put options. For a pair of European put and call options on the same underlying asset and with the same expiration date and strike price, we have

$$p = c - S + D + XB(\tau). \tag{1.2.17}$$

When the underlying asset is nondividend paying, we set $D = 0$.

The proof of the above put-call parity relation is quite straightforward. We consider the following two portfolios. The first portfolio involves long holding of a European call, a portfolio of bonds: τ_1-maturity discount bond with par $D_1, \cdots, \tau_n$-maturity discount bond with par D_n and τ-maturity discount bond with par X, and short selling of one unit of the asset. The second portfolio contains only one European put. The sum of the present values of the bonds in the first portfolio is

$$D_1 B(\tau_1) + \cdots + D_n B(\tau_n) + XB(\tau) = D + XB(\tau).$$

The bond par values are taken to match with the sizes of the dividends and they are used to compensate the dividends due to the short position of one unit of the asset. At expiry, both portfolios have the same value $\max(X - S_T, 0)$. Since both European options cannot be exercised prior to expiry, both portfolios have the same value throughout the life of the options. By equating the values of the two portfolios, we obtain the parity relation (1.2.17).

The above parity relation cannot be applied to a pair of American call and put options due to their early exercise feature. However, we can deduce the lower and upper bounds on the difference of the prices of American call and put options. First, we assume the underlying asset is nondividend paying. Since $P > p$ and $C = c$, we deduce from (1.2.17) (putting $D = 0$) that

$$C - P < S - XB(\tau),$$

giving the upper bound on $C - P$. Let us consider the following two portfolios: one contains a European call plus cash of amount X, and the other contains an American put together with one unit of underlying asset. The first portfolio can be shown to be dominant over the second portfolio, so we have

$$c + X > P + S.$$

Further, since $c = C$ when the asset does not pay dividends, the lower bound on $C - P$ is given by

$$S - X < C - P.$$

Combining the two bounds, the difference of the American call and put option values on a nondividend paying asset is bounded by

$$S - X < C - P < S - XB(\tau). \tag{1.2.18}$$

The right side inequality, $C - P < S - XB(\tau)$, also holds for options on a dividend paying asset since dividends decrease call value and increase put value. However, the left side inequality has to be modified as $S - D - X < C - P$ (see Problem 1.8). Combining the results, the difference of the American call and put option values on a dividend paying asset is bounded by

$$S - D - X < C - P < S - XB(\tau). \tag{1.2.19}$$

1.2.3 Foreign Currency Options

The above techniques of analysis are now extended to foreign currency options. Here, the underlying asset is a foreign currency and all prices are denominated in domestic currency. As an illustration, we take the domestic currency to be the U.S. dollar and the foreign currency to be the Japanese yen. In this case, the spot domestic currency price S of one unit of foreign currency refers to the spot value of one Japanese yen in U.S. dollars, say, ¥ 1 for U.S.$0.01. Now both domestic and foreign interest rates are involved. Let $B_f(\tau)$ denote the foreign currency price of a default-free zero coupon bond, which has unit par and time to maturity τ. Since the underlying asset, which is a foreign currency, earns the riskless foreign interest rate r_f continuously, it is analogous to an asset that pays continuous dividend yield. The rational boundaries for the European and American foreign currency option values have to be modified accordingly.

Lower and Upper Bounds on Foreign Currency Call and Put Values

First, we consider the lower bound on the value of a European foreign currency call. Consider the following two portfolios: Portfolio A contains the European foreign currency call with strike price X and a domestic discount bond with par value of X whose maturity date coincides with the expiration date of the call. Portfolio B contains a foreign discount bond with par value of unity in the foreign currency, which also matures on the expiration date of the call. Portfolio B is worth the foreign currency price of $B_f(\tau)$, so the domestic currency price of $SB_f(\tau)$. On expiry of the call, Portfolio B becomes one unit of foreign currency and this equals S_T in domestic currency. The value of Portfolio A equals $\max(S_T, X)$ in domestic currency, thus Portfolio A is dominant over Portfolio B. Together with the nonnegativity property of option value, we obtain

$$c \geq \max(SB_f(\tau) - XB(\tau), 0).$$

As mentioned earlier, premature exercise of the American call on a dividend paying asset may become optimal. Recall that a necessary (but not sufficient) condition for optimal early exercise is that the lower bound $SB_f(\tau) - XB(\tau)$ is less than the intrinsic value $S - X$. In the present context, the necessary condition is seen to be

$$SB_f(\tau) - XB(\tau) < S - X \quad \text{or} \quad S > X\frac{1 - B(\tau)}{1 - B_f(\tau)}. \tag{1.2.20}$$

When condition (1.2.20) is not satisfied, we then have $C > S - X$. The premature early exercise of the American foreign currency call would give $C = S - X$, resulting in a drop in value. Therefore, it is not optimal to exercise the American foreign currency call prematurely. In summary, the lower and upper bounds for the American and European foreign currency call values are given by

$$S \geq C \geq c \geq \max(SB_f(\tau) - XB(\tau), 0). \tag{1.2.21}$$

Using similar arguments, the necessary condition for the optimal early exercise of an American foreign currency put option is given by

$$S < X\frac{1 - B(\tau)}{1 - B_f(\tau)}. \tag{1.2.22}$$

The lower and upper bounds on the values of American and European foreign currency put options can be shown to be

$$X \geq P \geq p \geq \max(XB(\tau) - SB_f(\tau), 0). \tag{1.2.23}$$

The corresponding put-call parity relation for the European foreign currency put and call options is given by

$$p = c - SB_f(\tau) + XB(\tau), \tag{1.2.24}$$

and the bounds on the difference of the prices of American call and put options on a foreign currency are given by (see Problem 1.11)

$$SB_f(\tau) - X < C - P < S - XB(\tau). \tag{1.2.25}$$

In conclusion, we have deduced the rational boundaries for the option values of calls and puts and their put-call parity relations. The impact of the early exercise privilege and dividend payment on option values have also been analyzed. An important result is that it is never optimal to exercise prematurely an American call option on a nondividend paying asset. More comprehensive discussion of the analytic properties of option price functions can be found in the seminal paper by Merton (1973) and the review article by Smith (1976).

1.3 Forward and Futures Contracts

Recall that a forward contract is an agreement between two parties that the holder agrees to buy an asset from the writer at the delivery time T in the future for a predetermined delivery price K. Unlike an option contract where the holder pays the writer an up-front option premium, no up-front payment is involved when a forward contract is transacted. The delivery price of a forward is chosen so that the value of the forward contract to both parties is zero at the time when the contract is initiated. The *forward price* is defined as the delivery price which makes the initial value of the forward contract zero. The forward price in a new forward contract is liable to change due to the subsequent fluctuation of the price of the underlying asset while the delivery price of the already transacted forward contract is held fixed.

Suppose that on July 1 the forward price of silver with maturity date on October 31 is quoted at \$30. This means that \$30 is the price (paid upon delivery) at which the person in long (short) position of the forward contract agrees to buy (sell) the contracted amount and quality of silver on the maturity date. A week later (July 8), the quoted forward price of silver for October 31 delivery changes to a new value due to price fluctuation of silver during the week. Say, the forward price moves up to \$35. The forward contract entered on July 1 earlier now has positive *value* since the delivery price has been fixed at \$30 while the new forward price for the same maturity date has been increased to \$35. Imagine that while holding the earlier forward, the holder can short another forward on the same commodity and maturity date. The opposite positions of the two forward contracts will be exactly canceled off on the October 31 delivery date. The holder will pay \$30 to buy the asset but will receive \$35 from selling the asset. Hence, the holder will be secured to receive $\$35 - \$30 = \$5$ on the delivery date. Recall that the holder pays nothing on both July 1 and July 8 when the two forward contracts are transacted. Obviously, there is some value associated with the holding of the earlier forward contract. This value is related to the spot forward price and the fixed delivery price. While we have been using the terms "price" and "value" interchangeably for options, but "forward price" and "forward value" are different quantities for forward contracts.

1.3.1 Values and Prices of Forward Contracts

We would like to consider the pricing formulas for forward contracts under three separate cases of dividend behaviors of the underlying asset, namely, no dividend, known discrete dividends and known continuous dividend yields.

Nondividend Paying Asset

Let $f(S, \tau)$ and $F(S, \tau)$ denote, respectively, the value and the price of a forward contract with current asset value S and time to maturity τ, and let r denote the constant riskless interest rate. Consider a portfolio that contains one long forward contract and a bond with the same maturity date and par value as the delivery price. The bond price is $Ke^{-r\tau}$, where K is the delivery price at maturity. The other portfolio contains one unit of the underlying asset. At maturity, the par received from holding the bond could be used to pay for the purchase of one unit of the asset to honor the forward contract. Both portfolios become one unit of the asset at maturity. Assuming that the asset does not pay any dividend, by the principle of no arbitrage, both portfolios should have the same value at all times prior to maturity. It then follows that the value of the forward contract is given by

$$f = S - Ke^{-r\tau}. \tag{1.3.1}$$

Recall that we have defined the forward price to be the delivery price which makes the value of the forward contract zero. In (1.3.1), the value of K which makes $f = 0$ is given by $K = Se^{r\tau}$. The forward price is then $F = Se^{r\tau}$, which agrees with formula (1.1.1). Together with the put-call parity relation for a pair of European call and put options, we obtain

$$f = (F - K)e^{-r\tau} = c(S, \tau; K) - p(S, \tau; K), \tag{1.3.2}$$

where the strike prices of the call and put options are set equal to the delivery price of the forward contract. The put-call parity relation reveals that holding a call is equivalent to holding a put and a forward.

Discrete Dividend Paying Asset

Now, suppose the asset pays discrete dividends to the holder during the life of the forward contract. Let D denote the present value of all dividends paid by the asset within the life of the forward. To find the value of the forward contract, we modify the above second portfolio to contain one unit of the asset plus borrowing of D dollars. At maturity, the second portfolio again becomes worth one unit of the asset since the loan of D dollars will be repaid by the dividends received by holding the asset. Hence, the value of the forward contract on a discrete dividend paying asset is found to be

$$f = S - D - Ke^{-r\tau}.$$

By finding the value of K such that $f = 0$, we obtain the forward price to be given by

$$F = (S - D)e^{r\tau}. \tag{1.3.3}$$

Continuous Dividend Paying Asset

Next, suppose the asset pays a continuous dividend yield at the rate q. The dividend is paid continuously throughout the whole time period and the dividend amount over a differential time interval dt is $qS\,dt$, where S is the spot asset price. Under this dividend behavior, we choose the second portfolio to contain $e^{-q\tau}$ units of asset with all dividends being reinvested to acquire additional units of asset. At maturity, the second portfolio will be worth one unit of the asset since the number of units of asset can be considered to have the continuous compounded growth at the rate q. It follows from the equality of the values of the two portfolios that the value of the forward contract on a continuous dividend paying asset is

$$f = Se^{-q\tau} - Ke^{-r\tau},$$

and the corresponding forward price is

$$F = Se^{(r-q)\tau}. \tag{1.3.4}$$

Since an investor is not entitled to receive any dividends through holding a put, call or forward, the put-call parity relation (1.3.2) also holds for put, call and forward on assets that pay either discrete dividends or continuous dividend yield.

Interest Rate Parity Relation

When we consider forward contracts on foreign currencies, the value of the underlying asset S is the price in the domestic currency of one unit of the foreign currency. The foreign currency considered as an asset can earn interest at the foreign riskless rate r_f. This is equivalent to a continuous dividend yield at the rate r_f. Therefore, the delivery price of a forward contract on the domestic currency price of one unit of foreign currency is given by

$$F = Se^{(r-r_f)\tau}. \tag{1.3.5}$$

Equation (1.3.5) is called the *Interest Rate Parity Relation.*

Cost of Carry and Convenience Yield

For commodities like grain and livestock, there may be additional costs to hold the assets such as storage, insurance, spoilage, etc. In simple terms, these additional costs can be considered as negative dividends paid by the asset. Suppose we let U denote the present value of all additional costs that will be incurred during the life of the forward contract, then the forward price can be obtained by replacing $-D$ in (1.3.3) by U. The forward price is then given by

$$F = (S + U)e^{r\tau}. \tag{1.3.6}$$

If the additional holding costs incurred at any time is proportional to the price of the commodity, they can be considered as negative dividend yield. If u denotes the cost per annum as a proportion of the spot commodity price, then the forward price is

$$F = Se^{(r+u)\tau}, \tag{1.3.7}$$

which is obtained by replacing $-q$ in (1.3.4) by u.

We may interpret $r + u$ as the *cost of carry* that must be incurred to maintain the commodity inventory. The cost consists of two parts: one part is the cost of funds tied up in the asset which requires interest for borrowing and the other part is the holding costs due to storage, insurance, spoilage, etc. It is convenient to denote the cost of carry by b. When the underlying asset pays a continuous dividend yield at the rate q, then $b = r - q$. In general, the forward price is given by

$$F = Se^{b\tau}. \tag{1.3.8}$$

There may be some advantages to users who hold the commodity, like the avoidance of temporary shortages of supply and the ensurance of production process running. These holding advantages may be visualized as negative holding costs. Suppose the market forward price F is below the cost of ownership of the commodity

$Se^{(r+u)\tau}$, the difference gives a measure of the benefits realized from actual ownership. We define the convenience yield y (benefit per annum) as a proportion of the spot commodity price. In this way, y has the effect negative to that of u. By netting the costs and benefits, the forward price is then given by

$$F = Se^{(r+u-y)\tau}. \tag{1.3.9}$$

With the presence of convenience yield, F is seen to be less than $Se^{(r+u)\tau}$. This is due to the multiplicative factor $e^{-y\tau}$, whose magnitude is less than one.

1.3.2 Relation between Forward and Futures Prices

Forward contracts and futures are much alike, except that the former are traded over-the-counter and the latter are traded in exchanges. Since the exchanges would like to organize trading such that contract defaults are minimized, an investor who buy a futures in an exchange is requested to deposit funds in a *margin account* to safeguard against the possibility of default (the futures agreement is not honored at maturity). At the end of each trading day, the futures holder will pay to or receive from the writer the full amount of the change in the futures price from the previous day through the margin account. This process is called *marking to market the account*. Therefore, the payment required on the maturity date to buy the underlying asset is simply the spot price at that time. However, for a forward contract traded outside the exchanges, no money changes hands initially or during the life-time of the contract. Cash transactions occur only on the maturity date. Such difference in the payment schedules may lead to differences in the prices of a forward contract and a futures on the same underlying asset and date of maturity. This is attributed to the possibility of different interest rates applied on the intermediate payments. In Sect. 8.1, we show how the forward price and futures price differ when the interest rate is stochastic and exhibiting positive correlation with the underlying asset price process.

Here, we present the argument to illustrate the equality of the two prices when the interest rate is constant. First, consider one forward contract and one futures which both last for n days. Let F_i and G_i denote the forward price and the futures price at the end of the ith day, respectively, $i = 0, 1, \cdots, n$. We would like to show that $F_0 = G_0$. Let S_n denote the asset price at maturity. Let the constant interest rate per day be δ. Suppose we initiate the long position of one unit of the futures on day 0. The gain/loss of the futures on the ith day is $(G_i - G_{i-1})$ and this amount grows to the dollar value $(G_i - G_{i-1})\, e^{\delta(n-i)}$ at maturity, which is the end of the nth day $(n \geq i)$. Therefore, the value of this *one* long futures position at the end of the nth day is the summation of $(G_i - G_{i-1})\, e^{\delta(n-i)}$, where i runs from 1 to n. The sum can be expressed as

$$\sum_{i=1}^{n} (G_i - G_{i-1})\, e^{\delta(n-i)}.$$

The summation of gain/loss of each day reflects the daily settlement nature of a futures.

Instead of holding one unit of futures throughout the whole period, the investor now keeps changing the amount of futures to be held on each day. Suppose he holds α_i units at the end of the $(i-1)$th day, $i = 1, 2, \cdots, n$, α_i to be determined. Since there is no cost incurred when a futures is transacted, the investor's portfolio value at the end of the nth day becomes

$$\sum_{i=1}^{n} \alpha_i (G_i - G_{i-1})\, e^{\delta(n-i)}.$$

On the other hand, since the holder of one unit of the forward contract initiated on day 0 can purchase the underlying asset which is worth S_n using F_0 dollars at maturity, the value of the long position of one forward at maturity is $S_n - F_0$. Now, we consider the following two portfolios:

Portfolio A : long position of a bond with par value F_0 maturing on the nth day
long position of one unit of forward contract

Portfolio B : long position of a bond with par value G_0 maturing on the nth day
long position of $e^{-\delta(n-i)}$ units of futures held at the end of the $(i-1)$th day, $i = 1, 2, \cdots n$.

At maturity (end of the nth day), the values of the bond and the forward contract in Portfolio A become F_0 and $S_n - F_0$, respectively, so that the total value of the portfolio is S_n. For Portfolio B, the bond value is G_0 at maturity. The value of the long position of the futures (number of units of futures held is kept changing on each day) is obtained by setting $\alpha_i = e^{-\delta(n-i)}$. This gives

$$\sum_{i=1}^{n} e^{-\delta(n-i)}\, (G_i - G_{i-1})\, e^{\delta(n-i)} = \sum_{i=1}^{n} (G_i - G_{i-1}) = G_n - G_0.$$

Hence, the total value of Portfolio B at maturity is $G_0 + (G_n - G_0) = G_n$. Since the futures price must be equal to the asset price S_n at maturity, we have $G_n = S_n$. The two portfolios have the same value at maturity, while Portfolio A and Portfolio B require an initial investment of $F_0 e^{-\delta n}$ and $G_0 e^{-\delta n}$ dollars, respectively. In the absence of arbitrage opportunities, the initial values of the two portfolios must be the same. We then obtain $F_0 = G_0$, that is, the current forward and futures prices are equal.

1.4 Swap Contracts

A swap is a financial contract between two counterparties who agree to exchange one cash flow stream for another according to some prearranged rules. Two important types of swaps are considered in this section: interest rate swaps and currency swaps. Interest rate swaps have the effect of transforming a floating-rate loan into a fixed-rate loan or vice versa. A currency swap can be used to transform a loan in one currency into a loan in another currency. One may regard a swap as a *package of forward contracts*. It would be interesting to examine how two firms may benefit by entering into a swap and the financial rationales that determine the prearranged rules for the exchange of cash flows.

1.4.1 Interest Rate Swaps

The most common form of an interest rate swap is a fixed-for-floating swap, where a series of payments, calculated by applying a fixed rate of interest to a notional principal amount, are exchanged for a stream of payments calculated using a floating rate of interest. The exchange of cash flows in net amount occurs on designated swap dates during the life of the swap contract. In the simplest form, all payments are made in the same currency. The principal amount is said to be notional since no exchange of principal will occur and the principal is used only to compute the actual cash amounts to be exchanged periodically on the swap dates.

The floating rate in an interest rate swap is chosen from one of the money market rates, like the London interbank offer rate (LIBOR), Treasury bill rate, federal funds rate, etc. Among them, the most common choice is the LIBOR. It is the interest rate at which prime banks offer to pay on Eurodollar deposits available to other prime banks for a given maturity. A Eurodollar is a U.S. dollar deposited in a U.S. or foreign bank outside the United States. The LIBOR comes with different maturities, say, one-month LIBOR is the rate offered on one-month deposits, etc. In the floating-for-floating interest rate swaps, two different reference floating rates are used to calculate the exchange payments.

As an example, consider a five-year fixed-for-floating interest rate swap. The fixed rate payer agrees to pay 8% per year (quoted with semi-annual compounding) to the counterparty while the floating rate payer agrees to pay in return six-month LIBOR. Assume that payments are exchanged every six months throughout the life of the swap and the notional amount is \$10 million. This means for every six months, the fixed rate payer pays out fixed rate interest of amount \$10 million $\times$ 8% $\div 2 =$ \$0.4 million but receives floating rate interest of amount that equals \$10 million times half of the six-month LIBOR prevailing six months before the payment. For example, suppose April 1, 2008, is the initiation date of the swap and the prevailing six-month LIBOR on that date is 6.2%. The floating rate interest payment on the first swap date (scheduled on October 1, 2008) will be \$10 million $\times$ 6.2% $\div 2 =$ \$0.31 million. In this way, the fixed rate payer will pay a net amount of (\$0.4 $-$ \$0.31) million $=$ \$0.09 million to the floating rate payer on the first swap date.

The interest payments paid by the floating rate payer resemble those of a floating rate loan, where the interest rate is set at the beginning of the period over which the rate will be applied and the interest amount is paid at the end of the period. This class of swaps is known as plain vanilla interest rate swaps. Assuming no default of either swap counterparty, a plain vanilla interest rate swap can be characterized as the difference between a fixed rate bond and a floating rate bond. This property naturally leads to an efficient valuation approach for plain vanilla interest rate swaps.

Valuation of Plain Vanilla Interest Rate Swaps

Consider the fixed rate payer of the above five-year fixed-for-floating plain vanilla interest rate swap. The fixed rate payer will receive floating rate interest payments semi-annually according to the six-month LIBOR. This cash stream of interest payments will be identical to those generated by a floating rate bond having the same maturity, par value and reference interest rate as those of the swap. Unlike the holder of the floating rate bond, the fixed rate payer will not receive the notional principal on the maturity date of the swap. On the other hand, he or she will pay out, fixed rate interest rate payment semi-annually, like the issuer of a fixed rate bond with the same fixed interest rate, maturity and par value as those of the swap.

We observe that the position of the fixed rate payer of the plain vanilla interest swap can be replicated by long holding of the floating rate bond underlying the swap and short holding of the fixed rate bond underlying the swap. The two underlying bonds have the same maturity, par value and corresponding reference interest rates as those of the swap. Hence, the value of the swap to the fixed rate payer is the value of the underlying floating rate bond minus the value of the underlying fixed rate bond. Since the position of the floating rate payer of the fixed-for-floating swap is exactly opposite to that of the fixed rate payer, so the value of the swap to the floating rate payer is negative that of the fixed rate payer. In summary, we have

$$\begin{aligned} V_{fix} &= B_{f\ell} - B_{fix} \\ V_{f\ell} &= B_{fix} - B_{f\ell}, \end{aligned}$$

where V_{fix} and $V_{f\ell}$ denote the value of the interest rate swap to the fixed rate payer and floating rate payer, respectively; B_{fix} and $B_{f\ell}$ denote the value of the underlying fixed rate bond and floating rate bond, respectively.

Uses of Interest Rate Swaps in Asset and Liability Management

Financial institutions often use an interest rate swap to alter the cash flow characteristics of their assets or liabilities to meet certain management goals or lock in a spread. As an example, suppose a bank is holding an asset (say, a loan or a bond) that earns semi-annually a fixed rate of interest of 8% (annual rate). To fund the holding of this asset, the bank issues six-month certificates of deposit that pay six-month LIBOR plus 60 basis points (1 basis point = 0.01%). How can the bank lock in a spread over the cost of its funds? This can be achieved by converting the fixed rate interests generated from the asset into floating rate interest incomes. This type of transaction is called an *asset swap*, which consists of a simultaneous asset purchase and entry into an interest rate swap. Suppose that the following interest rate swap is available to the bank:

> Every six months the bank pays 7% (annual rate) and receives LIBOR.

By entering into this interest rate swap, for every six months, the bank receives $8\% - 7\% = 1\%$ of net fixed rate interest payments and pays (LIBOR + 60 bps) − LIBOR = 0.6% of net floating rate interest payments. In this way, the bank can lock in a spread of 40 basis points over the funding costs.

On the other hand, suppose the bank has issued a fixed rate loan that pays every six months at the annual rate of 7%. Through a *liability swap*, the bank can transform this fixed rate liability into a floating rate liability by serving as the floating rate payer in an interest rate swap. Say, for every six months, the bank pays LIBOR + 50 basis points and receives 7.2% (annual rate). Now, the bank then applies the loan capital to purchase a floating rate bond so that the floating rate coupons received may be used to cover the floating rate interest payments under the interest rate swap. Through simple calculations, if the floating coupon rate is higher than LIBOR + 30 basis points, then the bank again locks in a positive spread on funding costs.

1.4.2 Currency Swaps

A currency swap is used to transform a loan in one currency into a loan of another currency. Suppose a U.S. company wishes to borrow British sterling to finance a project in the United Kingdom. On the other hand, a British company wants to raise U.S. dollars. Both companies would suffer comparative disadvantages in raising foreign capitals as compared to raising domestic capitals in their own country. As an example, we consider the following fixed borrowing rates for the two companies on the two currencies.

The above borrowing rates indicate that the U.S. company has better creditworthiness so that it enjoys lower borrowing rates at both currencies as compared to the UK company. Note that the difference between the borrowing rates in U.S. dollars is 2% while that in UK sterling is only 1.2%. With a spread of $2\% - 1.2\% = 0.8\%$ on the borrowing rates in the two currencies, it seems possible to construct a currency swap so that both companies receive the desired types of capital and take advantage of the lower borrowing rates on their domestic currencies.

Let the current exchange rate be £1 = \$1.4, and assume the notional principals to be £1 million and \$1.4 million. First, both companies borrow the principals from their domestic borrowers in their respective currencies. That is, the U.S. company enters into a loan of \$1.4 million at the borrowing rate of 9.0% while the UK company enters into a loan of £1 million at the borrowing rate of 13.6%. Next, a currency swap is structured as follows. At initiation of the swap, the U.S. company exchanges the capital of \$1.4 million for £1 million with the UK company. In this way, both companies obtain the types of capital that they desire. Within the swap period, the

Table 1.3. Borrowing at fixed rates for the U.S. and UK companies

	U.S. dollars	UK sterling
U.S. company	9.0%	12.4%
UK company	11.0%	13.6%

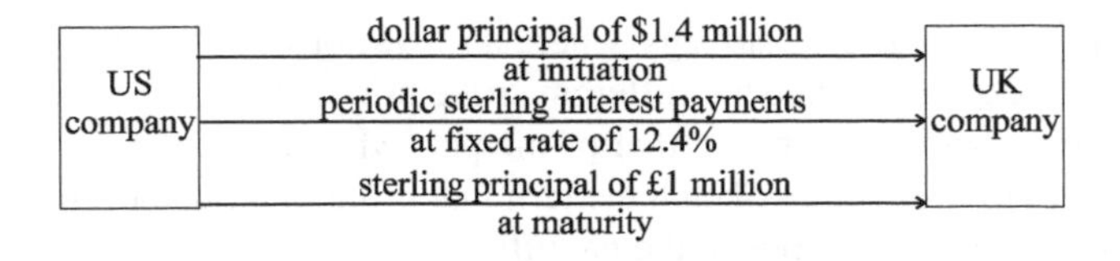

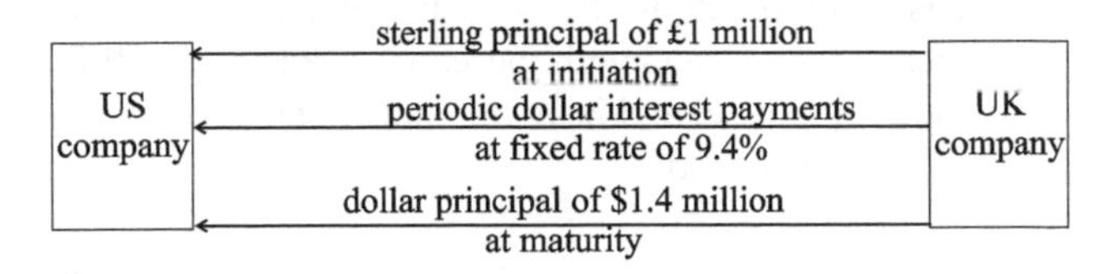

Fig. 1.5. Cash flow streams between the two counterparties in a currency swap.

US company pays periodically fixed sterling interest rate of 12.4% to the UK company, and in return, receives fixed dollar interest rate of 9.4% from the UK company. At maturity of the currency swap, the U.S. company returns the loan capital of £1 million to the UK company and receives $1.4 million back from the UK company. Both companies can pay back the loans to their respective domestic borrowers. The cash flow streams between the two companies are summarized in Fig. 1.5.

What would be the gains to both counterparties in the above currency swap? The U.S. company pays the same fixed sterling interest rate of 12.4%, but gains $9.4\% - 9.0\% = 0.4\%$ on the dollar interest rate. This is because it pays 9.0% to the domestic borrower but receives 9.4% from the UK company through the currency swap. On the other hand, the UK company pays only 9.4% on the dollar interest rate instead of 11.0%. This represents a gain on the spread of $11.0\% - 9.4\% = 1.6\%$ on the dollar interest rate, though it loses $13.6\% - 12.4\% = 1.2\%$ on the spread in the sterling interest rate. Note that the net gains and losses on the interest payments are in different currencies, so the parties in a currency swap face the exchange rate exposure.

1.5 Problems

1.1 How can we construct the portfolio of a butterfly spread that involves put options with different strike prices but the same date of expiration and on the same underlying asset? Draw the corresponding profit diagram of the spread at expiry.

1.2 A *strip* is a portfolio created by buying one call and writing two puts with the same strike price and expiration date. A *strap* is similar to a strip except it involves long holding of two calls and short selling of one put instead. Sketch the terminal profit diagrams for the strip and the strap and comment on their roles in monitoring risk exposure. How are they compared to a bottom straddle?

1.3 A *strangle* is a trading strategy where an investor buys a call and a put with the same expiration date but different strike prices. The strike price of the call may be higher or lower than that of the put (when the strike prices are equal, it reduces to a straddle). Sketch the terminal profit diagrams for both cases and discuss the characteristics of the payoffs at expiry.

1.4 A *box spread* is a combination of a bullish call spread with strike prices X_1 and X_2 and a bearish put spread with the same strike price. All four options are on the same underlying asset and have the same date of expiration. Discuss the characteristics of a box spread.

1.5 Suppose the strike prices X_1 and X_2 satisfy $X_2 > X_1$, show that for European calls on a nondividend paying asset, the difference in the call values satisfies

$$-B(\tau)(X_2 - X_1) \leq c(S, \tau; X_2) - c(S, \tau; X_1) \leq 0,$$

where $B(\tau)$ is the value of a pure discount bond with par value of unity and time to maturity τ. Furthermore, deduce that

$$-B(\tau) \leq \frac{\partial c}{\partial X}(S, \tau; X) \leq 0.$$

In other words, suppose the call price can be expressed as a differentiable function of the strike price, then the derivative must be nonpositive and not greater in absolute value than the price of a pure discount bond of the same maturity. Do the above results also hold for European/American calls on a dividend paying asset?

1.6 Show that a portfolio of holding various single-asset options with the same date of expiration is worth at least as much as a single option on the portfolio of the same number of units of each of the underlying assets. The single option is called a *basket option*. In mathematical representation, say for European call options, we have

$$\sum_{i=1}^{N} n_i c_i(S_i, \tau; X_i) \geq c\left(\sum_{i=1}^{N} n_i S_i, \tau; \sum_{i=1}^{N} n_i X_i\right), \qquad n_i > 0,$$

where N is the total number of options in the portfolio, and n_i is the number of units of asset i in the basket.

1.7 Show that the put prices (European and American) are convex functions of the asset price, that is,

$$p(\lambda S_1 + (1-\lambda)S_2, X) \leq \lambda p(S_1, X) + (1-\lambda)p(S_2, X), \quad 0 \leq \lambda \leq 1,$$

where S_1 and S_2 denote the asset prices and X denotes the strike price.

Hint: Let $S_1 = h_1 X$ and $S_2 = h_2 X$, and note that the put price function is homogeneous of degree one in the asset price and the strike price, the above inequality can be expressed as

$$[\lambda h_1 + (1-\lambda)h_2]p\left(X, \frac{X}{\lambda h_1 + (1-\lambda)h_2}\right)$$

$$\leq \lambda h_1 p\left(X, \frac{X}{h_1}\right) + (1-\lambda)h_2 p\left(X, \frac{X}{h_2}\right).$$

Apply the property that the put prices are convex functions of the strike price.

1.8 Consider the following two portfolios:

Portfolio A: One European call option plus X dollars of money market account.

Portfolio B: One American put option, one unit of the underlying asset and borrowing of loan amount D. The loan is in the form of a portfolio of bonds whose par values and dates of maturity match with the sizes and dates of the discrete dividends.

Assume the underlying asset pays dividends and D denotes the present value of the dividends paid by the underlying asset during the life of the option. Show that if the American put is not exercised early, Portfolio B is worth $\max(S_T, X)$, which is less than the value of Portfolio A. Even when the American put is exercised prior to expiry, show that Portfolio A is always worth more than Portfolio B at the moment of exercise. Hence, deduce that

$$S - D - X < C - P.$$

Hint: $c < C$ for calls on a dividend paying asset and the loan (bond) value in Portfolio A grows with time.

1.9 Deduce from the put-call parity relation that the price of a European put on a nondividend paying asset is bounded above by

$$p \leq XB(\tau).$$

Then deduce that the value of a perpetual European put option is zero. When does equality hold in the above inequality?

1.10 Consider a European call option on a foreign currency. Show that

$$c(S, \tau) \sim SB_f(\tau) - XB(\tau) \quad \text{as } S \to \infty.$$

Give a financial interpretation of the result. Deduce the conditions under which the value of a shorter-lived European foreign currency call option is worth more than that of the longer-lived counterpart.

Hint: Use the put-call parity relation (1.2.24). At exceedingly high exchange rates, the European call is almost sure to be in-the-money at expiry.

1.11 Show that the lower and upper bounds on the difference between the prices of the American call and put options on a foreign currency are given by

$$SB_f(\tau) - X < C - P < S - XB(\tau),$$

where $B_f(\tau)$ and $B(\tau)$ are bond prices in the foreign and domestic currencies, respectively, both with par value of unity in the respective currency and time to maturity τ, S is the spot domestic currency price of one unit of foreign currency.
Hint: To show the left inequality, consider the values of the following two portfolios: the first one contains a European currency call option plus X dollars of domestic currency, the second portfolio contains an American currency put option plus $B_f(\tau)$ units of foreign currency. To show the right inequality, we choose the first portfolio to contain an American currency call option plus $XB(\tau)$ dollars of domestic currency, and the second portfolio to contain a European currency put option plus one unit of the foreign currency.

1.12 Suppose the strike price is growing at the riskless interest rate, show that the price of an American put option is the same as that of the corresponding European counterpart.
Hint: Show that the early exercise privilege of the American put is rendered useless.

1.13 Consider a forward contract whose underlying asset has a holding cost of c_j paid at time t_j, $j = 1, 2, \cdots, M-1$, where time t_M is taken to be the maturity date of the forward. For notational simplicity, we take the initiation date of the swap contract to be time t_0. Assume that the asset can be sold short. Let S denote the spot price of the asset at the initiation date, and we use d_j to denote the discount factor at time t_j for cash received on the expiration date. Show that the forward price F of this forward contract is given by

$$F = \frac{S}{d_0} + \sum_{j=1}^{M-1} \frac{c_j}{d_j}.$$

1.14 Consider a one-year forward contract whose underlying asset is a coupon paying bond with maturity date beyond the forward's expiration date. Assume the bond pays coupon semi-annually at the coupon rate of 8%, and the face value of the bond is $100 (that is, each coupon payment is $4). The current market price of the bond is $94.6, and the previous coupon has just been paid. Taking the riskless interest rate to be at the constant value of 10% per annum, find the forward price of this bond forward.
Hint: The coupon payments may be considered as negative costs of carry.

1.15 Consider an interest rate swap of notional principal \$1 million and remaining life of nine months, the terms of the swap specify that six-month LIBOR is exchanged for the fixed rate of 10% per annum (quoted with semi-annual compounding). The market prices of unit par zero coupon bonds with maturity dates three months and nine months from now are \$0.972 and \$0.918, respectively, while the market price of unit par floating rate bond with maturity date three months from now is \$0.992. Find the value of the interest rate swap to the fixed-rate payer, assuming no default risk of the swap counterparty.

1.16 A financial institution X has entered into a five-year currency swap with another institution Y. The swap specifies that X receives fixed interest rate at 4% per annum in euros and pays fixed interest rate at 6% per annum in U.S. dollars. The principal amounts are 10 million U.S. dollars and 13 million euros, and interest payments are exchanged semi-annually. Suppose that Y defaults at the end of Year 3 after the initiation of the swap. Find the replacement cost to the counterparty X. Assume that the exchange rate at the time of default is \$1.32 per euro and the prevailing interest rates for all maturities for U.S. dollars and euros are 5.5% and 3.2%, respectively.

1.17 Suppose two financial institutions X and Y are faced with the following borrowing rates

	X	Y
U.S. dollars floating rate	LIBOR + 2.5%	LIBOR + 4.0%
British sterling fixed rate	4.0%	5.0%

Suppose X wants to borrow British sterling at a fixed rate and Y wants to borrow U.S. dollars at a floating rate. How can a currency swape be arranged that benefits both parties.

1.18 Consider an airlines company that has to purchase oil regularly (say, every three months) for its operations. To avoid the fluctuation of oil prices on the spot market, the company may wish to enter into a *commodity swap* with a financial institution. The following schematic diagram shows the flows of payment in the commodity swap:

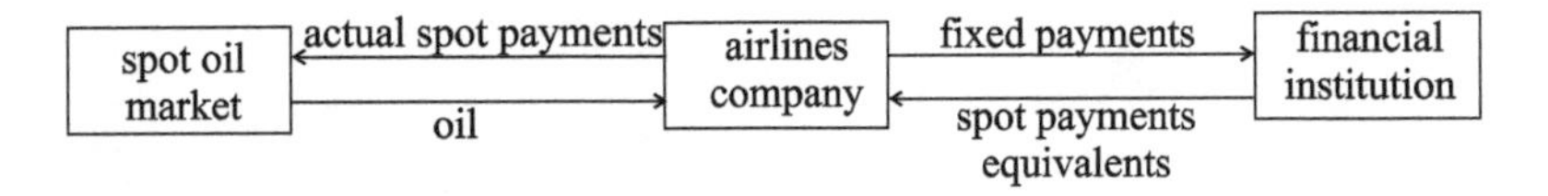

Under the terms of the commodity swap, the airline company receives spot price for a certain number units of oil at each swap date while paying a fixed amount K per unit. Let t_i, $i = 1, 2, \cdots, M$, denote the swap dates and d_i be the discount factor at the swap initiation date for cash received at t_i. Let F_i denote the forward price of one unit of oil to be received at time t_i, and K be the fixed payment per unit paid by the airline company to the swap counterparty. Suppose K is chosen such that the initial value of the commodity swap is zero, show that

$$K = \frac{\sum_{i=1}^{M} d_i F_i}{\sum_{i=1}^{M} d_i}.$$

That is, the fixed rate is a weighted average of the prices of the forward contracts maturing on the swap dates with the corresponding discount factors as weights.

1.19 This problem examines the role of a *financial intermediary* in arranging two separate interest rate swaps with two companies that would like to transform a floating rate loan into a fixed rate loan and vice versa. Consider the following situation:

Company A aims at transforming a fixed rate loan paying 6.2% per annum into a floating rate loan paying LIBOR + 0.2%.
Company B aims at transforming a floating rate loan paying LIBOR + 2.2% into a fixed rate loan paying 8.4% per annum.

Instead of having these two companies getting in touch directly to arrange an interest rate swap, how can a financial intermediary design separate interest swaps with the two companies and secure a profit on the spread of the borrowing rates?

2

Financial Economics and Stochastic Calculus

In the last chapter, we discussed how the application of the no arbitrage argument *enforces* the forward price of a forward contract. The enforceable forward price is not given by the expectation of the asset price at maturity of the forward contract. The adoption of replication of a derivative by marketed assets—together with the use of the no arbitrage argument—form the building blocks of derivative pricing models. For example, a call can be replicated by combining a put and a forward. More interestingly, a European option can be replicated *dynamically* by a portfolio containing the underlying asset and the riskless asset (in the form of a money market account). Assuming frictionless market and no premature termination of the option contract, if the option's payoff matches that of the replicating portfolio at maturity, one can show, by the no arbitrage argument, that the value of the option is then equal to the value of the replicating portfolio at all times throughout the life of the option. If every derivative can be replicated by a portfolio of assets available in the market, then the market is said to be complete. One then prices a derivative based on the prices of the marketed assets in the replicating portfolio.

In this chapter, we apply the theory of financial economics to show that the absence of arbitrage is equivalent to the existence of an equivalent martingale measure. This important result is coined the *Fundamental Theorem of Asset Pricing*. The term "martingale measure" is used because under this measure, all discounted price processes of risky assets are martingales. Also, this martingale measure is *equivalent* to the physical measure that gives the actual probability of occurrence of various states of the world. Further, if the market is complete (all contingent claims can be replicated), then the equivalent martingale measure is unique. It can be shown that the replication-based price of any contingent claim can be obtained by calculating the expected value of its discounted terminal payoff under the equivalent martingale probability measure (Harrison and Kreps, 1979). This approach has come to be known as risk neutral pricing. The term risk neutrality is used because all assets in the market offer the same return as the risk free asset under this probability, so an investor who is neutral to risk and faced with this probability would be indifferent among various assets. The concepts of *replicable contingent claims, absence of arbitrage* and *risk neutrality* form the cornerstones of modern option pricing theory.

In the first two sections, we limit our discussion to discrete time securities models. We start with the single period securities models in Sect. 2.1. We discuss the notions of the law of one price, nondominant trading strategy, linear pricing measure and absence of arbitrage. The use of the Separating Hyperplane Theorem leads to the identification of the risk neutral measure for valuation of contingent claims under the assumption of no arbitrage. In Sect. 2.2, we consider multiperiod securities models, starting with the construction of the information structures of securities models. Various notions in probability theory are presented, like filtrations, measurable random variables, conditional expectations and martingales. Under the multiperiod setting, the risk neutral probability measure is defined in terms of martingales. The highlight is the derivation of the *Fundamental Theorem of Asset Pricing*. More detailed exposition on the related concepts of financial economics can be found in Pliska (1997) and LeRoy and Werner (2001).

The price of a derivative has primary dependence on the stochastic process of the price of the underlying asset. In this text, most continuous asset price processes are modeled by the Ito processes. For equity prices, they are fairly described by the geometric Brownian processes, a popular class of Ito processes. In Sect. 2.3, we provide a brief exposition on the Brownian process. We start with the discrete random walk model and take the Brownian process as the continuous limit of the random walk process. The forward Fokker–Planck equation that governs the transition density function of a Brownian process is derived. In the last section, we introduce some basic tools in stochastic calculus, in particular, the notions of stochastic integrals and stochastic differentials. We explain the nondifferentiability of Brownian paths. We provide an intuitive proof of the Ito Lemma, which is an essential tool in performing calculus operations on functions of stochastic state variables. We also discuss the Feynman–Kac representation, Radon–Nikodym derivative and the Girsanov Theorem. The Girsanov Theorem provides an effective tool to transform Ito processes with general drifts into martingales. These preliminaries in stochastic calculus are essential to develop the option pricing theory and derive option price formulas in later chapters.

2.1 Single Period Securities Models

The no arbitrage approach is one of the cornerstones of pricing theory of financial derivatives. In simple language, arbitrage refers to the possibility of making an investment gain with no chance of loss (the rigorous definition of arbitrage will be given later). It is commonly assumed that there are no arbitrage opportunities in well-functioning and competitive financial markets.

In this section, we discuss various concepts of financial economics under the framework of single period securities models. Investment decisions on a finite set of securities are made at initial time $t = 0$ and the payoff is attained at terminal time $t = 1$. Though single period models may not quite reflect the realistic representation of the complex world of investment activities, a lot of fundamental concepts in financial economics can be revealed from the analysis of single period securities models.

Also, single period investment models approximate quite well the buy-and-hold investment strategies.

2.1.1 Dominant Trading Strategies and Linear Pricing Measures

In a single period securities model, the initial prices of M risky securities, denoted by $S_1(0), \cdots, S_M(0)$, are positive scalars that are known at $t = 0$. However, their values at $t = 1$ are random variables defined with respect to a sample space $\Omega = \{\omega_1, \omega_2, \cdots, \omega_K\}$ of K possible states of the world. At $t = 0$, the investors know the list of all possible outcomes of asset prices at $t = 1$, but which outcome does occur is revealed only at the end of the investment period. Further, a probability measure P satisfying $P(\omega) > 0$, for all $\omega \in \Omega$, is defined on Ω.

We use $\mathbf{S}$ to denote the price process $\{\mathbf{S}(t) : t = 0, 1\}$, where $\mathbf{S}(t)$ is the row vector $\mathbf{S}(t) = (S_1(t)\ S_2(t) \cdots S_M(t))$. The possible values of the asset price process at $t = 1$ are listed in the following $K \times M$ matrix

$$S(1; \Omega) = \begin{pmatrix} S_1(1;\omega_1) & S_2(1;\omega_1) & \cdots & S_M(1;\omega_1) \\ S_1(1;\omega_2) & S_2(1;\omega_2) & \cdots & S_M(1;\omega_2) \\ \cdots & \cdots & \cdots & \cdots \\ S_1(1;\omega_K) & S_2(1;\omega_K) & \cdots & S_M(1;\omega_K) \end{pmatrix}. \tag{2.1.1}$$

Since the assets are limited liability securities, the entries in $S(1; \Omega)$ are nonnegative scalars. We also assume the existence of a strictly positive riskless security or money market account, whose value is denoted by S_0. Without loss of generality, we take $S_0(0) = 1$ and the value at time 1 to be $S_0(1) = 1 + r$, where $r \geq 0$ is the deterministic interest rate over one period. The reciprocal of $S_0(1)$ is called the discount factor over the period. We define the discounted price process by

$$\mathbf{S}^*(t) = \mathbf{S}(t)/S_0(t), \quad t = 0, 1,$$

that is, we use the riskless security as the *numeraire* or *accounting unit*. Accordingly, the payoff matrix of the discounted price processes of the M risky assets and the riskless security can be expressed in the form

$$\widehat{S}^*(1; \Omega) = \begin{pmatrix} 1 & S_1^*(1;\omega_1) & \cdots & S_M^*(1;\omega_1) \\ 1 & S_1^*(1;\omega_2) & \cdots & S_M^*(1;\omega_2) \\ \cdots & \cdots & \cdots & \cdots \\ 1 & S_1^*(1;\omega_K) & \cdots & S_M^*(1;\omega_K) \end{pmatrix}. \tag{2.1.2}$$

The first column in $\widehat{S}^*(1; \Omega)$ (all entries are equal to one) represents the discounted payoff of the riskless security under all states of the world. Also, we define the vector of discounted price processes associated with the riskless security and the M risky securities by

$$\widehat{\mathbf{S}}^*(t) = (1 \quad S_1^*(t) \cdots S_M^*(t)), \qquad t = 0, 1.$$

An investor adopts a *trading strategy* by selecting a portfolio of the assets at time 0. The number of units of the mth asset held in the portfolio from $t = 0$ to $t = 1$

is denoted by $h_m, m = 0, 1, \cdots, M$. The scalars h_m can be positive (long holding), negative (short selling) or zero (no holding).

Let $V = \{V_t : t = 0, 1\}$ denote the value process that represents the total value of the portfolio over time. It is seen that

$$V_t = h_0 S_0(t) + \sum_{m=1}^{M} h_m S_m(t), \quad t = 0, 1. \tag{2.1.3}$$

The gain due to the investment on the mth risky security is given by $h_m[S_m(1) - S_m(0)] = h_m \Delta S_m$. Let G be the random variable that denotes the total gain generated by investing in the portfolio. We then have

$$G = h_0 r + \sum_{m=1}^{M} h_m \Delta S_m. \tag{2.1.4}$$

If there is no withdrawal or addition of funds within the investment horizon, then

$$V_1 = V_0 + G. \tag{2.1.5}$$

We define the discounted value process by $V_t^* = V_t / S_0(t)$ and discounted gain by $G^* = V_1^* - V_0^*$. It is seen that

$$V_t^* = h_0 + \sum_{m=1}^{M} h_m S_m^*(t), \quad t = 0, 1; \tag{2.1.6a}$$

$$G^* = V_1^* - V_0^* = \sum_{m=1}^{M} h_m \Delta S_m^*. \tag{2.1.6b}$$

Dominant Trading Strategies

Let $\mathcal{H}$ denote the trading strategy that involves the choice of the number of units of assets held in the portfolio. The trading strategy $\mathcal{H}$ is said to be *dominant* if there exists another trading strategy $\widehat{\mathcal{H}}$ such that

$$V_0 = \widehat{V}_0 \quad \text{and} \quad V_1(\omega) > \widehat{V}_1(\omega) \quad \text{for all } \omega \in \Omega. \tag{2.1.7}$$

Here, $\widehat{V}_0$ and $\widehat{V}_1$ denote the portfolio value of $\widehat{\mathcal{H}}$ at $t = 0$ and $t = 1$, respectively. Financially speaking, both strategies $\mathcal{H}$ and $\widehat{\mathcal{H}}$ start with the same initial investment amount but the dominant strategy $\mathcal{H}$ leads to a higher gain under all possible states of the world.

Suppose $\mathcal{H}$ dominates $\widehat{\mathcal{H}}$, we define a new trading strategy $\widetilde{\mathcal{H}} = \mathcal{H} - \widehat{\mathcal{H}}$. Let $\widetilde{V}_0$ and $\widetilde{V}_1$ denote the portfolio value of $\widetilde{\mathcal{H}}$ at $t = 0$ and $t = 1$, respectively. From (2.1.7), we then have $\widetilde{V}_0 = 0$ and $\widetilde{V}_1(\omega) > 0$ for all $\omega \in \Omega$. This trading strategy is dominant since it dominates the strategy which starts with zero value and does no investment at all. A securities model that allows the existence of a dominant trading strategy is not realistic since an investor starting with no money should not be guaranteed of ending up with positive returns by adopting a particular trading strategy. Equivalently, one can show that a dominant trading strategy is one that can transform strictly negative wealth at $t = 0$ into nonnegative wealth at $t = 1$ (see Problem 2.1). Later, we show how the nonexistence of dominant strategies is equivalent to the existence of a linear pricing measure.

Asset Span, Law of One Price and State Prices

Consider the following numerical example, where the number of possible states is taken to be three. First, we consider two risky securities whose discounted payoff vectors are $\mathbf{S}_1^*(1) = \begin{pmatrix} 1 \\ 2 \\ 3 \end{pmatrix}$ and $\mathbf{S}_2^*(1) = \begin{pmatrix} 3 \\ 1 \\ 2 \end{pmatrix}$. The payoff vectors are used to form the payoff matrix $S^*(1) = \begin{pmatrix} 1 & 3 \\ 2 & 1 \\ 3 & 2 \end{pmatrix}$. Let the current discounted prices be represented by the row vector $\mathbf{S}^*(0) = (1 \quad 2)$. We write $\mathbf{h}$ as the column vector whose entries are the weights of the securities in the portfolio. The current discounted portfolio value and the discounted portfolio payoff are given by $\mathbf{S}^*(0)\mathbf{h}$ and $S^*(1)\mathbf{h}$, respectively. As $S_0^*(0) = 1$, the current portfolio value and discounted portfolio value are the same.

The set of all portfolio payoffs via different holding of securities is called the *asset span* $\mathcal{S}$. The asset span is seen to be the column space of the payoff matrix $S^*(1)$. In this example, the asset span consists of all vectors of the form $h_1 \begin{pmatrix} 1 \\ 2 \\ 3 \end{pmatrix} + h_2 \begin{pmatrix} 3 \\ 1 \\ 2 \end{pmatrix}$, where h_1 and h_2 are scalars.

To these two securities in the portfolio, we may add a third security or even more securities. The newly added securities may or may not fall within the asset span. If the added security lies inside $\mathcal{S}$, then its payoff can be expressed as a linear combination of $\mathbf{S}_1^*(1)$ and $\mathbf{S}_2^*(1)$. In this case, it is said to be a *redundant security*. Since there are only three possible states, the dimension of the asset span cannot be more than three, that is, the maximal number of nonredundant securities is three. Suppose we add the third security whose discounted payoff is $\mathbf{S}_3^*(1) = \begin{pmatrix} 1 \\ 3 \\ 4 \end{pmatrix}$, it can be easily checked that it is a nonredundant security. The new asset span, which is the subspace in $\mathbb{R}^3$ spanned by $\mathbf{S}_1^*(1)$, $\mathbf{S}_2^*(1)$ and $\mathbf{S}_3^*(1)$, is the whole $\mathbb{R}^3$. Any further security added must be redundant since its discounted payoff vector must lie inside the new asset span. A securities model is said to be *complete* if every payoff vector lies inside the asset span. This occurs if and only if the dimension of the asset span equals the number of possible states. In this case, any new security added to the securities model must be a redundant security.

The law of one price states that all portfolios with the same payoff have the same price. Consider two portfolios with different portfolio weights $\mathbf{h}$ and $\mathbf{h}'$. Suppose these two portfolios have the same discounted payoff, that is, $S^*(1)\mathbf{h} = S^*(1)\mathbf{h}'$, then the law of one price infers that $\mathbf{S}^*(0)\mathbf{h} = \mathbf{S}^*(0)\mathbf{h}'$. It is quite straightforward to show that a sufficient condition for the law of one price to hold is that a portfolio with zero payoff must have zero price. This occurs if and only if the dimension of the null space of the payoff matrix $S^*(1)$ is zero. Also, if the law of one price fails, then it is possible to have two trading strategies $\mathbf{h}$ and $\mathbf{h}'$ such that $S^*(1)\mathbf{h} = S^*(1)\mathbf{h}'$ but $\mathbf{S}^*(0)\mathbf{h} > \mathbf{S}^*(0)\mathbf{h}'$. Let $G^*(\omega)$ and $G^{*\prime}(\omega)$ denote the respective discounted gain corresponding to the trading strategies $\mathbf{h}$ and $\mathbf{h}'$. We then have $G^{*\prime}(\omega) > G^*(\omega)$ for all $\omega \in \Omega$, so there exists a dominant trading strategy. Hence, the nonexistence of dominant trading strategy implies the law of one price. However, the converse statement does not hold (see Problem 2.4).

Given a discounted portfolio payoff $\mathbf{x}$ that lies inside the asset span, the payoff can be generated by some linear combination of the securities in the securities model. We have $\mathbf{x} = S^*(1)\mathbf{h}$ for some $\mathbf{h} \in \mathbb{R}^M$. The current discounted value of the portfolio is $\mathbf{S}^*(0)\mathbf{h}$, where $\mathbf{S}^*(0)$ is the discounted price vector. We may consider $\mathbf{S}^*(0)\mathbf{h}$ as a pricing functional $F(\mathbf{x})$ on the payoff $\mathbf{x}$. If the law of one price holds, then the pricing functional is single-valued. Furthermore, it can be shown to be a linear functional, that is,

$$F(\alpha_1\mathbf{x}_1 + \alpha_2\mathbf{x}_2) = \alpha_1 F(\mathbf{x}_1) + \alpha_2 F(\mathbf{x}_2) \tag{2.1.8}$$

for any scalars α_1 and α_2 and payoffs $\mathbf{x}_1$ and $\mathbf{x}_2$ (see Problem 2.5).

Let $\mathbf{e}_k$ denote the kth coordinate vector in the vector space $\mathbb{R}^K$, where $\mathbf{e}_k$ assumes the value 1 in the kth entry and zero in all other entries. The vector $\mathbf{e}_k$ can be considered as the discounted payoff vector of a security, and it is called the Arrow security of state k. This Arrow security has unit payoff when state k occurs and zero payoff otherwise. Suppose the securities model is complete and the law of one price holds, then the pricing functional F assigns unique value to each Arrow security. We write $s_k = F(\mathbf{e}_k)$, which is called the state price of state k.

Consider the risky security with discounted payoff at time $t = 1$ represented by

$$\mathbf{S}^*(1) = \begin{pmatrix} \alpha_1 \\ \vdots \\ \alpha_K \end{pmatrix},$$

then the current price of this risky security is given by

$$S^*(0) = F(\mathbf{S}^*(1)) = F\left(\sum_{k=1}^K \alpha_k \mathbf{e}_k\right) = \sum_{k=1}^K \alpha_k s_k.$$

Linear Pricing Measure

We consider securities models with the inclusion of the risk free security. A nonnegative row vector $\mathbf{q} = (q(\omega_1) \cdots q(\omega_K))$ is said to be a linear pricing measure if, for every trading strategy, the associated discounted portfolio values at $t = 0$ and $t = 1$ satisfy

$$V_0^* = \sum_{k=1}^K q(\omega_k) V_1^*(\omega_k). \tag{2.1.9}$$

The linear pricing measure exhibits the following properties. First, suppose we take the holding amount of each risky security to be zero, thereby $h_1 = h_2 = \cdots = h_M = 0$. With only the risk free asset in the portfolio, we have

$$V_0^* = h_0 = \sum_{k=1}^{K} q(\omega_k) h_0,$$

so that

$$\sum_{k=1}^{K} q(\omega_k) = 1. \tag{2.1.10}$$

Since we have taken $q(\omega_k) \geq 0, k = 1, \cdots, K$, and their sum is one, we may interpret $q(\omega_k)$ as a probability measure on the sample space Ω. Next, by taking the portfolio weights to be zero except for the mth security, we have

$$S_m^*(0) = \sum_{k=1}^{K} q(\omega_k) S_m^*(1; \omega_k), \quad m = 1, \cdots, M. \tag{2.1.11a}$$

The current discounted security price is given by the expectation of the discounted security payoff one period later under the linear pricing measure $q(\omega_k)$. Note that $q(\omega_k)$ is not related to the actual probability of occurrence of the state k. In matrix form, (2.1.11a) can be expressed as

$$\widehat{\mathbf{S}}^*(0) = \mathbf{q}\widehat{S}^*(1; \Omega), \quad \mathbf{q} \geq \mathbf{0}. \tag{2.1.11b}$$

As a numerical example, we consider a securities model with two risky securities and the risk free security, and there are three possible states. The current discounted price vector $\widehat{\mathbf{S}}^*(0)$ is $(1 \quad 4 \quad 2)$ and the discounted payoff matrix at $t = 1$ is $\widehat{S}^*(1) = \begin{pmatrix} 1 & 4 & 3 \\ 1 & 3 & 2 \\ 1 & 2 & 4 \end{pmatrix}$. Here, the law of one price holds since the only solution to $\widehat{S}^*(1)\mathbf{h} = \mathbf{0}$ is $\mathbf{h} = \mathbf{0}$. This is because the columns of $\widehat{S}^*(1)$ are independent so that the dimension of the null space of $\widehat{S}^*(1)$ is zero. We would like to see whether a linear pricing measure exists for the given securities model. By virtue of (2.1.10) and (2.1.11a), the linear pricing probabilities $q(\omega_1), q(\omega_2)$ and $q(\omega_3)$, if exist, should satisfy the following system of linear equations:

$$\begin{aligned} 1 &= q(\omega_1) + q(\omega_2) + q(\omega_3) \\ 4 &= 4q(\omega_1) + 3q(\omega_2) + 2q(\omega_3) \\ 2 &= 3q(\omega_1) + 2q(\omega_2) + 4q(\omega_3). \end{aligned}$$

Solving the above equations, we obtain $q(\omega_1) = q(\omega_2) = 2/3$ and $q(\omega_3) = -1/3$. Since not all the pricing probabilities are nonnegative, the linear pricing measure does not exist for this securities model.

Do dominant trading strategies exist for the above securities model? That is, can we find a trading strategy $(h_1 \; h_2)$ such that $V_0^* = 4h_1 + 2h_2 = 0$ but $V_1^*(\omega_k) > 0, k = 1, 2, 3$? This is equivalent to asking whether there exist h_1 and h_2 such that $4h_1 + 2h_2 = 0$ and

$$\begin{aligned} 4h_1 + 3h_2 &> 0 \\ 3h_1 + 2h_2 &> 0 \\ 2h_1 + 4h_2 &> 0. \end{aligned} \tag{2.1.12}$$

In Fig. 2.1, we show the region containing the set of points in the (h_1, h_2)-plane that satisfy inequalities (2.1.12). The region is found to be lying on the top right sides above the two bold lines: (i) $3h_1 + 2h_2 = 0$, $h_1 < 0$ and (ii) $2h_1 + 4h_2 = 0$, $h_1 > 0$. It is seen that all the points on the dotted half line: $4h_1 + 2h_2 = 0$, $h_1 < 0$ represent dominant trading strategies that start with zero wealth but end with positive wealth with certainty.

Suppose the initial discounted price vector is changed from $(1 \quad 4 \quad 2)$ to $(1 \quad 3 \quad 3)$, the new set of linear pricing probabilities will be determined by solving

$$
\begin{aligned}
1 &= q(\omega_1) + q(\omega_2) + q(\omega_3) \\
3 &= 4q(\omega_1) + 3q(\omega_2) + 2q(\omega_3) \\
3 &= 3q(\omega_1) + 2q(\omega_2) + 4q(\omega_3),
\end{aligned}
$$

which is seen to have the solution: $q(\omega_1) = q(\omega_2) = q(\omega_3) = 1/3$. Now, all the pricing probabilities have nonnegative values, and the row vector $\mathbf{q} = (1/3 \quad 1/3 \quad 1/3)$ represents a linear pricing measure. Referring to Fig. 2.1, we observe that the line $3h_1 + 3h_2 = 0$ always lies outside the region above the two bold lines. Hence, with respect to this new securities model, we cannot find $(h_1 \quad h_2)$ such that $3h_1 + 3h_2 = 0$ together with h_1 and h_2 satisfying inequalities (2.1.12). Since a linear pricing measure exists, by virtue of (2.1.11a), we expect that the initial price vector of the two risky securities: $(3 \quad 3)$ can be expressed as some linear combination of the three

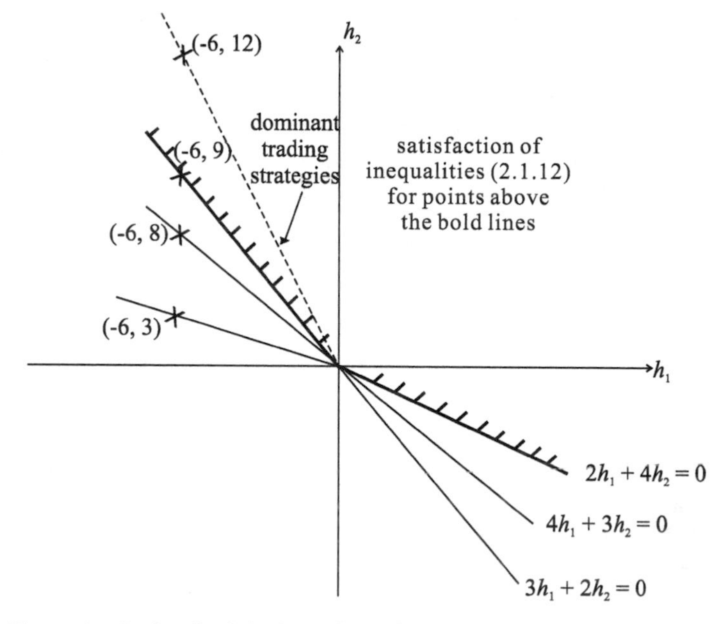

Fig. 2.1. The region in the $(h_1$-$h_2)$-plane above the two bold lines represents trading strategies that satisfy inequalities (2.1.12). The trading strategies that lie on the dotted line: $4h_1 + 2h_2 = 0$, $h_1 < 0$ are dominant trading strategies.

vectors: $(4 \quad 3)$, $(3 \quad 2)$ and $(2 \quad 4)$ with nonnegative weights. Actually, we have

$$(3 \quad 3) = \frac{1}{3}(4 \quad 3) + \frac{1}{3}(3 \quad 2) + \frac{1}{3}(2 \quad 4),$$

where the weights are the linear pricing probabilities.

The relation between the existence of a linear pricing measure and the nonexistence of dominant trading strategies is stated in the following theorem.

Theorem 2.1. *There exists a linear pricing measure if and only if there are no dominant trading strategies.*

The above linear pricing measure theorem can be seen to be a direct consequence of the Farkas Lemma.

Farkas Lemma. *There does not exist* $\mathbf{h} \in \mathbb{R}^{M+1}$ *such that*

$$\widehat{S}^*(1;\Omega)\mathbf{h} > \mathbf{0} \quad \text{and} \quad \widehat{\mathbf{S}}^*(0)\mathbf{h} = 0$$

if and only if there exists $\mathbf{q} \in \mathbb{R}^K$ *such that*

$$\widehat{\mathbf{S}}^*(0) = \mathbf{q}\widehat{S}^*(1;\Omega) \quad \text{and} \quad \mathbf{q} \geq \mathbf{0}.$$

As a remark, solution to the linear system (2.1.11b) exists if and only if $\widehat{\mathbf{S}}^*(0)$ lies in the row space of $\widehat{S}^*(1;\Omega)$. However, the solution vector $\mathbf{q}$ may not satisfy the nonnegativity property: $\mathbf{q} \geq \mathbf{0}$. Dominant trading strategies do not exist if and only if $\mathbf{q} \geq 0$. When the row vectors of $\widehat{S}^*(1;\Omega)$ are independent, if a solution $\mathbf{q}$ exists, then it must be unique.

2.1.2 Arbitrage Opportunities and Risk Neutral Probability Measures

Suppose $\mathbf{S}^*(0)$ in the above securities model is modified to $(3 \quad 2)$ and consider the trading strategy: $h_1 = -2$ and $h_2 = 3$. We observe that $V_0^* = 0$ and the possible discounted payoffs at $t = 1$ are: $V_1^*(\omega_1) = 1$, $V_1^*(\omega_2) = 0$ and $V_1^*(\omega_3) = 8$. This represents a trading strategy that starts with zero wealth, guarantees no loss, and ends up with a strictly positive wealth in some states (not necessarily in all states). The occurrence of such investment opportunity is called an arbitrage opportunity.

Formally, we define an *arbitrage opportunity* to be some trading strategy that has the following properties: (i) $V_0^* = 0$, (ii) $V_1^*(\omega) \geq 0$ and $E[V_1^*(\omega)] > 0$, where E is the expectation under the actual probability measure P, $P(\omega) > 0$. Note the difference between a dominant strategy and an arbitrage opportunity. Recall that a dominant trading strategy exists when a portfolio with initial zero wealth ends up with a *strictly* positive wealth in all states. Therefore, the existence of a dominant trading strategy implies the existence of an arbitrage opportunity, but the converse is not necessarily true. In other words, the absence of arbitrage implies the nonexistence of dominant trading strategy and in turn implying that the law of one price holds.

Existence of arbitrage opportunities is unreasonable from the economic standpoint. The natural question: What would be the necessary and sufficient condition for the nonexistence of arbitrage opportunities? The answer is related to the existence of a pricing measure, called the risk neutral probability measure. In financial markets with no arbitrage opportunities, we will show that every investor should use such risk neutral probability measure (though not necessarily unique) to find the fair value of a security or a portfolio, irrespective to the risk preference of the investor. That is, the fair value is independent of the probability values assigned to the occurrence of the states of the world by an individual investor.

Risk Neutral Probability Measure

The example just mentioned above represents the presence of an arbitrage opportunity but nonexistence of a dominant trading strategy [since $V_1^*(\omega) = 0$ for some ω]. The linear pricing measure vector is found to be $(0 \quad 1 \quad 0)$, where two of the linear pricing probabilities are zero. In order to exclude arbitrage opportunities, we need a bit stronger condition on the pricing probabilities, namely, the probabilities must be strictly positive.

A probability measure Q on Ω is said to be a risk neutral probability measure if it satisfies

(i) $Q(\omega) > 0$ for all $\omega \in \Omega$, and
(ii) $E_Q[\Delta S_m^*] = 0, m = 1, \cdots, M$,

where E_Q denotes the expectation under Q. Note that $E_Q[\Delta S_m^*] = 0$ is equivalent to

$$S_m^*(0) = \sum_{k=1}^{K} Q(\omega_k) S_m^*(1; \omega_k),$$

which takes a similar form as in (2.1.11a). Indeed, a linear pricing measure becomes a risk neutral probability measure if the probability masses are all positive. The strict positivity property of the risk neutral probability measure $Q(\omega)$ is more desirable since $Q(\omega)$ is seen to be "equivalent" to the actual probability measure $P(\omega)$, where $P(\omega) > 0$. That is P and Q may not agree on the assignment of probability values to individual events, but they always agree as to which events are possible or impossible. The notion of "equivalent probability measures" will be discussed in more detail in Sect. 2.2.1.

Fundamental Theorem of Asset Pricing (Single Period Models)

The existence of a risk neutral measure is directly related to the exclusion of arbitrage opportunities as stated in the following theorem.

Theorem 2.2. *No arbitrage opportunities exist if and only if there exists a risk neutral probability measure Q.*

The proof of Theorem 2.2 requires the Separating Hyperplane Theorem. A geometric intuition of the theorem is given here. First, we present the definitions of hyperplane and convex set in a vector space. Let $\mathbf{f}$ be a vector in $\mathbb{R}^n$. The hyperplane $H = [\mathbf{f}, \alpha]$ in $\mathbb{R}^n$ is defined to be the collection of those vectors $\mathbf{x}$ in $\mathbb{R}^n$ whose projection onto $\mathbf{f}$ has magnitude α. For example, the collection of vectors $\begin{pmatrix} x_1 \\ x_2 \\ x_3 \end{pmatrix}$ satisfying $x_1 + 2x_2 + 3x_3 = 2$ is a hyperplane in $\mathbb{R}^3$, where $\mathbf{f} = \begin{pmatrix} 1 \\ 2 \\ 3 \end{pmatrix}$ and $\alpha = 2$. A set C in $\mathbb{R}^n$ is said to be convex if for any pair of vectors $\mathbf{x}$ and $\mathbf{y}$ in C, all convex combinations of $\mathbf{x}$ and $\mathbf{y}$ represented by the form $\lambda\mathbf{x} + (1-\lambda)\mathbf{y}$, $0 \le \lambda \le 1$, also lie in C. For example, the set $C = \left\{ \begin{pmatrix} x_1 \\ x_2 \\ x_3 \end{pmatrix} : x_1 \ge 0, x_2 \ge 0, x_3 \ge 0 \right\}$ is a convex set in $\mathbb{R}^3$. The hyperplane $[\mathbf{f}, \alpha]$ separates the sets A and B in $\mathbb{R}^n$ if there exists α such that $\mathbf{f} \cdot \mathbf{x} \ge \alpha$ for all $\mathbf{x} \in A$ and $\mathbf{f} \cdot \mathbf{y} < \alpha$ for all $\mathbf{y} \in B$. For example, the hyperplane $\left[\begin{pmatrix} 1 \\ 1 \\ 1 \end{pmatrix}, 0 \right]$ separates the two disjoint convex sets $A = \left\{ \begin{pmatrix} x_1 \\ x_2 \\ x_3 \end{pmatrix} : x_1 \ge 0, x_2 \ge 0, x_3 \ge 0 \right\}$ and $B = \left\{ \begin{pmatrix} x_1 \\ x_2 \\ x_3 \end{pmatrix} : x_1 < 0, x_2 < 0, x_3 < 0 \right\}$ in $\mathbb{R}^3$.

The *Separating Hyperplane Theorem* states that if A and B are two nonempty disjoint convex sets in a vector space V, then they can be separated by a hyperplane. A pictorial interpretation of the Separating Hyperplane Theorem for the vector space $\mathbb{R}^2$ is shown in Fig. 2.2.

Proof of Theorem 2.2. "$\Leftarrow$ part". Assume that a risk neutral probability measure Q exists, that is, $\widehat{\mathbf{S}}^*(0) = \boldsymbol{\pi} \widehat{S}^*(1; \Omega)$, where $\boldsymbol{\pi} = (Q(\omega_1) \cdots Q(\omega_K))$. Consider a trading strategy $\mathbf{h} = (h_0 \; h_1 \; \cdots \; h_M)^T \in \mathbb{R}^{M+1}$ such that $\widehat{S}^*(1; \Omega)\mathbf{h} \ge 0$ in all $\omega \in \Omega$ and with strict inequality in some states. Now consider $\widehat{\mathbf{S}}^*(0)\mathbf{h} = \boldsymbol{\pi} \widehat{S}^*(1; \Omega)\mathbf{h}$, we can only have $\widehat{\mathbf{S}}^*(0)\mathbf{h} > 0$ since all entries in $\boldsymbol{\pi}$ are strictly positive and entries in $\widehat{S}^*(1; \Omega)\mathbf{h}$ are either zero or strictly positive. Hence, no arbitrage opportunities exist.

"$\Rightarrow$ part". First, we define the subset U in $\mathbb{R}^{K+1}$ which consists of vectors of the form $\begin{pmatrix} -\widehat{\mathbf{S}}^*(0)\mathbf{h} \\ \widehat{\mathbf{S}}^*(1;\omega_1)\mathbf{h} \\ \vdots \\ \widehat{\mathbf{S}}^*(1;\omega_K)\mathbf{h} \end{pmatrix}$, where $\widehat{\mathbf{S}}^*(1; \omega_k)$ is the kth row in $\widehat{S}^*(1; \Omega)$ and $\mathbf{h} \in \mathbb{R}^{M+1}$ represents

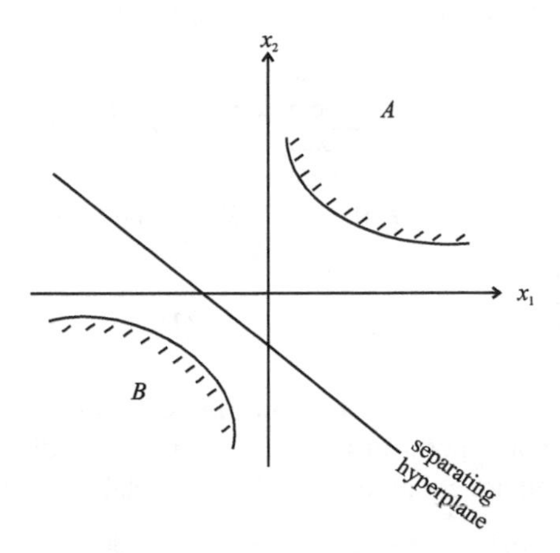

Fig. 2.2. The hyperplane (represented by a line in $\mathbb{R}^2$) separates the two convex sets A and B in $\mathbb{R}^2$.

a trading strategy. Since the sum of any two trading strategies is a trading strategy and any scalar multiple of a trading strategy is also a trading strategy, the subset U is seen to be a subspace in $\mathbb{R}^{K+1}$. Note that U contains the zero vector in $\mathbb{R}^{K+1}$ and obviously U is also convex. Consider another subset $\mathbb{R}_+^{K+1}$ defined by

$$\mathbb{R}_+^{K+1} = \{(x_0\, x_1 \cdots x_K)^T \in \mathbb{R}^{K+1} : x_i \geq 0 \quad \text{for all} \quad 0 \leq i \leq K\},$$

which is a convex set in $\mathbb{R}^{K+1}$. We claim that the nonexistence of arbitrage opportunities implies that U and $\mathbb{R}_+^{K+1}$ can have only the zero vector in common.

Assume the contrary, suppose there exists a nonzero vector $\mathbf{x} \in U \cap \mathbb{R}_+^{K+1}$. Since there is a trading strategy vector $\mathbf{h}$ associated with every vector in U, it suffices to show that the trading strategy $\mathbf{h}$ associated with $\mathbf{x}$ always represents an arbitrage opportunity. We consider the following two cases: $-\widehat{\mathbf{S}}^*(0)\mathbf{h} = 0$ or $-\widehat{\mathbf{S}}^*(0)\mathbf{h} > 0$.

(i) When $\widehat{\mathbf{S}}^*(0)\mathbf{h} = 0$, since $\mathbf{x} \neq \mathbf{0}$ and $\mathbf{x} \in \mathbb{R}_+^{K+1}$, then the entries $\widehat{\mathbf{S}}(1;\omega_k)\mathbf{h}$, $k = 1, 2, \cdots K$, must be all greater than or equal to zero, with at least one strict inequality. In this case, $\mathbf{h}$ is seen to represent an arbitrage opportunity.

(ii) When $\widehat{\mathbf{S}}^*(0)\mathbf{h} < 0$, all the entries $\widehat{\mathbf{S}}(1;\omega_k)\mathbf{h}$, $k = 1, 2, \cdots, K$ must be all nonnegative. Consequently, $\mathbf{h}$ represents a dominant trading strategy (see Problem 2.1) and in turn $\mathbf{h}$ is an arbitrage opportunity.

Since $U \cap \mathbb{R}_+^{K+1} = \{\mathbf{0}\}$, U and $\mathbb{R}_+^{K+1}\backslash\{\mathbf{0}\}$ are disjoint convex subsets in $\mathbb{R}^{K+1}$. By the Separating Hyperplane Theorem, there exists a hyperplane that separates these two disjoint nonempty convex sets. Let $\mathbf{f} \in \mathbb{R}^{K+1}$ be the normal to this hyperplane, then we have $\mathbf{f}\cdot\mathbf{x} > \mathbf{f}\cdot\mathbf{y}$, where $\mathbf{x} \in \mathbb{R}_+^{K+1}\backslash\{\mathbf{0}\}$ and $\mathbf{y} \in U$. [*Remark*: We may have $\mathbf{f}\cdot\mathbf{x} < \mathbf{f}\cdot\mathbf{y}$, depending on the orientation of the normal. However, the final conclusion remains unchanged.] Since U is a linear subspace so that a negative multiple of $\mathbf{y} \in U$ also belongs to U, the condition $\mathbf{f}\cdot\mathbf{x} > \mathbf{f}\cdot\mathbf{y}$ holds only if $\mathbf{f}\cdot\mathbf{y} = 0$ for all $\mathbf{y} \in U$. We then have $\mathbf{f}\cdot\mathbf{x} > 0$ for all $\mathbf{x}$ in $\mathbb{R}_+^{K+1}\backslash\{\mathbf{0}\}$. This requires all entries in $\mathbf{f}$ to be strictly positive. If otherwise, suppose the ith component of $\mathbf{f}$ is nonpositive, then we choose $\mathbf{x}$ to be the coordinate vector with only the ith component equals one while all other components are zero. This leads to a contradiction that $\mathbf{f}\cdot\mathbf{x} > 0$ for all $\mathbf{x}$. Also, from $\mathbf{f}\cdot\mathbf{y} = 0$, we have

$$-f_0\widehat{\mathbf{S}}^*(0)\mathbf{h} + \sum_{k=1}^{K} f_k\widehat{\mathbf{S}}^*(1;\omega_k)\mathbf{h} = 0$$

for all $\mathbf{h} \in \mathbb{R}^{M+1}$, where f_j, $j = 0, 1, \cdots, K$ are the entries of $\mathbf{f}$. We then deduce that

$$\widehat{\mathbf{S}}^*(0) = \sum_{k=1}^{K} Q(\omega_k)\widehat{\mathbf{S}}^*(1;\omega_k), \text{ where } Q(\omega_k) = f_k/f_0. \tag{2.1.13a}$$

Finally, we consider the first component in the vectors on both sides of the above equation. They both correspond to the current price and discounted payoff of the riskless security, and all are equal to one. We then obtain

$$1 = \sum_{k=1}^{K} Q(\omega_k).$$

Here, we obtain the risk neutral probabilities $Q(\omega_k), k = 1, \cdots, K$, whose sum is equal to one and they are all strictly positive since $f_j > 0, j = 0, 1, \cdots, K$.

Remarks.

1. Corresponding to each risky asset, (2.1.13a) dictates that

$$S_m(0) = S_m^*(0) = \sum_{k=1}^{K} Q(\omega_k) S_m^*(1; \omega_k), \quad m = 1, 2, \cdots, M. \tag{2.1.13b}$$

   Hence, the current price of any risky security is given by the expectation of the discounted payoff under the risk neutral measure Q.
2. The risk neutral probabilities $Q(\omega_k)$ are related to the components of the normal to the separating hyperplane. The existence of Q arises from the existence of the hyperplane (not necessarily unique). Since f_0 and other components of $\mathbf{f}$ always have the same sign, the positivity of $Q(\omega_k)$ then follows (independent of the choice of the orientation of the normal to the hyperplane).

Set of Risk Neutral Measures

Consider the earlier securities model with the risk free security and only one risky security, where $\widehat{S}(1; \Omega) = \begin{pmatrix} 1 & 4 \\ 1 & 3 \\ 1 & 2 \end{pmatrix}$ and $\widehat{\mathbf{S}}(0) = (1 \quad 3)$. The risk neutral probability measure $\boldsymbol{\pi} = (Q(\omega_1) \quad Q(\omega_2) \quad Q(\omega_3))$, if exists, is determined by the following system of equations

$$(Q(\omega_1) \quad Q(\omega_2) \quad Q(\omega_3)) \begin{pmatrix} 1 & 4 \\ 1 & 3 \\ 1 & 2 \end{pmatrix} = (1 \quad 3). \tag{2.1.14}$$

Since there are more unknowns than the number of equations, the solution is not unique. The solution is found to be $\boldsymbol{\pi} = (\lambda \quad 1-2\lambda \quad \lambda)$, where λ is a free parameter. In order that all risk neutral probabilities are all strictly positive, we must have $0 < \lambda < 1/2$.

Suppose we add another risky security with discounted payoff $\mathbf{S}_2^*(1) = \begin{pmatrix} 3 \\ 2 \\ 4 \end{pmatrix}$ and current discounted value $S_2^*(0) = 3$. With this new addition, the securities model becomes complete (the asset span of the two risky securities and the risk free security is the whole $\mathbb{R}^3$ space). With the new equation $3Q(\omega_1) + 2Q(\omega_2) + 4Q(\omega_3) = 3$ added to the system (2.1.14), this new securities model is seen to have the unique risk neutral measure $(1/3 \quad 1/3 \quad 1/3)$. The uniqueness of the risk neutral measure stems from the completeness of the securities model (see Problem 2.14).

Let W be a subspace in $\mathbb{R}^K$ which consists of discounted gains corresponding to some trading strategy $\mathbf{h}$. In the above securities model, the discounted gains of the first and second risky securities are $\begin{pmatrix} 4 \\ 3 \\ 2 \end{pmatrix} - \begin{pmatrix} 3 \\ 3 \\ 3 \end{pmatrix} = \begin{pmatrix} 1 \\ 0 \\ -1 \end{pmatrix}$ and $\begin{pmatrix} 3 \\ 2 \\ 4 \end{pmatrix} - \begin{pmatrix} 3 \\ 3 \\ 3 \end{pmatrix} = \begin{pmatrix} 0 \\ -1 \\ 1 \end{pmatrix}$, respectively. Therefore, the corresponding discounted gain subspace is given by

$$W = \left\{ h_1 \begin{pmatrix} 1 \\ 0 \\ -1 \end{pmatrix} + h_2 \begin{pmatrix} 0 \\ -1 \\ 1 \end{pmatrix}, \text{ where } h_1 \text{ and } h_2 \text{ are scalars} \right\}.$$

For any risk neutral probability measure Q, we have

$$\begin{aligned} E_Q G^* &= \sum_{k=1}^{K} Q(\omega_k) \left[\sum_{m=1}^{M} h_m \Delta S_m^*(\omega_k) \right] \\ &= \sum_{m=1}^{M} h_m E_Q[\Delta S_m^*] = 0, \end{aligned} \tag{2.1.15}$$

where $\Delta S_m^*(\omega_k)$ is the discounted gain on the mth risky security when the state ω_k occurs. Therefore, the risk neutral probability vector $\boldsymbol{\pi}$ must lie in the orthogonal complement $W^\perp$. Since the sum of risk neutral probabilities must be one and all probability values must be positive, the risk neutral probability vector $\boldsymbol{\pi}$ must lie in the following subset

$$P^+ = \{\mathbf{y} \in \mathbb{R}^K : y_1 + y_2 + \cdots + y_K = 1 \quad \text{and} \quad y_k > 0, k = 1, \cdots K\}.$$

Let R denote the set of all risk neutral measures. Combining the above results, we see that

$$R = P^+ \cap W^\perp. \tag{2.1.16}$$

In the above numerical example, $W^\perp$ is the line through the origin in $\mathbb{R}^3$ which is perpendicular to $(1 \quad 0 \quad -1)^T$ and $(0 \quad -1 \quad 1)^T$. The line should assume the form $\lambda(1 \quad 1 \quad 1)^T$ for some scalar λ. Together with the constraints that sum of components equals one and each component is positive, we obtain the risk neutral probability vector $\boldsymbol{\pi} = (1/3 \quad 1/3 \quad 1/3)$.

2.1.3 Valuation of Contingent Claims

A contingent claim can be considered as a random variable Y that represents the terminal payoff whose value depends on the occurrence of a particular state ω_k, where $\omega_k \in \Omega$. Suppose the holder of the contingent claim is promised to receive the preset payoff: How much should the writer charge at $t = 0$ when selling the contingent claim so that the price is *fair* to both parties?

Consider the securities model with the risk free security whose values at $t = 0$ and $t = 1$ are $S_0(0) = 1$ and $S_0(1) = 1.1$, respectively, and a risky security with $S_1(0) = 3$ and $\mathbf{S}_1(1) = \begin{pmatrix} 4.4 \\ 3.3 \\ 2.2 \end{pmatrix}$. The set of $t = 1$ payoffs that can be generated by a trading strategy is given by $h_0 \begin{pmatrix} 1.1 \\ 1.1 \\ 1.1 \end{pmatrix} + h_1 \begin{pmatrix} 4.4 \\ 3.3 \\ 2.2 \end{pmatrix}$ for some scalars h_0 and h_1. For example, the contingent claim $\begin{pmatrix} 5.5 \\ 4.4 \\ 3.3 \end{pmatrix}$ can be generated by the trading strategy:

$h_0 = 1$ and $h_1 = 1$, while the other contingent claim $\begin{pmatrix} 5.5 \\ 4.0 \\ 3.3 \end{pmatrix}$ cannot be generated by any trading strategy associated with the given securities model. A contingent claim Y is said to be *attainable* if there exists some trading strategy $\mathbf{h}$ for constructing the *replicating portfolio* such that the portfolio value V_1 equals Y for all possible states occurring at $t = 1$.

What should be the price at $t = 0$ of the attainable contingent claim $\begin{pmatrix} 5.5 \\ 4.4 \\ 3.3 \end{pmatrix}$? One may propose that the price at $t = 0$ of the replicating portfolio is given by $V_0 = h_0 S_0(0) + h_1 S_1(0) = 1 \times 1 + 1 \times 3 = 4$. As discussed in Sect. 2.1.2, suppose there are no arbitrage opportunities (equivalent to the existence of a risk neutral probability measure), then the law of one price holds and so V_0 is unique. The price at $t = 0$ of the contingent claim Y is simply V_0, the price that is implied by the arbitrage pricing theory. If otherwise, suppose the price p of the contingent claim at $t = 0$ is greater than V_0, an arbitrageur can lock in a risk free profit of amount $p - V_0$ by short selling the contingent claim and buying the replicating portfolio. The arbitrage strategy is reversed if $p < V_0$. In this securities model, we have shown earlier that risk neutral probability measures do exist (though are not unique). However, the initial price of the contingent claim $\begin{pmatrix} 5.5 \\ 4.4 \\ 3.3 \end{pmatrix}$ is unique and it is found to be $V_0 = 4$.

Risk Neutral Valuation Principle

Given an attainable contingent claim Y that can be generated by a certain trading strategy, the associated discounted gain G^* of the trading strategy is given by $G^* = \sum_{m=1}^{M} h_m \Delta S_m^*$. Assuming that a risk neutral probability measure Q associated with the securities model exists, and using the relations: $V_0^* = V_1^* - G^*$, $E_Q[G^*] = 0$ and $V_1^* = Y/S_0(1)$, we obtain

$$V_0 = E_Q V_0^* = E_Q[V_1^* - G^*] = E_Q[Y/S_0(1)]. \tag{2.1.17}$$

Recall that the existence of the risk neutral probability measure implies the law of one price. Does $E_Q[Y/S_0(1)]$ assume the same value for every risk neutral probability measure Q? This must be true by virtue of the law of one price since we cannot have two different values for V_0 corresponding to the same contingent claim Y. The *risk neutral valuation principle* can be stated as follows:

> The price at $t = 0$ of an attainable claim Y is given by the expectation under any risk neutral measure Q of the discounted value of the contingent claim.

Actually, one can show a rather strong result: If $E_Q[Y/S_0(1)]$ takes the same value for every risk neutral measure Q, then the contingent claim Y is attainable [for its proof, see Pliska (1997)].

Readers are reminded that if the law of one price does not hold for a given securities model, we cannot define a unique price for an attainable contingent claim (see Problem 2.14).

State Prices

Consider the discounted value of a contingent claim $Y^* = Y/S_0(1)$, which equals one if $\omega = \omega_k$ for some $\omega_k \in \Omega$ and zero otherwise. This is just the Arrow security $\mathbf{e}_k$ corresponding to the state ω_k. We then have

$$E_Q[Y/S_0(1)] = \boldsymbol{\pi}\mathbf{e}_k = Q(\omega_k). \tag{2.1.18}$$

The price of the Arrow security with the discounted payoff $\mathbf{e}_k$ is called the state price for state $\omega_k \in \Omega$. The above result shows that the state price for ω_k is equal to the risk neutral probability for the same state.

Any contingent claim Y can be written as a linear combination of these basic Arrow securities. Suppose $Y^* = Y/S_0(1) = \sum_{k=1}^K \alpha_k \mathbf{e}_k$, then the price at $t = 0$ of the contingent claim is equal to $\sum_{k=1}^K \alpha_k Q(\omega_k)$. For example, suppose

$$\mathbf{Y}^* = \begin{pmatrix} 5 \\ 4 \\ 3 \end{pmatrix} \quad \text{and} \quad \widehat{S}^*(1;\Omega) = \begin{pmatrix} 1 & 4 \\ 1 & 3 \\ 1 & 2 \end{pmatrix},$$

we have seen that the risk neutral probability is given by

$$\boldsymbol{\pi} = (\lambda \quad 1-2\lambda \quad \lambda), \text{ where } 0 < \lambda < 1/2.$$

The price at $t = 0$ of the contingent claim is given by

$$V_0 = \boldsymbol{\pi}\mathbf{Y}^* = 5\lambda + 4(1-2\lambda) + 3\lambda = 4,$$

which is independent of λ. This verifies the earlier claim that $E_Q[Y/S_0(1)]$ assumes the same value for any risk neutral measure Q.

Complete Markets

Recall that a securities model is complete if every contingent claim Y lies in the asset span, that is, Y can be replicated by a portfolio generated using some trading strategy. Consider the augmented terminal payoff matrix of dimension $K \times (M+1)$

$$\widehat{S}(1;\Omega) = \begin{pmatrix} S_0(1;\omega_1) & S_1(1;\omega_1) & \cdots & S_M(1;\omega_1) \\ \vdots & \vdots & & \vdots \\ S_0(1;\omega_K) & S_1(1;\omega_K) & \cdots & S_M(1;\omega_K) \end{pmatrix},$$

if the columns of $\widehat{S}(1;\Omega)$ span the whole $\mathbb{R}^K$, then Y always lies in the asset span. Since the dimension of the column space of $\widehat{S}(1;\Omega)$ cannot be greater than $M+1$, a necessary condition for market completeness is given by $M+1 \geq K$. Under market completeness, if the set of risk neutral probability measures is nonempty, then it must be a singleton (see Problem 2.14). Furthermore, when $\widehat{S}(1;\Omega)$ has independent columns and the asset span is the whole $\mathbb{R}^K$, then $M + 1 = K$. In this case, all contingent claims are attainable and the trading strategy that generates Y must be unique since there are no redundant securities. Hence, we have a unique price for any contingent claim. On the other hand, when the asset span is the whole $\mathbb{R}^K$ but some securities are redundant, the trading strategy that generates Y would not be unique. Assuming that a risk neutral measure exists, the price at $t = 0$ of the contingent claim must be unique under arbitrage pricing, independent of the chosen trading strategy. This is a consequence of the law of one price, which is implied by the existence of the risk neutral measure. These results illustrate that nonexistence of redundant securities is a sufficient but not necessary condition for the law of one price.

As a numerical example, we consider the securities model defined by

$$\widehat{\mathbf{S}}^*(0) = (1 \quad 3) \quad \text{and} \quad \widehat{S}^*(1;\Omega) = \begin{pmatrix} 1 & 4 \\ 1 & 5 \\ 1 & 6 \end{pmatrix}.$$

Since there is no redundant security, so the law of one price holds. Suppose the securities model is modified by adding another risky security so that

$$\widehat{\mathbf{S}}^*(0) = (1 \quad 3 \quad 4) \quad \text{and} \quad \widehat{S}^*(1;\Omega) = \begin{pmatrix} 1 & 4 & 5 \\ 1 & 5 & 6 \\ 1 & 6 & 7 \end{pmatrix}.$$

The last risky security is seen to be redundant (the third column is the sum of the first and second columns). However, the law of one price still holds since

$$\mathbf{S}_2^*(1) = \mathbf{S}_0^*(1) + \mathbf{S}_1^*(1) \quad \text{while} \quad S_2(0) = S_0(0) + S_1(0).$$

Actually, one can see that $\widehat{\mathbf{S}}^*(0)$ lies in the row space of $\widehat{S}^*(1;\Omega)$.

When the dimension of the column space $\widehat{S}(1;\Omega)$ is less than K, then not all contingent claims can be attainable. In this case, a nonattainable contingent claim cannot be priced using arbitrage pricing theory. However, we may specify an interval $(V_-(Y), V_+(Y))$ where a reasonable price at $t = 0$ of the contingent claim should lie. The lower and upper bounds are given by

$$V_-(Y) = \sup\{E_Q[\widetilde{Y}/S_0(1)] : \widetilde{Y} \leq Y \text{ and } \widetilde{Y} \text{ is attainable}\} \tag{2.1.19a}$$
$$V_+(Y) = \inf\{E_Q[\widetilde{Y}/S_0(1)] : \widetilde{Y} \geq Y \text{ and } \widetilde{Y} \text{ is attainable}\}. \tag{2.1.19b}$$

Here, $V_+(Y)$ is the minimum value among all prices of attainable contingent claims that dominate the nonattainable claim Y, while $V_-(Y)$ is the maximum value among all prices of attainable contingent claims that are dominated by Y. Suppose $V(Y) > V_+(Y)$, an arbitrageur can lock in a riskless profit by selling the contingent claim to receive $V(Y)$ and use $V_+(Y)$ to construct the replicating portfolio that generates the attainable $\widetilde{Y}$ as defined in (2.1.19b). The upfront positive gain is $V(Y) - V_+(Y)$. At $t = 1$, the payoff from the replicating portfolio always dominates that of Y so that no loss at expiry is also ensured. Similarly, an arbitrageur can lock in a riskless profit when $V(Y) < V_{-}(Y)$ by buying the contingent claim and short holding the replicating portfolio that generates the attainable $\widetilde{Y}$ as defined in (2.1.19a).

Summary

1. The relations between the law of one price, absence of dominant trading strategies and absence of arbitrage opportunities are:

   absence of arbitrage opportunities
   ⇒ absence of dominant trading strategies
   ⇒ law of one price.

   while

   law of one price
   ⇔ single-valuedness of linear pricing functional.

2. Theorems 2.1 and 2.2 show that

   absence of arbitrage opportunities ⇔ existence of risk neutral measure

   absence of dominant trading strategies ⇔ existence of linear pricing measure.

3. The state prices are nonnegative when a linear pricing measure exists and they become strictly positive when a risk neutral measure exists.
4. Under the absence of arbitrage opportunities, the risk neutral valuation principle can be applied to find the fair price of a contingent claim.

2.1.4 Principles of Binomial Option Pricing Model

We would like to illustrate the risk neutral valuation principle to price a call option using the renowned binomial option pricing model. In the binomial model, the asset price movement is simulated by a discrete binomial random walk model (see Sect. 2.3.1 for a more detailed discussion on the random walk models). Here, we limit our discussion to the one-period binomial model and defer the analysis of the multiperiod binomial model in Sect. 2.2.4. We show that the call option price obtained from the binomial model depends only on the riskless interest rate but independent on the actual expected rate of return of the asset price.

Formulation of the Replicating Portfolio

We follow the derivation of the discrete binomial model presented by Cox, Ross and Rubinstein (1979). They showed that by buying the asset and borrowing cash (in the form of a riskless money market account) in appropriate proportions, one can replicate the position of a call option. Let S denote the current asset price. Under the binomial random walk model, the asset price after one period Δt will be either uS or dS with probability q and $1-q$, respectively (see Fig. 2.3). We assume $u > 1 > d$ so that uS and dS represent the up-move and down-move of the asset price, respectively. The proportional jump parameters u and d are related to the asset price

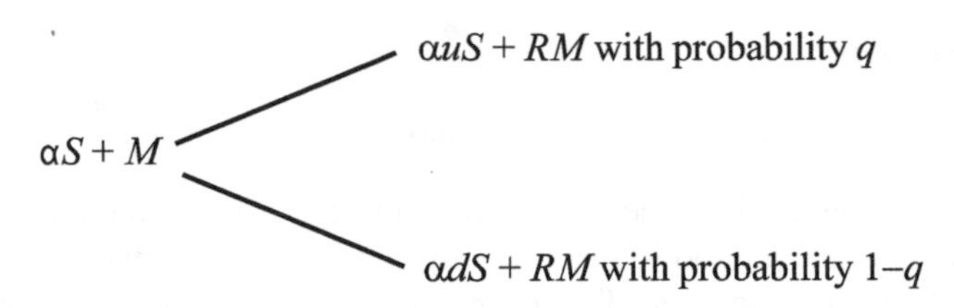

Fig. 2.3. Evolution of the asset price S and money market account M after one time period under the binomial model. The risky asset value may either go up to uS or go down to dS, while the riskless money market account M grows to RM.

dynamics, the detailed discussion of which will be relegated to Sect. 6.1.1. Let R denote the growth factor of the money market account over one period so that \$1 invested in the riskless money market account will grow to \$$R$ after one period. In order to ensure absence of arbitrage opportunities, we must have $u > R > d$ (see Problem 2.16).

Suppose we form a portfolio that consists of α units of asset and cash amount M in the form of riskless money market account. After one period of time Δt, the value of the portfolio becomes (see Fig. 2.3)

$$\begin{cases} \alpha uS + RM & \text{with probability } q \\ \alpha dS + RM & \text{with probability } 1-q. \end{cases}$$

The portfolio is used to replicate the long position of a call option on a nondividend paying asset. As there are two possible states of the world: asset price goes up or down, the call is thus a contingent claim. Suppose the current time is only one period Δt prior to expiration. Let c denote the current call price, and c_u and c_d denote the call price after one period (which is the expiration time) corresponding to the up-move and down-move of the asset price, respectively. Let X denote the strike price of the call. The payoff of the call at expiry is given by

$$\begin{cases} c_u = \max(uS - X, 0) & \text{with probability } q \\ c_d = \max(dS - X, 0) & \text{with probability } 1-q. \end{cases}$$

The above portfolio containing the risky asset and money market account is said to replicate the long position of the call if and only if the values of the portfolio and the call option match for each possible outcome, that is,

$$\alpha uS + RM = c_u \quad \text{and} \quad \alpha dS + RM = c_d.$$

The unknowns are α and M in the above linear system of equations. It occurs that the number of unknowns (related to the number of units of asset and cash amount) and the number of equations (two possible states of the world under the binomial model) are equal. Solving the equations, we obtain

$$\alpha = \frac{c_u - c_d}{(u-d)S} \geq 0, \qquad M = \frac{uc_d - dc_u}{(u-d)R} \leq 0. \tag{2.1.20}$$

It is easy to establish

$$u \max(dS - X, 0) - d \max(uS - X, 0) \leq 0,$$

so M is always nonpositive. The replicating portfolio involves buying the asset and borrowing cash in the proportions as given by (2.1.20). The number of units of asset held is seen to be the ratio of the difference of call values $c_u - c_d$ to the difference of asset values $uS - dS$.

Under the one-period binomial model of asset price dynamics, we observe that the call option can be replicated by a portfolio of basic securities: risky asset and riskless money market account.

Binomial Option Pricing Formula

By the principle of no arbitrage, the current value of the call must be the same as that of the replicating portfolio. What happens if otherwise? Suppose the current value of the call is less than the portfolio value, then we could make a riskless profit by buying the cheaper call and selling the more expensive portfolio. The net gain from the above two transactions is secured since the portfolio value and call value cancel each other at a later period. The argument can be reversed if the call is worth more than the portfolio. Therefore, the current value of the call is given by the current value of the portfolio, that is,

$$c = \alpha S + M = \frac{\frac{R-d}{u-d} c_u + \frac{u-R}{u-d} c_d}{R} = \frac{p c_u + (1-p) c_d}{R}, \tag{2.1.21}$$

where $p = \frac{R-d}{u-d}$. Note that the probability q, which is the subjective probability about upward or downward movement of the asset price, does not appear in the call value formula (2.1.21). The parameter p can be shown to be $0 < p < 1$ since $u > R > d$, so p can be interpreted as a probability. Furthermore, from the relation

$$puS + (1-p)dS = \frac{R-d}{u-d} uS + \frac{u-R}{u-d} dS = RS, \tag{2.1.22}$$

one can interpret the result as follows: the expected rate of return on the asset with p as the probability of upside move is just equal to the riskless interest rate. Let $S^{\Delta t}$ be the random variable that denotes the asset price at one period later. We may express (2.1.22) as

$$S = \frac{1}{R} E^*[S^{\Delta t} | S], \tag{2.1.23}$$

where E^* is expectation under this probability measure. According to the definition given in Sect. 2.1.2 [see (2.1.13b)], we may view p as the *risk neutral probability*. Similarly, the binomial formula (2.1.21) for the call value can be expressed as

$$c = \frac{1}{R} E^*\left[c^{\Delta t} | S\right], \tag{2.1.24}$$

where c denotes the call value at the current time, and $c^{\Delta t}$ denotes the random variable representing the call value at one period later.

As a summary, when the call option can be replicated by existing marketed assets, its current value is given by the expectation of the discounted terminal payoff under the risk neutral measure. Under the no arbitrage pricing framework, assuming the existence of the risk neutral probability values $Q(\omega_u)$ and $Q(\omega_d)$ corresponding to the up-state ω_u and down-state ω_d, these probability values can be found by solving [see (2.1.23)]

$$S = Q(\omega_u)\frac{uS}{R} + Q(\omega_d)\frac{dS}{R} \quad \text{and} \quad Q(\omega_u) + Q(\omega_d) = 1.$$

This gives

$$Q(\omega_u) = 1 - Q(\omega_d) = \frac{R-d}{u-d} = p.$$

The current call value is then obtained from the corresponding discounted expectation formula:

$$c = Q(\omega_u)\frac{c_u}{R} + Q(\omega_d)\frac{c_d}{R} = \frac{pc_u + (1-p)c_d}{R},$$

giving the same result as in (2.1.21).

Besides applying the principle of replication of contingent claims, the binomial option pricing formula can also be derived using the riskless hedging principle or via the concept of state prices (see Problems 2.17 and 2.18).

2.2 Filtrations, Martingales and Multiperiod Models

In this section, we extend our discussion of securities models to multiperiod, where there are $T+1$ trading dates: $t = 0, 1, \cdots, T, T > 1$. Similar to an one-period model, we have a finite sample space Ω of K elements, $\Omega = \{\omega_1, \omega_2, \cdots, \omega_K\}$, which represents the possible states of the world. There is a probability measure P defined on the sample space with $P(\omega) > 0$ for all $\omega \in \Omega$. The securities model consists of M risky securities whose price processes are nonnegative stochastic processes, as denoted by $S_m = \{S_m(t); t = 0, 1, \cdots, T\}, m = 1, \cdots, M$. In addition, there is a risk free security whose price process $S_0(t)$ is deterministic, with $S_0(t)$ being strictly positive and possibly nondecreasing in t. We may consider $S_0(t)$ as a money market account, and the quantity $r_t = \frac{S_0(t)-S_0(t-1)}{S_0(t-1)}, t = 1, \cdots, T$, is visualized as the interest rate over the time interval $(t-1, t)$.

In this section, we would like to show that the concepts of arbitrage opportunity and risk neutral valuation can be carried over from single-period models to multiperiod models. However, it is necessary to specify how the investors learn about the true state of the world on intermediate trading dates in a multi-period model. Accordingly, we construct an information structure that models how information is revealed to investors in terms of the subsets of the sample space Ω. We show how the information structure can be described by a filtration and understand how security price processes can be adapted to a given filtration. Then we introduce martingales that are adapted stochastic processes modeled as "fair gambling" under a given filtration and probability measure. We also discuss the notions of stopping rule, stopping time and stopped process. The renowned Doob Optimal Sampling Theorem states that a stopped martingale remains a martingale. The highlight of this section is the multiperiod version of the *Fundamental Theorem of Asset Pricing*. The last part of this section is devoted to the multiperiod binomial models for pricing options.

2.2.1 Information Structures and Filtrations

Consider the sample space $\Omega = \{\omega_1, \omega_2, \cdots, \omega_{10}\}$ with 10 elements. We can construct various partitions of the set Ω. A *partition* of Ω is a collection $\mathcal{P} = \{B_1, B_2, \cdots B_n\}$ such that B_j, $j = 1, \cdots, n$, are subsets of Ω and $B_i \cap B_j = \phi, i \neq j$, and $\bigcup_{j=1}^n B_j = \Omega$. Each of these sets $B_1, \cdots, B_n$ is called an *atom* of the partition. For example, we may form the partitions of Ω as

$$\begin{aligned}
\mathcal{P}_0 &= \{\Omega\} \\
\mathcal{P}_1 &= \{\{\omega_1, \omega_2, \omega_3, \omega_4\}, \{\omega_5, \omega_6, \omega_7, \omega_8, \omega_9, \omega_{10}\}\} \\
\mathcal{P}_2 &= \{\{\omega_1, \omega_2\}, \{\omega_3, \omega_4\}, \{\omega_5, \omega_6\}, \{\omega_7, \omega_8, \omega_9\}, \{\omega_{10}\}\} \\
\mathcal{P}_3 &= \{\{\omega_1\}, \{\omega_2\}, \{\omega_3\}, \{\omega_4\}, \{\omega_5\}, \{\omega_6\}, \{\omega_7\}, \{\omega_8\}, \{\omega_9\}, \{\omega_{10}\}\}.
\end{aligned}$$

We have defined a finite sequence of partitions of Ω with the property that they are nested with successive refinements of one another. Each set belonging to $\mathcal{P}_k$ splits into smaller sets which are atoms of $\mathcal{P}_{k+1}$.

Consider a three-period securities model that consists of the above sequence of successively finer partitions: $\{\mathcal{P}_k : k = 0, 1, 2, 3\}$. The pair $(\Omega, \mathcal{P}_k)$ is called a *filtered space*, which consists of a sample space Ω and a sequence of partitions $\mathcal{P}_k$ of Ω. The filtered space is used to model the unfolding of information through time. At time $t = 0$, the investors know only the set of all possible states of the world, so $\mathcal{P}_0 = \{\Omega\}$. At time $t = 1$, the investors get a bit more information: the actual state ω is in either $\{\omega_1, \omega_2, \omega_3, \omega_4\}$ or $\{\omega_5, \omega_5, \omega_7, \omega_8, \omega_9, \omega_{10}\}$. In the next trading date $t = 2$, more information is revealed, say, ω is in the set $\{\omega_7, \omega_8, \omega_9\}$. On the last date, $t = 3$, we have $\mathcal{P}_3 = \{\{\omega_i\}, i = 1, \cdots, 10\}$. Each set of $\mathcal{P}_3$ consists of a single element of Ω, so the investors have full information regarding which particular state has occurred. The information submodel of this three-period securities model can be represented by the information tree shown in Fig. 2.4.

Algebra

Let Ω be a finite set and $\mathcal{F}$ be a collection of subsets of Ω. The collection $\mathcal{F}$ is an *algebra* on Ω if

(i) $\Omega \in \mathcal{F}$
(ii) $B \in \mathcal{F} \Rightarrow B^c \in \mathcal{F}$
(iii) B_1 and $B_2 \in \mathcal{F} \Rightarrow B_1 \cup B_2 \in \mathcal{F}$.

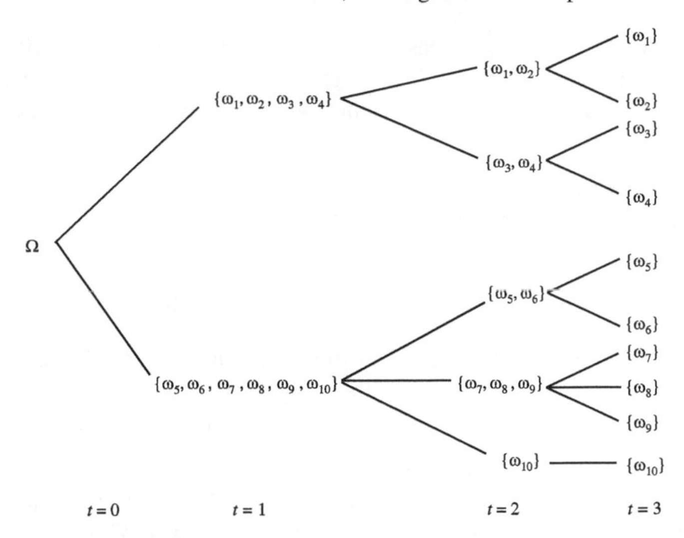

Fig. 2.4. Information tree of a three-period securities model with 10 possible states. The partitions form a sequence of successively finer partitions.

Given an algebra $\mathcal{F}$ on Ω, one can always find a unique collection of disjoint subsets B_n such that each $B_n \in \mathcal{F}$ and the union of all of these subsets gives Ω. The algebra $\mathcal{F}$ generated by a partition $\mathcal{P} = \{B_1, \cdots, B_n\}$ is a set of subsets of Ω. Actually, when Ω is a finite sample space, there is a one-to-one correspondence between partitions of Ω and algebras on Ω. The information structure defined by a sequence of partitions can be visualized as a sequence of algebras. We define a *filtration* $\mathbb{F} = \{\mathcal{F}_k; k = 0, 1, \cdots, T\}$ to be a nested sequence of algebras satisfying $\mathcal{F}_k \subseteq \mathcal{F}_{k+1}$.

As an example, given the algebra $\mathcal{F} = \{\phi, \{\omega_1\}, \{\omega_2, \omega_3\}, \{\omega_4\}, \{\omega_1, \omega_2, \omega_3\}, \{\omega_2, \omega_3, \omega_4\}, \{\omega_1, \omega_4\}, \{\omega_1, \omega_2, \omega_3, \omega_4\}\}$, the corresponding partition $\mathcal{P}$ is found to be $\{\{\omega_1\}, \{\omega_2, \omega_3\}, \{\omega_4\}\}$. The atoms of $\mathcal{P}$ are $B_1 = \{\omega_1\}$, $B_2 = \{\omega_2, \omega_3\}$ and $B_3 = \{\omega_4\}$. A nonempty event whose occurrence, shown through the revelation of $\mathcal{P}$, would be an union of atoms in $\mathcal{P}$. Take the event $A = \{\omega_1, \omega_2, \omega_3\}$, which is the union of B_1 and B_2. Given that $B_2 = \{\omega_2, \omega_3\}$ of $\mathcal{P}$ has occurred, we can decide whether A or its complement A^c has occurred. However, for another event $\widetilde{A} = \{\omega_1, \omega_2\}$, even though we know that B_2 has occurred, we cannot determine whether $\widetilde{A}$ or $\widetilde{A}^c$ has occurred.

Next, we define a probability measure P defined on an algebra $\mathcal{F}$. The probability measure P is a function

$$P : \mathcal{F} \to [0, 1]$$

such that

1. $P(\Omega) = 1$.
2. If $B_1, B_2, \cdots$ are pairwise disjoint sets belonging to $\mathcal{F}$, then

$$P(B_1 \cup B_2 \cup \cdots) = P(B_1) + P(B_2) + \cdots.$$

Equipped with a probability measure, the elements of $\mathcal{F}$ are called measurable events. Given the sample space Ω, an algebra $\mathcal{F}$ and a probability measure P defined on Ω, the triplet $(\Omega, \mathcal{F}, P)$ together with the filtration $\mathbb{F}$ is called a *filtered probability space*.

Equivalent Measures

Given two probability measures P and P' defined on the same measurable space $(\Omega, \mathcal{F})$, suppose that

$$P(\omega) > 0 \quad \Longleftrightarrow \quad P'(\omega) > 0, \quad \text{for all } \omega \in \Omega,$$

then P and P' are said to be equivalent measures. In other words, though the two equivalent measures may not agree on the assignment of probability values to individual events, but they always agree as to which events are possible or impossible.

Measurability of Random Variables

Consider an algebra $\mathcal{F}$ generated by a partition $\mathcal{P} = \{B_1, \cdots, B_n\}$, a random variable X is said to be measurable with respect to $\mathcal{F}$ (denoted by $X \in \mathcal{F}$) if $X(\omega)$ is constant for all $\omega \in B_i$, B_i is any atom in $\mathcal{P}$. For example, consider the algebra $\mathcal{F}_1$ generated by $\mathcal{P}_1 = \{\{\omega_1, \omega_2, \omega_3, \omega_4\}, \{\omega_5, \omega_6, \omega_7, \omega_8, \omega_9, \omega_{10}\}\}$. If $X(\omega_1) = 3$ and $X(\omega_4) = 5$, then X is not measurable with respect to $\mathcal{F}_1$ since ω_1 and ω_4 belong to the same atom but $X(\omega_1)$ and $X(\omega_4)$ have different values.

Consider an example where $\mathcal{P} = \{\{\omega_1, \omega_2\}, \{\omega_3, \omega_4\}, \{\omega_5\}\}$ and X is measurable with respect to the algebra $\mathcal{F}$ generated by $\mathcal{P}$. Let $X(\omega_1) = X(\omega_2) = 3$, $X(\omega_3) = X(\omega_4) = 5$ and $X(\omega_5) = 7$. Suppose the random experiment associated with the random variable X is performed, giving $X = 5$. This tells the information that the event $\{\omega_3, \omega_4\}$ has occurred. In this sense, the information of outcome from the random experiment is revealed through the random variable X. We say that $\mathcal{F}$ is being generated by X.

A stochastic process $S_m = \{S_m(t); t = 0, 1, \cdots, T\}$ is said to be *adapted to the filtration* $\mathbb{F} = \{\mathcal{F}_t; t = 0, 1, \cdots, T\}$ if the random variables $S_m(t)$ is $\mathcal{F}_t$-measurable for each $t = 0, 1, \cdots, T$. For the money market account process $S_0(t)$, the interest rate is normally known at the beginning of the period so that $S_0(t)$ is $\mathcal{F}_{t-1}$-measurable, $t = 1, \cdots, T$. In this case, we say that the process $S_0(t)$ is *predictable*.

2.2.2 Conditional Expectations and Martingales

Consider the filtered probability space defined by the triplet $(\Omega, \mathcal{F}, P)$ together with the filtration $\mathbb{F}$. Recall that a random variable is a mapping $\omega \to X(\omega)$ that assigns a real number $X(\omega)$ to each $\omega \in \Omega$. A random variable is said to be simple if X can be decomposed into the form

$$X(\omega) = \sum_{j=1}^{n} a_j \mathbf{1}_{B_j}(\omega), \tag{2.2.1}$$

where $\{B_1, \cdots, B_n\}$ is a finite partition of Ω with each $B_j \in \mathcal{F}$ and the indicator of B_j is defined by

$$\mathbf{1}_{B_j}(\omega) = \begin{cases} 1 & \text{if } \omega \in B_j \\ 0 & \text{if otherwise.} \end{cases}$$

The expectation of X with respect to the probability measure P is defined by

$$E[X] = \sum_{j=1}^{n} a_j E[\mathbf{1}_{B_j}(\omega)] = \sum_{j=1}^{n} a_j P(B_j), \tag{2.2.2}$$

where $P(B_j)$ is the probability that a state ω contained in B_j occurs. The conditional expectation of X given that event B has occurred is defined to be

$$\begin{aligned} E[X|B] &= \sum_x x P(X = x|B) \\ &= \sum_x x P(X = x, B)/P(B) \\ &= \frac{1}{P(B)} \sum_{\omega \in B} X(\omega) P(\omega). \end{aligned} \tag{2.2.3}$$

As a numerical example, consider the sample space $\Omega = \{\omega_1, \omega_2, \omega_3, \omega_4\}$ and the algebra is generated by the partition $\mathcal{P} = \{\{\omega_1, \omega_2\}, \{\omega_3, \omega_4\}\}$. The probabilities of occurrence of the states are given by $P(\omega_1) = 0.2$, $P(\omega_2) = 0.3$, $P(\omega_3) = 0.35$ and $P(\omega_4) = 0.15$. Consider the two-period price process S whose values are given by

$$\begin{array}{llll} S(1;\omega_1) = 3, & S(1;\omega_2) = 3, & S(1;\omega_3) = 5, & S(1;\omega_4) = 5, \\ S(2;\omega_1) = 4, & S(2;\omega_2) = 2, & S(2;\omega_3) = 4, & S(2;\omega_4) = 6. \end{array}$$

The corresponding tree representation is shown in Fig. 2.5.

The conditional expectations

$$E[S(2)|S(1) = 3] \text{ and } E[S(2)|S(1) = 5]$$

are computed using (2.2.3) as follows:

$$\begin{aligned} E[S(2)|S(1) = 3] &= \frac{S(2;\omega_1)P(\omega_1) + S(2;\omega_2)P(\omega_2)}{P(\omega_1) + P(\omega_2)} \\ &= (4 \times 0.2 + 2 \times 0.3)/0.5 = 2.8; \\ E[S(2)|S(1) = 5] &= \frac{S(2;\omega_3)P(\omega_3) + S(2;\omega_4)P(\omega_4)}{P(\omega_3) + P(\omega_4)} \\ &= (4 \times 0.35 + 6 \times 0.15)/0.5 = 4.6. \end{aligned}$$

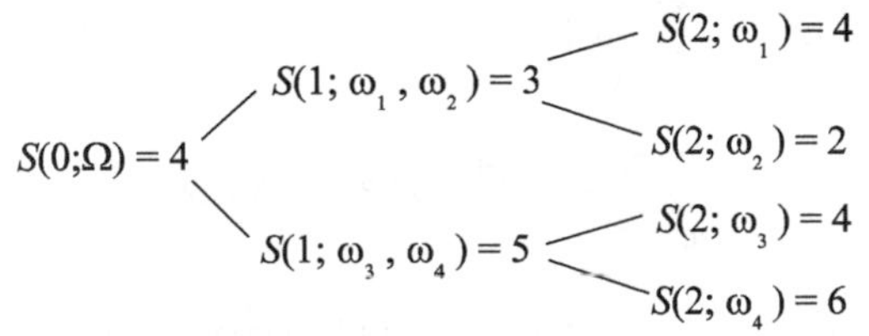

$S(0;\Omega) = 4$
$S(1;\omega_1,\omega_2) = 3$
$S(2;\omega_1) = 4$
$S(2;\omega_2) = 2$
$S(1;\omega_3,\omega_4) = 5$
$S(2;\omega_3) = 4$
$S(2;\omega_4) = 6$

Fig. 2.5. The tree representation of an asset price process in a two-period securities model.

Interpretation of $E[X|\mathcal{F}]$
It is quite often that we would like to consider all conditional expectations of the form $E[X|B]$ where the event B runs through the algebra $\mathcal{F}$. Let B_j, $j = 1, 2, \cdots, n$, be the atoms of the algebra $\mathcal{F}$. We define the quantity $E[X|\mathcal{F}]$ by

$$E[X|\mathcal{F}] = \sum_{j=1}^{n} E[X|B_j]\mathbf{1}_{B_j}. \tag{2.2.4}$$

We see that $E[X|\mathcal{F}]$ is actually a random variable that is measurable with respect to the algebra $\mathcal{F}$. In the above numerical example, we have $\mathcal{F}_1 = \{\phi, \{\omega_1, \omega_2\}, \{\omega_3, \omega_4\}, \Omega\}$, and the atoms of $\mathcal{F}_1$ are $B_1 = \{\omega_1, \omega_2\}$ and $B_2 = \{\omega_3, \omega_4\}$. Since we have

$$E[S(2)|S(1) = 3] = 2.8 \quad \text{and} \quad E[S(2)|S(1) = 5] = 4.6,$$

so that

$$E[S(2)|\mathcal{F}_1] = 2.8\mathbf{1}_{B_1} + 4.6\mathbf{1}_{B_2}.$$

Tower Property
Since $E[X|\mathcal{F}]$ is a random variable, we may compute its expectation. We find that

$$\begin{aligned} E[E[X|\mathcal{F}]] &= \sum_{B\in\mathcal{F}} E[X|B]P(B) = \sum_{B\in\mathcal{F}}\sum_{\omega\in B} X(\omega)(P[\omega]/P(B))P(B) \\ &= \sum_{B\in\mathcal{F}}\sum_{\omega\in B} X(\omega)P(\omega) = E[X]. \end{aligned} \tag{2.2.5}$$

The above result can be generalized as follows. If $\mathcal{F}_1 \subset \mathcal{F}_2$, then

$$E[E[X|\mathcal{F}_2]|\mathcal{F}_1] = E[X|\mathcal{F}_1]. \tag{2.2.6}$$

If we condition first on the information up to $\mathcal{F}_2$ and later on the information $\mathcal{F}_1$ at an earlier time, then it is the same as conditioning originally on $\mathcal{F}_1$. This is called the *tower property* of conditional expectations.

Suppose that the random variable X is $\mathcal{F}$-measurable, we would like to show $E[XY|\mathcal{F}] = XE[Y|\mathcal{F}]$ for any random variable Y. Using (2.2.1), we may write $X = \sum_{B_j\in\mathcal{P}} a_j\mathbf{1}_{B_j}$, where $\mathcal{P}$ is the partition corresponding to the algebra $\mathcal{F}$. By (2.2.4), we obtain

$$\begin{aligned} E[XY|\mathcal{F}] &= \sum_{B_j\in\mathcal{P}} E[XY|B_j]\mathbf{1}_{B_j} = \sum_{B_j\in\mathcal{P}} E[a_jY|B_j]\mathbf{1}_{B_j} \\ &= \sum_{B_j\in\mathcal{P}} a_jE[Y|B_j]\mathbf{1}_{B_j} = XE[Y|\mathcal{F}]. \end{aligned} \tag{2.2.7}$$

When we take the conditional expectation with respect to the filtration $\mathcal{F}$, we can treat X as constant if X is known with regard to the information provided by $\mathcal{F}$. The proofs of other properties on conditional expectations are relegated as exercises (see Problem 2.20).

Martingales

The term martingale has its origin in gambling. It refers to the gambling tactic of doubling the stake when losing in order to recoup oneself. In the studies of stochastic processes, martingales are defined in relation to an adapted stochastic process.

Consider a filtered probability space with filtration $\mathbb{F} = \{\mathcal{F}_t; t = 0, 1, \cdots, T\}$. An adapted stochastic process $S = \{S(t); t = 0, 1 \cdots, T\}$ is said to be martingale if it observes

$$E[S(u)|\mathcal{F}_t] = S(t) \quad \text{for } 0 \leq t \leq u \leq T. \tag{2.2.8}$$

We define an adapted stochastic process S to be a supermartingale if

$$E[S(u)|\mathcal{F}_t] \leq S(t) \quad \text{for } 0 \leq t \leq u \leq T; \tag{2.2.9a}$$

and a submartingale if

$$E[S(u)|\mathcal{F}_t] \geq S(t) \quad \text{for } 0 \leq t \leq u \leq T. \tag{2.2.9b}$$

It is straightforward to deduce the following properties:

1. All martingales are supermartingales, but not vice versa. The same observation is applied to submartingales.
2. An adapted stochastic process S is a submartingale if and only if $-S$ is a supermartingale; S is a martingale if and only if it is both a supermartingale and a submartingale.

Martingales are related to models of fair gambling. For example, let X_n represent the amount of money a player possesses at stage n of the game. The martingale property means that the expected amount of the player would have at stage $n + 1$ given that $X_n = \alpha_n$, is equal to α_n, regardless of his past history of fortune. A supermartingale (submartingale) can be used to model an unfavorable (favorable) game since the gambler is more likely to lose than to win (win than to lose).

It must be emphasized that a martingale is defined with respect to a filtration (information set) and a probability measure. The risk neutral valuation approach in option pricing theory is closely related to the theory of martingales. In Sect. 2.2.4, we show that the necessary and sufficient condition for the exclusion of arbitrage opportunities in a securities model is the existence of a risk neutral pricing measure constructed from the martingale property of the asset price processes.

Martingale Transforms

Suppose S is a martingale and H is a predictable process with respect to the filtration $\mathbb{F} = \{\mathcal{F}_t; t = 0, 1, \cdots, T\}$, we define the process

$$G_t = \sum_{u=1}^{t} H_u \Delta S_u, \tag{2.2.10}$$

where $\Delta S_u = S_u - S_{u-1}$. One then deduces that $\Delta G_u = G_u - G_{u-1} = H_u \Delta S_u$. If S and H represent the asset price process and trading strategy, respectively, then G can be visualized as the gain process. Note that the trading strategy H is a predictable process, that is, H_t is $\mathcal{F}_{t-1}$-measurable. This is because the number of units held for each security is determined at the beginning of the trading period by taking into account all the information available up to that time.

We call G to be the martingale transform of S by H, as G itself is also a martingale. To show the claim, it suffices to show that $E[G_{t+s}|\mathcal{F}_t] = G_t, t \geq 0, s \geq 0$. We consider

$$\begin{aligned} E[G_{t+s}|\mathcal{F}_t] &= E[G_{t+s} - G_t + G_t|\mathcal{F}_t] \\ &= E[H_{t+1}\Delta S_{t+1} + \cdots + H_{t+s}\Delta S_{t+s}|\mathcal{F}_t] + E[G_t|\mathcal{F}_t] \\ &= E[H_{t+1}\Delta S_{t+1}|\mathcal{F}_t] + \cdots + E[H_{t+s}\Delta S_{t+s}|\mathcal{F}_t] + G_t. \end{aligned}$$

Consider the typical term $E[H_{t+u}\Delta S_{t+u}|\mathcal{F}_t]$, by the tower property of conditional expectations, we can express it as $E[E[H_{t+u}\Delta S_{t+u}|\mathcal{F}_{t+u-1}]|\mathcal{F}_t]$. Further, since H_{t+u} is $\mathcal{F}_{t+u-1}$-measurable and S is a martingale, by virtue of (2.2.7)–(2.2.8), we have

$$E[H_{t+u}\Delta S_{t+u}|\mathcal{F}_{t+u-1}] = H_{t+u}E[\Delta S_{t+u}|\mathcal{F}_{t+u-1}] = 0.$$

Collecting all the calculations, we obtain the desired result.

2.2.3 Stopping Times and Stopped Processes

Given a filtered probability space $(\Omega, \mathcal{F}, P)$ and an adapted process X_t, we consider a game in which the player has the option either to continue the game or quit to receive the reward X_t. A stopping rule is defined such that the game player knows at each time t whether to continue or quit the game, given the information available at that time. A stopping time τ is a random variable: $\Omega \to \{0, 1, \cdots, T\}$ such that

$$\{\tau = t\} = \{\tau(\omega) = t; \omega \in \Omega\} \in \mathcal{F}_t. \tag{2.2.11}$$

That is, conditional on the information $\mathcal{F}_t$ at time t, one can determine whether the event $\{\tau = t\}$ has occurred or not. It can be shown that τ is a stopping time if and only if $\{\tau \leq t\} \in \mathcal{F}_t$ (see Problem 2.23).

For example, consider the adapted process S_t defined in the two-period model in Fig. 2.5. We define τ to be the first time that S_t assumes the value 3. That is,

$$\tau = \inf\{t \geq 0 : S_t = 3\}.$$

This is seen to be a stopping time since we can determine whether the event $\{\tau = t\}$ occurs conditional on $\mathcal{F}_t$, $t = 0, 1, 2$. On the other hand, the random time defined by

$$\tau = \sup\{t \geq 0 : S_t = 3\}$$

is not a stopping time since this random time depends on knowledge about the future.

Stopped (Sampled) Processes
Given an adapted process S_t, the stopped (sampled) process $S_t^\tau(\omega)$ with reference to the stopping time τ is defined by

$$S_t^\tau(\omega) = \begin{cases} S_t(\omega) & \text{if } t \leq \tau(\omega) \\ S_{\tau(\omega)}(\omega) & \text{if } t \geq \tau(\omega). \end{cases} \tag{2.2.12}$$

Under the discrete multiperiod model, $S_t^\tau(\omega)$ can be expressed as

$$S_t^\tau(\omega) = \mathbf{1}_{\{\tau \geq t\}} S_t + \sum_{u=0}^{t-1} \mathbf{1}_{\{\tau = u\}} S_u.$$

Since $\mathbf{1}_{\{\tau \geq t\}} S_t$ and $\mathbf{1}_{\{\tau = u\}} S_u$, $u = 0, 1, \cdots, t-1$ are $\mathcal{F}_t$-measurable, so the stopped process $S_t^\tau(\omega)$ is also adapted. More interestingly, if we stop a martingale by a stopping rule, the stopped process remains a martingale. That is, suppose M_t is a martingale, then

$$E\big[M_{t+s}^\tau\big] = M_t, \quad s = 1, 2, \cdots. \tag{2.2.13}$$

This result is known as the *Doob Optional Sampling Theorem.* Actually, the theorem remains valid even if we replace martingale by supermartingale or submartingale.

In the proof procedure, it is easier to show the validity of the theorem for submartingales or supermartingales. Once the results for submartingales and supermartingales have been established, and noting that a martingale is both a submartingale or supermartingale, the result for martingales then holds.

Let X_t be a submartingale and observe that $\{\tau = s\}$, $s = 0, 1, \cdots, t$, and $\{\tau \geq t+1\}$ are $\mathcal{F}_t$-measurable so that

$$E[\mathbf{1}_{\{\tau = s\}} X_s | \mathcal{F}_t] = \mathbf{1}_{\{\tau = s\}} X_s, \quad s = 0, 1, \cdots, t.$$

By virtue of the submartingale property, we have

$$E\big[\mathbf{1}_{\{\tau \geq t+1\}} X_{t+1} | \mathcal{F}_t\big] = \mathbf{1}_{\{\tau \geq t+1\}} E[X_{t+1} | \mathcal{F}_t] \geq \mathbf{1}_{\{\tau \geq t+1\}} X_t.$$

Next, we consider

$$\begin{aligned} E[X_{t+1}^\tau | \mathcal{F}_t] &= E\big[\mathbf{1}_{\{\tau \geq t+1\}} X_{t+1} | \mathcal{F}_t\big] + \sum_{s=0}^{t} E\big[\mathbf{1}_{\{\tau = s\}} X_s | \mathcal{F}_t\big] \\ &\geq \mathbf{1}_{\{\tau \geq t+1\}} X_t + \sum_{s=0}^{t} \mathbf{1}_{\{\tau = s\}} X_s = X_t^\tau, \end{aligned}$$

so the stopped submartingale remains a submartingale. Hence, the result for a submartingale is established. A similar proof can be extended to supermartingales.

The terminal value of the stopped process is seen to be $S_{\tau(\omega)}(\omega)$, $\omega \in \Omega$. An optimal stopping rule is defined to be the optimal choice of the stopping time such that the expected terminal value is maximized. Accordingly, a stopping time τ^* is said to be optimal if

$$E[S_{\tau^*}] = \max_{\tau \in \{0,1,\cdots,T\}} E[S_\tau]. \tag{2.2.14}$$

The optimal early exercise time of an American option is related to the notion of an optimal stopping time, the details of which can be found in Sect. 5.2.

2.2.4 Multiperiod Securities Models

We are now equipped with the knowledge of filtrations, adapted stochastic processes and martingales. Next, we discuss the fundamentals of financial economics of the multiperiod securities models. In particular, we consider the relation between absence of arbitrage opportunities and existence of the martingale measure (risk neutral probability measure).

We start with the prescription of a discrete n-period securities model with M risky securities. Like the discrete single-period model, there is a sample space $\Omega = \{\omega_1, \omega_2, \cdots, \omega_K\}$ of K possible states of the world. The asset price process is the row vector $\mathbf{S}(t) = (S_1(t)\ S_2(t) \cdots S_M(t))$ whose components are the security prices, $t = 0, 1, \cdots, n$. Also, there is a money market account process $S_0(t)$ whose value is given by

$$S_0(t) = (1 + r_1)(1 + r_2) \cdots (1 + r_t),$$

where r_u is the interest rate applied over one time period $(u-1, u)$, $u = 1, \cdots, t$. It is commonly assumed that r_t is known at the beginning of the period $(t-1, t)$ so that r_t is $\mathcal{F}_{t-1}$-measurable. A trading strategy is the rule taken by an investor that specifies the investor's position in each security at each time and in each state of the world based on the available information as prescribed by the filtration. Hence, one can visualize a trading strategy as an adapted stochastic process. We prescribe a trading strategy by a vector stochastic process $\mathbf{h}(t) = (h_0(t)\ h_1(t)\ h_2(t) \cdots h_M(t))^{\mathrm{T}}$, $t = 1, 2, \cdots, n$ (represented as a column vector), where $h_m(t)$ is the number of units held in the portfolio for the mth security from time $t-1$ to time t. Thus, $h_m(t)$ is $\mathcal{F}_{t-1}$-measurable, $m = 0, 1, \cdots, M$.

The value of the portfolio is a stochastic process given by

$$V(t) = h_0(t)S_0(t) + \sum_{m=1}^{M} h_m(t)S_m(t), \quad t = 1, 2, \cdots, n, \tag{2.2.15}$$

which gives the portfolio value at the moment right after the asset prices are observed but before changes in portfolio weights are made.

We write $\Delta S_m(t) = S_m(t) - S_m(t-1)$ as the change in value of one unit of the mth security between times $t-1$ and t. The cumulative gain associated with investing in the mth security from time zero to time t is given by

$$\sum_{u=1}^{t} h_m(u)\Delta S_m(u).$$

We define the portfolio gain process $G(t)$ to be the total cumulative gain in holding the portfolio consisting of the M risky securities and the money market account up to time t. The value of $G(t)$ is found to be

$$G(t) = \sum_{u=1}^{t} h_0(u) \Delta S_0(u) + \sum_{m=1}^{M} \sum_{u=1}^{t} h_m(u) \Delta S_m(u), \quad t = 1, 2, \cdots, n.$$

If we define the discounted price process $S_m^*(t)$ by

$$S_m^*(t) = S_m(t)/S_0(t), \quad t = 0, 1, \cdots, n, \text{ and } m = 1, 2, \cdots, M,$$

and write $\Delta S_m^*(t) = S_m^*(t) - S_m^*(t-1)$, then the discounted value process $V^*(t)$ and discounted gain process $G^*(t)$ are given by

$$V^*(t) = h_0(t) + \sum_{m=1}^{M} h_m(t) S_m^*(t), \quad t = 1, 2, \cdots n, \tag{2.2.16a}$$

$$G^*(t) = \sum_{m=1}^{M} \sum_{u=1}^{t} h_m(u) \Delta S_m^*(u), \quad t = 1, 2, \cdots, n. \tag{2.2.16b}$$

Once the asset prices, $S_m(t), m = 1, 2, \cdots, M$, are revealed to the investor, he changes the trading strategy from $\mathbf{h}(t)$ to $\mathbf{h}(t+1)$ in response to the arrival of the new information. Let t^+ denote the moment right after the portfolio rebalancing at time t. Since the portfolio holding of assets changes from $\mathbf{h}(t)$ to $\mathbf{h}(t+1)$, the new portfolio value at time t^+ becomes

$$V(t^+) = h_0(t+1) S_0(t) + \sum_{m=1}^{M} h_m(t+1) S_m(t). \tag{2.2.17}$$

Suppose we adopt the self-financing trading strategy such that the purchase of additional units of one particular security is financed by the sales of other securities within the portfolio, then $V(t) = V(t^+)$ since there is no addition or withdrawal of fund from the portfolio. By combining (2.2.15) and (2.2.17), the portfolio rebalancing from $\mathbf{h}(t)$ to $\mathbf{h}(t+1)$ under the self-financing condition must observe

$$[h_0(t+1) - h_0(t)] S_0(t) + \sum_{m=1}^{M} [h_m(t+1) - h_m(t)] S_m(t) = 0. \tag{2.2.18}$$

If there were no addition or withdrawal of funds at all trading times, then the cumulative change of portfolio value $V(t) - V(0)$ should be equal to the gain $G(t)$ associated with price changes of the securities on all trading dates. Hence, a trading strategy H is self-financing if and only if

$$\begin{aligned} V(t) &= V(0) + G(t) \\ &= V(0) + \sum_{u=1}^{t} h_0(u) \Delta S_0(u) + \sum_{u=1}^{t} \sum_{m=1}^{M} h_m(u) \Delta S_m(u). \end{aligned} \tag{2.2.19a}$$

In a similar manner, we can use (2.2.16a,b) to show that H is self-financing if and only if

$$V^*(t) = V^*(0) + G^*(t). \tag{2.2.19b}$$

No Arbitrage Principle

The definition of an arbitrage opportunity for the single period securities model (see Sect. 2.1.2) is extended to the multiperiod models. A trading strategy H represents an arbitrage opportunity if and only if the value process $V(t)$ and H satisfy the following properties:

(i) $V(0) = 0$,
(ii) $V(T) \geq 0$ and $E[V(T)] > 0$, and
(iii) H is self-financing.

Here, E is the expectation under the actual probability measure. Equivalently, the self-financing trading strategy H is an arbitrage opportunity if and only if (i) $G^*(T) \geq 0$ and (ii) $E[G^*(T)] > 0$. Like that in the single-period models, we expect that an arbitrage opportunity does not exist if and only if there exists a risk neutral probability measure. In the multiperiod models, risk neutral probabilities are defined in terms of martingales.

Martingale Measure

The measure Q is called a martingale measure (or called a risk neutral probability measure) if it has the following properties:

1. $Q(\omega) > 0$ for all $\omega \in \Omega$.
2. Every discounted price process S_m^* in the securities model is a martingale under $Q, m = 1, 2, \cdots, M$, that is,

$$E_Q[S_m^*(u)|\mathcal{F}_t] = S_m^*(t) \quad \text{for } 0 \leq t \leq u \leq T.$$

We call the discounted price process $S_m^*(t)$ a Q-martingale.

Calculations of Martingale Probability Values

As a numerical example, we determine the martingale measure Q associated with the two-period securities model shown in Fig. 2.5. Let $r \geq 0$ be the constant riskless interest rate over one period, and write $Q(\omega_j)$ as the martingale measure associated with the state ω_j, $j = 1, 2, 3, 4$. By invoking the martingale property of S_t, we obtain the following equations for $Q(\omega_1), \cdots, Q(\omega_4)$:

(i) $t = 0$ and $u = 1$

$$4 = \frac{3}{1+r}[Q(\omega_1) + Q(\omega_2)] + \frac{5}{1+r}[Q(\omega_3) + Q(\omega_4)]. \tag{2.2.20a}$$

(ii) $t = 0$ and $u = 2$

$$\begin{aligned} 4 = {} & \frac{4}{(1+r)^2}Q(\omega_1) + \frac{2}{(1+r)^2}Q(\omega_2) \\ & + \frac{4}{(1+r)^2}Q(\omega_3) + \frac{6}{(1+r)^2}Q(\omega_4). \end{aligned} \tag{2.2.20b}$$

(iii) $t = 1$ and $u = 2$

$$3 = \frac{4}{1+r}\frac{Q(\omega_1)}{Q(\omega_1)+Q(\omega_2)} + \frac{2}{1+r}\frac{Q(\omega_2)}{Q(\omega_1)+Q(\omega_2)} \tag{2.2.20c}$$

$$5 = \frac{4}{1+r}\frac{Q(\omega_3)}{Q(\omega_3)+Q(\omega_4)} + \frac{6}{1+r}\frac{Q(\omega_4)}{Q(\omega_3)+Q(\omega_4)}. \tag{2.2.20d}$$

It may be quite tedious to solve the above equations simultaneously. The calculation procedure can be simplified by observing that $Q(\omega_j)$ is given by the product of the conditional probabilities along the path from the node at $t = 0$ to the node ω_j at $t = 2$. First, we start with the conditional probability p associated with the upper branch $\{\omega_1, \omega_2\}$. The corresponding conditional probability p is given by

$$4 = \frac{3}{1+r}p + \frac{5}{1+r}(1-p)$$

so that $p = \frac{1-4r}{2}$. Similarly, the conditional probability p' associated with the branch $\{\omega_1\}$ from the node $\{\omega_1, \omega_2\}$ is given by

$$3 = \frac{4}{1+r}p' + \frac{2}{1+r}(1-p')$$

giving $p' = \frac{1-3r}{2}$. In a similar manner, the conditional probability p'' associated with $\{\omega_3\}$ from $\{\omega_3, \omega_4\}$ is found to be $\frac{1-5r}{2}$. The martingale probabilities are then found to be

$$\begin{aligned}
Q(\omega_1) &= pp' = \frac{1-4r}{2}\frac{1-3r}{2}, \\
Q(\omega_2) &= p(1-p') = \frac{1-4r}{2}\frac{1+3r}{2}, \\
Q(\omega_3) &= (1-p)p'' = \frac{1+4r}{2}\frac{1-5r}{2}, \\
Q(\omega_4) &= (1-p)(1-p'') = \frac{1+4r}{2}\frac{1+5r}{2}.
\end{aligned} \tag{2.2.21}$$

These martingale probabilities can be shown to satisfy (2.2.20a–d). In order that the martingale probabilities remain positive, we have to impose the restriction: $r < 0.2$. It can be shown that an arbitrage opportunity exists for the securities model when $r \geq 0.2$ (see Problem 2.25).

As a remark, an arbitrage opportunity in any underlying single period would lead to an arbitrage opportunity in the overall multiperiod model. This is because one can follow the arbitrage trading strategy in that particular single period and do nothing in all other periods, thus arbitrage arises in the multiperiod model.

Martingale Property of Value Processes

Suppose H is a self-financing trading strategy and Q is a martingale measure with respect to a filtration $\mathcal{F}$, then the value process $V(t)$ is a Q-martingale. To show the claim, since H is self-financing, we apply (2.2.19b) to obtain

$$V^*(t+1) - V^*(t) = G^*(t+1) - G^*(t)$$
$$= [\mathbf{S}^*(t+1) - \mathbf{S}^*(t)]\mathbf{h}(t+1).$$

As H is a predictable process, $V^*(t)$ is the martingale transform of the Q-martingale $\mathbf{S}^*(t)$. Hence, $V^*(t)$ itself is also a Q-martingale.

Fundamental Theorem of Asset Pricing (Multiperiod Models)
The above result can be applied to show that the existence of the martingale measure Q implies the nonexistence of arbitrage opportunities. To prove the claim, suppose H is a self-financing trading strategy with $V^*(T) \geq 0$ and $E[V^*(T)] > 0$. Here, E is the expectation under the actual probability measure P, with $P(\omega) > 0$. That is, $V^*(T)$ is strictly positive for some states of the world. As $Q(\omega) > 0$, we then have $E_Q[V^*(T)] > 0$. However, since $V^*(T)$ is a Q-martingale so that $V^*(0) = E_Q[V^*(T)]$, and by virtue of $E_Q[V^*(T)] > 0$, we always have $V^*(0) > 0$. It is then impossible to have $V^*(T) \geq 0$ and $E[V^*(T)] > 0$ while $V^*(0) = 0$. Hence, the self-financing strategy H cannot be an arbitrage opportunity.

The converse of the above claim remains valid, that is, the nonexistence of arbitrage opportunities implies the existence of a martingale measure. The intuition behind the proof can be outlined as follows. If there are no arbitrage opportunities in the multiperiod model, then there will be no arbitrage opportunities in any underlying single period. Since each single period does not admit arbitrage opportunities, one can construct the one-period risk neutral conditional probabilities. The martingale probability measure $Q(\omega)$ is then obtained by multiplying all the risk neutral conditional probabilities along the path from the node at $t = 0$ to the terminal node (T, ω). The construction of a rigorous proof based on the above arguments is quite technical, the details of which can be found in Harrison and Kreps (1979) and Bingham and Kiesel (2004).

We summarize the above results into the following theorem.

Theorem 2.3. *A multiperiod securities model is arbitrage free if and only if there exists a probability measure Q such that the discounted asset price processes are Q-martingales.*

Valuation of Contingent Claims
Most of the results on valuation of contingent claims in single period models can be extended to multiperiod models. First, the martingale measure is unique if and only if the multiperiod securities model is complete. Here, completeness implies that all contingent claims ($\mathcal{F}_T$-measurable random variables) can be replicated by a self-financing trading strategy. In an arbitrage free complete market, the arbitrage price of an attainable contingent claim is then given by the discounted expectation under the martingale measure of the value of the portfolio that replicates the claim. Let Y denote the contingent claim at maturity T and $V(t)$ denote the arbitrage price of the contingent claim at time t, $t < T$. Using the Q-martingale property, we then have

$$V^*(t) = E_Q[V^*(T)|\mathcal{F}_t] = E_Q[Y^*|\mathcal{F}_t].$$

Assuming deterministic interest rates, we obtain

$$V(t) = \frac{S_0(t)}{S_0(T)} E_Q[Y|\mathcal{F}_t], \tag{2.2.22}$$

where $S_0(t)$ is the time-t price of the riskless asset and the ratio $S_0(t)/S_0(T)$ is the discount factor over the period from t to T.

2.2.5 Multiperiod Binomial Models

We extend the one-period binomial model to its multiperiod version. We start with the two-period binomial model. The corresponding dynamics of the binomial process for the asset price and the call price are shown in Fig. 2.6. The jump ratios of the asset price, u and d, are taken to have the same value for all binomial steps.

Let c_{uu} denote the call value at two periods beyond the current time with two consecutive upward moves of the asset price and similar notational interpretation for c_{ud} and c_{dd}. Based on a similar relation as depicted in (2.1.21), the call values c_u and c_d are related to c_{uu}, c_{ud} and c_{dd} as follows:

$$c_u = \frac{pc_{uu} + (1-p)c_{ud}}{R} \quad \text{and} \quad c_d = \frac{pc_{ud} + (1-p)c_{dd}}{R},$$

where $p = \frac{R-d}{u-d}$ and $R = e^{r\Delta t}$. Subsequently, by substituting the above results into (2.1.21), the call value at the current time which is two periods from expiry is found to be

$$c = \frac{p^2 c_{uu} + 2p(1-p)c_{ud} + (1-p)^2 c_{dd}}{R^2},$$

where the corresponding terminal payoff values are given by

$$c_{uu} = \max(u^2S - X, 0),\ c_{ud} = \max(udS - X, 0),\ c_{dd} = \max(d^2S - X, 0).$$

Note that the coefficients $p^2, 2p(1-p)$ and $(1-p)^2$ represent the respective risk neutral probability of having two up-jumps, one up-jump and onc down-jump, and two down-jumps in the two-step binomial asset price process.

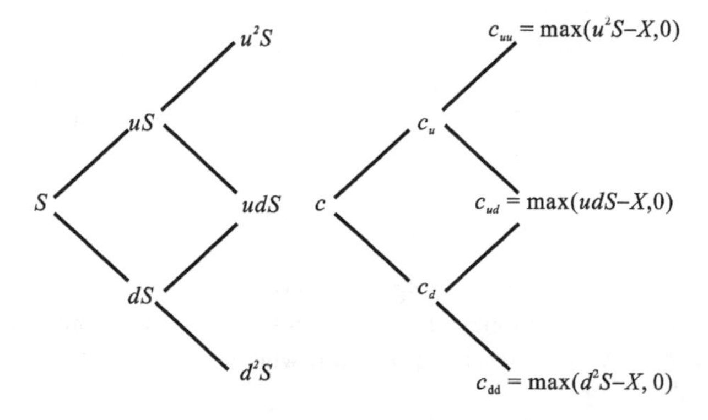

Fig. 2.6. Dynamics of the asset price and call price in a two-period binomial model.

The extension of the binomial model to the n-period case should be quite straightforward. With n binomial steps, the risk neutral probability of having j up-jumps and $n-j$ down-jumps is given by $C_j^n p^j(1-p)^{n-j}$, where $C_j^n = \frac{n!}{j!(n-j)!}$ is the number of choices of choosing j up-jumps from the n binomial steps. The corresponding terminal payoff when j up-jumps and $n-j$ down-jumps occur is seen to be $\max(u^j d^{n-j} S - X, 0)$. The call value obtained from the n-period binomial model is given by

$$c = \frac{\sum_{j=0}^{n} C_j^n p^j (1-p)^{n-j} \max(u^j d^{n-j} S - X, 0)}{R^n}. \tag{2.2.23}$$

We define k to be the smallest nonnegative integer such that $u^k d^{n-k} S \geq X$, that is, $k \geq \frac{\ln \frac{X}{Sd^n}}{\ln \frac{u}{d}}$. Accordingly, we have

$$\max(u^j d^{n-j} S - X, 0) = \begin{cases} 0 & \text{when} \quad j < k \\ u^j d^{n-j} S - X & \text{when} \quad j \geq k. \end{cases} \tag{2.2.24}$$

The integer k gives the minimum number of upward moves required for the asset price in the multiplicative binomial process in order that the call expires in-the-money. The call formula in (2.2.23) can then be simplified as

$$c = S \sum_{j=k}^{n} C_j^n p^j (1-p)^{n-j} \frac{u^j d^{n-j}}{R^n} - X R^{-n} \sum_{j=k}^{n} C_j^n p^j (1-p)^{n-j}. \tag{2.2.25}$$

The last term in above equation can be interpreted as the expectation value under the risk neutral measure of the payment made by the holder at expiration discounted by the factor R^{-n}, and $\sum_{j=k}^{n} C_j^n p^j (1-p)^{n-j}$ is seen to be the risk neutral probability that the call expires in-the-money. The above probability value is related to the *complementary binomial distribution function* defined by

$$\Phi(n, k, p) = \sum_{j=k}^{n} C_j^n p^j (1-p)^{n-j}. \tag{2.2.26}$$

Note that $\Phi(n, k, p)$ gives the probability for having at least k successes in n trials of a binomial experiment, where p is the probability of success in each trial. Further, if we write $p' = \frac{up}{R}$ so that $1 - p' = \frac{d(1-p)}{R}$, then the call price formula for the n-period binomial model can be expressed as

$$c = S\Phi(n, k, p') - XR^{-n}\Phi(n, k, p). \tag{2.2.27}$$

The first term gives the discounted expectation of the asset price at expiration given that the call expires in-the-money and the second term gives the present value of the expected cost incurred by exercising the call, where the expectation is taken under the risk neutral measure.

Using the argument of discounted expectation of the payoff of a contingent claim under the risk neutral measure, the call price for the n-period binomial model can be expressed in the following canonical form

$$c = \frac{1}{R^n} E^*[c_T] = \frac{1}{R^n} E^*[\max(S_T - X, 0)], \quad T = t + n\Delta t, \tag{2.2.28}$$

where c_T is the payoff, as defined by $\max(S_T - X, 0)$, of the call at expiration time T and $\frac{1}{R^n}$ is the discount factor over n periods. Here, the expectation operator E^* is taken under the risk neutral measure rather than the true probability measure associated with the actual (subjective) asset price process.

Numerical Implementation

The n-period binomial model can be represented schematically by a n-step tree structure (see Fig. 2.7 for a three-step tree). The binomial tree will be symmetrical about S if $ud = 1$, skewed upward if $ud > 1$ and skewed downward if $ud < 1$. At the time level that is m time steps marching forward from the current time in the binomial tree, there are $m + 1$ nodes. The asset price at the node obtained by j upward moves and $m - j$ downward moves equals $Su^j d^{m-j}$, $j = 0, 1, \cdots, m$. The possible option values at expiration are known since the payoff function at expiry is defined in the option contract. Rather than using the multiplicative binomial formula (2.2.25), the following stepwise backward induction procedure is more effective in numerical implementation. First, we compute option values at the nodes that are one time step from expiration using the binomial formula (2.1.21). Once option values at one time step from expiration are known, we proceed two time steps from expiration and repeat the same numerical procedure. After performing n backward steps in the tree, we come to the starting node (tip of the tree) at which the option value is desired.

As a numerical example, suppose we have chosen the following values for the binomial parameters: $u = 1.25$, $d = 0.8$, and the discount factor for one period $= 1/R = 0.95$. According to (2.1.21), we have

$$p = \frac{R - d}{u - d} = \left(\frac{1}{0.95} - 0.8\right) \Big/ (1.25 - 0.8) = 0.5614.$$

The strike price of the call is taken to be 70 and the asset price S at the current time is 120. The binomial tree with three time steps is illustrated in Fig. 2.7. The upper and lower figures at the nodes denote the asset prices and option values, respectively. For example, the option values at nodes P and Q are, respectively, $\max(150 - 70, 0) = 80$ and $\max(96 - 70, 0) = 26$. The option value at node Y is computed by

$$\begin{aligned} c_Y &= \frac{1}{R}[pc_P + (1 - p)c_Q] \\ &= 0.95(0.5614 \times 80 + 0.4386 \times 26) \\ &= 53.50 \text{ (2 decimal places).} \end{aligned}$$

Working three steps backward from the expiration time to the current time, the current option value at $S = 120$ is found to be 60.61 (see Fig. 2.7).

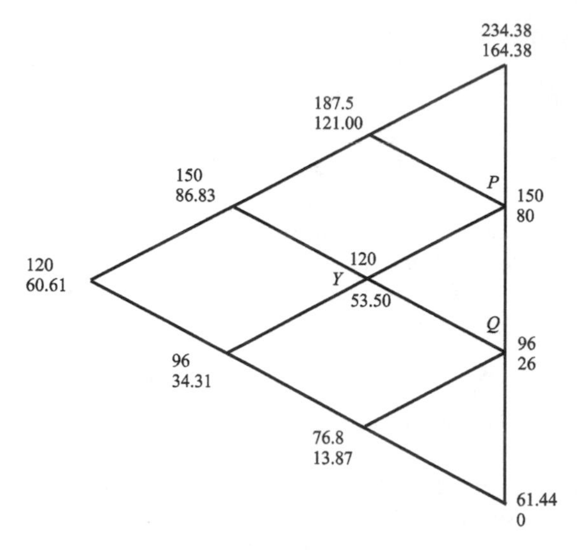

Fig. 2.7. Illustration of the binomial calculations with three time steps for a European call with strike price $X = 70$. The top figures are asset prices and the bottom figures are option values.

2.3 Asset Price Dynamics and Stochastic Processes

In this section, we discuss the stochastic models for the simulation of the asset price movement. The asset price movement is said to follow a *stochastic process* if its value changes over time in an uncertain manner. The study of stochastic processes is concerned with the investigation of the structure of families of random variables X_t, where t is a parameter (t is usually interpreted as the time parameter) running over some index set $\mathcal{T}$. If the index set $\mathcal{T}$ is discrete, then the stochastic process $\{X_t, t \in \mathcal{T}\}$ is called a discrete stochastic process, and for a continuous index set, $\{X_t, t \in \mathcal{T}\}$ becomes a continuous stochastic process. In other words, a discrete-time stochastic process for the asset price is one where the asset price can change at some discrete fixed times. On the other hand, the asset price which follows a continuous-time stochastic process can change its value at any time. Further, the value taken by the random variable X_t can be either discrete or continuous, and the corresponding stochastic process is called discrete-valued or continuous-valued, respectively. In reality, stock prices can change only at discrete values and during periods when the stock exchange is open. In order that analytic tools in stochastic calculus can be employed, the asset price processes are assumed to be continuous-valued continuous-time stochastic processes in later chapters.

A *Markovian process* is a stochastic process that, given the value of X_s, the value of $X_t, t > s$, depends only on X_s but not on the values taken by $X_u, u < s$. If the asset price follows a Markovian process, then only the present asset price is relevant for predicting its future values. This Markovian property of an asset price process is consistent with the *weak form of market efficiency*, which assumes that the present value of an asset price already impounds all information in its past prices and the particular path taken by the asset price to reach the present value is irrelevant. If the past history is indeed relevant, that is, a particular pattern might have a higher chance of price increases, then investors would bid up the asset price once such a pattern occurs and the profitable advantage would be eliminated.

We start with the discussion of the discrete random walk model and subsequently deduce its continuum limit. We obtain the Fokker–Planck equation that governs the probability density function of the continuous random walk motion. We then present the formal definition of a Brownian process and discuss some of the properties of Brownian processes.

2.3.1 Random Walk Models

We describe the unrestricted, one-dimensional discrete random walk and consider the continuum limit of the discrete random walk problem to yield the continuous random walk model. Suppose a particle starts at the origin of the x-axis and it jumps either to the left or the right of the same length δ. We define x_i to be the random variable which takes the value δ or $-\delta$ when the particle at the ith step moves to the right or the left, respectively. Assume that the jump probabilities are stationary, that is, these probabilities are the same at all times. We then write the probabilities as

$$P(x_i = \delta) = p, \quad P(x_i = -\delta) = q, \tag{2.3.1}$$

where $p+q = 1$, p and q are independent of i. The individual jumps are assumed to be independent of each other so that x_i, $i = 1, 2, \cdots$, are independent. This discrete random walk problem is seen to be a discrete Markovian process (see Fig. 2.8).

Define the discrete sum process

$$X_n = x_1 + x_2 + \cdots + x_n, \tag{2.3.2}$$

which gives the position of the particle at the end of the nth step. Since the expected value of x_i is

$$E[x_i] = \delta p - \delta q = (p - q)\delta, \quad i = 1, 2, \cdots, n,$$

therefore

$$E[X_n] = E\left[\sum_{i=1}^{n} x_i\right] = \sum_{i=1}^{n} E[x_i] = (p - q)\delta n. \tag{2.3.3}$$

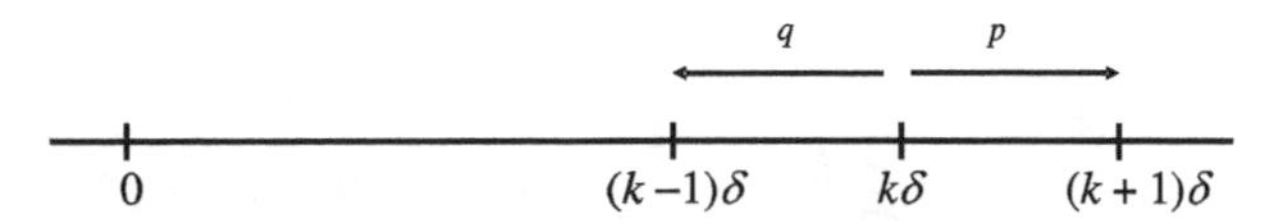

Fig. 2.8. A graphical representation of the discrete random walk model. Suppose the particle is at position $x = k\delta$ after $i - 1$ steps, $|k| \leq i - 1$. In the ith step, it moves to the right with probability p or the left with probability q.

As x_i's are independent, we have

$$\mathrm{var}(X_n) = n\ \mathrm{var}(x_i).$$

The variance of x_i is

$$\mathrm{var}(x_i) = [\delta^2 p + (-\delta)^2 q] - (E[x_i])^2 = \delta^2 - (p-q)^2\delta^2 = 4pq\delta^2$$

so that

$$\mathrm{var}(X_n) = 4pq\delta^2 n. \tag{2.3.4}$$

We call $X_{n+1} - X_n$ an increment of the discrete random walk model. Since X_n is a sum process of independent and identically distributed (iid) random variables, it observes the properties of stationary and independent increments.

Continuum Limit

Next, we take the continuum limit of an infinitesimally small step size of the above discrete model to yield the continuous random walk model. Suppose there are r steps per unit time, then according to (2.3.3)–(2.3.4), the mean displacement of the particle per unit time μ is $(p-q)\delta r$ and the variance of the observed displacement around the mean position per unit time σ^2 is $4pq\delta^2 r$. Let $\lambda = 1/r$, which is the time interval between two successive steps, and let $u(x,t)$ denote the probability that the particle takes the position x at time t. Now, we write $X_n = x$ and $n\lambda = t$ so that

$$u(x,t) = P(X_n = x) \quad \text{at } t = n\lambda. \tag{2.3.5}$$

To arrive at the position x at time $t+\lambda$, the particle must be either at $x-\delta$ or $x+\delta$ at time t. With probability p (or q), the particle at $x-\delta$ (or $x+\delta$) moves to x in the next time step. Therefore, the probability function $u(x,t)$ satisfies the recurrence relation:

$$u(x, t+\lambda) = pu(x-\delta, t) + qu(x+\delta, t). \tag{2.3.6}$$

In the continuum limit, we take $\delta \to 0$ and $r \to \infty$ so that $\lambda \to 0$. Now, consider the Taylor expansion of relation (2.3.6):

$$\begin{aligned} u(x,t) + \lambda\frac{\partial u}{\partial t}(x,t) + O(\lambda^2) = {} & p\left[u(x,t) - \delta\frac{\partial u}{\partial x}(x,t) + \frac{\delta^2}{2}\frac{\partial^2 u}{\partial x^2}(x,t) + O(\delta^3)\right] \\ & + q\left[u(x,t) + \delta\frac{\partial u}{\partial x}(x,t) + \frac{\delta^2}{2}\frac{\partial^2 u}{\partial x^2}(x,t) + O(\delta^3)\right], \end{aligned}$$

and upon simplification, we obtain

$$\frac{\partial u}{\partial t} = \left[(q-p)\frac{\delta}{\lambda}\right]\frac{\partial u}{\partial x} + \frac{1}{2}\left(\frac{\delta^2}{\lambda}\right)\frac{\partial^2 u}{\partial x^2} + O(\lambda) + O\left((q-p)\frac{\delta^3}{\lambda}\right). \tag{2.3.7}$$

We take the limits $\delta, \lambda \to 0$ in the manner that the mean displacement and variance per unit time are given by

$$(p-q)\frac{\delta}{\lambda} = \mu \text{ and } 4pq\frac{\delta^2}{\lambda} = \sigma^2, \tag{2.3.8}$$

where μ and σ^2 are finite quantities. The discrete random walk model fails to make sense if p and q are infinitesimal quantities. In other words, we must observe $p = O(1), q = O(1)$ and $p + q = 1$. Consequently, we can deduce from $4pq\frac{\delta^2}{\lambda} = \sigma^2$ that $\frac{\delta^2}{\lambda} = O(1)$ or $\frac{\delta}{\lambda} = O(\frac{1}{\delta})$. Also, as deduced from $(p-q)\frac{\delta}{\lambda} = \mu$ and $p+q = 1$, the asymptotic expansion up to $O(\delta)$ of p and q must take the following forms:

$$p \approx \frac{1}{2}(1 + k\delta) \text{ and } q \approx \frac{1}{2}(1 - k\delta)$$

for some k to be determined. We then have $4pq \approx 1$ and so

$$\lim_{\delta,\lambda \to 0} \frac{\delta^2}{\lambda} = \sigma^2. \tag{2.3.9}$$

Lastly, from $(p-q)\frac{\delta}{\lambda} = \mu$ and condition (2.3.9), one deduces that $p - q \approx \frac{\mu}{\sigma^2}\delta$ and so $k = \frac{\mu}{\sigma^2}$. The asymptotic expansion of p and q are then found to be

$$p \approx \frac{1}{2}\left(1 + \frac{\mu}{\sigma^2}\delta\right) \text{ and } q \approx \frac{1}{2}\left(1 - \frac{\mu}{\sigma^2}\delta\right). \tag{2.3.10}$$

Note that $p \to \frac{1}{2}$ and $q \to \frac{1}{2}$ when taking the asymptotic limit $\delta \to 0$. If this is not the case, then the drift rate would become infinite. Since $\frac{\delta^2}{\lambda} = O(1)$, the last term in (2.3.7) becomes $O((q-p)\frac{\delta^3}{\lambda}) = O(\lambda)$. Consequently, by taking the limits $\delta, \lambda \to 0$ in (2.3.7), we obtain the following partial differential equation

$$\frac{\partial u}{\partial t} = -\mu\frac{\partial u}{\partial x} + \frac{\sigma^2}{2}\frac{\partial^2 u}{\partial x^2} \tag{2.3.11}$$

for the probability density function $u(x,t)$ of the continuous random walk motion with drift.

The above differential equation is called the *forward Fokker–Planck equation.* The drift rate is μ and the diffusion rate is σ^2. In time t, the mean displacement of the particle is μt and the variance of the observed displacement around the mean position is $\sigma^2 t$.

From the Central Limit Theorem in probability theory, one can show that the continuum limit of the probability density of the discrete random variable X_n defined in (2.3.2) tends to that of a normal random variable with the same mean and variance. The probability density function of the normal random variable X with mean μt and variance $\sigma^2 t$ is given by

$$f_X(x,t) = \frac{1}{\sqrt{2\pi\sigma^2 t}} \exp\left(-\frac{(x-\mu t)^2}{2\sigma^2 t}\right). \tag{2.3.12}$$

From the partial differential equation theory, $f_X(x,t)$ can be shown to satisfy the following initial value problem:

$$\frac{\partial u}{\partial t} = -\mu \frac{\partial u}{\partial x} + \frac{\sigma^2}{2} \frac{\partial^2 u}{\partial x^2}, \quad -\infty < x < \infty, \ t > 0, \tag{2.3.13}$$

with initial condition: $u(x, 0^+) = \delta(x)$, where $u(x, 0^+)$ signifies $\lim_{t \to 0^+} u(x, t)$. Here, $\delta(x)$ represents the Dirac function with the following properties:

$$\delta(x) = \begin{cases} 0 & \text{if } x \neq 0 \\ \infty & \text{if } x = 0 \end{cases} \quad \text{and} \quad \int_{-\infty}^{\infty} \delta(x)\, dx = 1.$$

The above result has the following probabilistic interpretation. Conditional on the event that the particle starts at the position $x = 0$ initially, $f_X(x, t)\Delta x$ gives the probability that the particle stays within $[x, x + \Delta x]$ at some future time t. This is why $f_X(x, t)$ is usually called the *transition density function*. The initial condition: $u(x, 0^+) = \delta(x)$ indicates that the particle starts at $x = 0$ almost surely. Also, the continuous random walk model inherits the properties of stationary and independent increments from the discrete random walk model.

2.3.2 Brownian Processes

The *Brownian motion* refers to the ceaseless, irregular random motion of small particles immersed in a liquid or gas, as observed by R. Brown in 1827. The phenomena can be explained by the perpetual collisions of the particles with the molecules of the surrounding medium. The stochastic process associated with the Brownian motion is called the *Brownian process* or the *Wiener process*.

The formal definition of a Brownian process with drift is presented below.

Definition. The *Brownian process with drift* is a stochastic process $\{X(t); t \geq 0\}$ with the following properties:

(i) Every increment $X(t + s) - X(s)$ is normally distributed with mean μt and variance $\sigma^2 t$; μ and σ are fixed parameters.
(ii) For every $t_1 < t_2 < \cdots < t_n$, the increments $X(t_2) - X(t_1), \cdots, X(t_n) - X(t_{n-1})$ are independent random variables with distributions given in (i).
(iii) $X(0) = 0$ and the sample paths of $X(t)$ are continuous.

Note that $X(t + s) - X(s)$ is independent of the past history of the random path, that is, the knowledge of $X(\tau)$ for $\tau < s$ has no effect on the probability distribution for $X(t + s) - X(s)$. This is precisely the Markovian character of the Brownian process.

Standard Brownian Process
For the particular case $\mu = 0$ and $\sigma^2 = 1$, the Brownian process is called the *standard Brownian process* (or *standard Wiener process*). The corresponding probability distribution for the standard Brownian process $\{Z(t); t \geq 0\}$ is given by [see (2.3.12)]

$$\begin{aligned}
P(Z(t) \le z | Z(t_0) = z_0) &= P(Z(t) - Z(t_0) \le z - z_0) \\
&= \frac{1}{\sqrt{2\pi(t-t_0)}} \int_{-\infty}^{z-z_0} \exp\left(-\frac{x^2}{2(t-t_0)}\right) dx \\
&= N\left(\frac{z-z_0}{\sqrt{t-t_0}}\right), \qquad (2.3.14)
\end{aligned}$$

where

$$N(x) = \frac{1}{\sqrt{2\pi}} \int_{-\infty}^{x} e^{-t^2/2}\, dt$$

is the cumulative normal distribution function. With zero mean and unit variance, the density function of the standard normal random variable is given by

$$n(x) = \frac{1}{\sqrt{2\pi}} e^{-x^2/2}.$$

Some Useful Properties

(a) $E[Z(t)^2] = \text{var}(Z(t)) + E[Z(t)]^2 = t.$
(b) $E[Z(t)Z(s)] = \min(t, s).$

To show the result in (b), we assume $t > s$ (without loss of generality) and consider

$$\begin{aligned}
E[Z(t)Z(s)] &= E[\{Z(t) - Z(s)\}Z(s) + Z(s)^2] \\
&= E[\{Z(t) - Z(s)\}Z(s)] + E[Z(s)^2].
\end{aligned}$$

Since $Z(t) - Z(s)$ and $Z(s)$ are independent and both $Z(t) - Z(s)$ and $Z(s)$ have zero mean, so

$$E[Z(t)Z(s)] = E[Z(s)^2] = s = \min(t, s). \qquad (2.3.15)$$

Overlapping Brownian Increments
When $t > s$, the correlation coefficient ρ between the two overlapping Brownian increments $Z(t)$ and $Z(s)$ is given by

$$\rho = \frac{E[Z(t)Z(s)]}{\sqrt{\text{var}(Z(t))}\sqrt{\text{var}(Z(s))}} = \frac{s}{\sqrt{st}} = \sqrt{\frac{s}{t}}. \qquad (2.3.16)$$

The Brownian increments $Z(t)$ and $Z(s)$ are bivariate normally distributed with zero mean, variance t and s, respectively, and their correlation coefficient is $\sqrt{s/t}$. If we define $X_1 = Z(t)/\sqrt{t}$ and $X_2 = Z(s)/\sqrt{s}$, then X_1 and X_2 become standard normal random variables. The joint distribution of $Z(t)$ and $Z(s)$ is given by

$$\begin{aligned}
P(Z(t) \le z_t, Z(s) \le z_s) &= P(X_1 \le z_t/\sqrt{t}, X_2 \le z_s/\sqrt{s}) \\
&= N_2(z_t/\sqrt{t}, z_s/\sqrt{s}; \sqrt{s/t}), \qquad (2.3.17)
\end{aligned}$$

where the bivariate normal distribution function is defined by

$$N_2(x_1, x_2; \rho) = \int_{-\infty}^{x_2} \int_{-\infty}^{x_1} \frac{1}{2\pi\sqrt{1-\rho^2}} \exp\left(-\frac{\xi_1^2 - 2\rho\xi_1\xi_2 + \xi_2^2}{2(1-\rho^2)}\right) d\xi_1 d\xi_2.$$

Geometric Brownian Process

Let $X(t)$ denote the Brownian process with drift parameter $\mu \geq 0$ and variance parameter σ^2. The stochastic process defined by

$$Y(t) = e^{X(t)}, \quad t \geq 0, \tag{2.3.18}$$

is called the *Geometric Brownian process*. Obviously, the value taken by $Y(t)$ is nonnegative. Since $X(t) = \ln Y(t)$ is a Brownian process, by properties (i) and (ii) we deduce that $\ln Y(t) - \ln Y(0)$ is normally distributed with mean μt and variance $\sigma^2 t$. For common usage, $\frac{Y(t)}{Y(0)}$ is said to be log-normally distributed. From the density function of $X(t)$ given in (2.3.12), the density function of $\frac{Y(t)}{Y(0)}$ is deduced to be

$$f_Y(y, t) = \frac{1}{y\sqrt{2\pi\sigma^2 t}} \exp\left(-\frac{(\ln y - \mu t)^2}{2\sigma^2 t}\right). \tag{2.3.19}$$

The mean of $Y(t)$ conditional on $Y(0) = y_0$ is found to be

$$\begin{aligned}
&E[Y(t)|Y(0) = y_0] \\
&= y_0 \int_0^\infty y f_Y(y, t)\, dy \\
&= y_0 \int_{-\infty}^\infty \frac{e^x}{\sqrt{2\pi\sigma^2 t}} \exp\left(-\frac{(x - \mu t)^2}{2\sigma^2 t}\right) dx, \quad x = \ln y, \\
&= y_0 \int_{-\infty}^\infty \frac{1}{\sqrt{2\pi\sigma^2 t}} \exp\left(-\frac{[x - (\mu t + \sigma^2 t)]^2 - 2\mu t\sigma^2 t - \sigma^4 t^2}{2\sigma^2 t}\right) dx \\
&= y_0 \exp\left(\mu t + \frac{\sigma^2 t}{2}\right).
\end{aligned} \tag{2.3.20}$$

Similarly, the variance of $Y(t)$ conditional on $Y(0) = y_0$ is found to be

$$\begin{aligned}
&\mathrm{var}(Y(t)|Y(0) = y_0) \\
&= y_0^2 \int_0^\infty y^2 f_Y(y, t)\, dy - \left[y_0 \exp\left(\mu t + \frac{\sigma^2 t}{2}\right)\right]^2 \\
&= y_0^2 \left\{ \int_{-\infty}^\infty \frac{1}{\sqrt{2\pi\sigma^2 t}} \exp\left(-\frac{[x - (\mu t + 2\sigma^2 t)]^2 - 4\mu t\sigma^2 t - 4\sigma^4 t^2}{2\sigma^2 t}\right) dx \right. \\
&\qquad \left. - \left[\exp\left(\mu t + \frac{\sigma^2 t}{2}\right)\right]^2 \right\} \\
&= y_0^2 \exp(2\mu t + \sigma^2 t)[\exp(\sigma^2 t) - 1].
\end{aligned} \tag{2.3.21}$$

Given the set of discrete times $t_1 < t_2 < \cdots < t_n$, the successive ratios $Y(t_2)/Y(t_1), \cdots, Y(t_n)/Y(t_{n-1})$ are independent random variables, that is, the percentage changes over nonoverlapping time intervals are independent.

2.4 Stochastic Calculus: Ito's Lemma and Girsanov's Theorem

The price of a derivative is a function of the underlying asset price where the asset price process is modeled by a stochastic process. In order to construct pricing models for derivatives, it is necessary to develop calculus tools that allow us to perform mathematical operations, like composition, differentiation, integration, etc. on functions of stochastic random variables. In this section, we define stochastic integrals and stochastic differentials of functions that involve the Brownian random variables. In particular, we develop the Ito differentiation rule that computes the differentials of functions of stochastic state variables. We also derive the Feynman–Kac representation formula, which gives a stochastic representation of the solution of a parabolic partial differential equation. We then discuss the notion of Radon–Nikodym derivatives and the Girsanov Theorem that effect the change of equivalent probability measures.

2.4.1 Stochastic Integrals

Brownian processes are the continuous limit of discrete random walk models. Intuitively, one may visualize Brownian paths to be continuous (though a rigorous mathematical proof of the continuity property is not trivial). However, Brownian paths are seen to be nonsmooth. In fact, they are not differentiable. The nondifferentiability property can be shown by proving the finiteness of the quadratic variation of a Brownian process. This stems from the result in calculus that differentiability implies vanishing of the quadratic variation of the function.

Quadratic Variation of a Brownian Process

Suppose we form a partition π of the time interval $[0, T]$ by the discrete points

$$0 = t_0 < t_1 < \cdots < t_n = T,$$

and let $\delta t_{max} = \max_k (t_k - t_{k-1})$. We write $\Delta t_k = t_k - t_{k-1}$, and define the corresponding quadratic variation of the standard Brownian process $Z(t)$ by

$$Q_\pi = \sum_{k=1}^{n} [Z(t_k) - Z(t_{k-1})]^2. \tag{2.4.1}$$

The quadratic variation of $Z(t)$ over $[0, T]$ is nonzero and its value is given by

$$Q_{[0,T]} = \lim_{\delta t_{max} \to 0} Q_\pi = T. \tag{2.4.2}$$

To prove the above claim, it suffices to show that

$$\lim_{\delta t_{max} \to 0} E[Q_\pi] = T \quad \text{and} \quad \lim_{\delta t_{max} \to 0} \text{var}(Q_\pi - T) = 0. \tag{2.4.3}$$

First, we consider

$$
\begin{aligned}
&E[Q_\pi] \\
&= \sum_{k=1}^{n} E[\{Z(t_k) - Z(t_{k-1})\}^2] \\
&= \sum_{k=1}^{n} \mathrm{var}(Z(t_k) - Z(t_{k-1})) \quad \text{since } Z(t_k) - Z(t_{k-1}) \text{ has zero mean} \\
&= \mathrm{var}(Z(t_n) - Z(t_0)) \quad \text{since } Z(t_k) - Z(t_{k-1}), k = 1, \cdots, n \text{ are independent} \\
&= t_n - t_0 = T
\end{aligned} \tag{2.4.4}
$$

so that the first result in (2.4.3) is established. Next, we consider

$$
\mathrm{var}(Q_\pi - T) = E\left[\sum_{k=1}^{n}\sum_{\ell=1}^{n}\{[Z(t_k) - Z(t_{k-1})]^2 - \Delta t_k\} \{[Z(t_\ell) - Z(t_{\ell-1})]^2 - \Delta t_\ell\}\right].
$$

Since the increments $[Z(t_k) - Z(t_{k-1})], k = 1, \cdots, n$, are independent, only those terms corresponding to $k = \ell$ in the above series survive. Hence, we have

$$
\begin{aligned}
\mathrm{var}(Q_\pi - T) &= E\left[\sum_{k=1}^{n}\{[Z(t_k) - Z(t_{k-1})]^2 - \Delta t_k\}^2\right] \\
&= \sum_{k=1}^{n} E\left[\{Z(t_k) - Z(t_{k-1})\}^4\right] \\
&\qquad -2\Delta t_k \sum_{k=1}^{n} E\left[\{Z(t_k) - Z(t_{k-1})\}^2\right] + \Delta t_k^2.
\end{aligned}
$$

Since $Z(t_k) - Z(t_{k-1})$ is normally distributed with zero mean and variance Δt_k, its fourth-order moment is known to be (see Problem 2.28)

$$
E[\{Z(t_k) - Z(t_{k-1})\}^4] = 3\Delta t_k^2,
$$

so

$$
\mathrm{var}(Q_\pi - T) = \sum_{k=1}^{n}[3\Delta t_k^2 - 2\Delta t_k^2 + \Delta t_k^2] = 2\sum_{k=1}^{n}\Delta t_k^2. \tag{2.4.5}
$$

In taking the limit $\delta t_{max} \to 0$, we observe that $\mathrm{var}(Q_\pi - T) \to 0$, thus we obtain the second result in (2.4.3). By virtue of $\lim_{n\to\infty} \mathrm{var}(Q_\pi - T) = 0$, we say that T is the *mean square limit* of Q_π.

Remarks.

1. In general, the quadratic variation of the Brownian process with variance rate σ^2 over the time interval $[t_1, t_2]$ is given by

$$
Q_{[t_1,t_2]} = \sigma^2(t_2 - t_1). \tag{2.4.6}
$$

2. If we write $dZ(t) = Z(t) - Z(t - dt)$, where $dt \to 0$, then we can deduce from the above calculations that

$$E[dZ(t)^2] = dt \quad \text{and} \quad \text{var}(dZ(t)^2) = 2\,dt^2. \tag{2.4.7}$$

Since dt^2 is a higher order infinitesimally small quantity, we may claim that the random quantity $dZ(t)^2$ converges in the *mean square sense* to the deterministic quantity dt.

Definition of Stochastic Integration

Let $f(t)$ be an arbitrary function of t and $Z(t)$ be the standard Brownian process. First, we consider the definition of the stochastic integral $\int_0^T f(t)\,dZ(t)$ as the limit of the following partial sums (defined in the usual Riemann–Stieltjes sense):

$$\int_0^T f(t)\,dZ(t) = \lim_{n\to\infty} \sum_{k=1}^n f(\xi_k)[Z(t_k) - Z(t_{k-1})], \tag{2.4.8}$$

where the discrete points $0 < t_0 < t_1 < \cdots < t_n = T$ form a partition of the interval $[0, T]$ and ξ_k is some immediate point between t_{k-1} and t_k. The limit is taken in the mean square sense. Unfortunately, the limit depends on how the immediate points are chosen. For example, suppose we take $f(t) = Z(t)$ and choose $\xi_k = \alpha t_k + (1-\alpha)t_{k-1}, 0 < \alpha < 1$, for all k. We consider

$$\begin{aligned}
&E\left[\sum_{k=1}^n Z(\xi_k)[Z(t_k) - Z(t_{k-1})\right] \\
&= \sum_{k=1}^n E\big[Z(\xi_k)Z(t_k) - Z(\xi_k)Z(t_{k-1})\big] \\
&= \sum_{k=1}^n [\min(\xi_k, t_k) - \min(\xi_k, t_{k-1})] \qquad \text{[see (2.3.15)]} \\
&= \sum_{k=1}^n (\xi_k - t_{k-1}) = \alpha \sum_{k=1}^n (t_k - t_{k-1}) = \alpha T,
\end{aligned} \tag{2.4.9}$$

so that the expected value of the stochastic integral depends on the choice of the immediate points ξ_k chosen in $[t_{k-1}, t_k]$, $k = 1, 2, \cdots, n$.

A function is said to be *nonanticipative* with respect to the Brownian process $Z(t)$ if the value of the function at time t is determined by the path history of $Z(t)$ up to time t. In finance, the investor's action is nonanticipative in nature since he makes the investment decision before the asset prices move. Accordingly, the stochastic integration is defined by taking $\xi_k = t_{k-1}$ (left-hand point in each subinterval). The Ito definition of a stochastic integral is given by

$$\int_0^T f(t)\,dZ(t) = \lim_{n\to\infty} \sum_{k=1}^n f(t_{k-1})[Z(t_k) - Z(t_{k-1})], \tag{2.4.10}$$

where the limit is taken in the mean square sense and $f(t)$ is nonanticipative with respect to $Z(t)$.

As an example, we consider the evaluation of the Ito stochastic integral $\int_0^T Z(t)\,dZ(t)$. A naive evaluation according to the usual integration rule gives

$$\int_0^T Z(t)\,dZ(t) = \frac{1}{2}\int_0^T \frac{d}{dt}[Z(t)]^2\,dt = \frac{Z(T)^2 - Z(0)^2}{2},$$

which unfortunately gives a wrong result (see the explanation below). According to the definition in (2.4.10), we have

$$\begin{aligned}
\int_0^T Z(t)\,dZ(t) &= \lim_{n\to\infty}\sum_{k=1}^{n} Z(t_{k-1})[Z(t_k) - Z(t_{k-1})] \\
&= \lim_{n\to\infty}\frac{1}{2}\sum_{k=1}^{n}(\{Z(t_{k-1}) + [Z(t_k) - Z(t_{k-1})]\}^2 \\
&\qquad - Z(t_{k-1})^2 - [Z(t_k) - Z(t_{k-1})]^2) \\
&= \frac{1}{2}\lim_{n\to\infty}[Z(t_n)^2 - Z(t_0)^2] \\
&\qquad - \frac{1}{2}\lim_{n\to\infty}\sum_{k=1}^{n}[Z(t_k) - Z(t_{k-1})]^2 \\
&= \frac{Z(T)^2 - Z(0)^2}{2} - \frac{T}{2} \qquad \text{[by (2.4.3)].}
\end{aligned} \tag{2.4.11}$$

Rearranging the terms, we may rewrite the above result as

$$2\int_0^T Z(t)\,dZ(t) + \int_0^T dt = \int_0^T \frac{d}{dt}[Z(t)]^2\,dt, \tag{2.4.12a}$$

or in differential form,

$$2Z(t)\,dZ(t) + dt = d[Z(t)]^2. \tag{2.4.12b}$$

Unlike the usual differential rule, we have the extra term dt. This arises from the finiteness of the quadratic variation of the Brownian process since $|Z(t_k) - Z(t_{k-1})|$ is of order $\sqrt{\Delta t_k}$ and $\lim_{n\to\infty}\sum_{k=1}^{n}[Z(t_k) - Z(t_{k-1})]^2$ remains finite on taking the limit. Apparently, it is necessary to develop a new set of differential rules that deal with the computation of differentials of stochastic functions.

2.4.2 Ito's Lemma and Stochastic Differentials

Once we have defined stochastic integrals, we can give a formal definition of a class of continuous stochastic processes, called the Ito processes. Let $\mathcal{F}_t$ be the natural filtration generated by the standard Brownian process $Z(t)$ through the observation of the trajectory of $Z(t)$. Let $\mu(t)$ and $\sigma(t)$ be adapted to $\mathcal{F}_t$ with $\int_0^T |\mu(t)|\,dt < \infty$ and $\int_0^T \sigma^2(t)\,dt < \infty$ (almost surely) for all T, then the process $X(t)$ defined by

$$X(t) = X(0) + \int_0^t \mu(s)\,ds + \int_0^t \sigma(s)\,dZ(s), \tag{2.4.13}$$

is called an *Ito process*. The differential form of the above equation is given by

$$dX(t) = \mu(t)\,dt + \sigma(t)\,dZ(t). \tag{2.4.14}$$

Ito's Lemma

Suppose $f(x,t)$ is a deterministic twice continuously differentiable function and the stochastic process Y is defined by $Y = f(X,t)$, where $X(t)$ is an Ito process whose dynamics are governed by (2.4.14). How do we compute the differential $dY(t)$? We have seen the justification by why $dZ(t)^2$ converges in the mean square sense to dt [see (2.4.7)]. Hence, the second-order term dX^2 also contributes to the differential dY. The *Ito formula* of computing the differential of the stochastic function $f(X,t)$ is given by

$$\begin{aligned} dY = &\left[\frac{\partial f}{\partial t}(X,t) + \mu(t)\frac{\partial f}{\partial x}(X,t) + \frac{\sigma^2(t)}{2}\frac{\partial^2 f}{\partial x^2}\right]dt \\ &+ \sigma(t)\frac{\partial f}{\partial x}(X,t)\,dZ. \end{aligned} \tag{2.4.15}$$

The rigorous proof of the Ito formula is quite technical, so only a heuristic proof is provided below. We expand ΔY by the Taylor series up to the second-order terms as follows:

$$\begin{aligned} \Delta Y = &\frac{\partial f}{\partial t}\Delta t + \frac{\partial f}{\partial x}\Delta X \\ &+ \frac{1}{2}\left(\frac{\partial^2 f}{\partial t^2}\Delta t^2 + 2\frac{\partial^2 f}{\partial x \partial t}\Delta X \Delta t + \frac{\partial^2 f}{\partial x^2}\Delta X^2\right) + O(\Delta X^3, \Delta t^3). \end{aligned}$$

In the limits $\Delta X \to 0$ and $\Delta t \to 0$, we apply the multiplication rules where $dX^2 = \sigma^2(t)\,dt$, $dXdt = 0$ and $dt^2 = 0$ so that

$$dY = \frac{\partial f}{\partial t}dt + \frac{\partial f}{\partial x}dX + \frac{\sigma^2(t)}{2}\frac{\partial^2 f}{\partial x^2}dt.$$

Writing out in full in terms of dZ and dt, we obtain the Ito formula (2.4.15).

As a simple verification, when we apply the Ito formula to $f = Z^2$, we obtain the result in (2.4.12b) immediately. As an additional example, we consider the exponential Brownian function

$$S(t) = S_0 e^{(r - \frac{\sigma^2}{2})t + \sigma Z(t)}. \tag{2.4.16}$$

Suppose we write

$$X(t) = \left(r - \frac{\sigma^2}{2}\right)t + \sigma Z(t)$$

so that

$$X(t) = \ln \frac{S(t)}{S_0} \quad \text{or} \quad S(t) = S_0 e^{X(t)}.$$

Now, the respective partial derivatives of S are

$$\frac{\partial S}{\partial t} = 0, \quad \frac{\partial S}{\partial X} = S \quad \text{and} \quad \frac{\partial^2 S}{\partial X^2} = S.$$

By the Ito lemma, we obtain

$$dS(t) = \left(r - \frac{\sigma^2}{2} + \frac{\sigma^2}{2}\right) S(t)\, dt + \sigma S(t)\, dZ(t)$$

or

$$\frac{dS(t)}{S(t)} = r\, dt + \sigma\, dZ(t), \text{ with } S(0) = S_0. \tag{2.4.17}$$

Conversely, we observe that $S(t)$ defined in (2.4.16) is the solution to the stochastic differential equation (2.4.17). Since $E[X(t)] = (r - \frac{\sigma^2}{2})t$ and $\text{var}(X(t)) = \sigma^2 t$, the mean and variance of $\ln \frac{S(t)}{S_0}$ are found to be $(r - \frac{\sigma^2}{2})t$ and $\sigma^2 t$, respectively.

Multidimensional Version of Ito's Lemma

Suppose $f(x_1, \cdots, x_n, t)$ is a multidimensional twice continuously differentiable function and the stochastic process Y_n is defined by

$$Y_n = f(X_1, \cdots, X_n, t), \tag{2.4.18a}$$

where the process $X_j(t)$ follows the Ito process

$$dX_j(t) = \mu_j(t)\, dt + \sigma_j(t)\, dZ_j(t), \quad j = 1, 2, \cdots, n. \tag{2.4.18b}$$

The standard Brownian processes $Z_j(t)$ and $Z_k(t)$ are assumed to be correlated with correlation coefficient ρ_{jk} so that $dZ_j\, dZ_k = \rho_{jk}\, dt$. In a similar manner, we expand ΔY_n up to the second-order terms in ΔX_j:

$$\begin{aligned}\Delta Y_n = &\frac{\partial f}{\partial t}(X_1, \cdots, X_n, t)\, \Delta t + \sum_{j=1}^{n} \frac{\partial f}{\partial x_j}(X_1, \cdots, X_n, t)\, \Delta X_j \\ &+ \frac{1}{2} \sum_{j=1}^{n} \sum_{k=1}^{n} \frac{\partial^2 f}{\partial x_j \partial x_k}(X_1, \cdots, X_n, t)\, \Delta X_j\, \Delta X_k \\ &+ O(\Delta t \Delta X_j) + O(\Delta t^2).\end{aligned}$$

In the limits $\Delta X_j \to 0$, $j = 1, 2, \cdots, n$, and $\Delta t \to 0$, we neglect the higher order terms in $O(\Delta t \Delta X_j)$ and $O(\Delta t^2)$ and observe $dX_j\, dX_k = \sigma_j(t)\sigma_k(t)\rho_{jk}\, dt$. We then obtain the following multidimensional version of the Ito lemma:

$$dY_n = \left[\frac{\partial f}{\partial t}(X_1, \cdots, X_n, t) + \sum_{j=1}^{n} \mu_j(t) \frac{\partial f}{\partial x_j}(X_1, \cdots, X_n, t) + \frac{1}{2}\sum_{j=1}^{n}\sum_{k=1}^{n} \sigma_j(t)\sigma_k(t)\rho_{jk} \frac{\partial^2 f}{\partial x_j \partial x_k}(X_1, \cdots, X_n, t)\right] dt + \sum_{j=1}^{n} \sigma_j(t) \frac{\partial f}{\partial x_j}(X_1, \cdots, X_n, t)\, dZ_j. \tag{2.4.19}$$

2.4.3 Ito's Processes and Feynman–Kac Representation Formula

Consider an Ito process defined either in the differential form

$$dY(t) = \mu(t)\, dt + \sigma(t)\, dZ(t), \tag{2.4.20a}$$

or in the integral form

$$Y(t) = Y(0) + \int_0^t \mu(s)\, ds + \int_0^t \sigma(s)\, dZ(s) \tag{2.4.20b}$$

with drift term $\mu(t)$. We let

$$M(t) = \int_0^t \sigma(s)\, dZ(s)$$

and note that

$$M(T) = M(t) + \int_t^T \sigma(s)\, dZ(s), \qquad t < T.$$

Suppose we take the conditional expectation of $M(T)$ given the history of the Brownian path up to the time t, we obtain

$$E_t[M(T)] = M(t) \tag{2.4.21}$$

since the stochastic integral in (2.4.21) has zero conditional expectation. Hence, $M(t)$ is a martingale. However, $Y(t)$ is not a martingale if $\mu(t)$ is nonzero.

As an additional example, we consider the following stochastic differential equation

$$\frac{dS(t)}{S(t)} = \sigma\, dZ(t) \quad \text{with } S(0) = S_0 \tag{2.4.22}$$

whose integral form can be formally expressed as

$$S(t) = S_0 + \int_0^t \sigma S(u) Z(u)\, du.$$

As deduced from the result in (2.4.16), the closed form solution to the above stochastic differential equation is given by

$$S(t) = S_0 e^{-\frac{\sigma^2}{2}t + \sigma Z(t)}.$$

We would like to verify that $S(t)$ is a martingale using the first principle. For $u < t$, we consider the expectation of $S(t)$ conditional on the filtration $\mathcal{F}_u$:

$$\begin{aligned}
&E\left[S_0 \exp\left(-\frac{\sigma^2}{2}t + \sigma Z(t)\right)\Big|\mathcal{F}_u\right] \\
&= E\left[S_0 \exp\left(-\frac{\sigma^2}{2}u + \sigma Z(u)\right)\exp(\sigma(Z(t) - Z(u)))\exp\left(-\frac{\sigma^2}{2}(t-u)\right)\Big|\mathcal{F}_u\right] \\
&= S_0 \exp\left(-\frac{\sigma^2}{2}u + \sigma Z(u)\right)\exp\left(-\frac{\sigma^2}{2}(t-u)\right)E\left[\exp(\sigma(Z(t) - Z(u)))\Big|\mathcal{F}_u\right].
\end{aligned}$$

Conditional on $\mathcal{F}_u$, the Brownian increment $Z(t) - Z(u)$ is normal with variance $t - u$. Recall that a random variable X is normal with mean m_X and variance σ_X^2 if and only if the moment generating function of X is given by

$$E[\exp(\alpha X)] = \exp\left(\alpha m_X + \frac{\alpha^2}{2}\sigma_X^2\right) \tag{2.4.23}$$

for any real value of α (see Problem 2.28). We then obtain

$$E\big[\exp(\sigma(Z(t) - Z(u))]|\mathcal{F}_u\big] = \exp\left(\frac{\sigma^2}{2}(t-u)\right)$$

so that

$$\begin{aligned}
&E\left[S_0 \exp\left(-\frac{\sigma^2}{2}t + \sigma Z(t)\right)\Big|\mathcal{F}_u\right] \\
&= S_0 \exp\left(-\frac{\sigma^2}{2}u + \sigma Z(u)\right), \quad u < t,
\end{aligned} \tag{2.4.24}$$

hence $S(t)$ is a martingale.

Suppose we consider the more general case of an Ito process $X(s)$ whose dynamics is governed by the stochastic differential equation

$$dX(s) = \mu(X(s), s)\,ds + \sigma(X(s), s)\,dZ(s), \quad t \le s \le T, \tag{2.4.25}$$

with initial condition: $X(t) = x$. Consider a smooth function $F(X(t), t)$, by virtue of the Ito lemma, the differential of which is given by

$$dF = \left[\frac{\partial F}{\partial t} + \mu(X, t)\frac{\partial F}{\partial X} + \frac{\sigma^2(X, t)}{2}\frac{\partial^2 F}{\partial X^2}\right]dt + \sigma\frac{\partial F}{\partial X}\,dZ. \tag{2.4.26}$$

We define the infinitesimal generator $\mathcal{A}$ associated with the Ito process $X(t)$ by

$$\mathcal{A} = \mu(X,t)\frac{\partial}{\partial X} + \frac{\sigma^2(X,t)}{2}\frac{\partial^2}{\partial X^2}. \tag{2.4.27}$$

Suppose F satisfies the parabolic partial differential equation

$$\frac{\partial F}{\partial t} + \mathcal{A}F = 0 \tag{2.4.28}$$

with terminal condition: $F(X(T),T) = h(X(T))$, then dF becomes

$$dF = \sigma\frac{\partial F}{\partial X}\,dZ.$$

Supposing that $\sigma\frac{\partial F}{\partial X}$ is nonanticipative with the Brownian process $Z(t)$, we can express the above stochastic differential equation into the following integral form

$$F(X(s),s) = F(X(t),t) + \int_t^s \sigma(X(u),u)\frac{\partial F}{\partial X}(X(u),u)\,dZ(u). \tag{2.4.29}$$

The stochastic integral can be viewed as a sum of inhomogeneous consecutive Gaussian increments with zero mean, hence it has zero conditional expectation. By taking the conditional expectation and setting $s = T$ and $F(X(T),T) = h(X(T))$, we then obtain the following Feynman–Kac representation formula

$$F(x,t) = E_{x,t}[h(X(T))], \quad t < T, \tag{2.4.30}$$

where $F(x,t)$ satisfies the partial differential equation (2.4.23) and $E_{x,t}$ refers to expectation taken conditional on $X(t) = x$ and $\mathcal{F}_t$. The process $X(s)$ is initialized at the fixed point x at time t and it follows the Ito process defined in (2.4.20).

2.4.4 Change of Measure: Radon–Nikodym Derivative and Girsanov's Theorem

Under the risk neutral measure, the discounted price of the underlying asset becomes a martingale. The effective valuation of contingent claims under the risk neutral measure often requires the transformation of an underlying price process with drift into a martingale, but under a different measure. The transformation can be performed effectively using Girsanov's Theorem. Before stating the theorem, we discuss the Radon–Nikodym derivative which relates the transformation between two equivalent probability measures.

Let us consider the standard Brownian process $Z_P(t)$ under the measure P. Adding the drift μt to $Z_P(t)$, μ is a constant, we write

$$Z_{\widetilde{P}}(t) = Z_P(t) + \mu t. \tag{2.4.31}$$

Here, $Z_{\widetilde{P}}(t)$ is a Brownian process with drift under P. How can we change from measure P to another measure $\widetilde{P}$ so that $Z_{\widetilde{P}}(t)$ becomes a Brownian process with zero drift under $\widetilde{P}$? Formally, we multiply dP by a factor $\frac{d\widetilde{P}}{dP}$ to give $d\widetilde{P}$. The factor

$\frac{d\widetilde{P}}{dP}$ is called the *Radon–Nikodym derivative*. It is postulated that the corresponding Radon–Nikodym derivative for this case is given by

$$\frac{d\widetilde{P}}{dP} = \exp\left(-\mu Z_P(t) - \frac{\mu^2}{2}t\right). \tag{2.4.32}$$

For a fixed time horizon T, $Z_P(T)$ is known to have zero mean and variance T under P, where

$$P\left(\xi - \frac{d\xi}{2} < Z_P(T) < \xi + \frac{d\xi}{2}\right) = \frac{1}{\sqrt{2\pi T}}\, e^{-\xi^2/2T}\, d\xi. \tag{2.4.33}$$

To show the validity of the claim, it suffices to show that $Z_{\widetilde{P}}(T)$ is normal with zero mean and variance T under the measure $\widetilde{P}$ by looking at the corresponding moment generating function of $Z_{\widetilde{P}}(T)$. We consider

$$\begin{aligned} E_{\widetilde{P}}\big[\exp\big(\alpha Z_{\widetilde{P}}(T)\big)\big] &= E_P\left[\frac{d\widetilde{P}}{dP}\exp\big(\alpha Z_P(T) + \alpha\mu T\big)\right] \\ &= E_P\left[\exp\big((\alpha-\mu)Z_P(T)\big)\exp\left(\alpha\mu T - \frac{\mu^2}{2}T\right)\right] \\ &= \exp\left(\frac{(\alpha-\mu)^2}{2}T + \alpha\mu T - \frac{\mu^2}{2}T\right) \\ &= \exp\left(\frac{\alpha^2}{2}T\right), \end{aligned} \tag{2.4.34}$$

which is valid for any real value of α. By virtue of (2.4.23), we obtain the desired result.

The Girsanov Theorem presented below provides the procedure of finding the Radon–Nikodym derivative for the general case when the drift rate is not constant.

Girsanov Theorem

What would be the Radon–Nikodym derivative when the drift rate is taken to be a $\mathcal{F}_t$-adapted stochastic process. We state without proof a version of the Girsanov Theorem, which is a useful tool to effect a change of measure on an Ito process. The application of the Girsanov Theorem in the determination of an equivalent martingale measure for pricing contingent claims will be demonstrated in Sect. 3.2.

Theorem 2.4. *Let $Z_P(t)$ be a Brownian process under the measure P (called a P-Brownian process). Let $\mathcal{F}_t, t \geq 0$, be the natural filtration generated by $Z(t)$. Consider a $\mathcal{F}_t$-adapted stochastic process $\gamma(t)$ which satisfies the Novikov condition*

$$E[e^{\int_0^t \frac{1}{2}\gamma(s)^2\, ds}] < \infty, \tag{2.4.35}$$

and consider the Radon–Nikodym derivative

$$\frac{d\widetilde{P}}{dP} = \rho(t), \tag{2.4.36a}$$

where

$$\rho(t) = \exp\left(\int_0^t -\gamma(s)\, dZ_P(s) - \frac{1}{2}\int_0^t \gamma(s)^2\, ds\right). \tag{2.4.36b}$$

Under the measure $\widetilde{P}$, the Ito process

$$Z_{\widetilde{P}}(t) = Z_P(t) + \int_0^t \gamma(s)\, ds \tag{2.4.37}$$

is a $\widetilde{P}$-Brownian process.

2.5 Problems

2.1 Show that a dominant trading strategy exists if and only if there exists a trading strategy satisfying $V_0 < 0$ and $V_1(\omega) \geq 0$ for all $\omega \in \Omega$.
Hint: Consider the dominant trading strategy $\mathcal{H} = (h_0\ h_1 \cdots h_M)^T$ satisfying $V_0 = 0$ and $V_1(\omega) > 0$ for all $\omega \in \Omega$. Take $G^*_{\min} = \min_{\omega} G^*(\omega) > 0$ and define a new trading strategy with $\widehat{h}_m = h_m, m = 1, \cdots, M$ and $\widehat{h}_0 = -G^*_{\min} - \sum_{m=1}^{M} h_m S^*_m(0)$.

2.2 Consider a portfolio with one risky security and a risk free security. Suppose the price of the risky asset at time 0 is 4 and the possible values of the $t = 1$ price are $1.1, 2.2$ and 3.3 (three possible states of the world at the end of a single trading period). Let the risk free interest rate r be 0.1 and take the price of the risk free security at $t = 0$ to be unity.
(a) Show that the trading strategy: $h_0 = 4$ and $h_1 = -1$ is a dominant trading strategy that starts with zero wealth and ends with positive wealth with certainty.
(b) Find the discounted gain G^* over the single trading period.
(c) Find a trading strategy that starts with negative wealth and ends with non-negative wealth with certainty.

2.3 Show that if the law of one price does not hold, then every payoff in the asset span can be bought at any price.

2.4 Construct a securities model such that it satisfies the law of one price but admits a dominant trading strategy.
Hint: Construct a securities model where the initial price vector lies in the row space of the discounted terminal payoff matrix $\widehat{S}^*(1)$ while the nonnegativity of the linear measure does not hold.

2.5 Define the pricing functional $F(\mathbf{x})$ on the asset span $\mathcal{S}$ by $F(\mathbf{x}) = \{y : y = \mathbf{S}^*(0)\mathbf{h}$ for some $\mathbf{h}$ such that $\mathbf{x} = S^*(1)\mathbf{h}$, where $\mathbf{x} \in \mathcal{S}\}$. Show that if the law of one price holds, then F is a *linear* functional.

2.6 Given the discounted terminal payoff matrix

$$\widehat{S}^*(1;\Omega) = \begin{pmatrix} 1 & 3 & 5 \\ 1 & 2 & 4 \\ 1 & 1 & 3 \end{pmatrix},$$

and the current price vector $\widehat{\mathbf{S}}^*(0) = (1 \quad 3 \quad 4)$.

(a) By presenting a counter example, show that the law of one price does not hold for this one-period securities model.

(b) How can we modify the current price vector such that the law of one price holds under the modified model?

2.7 Given the discounted terminal payoff matrix

$$\widehat{S}^*(1;\Omega) = \begin{pmatrix} 1 & 3 & 2 \\ 1 & 1 & 3 \\ 1 & 2 & 4 \end{pmatrix},$$

and the current price vector $\widehat{\mathbf{S}}^*(0) = (1 \quad 2 \quad 3)$, find the state price of the Arrow security with discounted payoff $\mathbf{e}_k$, $k = 1, 2, 3$.
Does the securities model admit dominant trading strategies? If so, find an example where one trading strategy dominates the other.

2.8 Consider the securities model with

$$\widehat{\mathbf{S}}^*(0) = (1 \quad 2 \quad 3 \quad k) \quad \text{and} \quad \widehat{S}^*(1;\Omega) = \begin{pmatrix} 1 & 2 & 6 & 9 \\ 1 & 3 & 3 & 7 \\ 1 & 6 & 12 & 19 \end{pmatrix},$$

determine the value of k such that the law of one price holds. Taking this particular value of k in $\widehat{\mathbf{S}}^*(0)$, does the securities model admit dominant trading strategies. If yes, find one such dominant trading strategy.

2.9 Show that if there exists a dominant trading strategy, then there exists an arbitrage opportunity. How to construct a securities model such that there exists an arbitrage opportunity but dominant trading strategy does not exist?

2.10 Show that $\mathbf{h}$ is an arbitrage if and only if the discounted gain G^* satisfies (i) $G^* \geq 0$ and (ii) $E[G^*] > 0$. Here, E is the expectation under the actual probability measure P, $P(\omega) > 0$.

2.11 Suppose a betting game has three possible outcomes. If a gambler bets on outcome i, then he receives a net gain of d_i dollars for one dollar betted, $i = 1, 2, 3$. The payoff matrix thus takes the form (consideration of discounting is not necessary in a betting game)

$$S(1;\Omega) = \begin{pmatrix} d_1 + 1 & 0 & 0 \\ 0 & d_2 + 1 & 0 \\ 0 & 0 & d_3 + 1 \end{pmatrix}.$$

Find the condition on d_i such that a risk neutral probability measure exists for the above betting game (visualized as an investment model).

2.12 Consider the following securities model

$$S^*(1;\Omega) = \begin{pmatrix} 3 & 4 \\ 2 & 5 \\ 2 & 4 \end{pmatrix}, \quad \mathbf{S}^*(0) = (2 \quad 4),$$

do risk neutral measures exist? If not, explain why. If yes, find the set of all risk neutral measures.

2.13 Consider the following securities model with discounted payoffs of the securities at $t = 1$ given by the discounted terminal payoff matrix

$$\widehat{S}^*(1;\Omega) = \begin{pmatrix} 1 & 2 & 3 & 4 \\ 1 & 3 & 4 & 5 \\ 1 & 5 & 6 & 7 \end{pmatrix}.$$

Let the initial price vector $\widehat{\mathbf{S}}^*(0)$ be $(1 \quad 3 \quad 5 \quad 9)$. Does the law of one price hold for this securities model? Show that the contingent claim with discounted payoff $\begin{pmatrix} 6 \\ 8 \\ 12 \end{pmatrix}$ is attainable and find the set of all possible trading securities that generate the payoff. Can we find the price at $t = 0$ of this contingent claim?
Hint: Write the discounted payoff vectors of the securities as

$$\mathbf{S}_0^*(1) = \begin{pmatrix} 1 \\ 1 \\ 1 \end{pmatrix}, \mathbf{S}_1^*(1) = \begin{pmatrix} 2 \\ 3 \\ 5 \end{pmatrix}, \mathbf{S}_2^*(1) = \begin{pmatrix} 3 \\ 4 \\ 6 \end{pmatrix}, \mathbf{S}_3^*(1) = \begin{pmatrix} 4 \\ 5 \\ 7 \end{pmatrix}.$$

Note that

$$\mathbf{S}_2^*(1) = \mathbf{S}_0^*(1) + \mathbf{S}_1^*(1) \quad \text{and} \quad \mathbf{S}_3^*(1) = \mathbf{S}_0^*(1) + \mathbf{S}_2^*(1),$$

but the initial prices observe

$$S_2^*(0) \neq S_0^*(0) + S_1^*(0) \quad \text{and} \quad S_3^*(0) \neq S_0^*(0) + S_2^*(0).$$

2.14 Suppose the set of risk neutral measures for a given securities model is nonempty. Show that if the securities model is complete, then the set of risk neutral measures must be singleton.
Hint: Under market completeness, column rank of $\widehat{S}(1;\Omega)$ equals number of states. Since column rank = row rank, all rows of $\widehat{S}^*(1;\Omega)$ are independent.

2.15 Let P be the true probability measure, where $P(\omega)$ denotes the actual probability that the state ω occurs. Define the *state price density* by the random variable

$L(\omega) = Q(\omega)/P(\omega)$, where Q is a risk neutral measure. Use R_m to denote the return of the risky security m, where $R_m = [S_m(1) - S_m(0)]/S_m(0), m = 1, \cdots, M$. Show that $E_Q[R_m] = r, m = 1, \cdots, M$, where r is the interest over one period. Let $E_P[R_m]$ denote the expectation of R_m under the actual probability measure P, show that

$$E_P[R_m] - r = -\text{cov}(R_m, L),$$

where cov denotes the covariance operator.

2.16 Suppose $u > d > R$ in the discrete binomial model. Show that an investor can lock in a riskless profit by borrowing cash as much as possible to purchase the asset, and selling the asset after one period and returning the loan. When $R > u > d$, what should be the corresponding strategy in order to take arbitrage?

2.17 We can also derive the binomial formula using the *riskless hedging principle* (see Sect. 3.1.1). Suppose we have a call that is one period from expiry and we would like to create a perfectly hedged portfolio with a long position of one unit of the underlying asset and a short position of m units of call. Let c_u and c_d denote the payoff of the call at expiry corresponding to the upward and downward movement of the asset price, respectively. Show that the number of calls to be sold short in the portfolio should be

$$m = \frac{S(u-d)}{c_u - c_d}$$

in order that the portfolio is perfectly hedged. The hedged portfolio should earn the risk-free interest rate. Let R denote the growth factor of a perfectly hedged riskfree portfolio over one period. Show that the binomial option pricing formula for the call as deduced from the riskless hedging principle is given by

$$c = \frac{pc_u + (1-p)c_d}{R} \quad \text{where} \quad p = \frac{R-d}{u-d}.$$

2.18 Let Π_u and Π_d denote the state prices corresponding to the states of asset value going up and going down, respectively. The state prices can also be interpreted as state contingent discount rates. If no arbitrage opportunities are available, then all securities (including the bond, the asset and the call option) must have returns with the same state contingent discount rates Π_u and Π_d. Hence, the respective relations for the money market account, asset price and call option value with Π_u and Π_d are given by

$$\begin{aligned} 1 &= \Pi_u R + \Pi_d R \\ S &= \Pi_u u S + \Pi_d d S \\ c &= \Pi_u c_u + \Pi_d c_d. \end{aligned}$$

By solving for Π_u and Π_d from the first two equations and substituting the solutions into the third equation, show that the binomial call price formula over one period is given by

$$c = \frac{pc_u + (1-p)c_d}{R} \quad \text{where} \quad p = \frac{R-d}{u-d}.$$

2.19 Consider the sample space $\Omega = \{-3, -2, -1, 1, 2, 3\}$ and the algebra $\mathcal{F} = \{\phi, \{-3, -2\}, \{-1, 1\}, \{2, 3\}, \{-3, -2, -1, 1\}, \{-3, -2, 2, 3\}, \{-1, 1, 2, 3\}, \Omega\}$. For each of the following random variables, determine whether it is $\mathcal{F}$-measurable:
(i) $X(\omega) = \omega^2$, (ii) $X(\omega) = \max(\omega, 2)$.
Find a random variable that is $\mathcal{F}$-measurable.

2.20 Let $X, X_1, \cdots, X_n$ be random variables defined on the filtered probability space $(\Omega, \mathcal{F}, P)$. Prove the following properties on conditional expectations:
(a) $E[XI_B] = E[I_B E[X|\mathcal{F}]]$ for all $B \in \mathcal{F}$,
(b) $E[\max(X_1, \cdots, X_n)|\mathcal{F}] \geq \max(E[X_1|\mathcal{F}], \cdots, E[X_n|\mathcal{F}])$.

2.21 Let $X = \{X_t; t = 0, 1, \cdots, T\}$ be a stochastic process adapted to the filtration $\mathbb{F} = \{\mathcal{F}_t; t = 0, 1, \cdots, T\}$. Does the property: $E[X_{t+1} - X_t|\mathcal{F}_t] = 0, t = 0, 1, \cdots, T-1$ imply that X is a martingale?

2.22 Consider the binomial experiment with the probability of success p, $0 < p < 1$. We let N_k denote the number of successes after k independent trials. Define the discrete process Y_k by $N_k - kp$, the excess number of successes above the mean kp. Show that Y_k is a martingale.

2.23 Show that τ is a stopping time if and only if $\{\tau \leq t\} \in \mathcal{F}_t$.
Hint: $\{\tau \leq t\} = \{\tau = 0\} \cup \{\tau = 1\} \cup \cdots \cup \{\tau = t\}$ and
$\{\tau = t\} = \{\tau \leq t\} \cap \{\tau \leq t-1\}^C$.

2.24 Suppose τ_1 and τ_2 are stopping times, show that $\max(\tau_1, \tau_2)$ and $\min(\tau_1, \tau_2)$ are also stopping times.

2.25 Consider the two-period securities model shown in Fig. 2.5. Suppose the riskless interest rate r violates the restriction $r < 0.2$, say, $r = 0.3$. Construct an arbitrage opportunity associated with the securities model.

2.26 Deduce the price formula for a European put option with terminal payoff: $\max(X - S, 0)$ for the n-period binomial model.

2.27 Suppose the particle starts initially at $x = a_0$ in the continuous random walk model (see Sect. 2.3.1), find the probability that the particle stays above $x = a$ at time t.

2.28 Let X be a normal random variable with mean m_X and variance σ_X^2. Show that the higher central moments of the normal random variable are given by

$$E[(X - m_X)^n] = \begin{cases} 0, & n \text{ odd} \\ (n-1)(n-3)\cdots 3\cdot 1\sigma_X^n, & n \text{ even.} \end{cases}$$

For the log-normal random variable $Z = \exp(\alpha X)$, α is a real constant, show that

$$E[Z] = \exp\left(\alpha m_X + \frac{\alpha^2}{2}\sigma_X^2\right).$$

2.29 Suppose $\{X(t), t \geq 0\}$ is the standard Brownian process, its corresponding reflected Brownian process is defined by

$$Y(t) = |X(t)|, \quad t \geq 0.$$

Show that $Y(t)$ is also Markovian and its mean and variance are, respectively,

$$E[Y] = \sqrt{\frac{2t}{\pi}}$$

and

$$\text{var}(Y) = \left(1 - \frac{2}{\pi}\right)t.$$

2.30 Suppose $Z(t)$ is the standard Brownian process, show that the following processes defined by

$$X_1(t) = kZ(t/k^2), \quad k > 0,$$

$$X_2(t) = \begin{cases} tZ(\frac{1}{t}) & \text{for } t > 0 \\ 0 & \text{for } t = 0, \end{cases}$$

and

$$X_3(t) = Z(t+h) - Z(h), \quad h > 0,$$

are also Brownian processes.

Hint: To show that $X_i(t)$ is a Brownian process, $i = 1, 2, 3$, it suffices to show that

$$X_i(t+s) - X_i(s)$$

is normally distributed with zero mean, and

$$E[[X_i(t+s) - X_i(s)]^2] = t.$$

Also, the increments over disjoint time intervals are independent and $X_i(0) = 0$.

2.31 Consider the Brownian process with drift defined by

$$X(t) = \mu t + \sigma Z(t), \quad X(0) = 0,$$

where $Z(t)$ is the standard Brownian process, find $E[X(t)|X(t_0)]$, $\text{var}(X(t)|X(t_0))$ and $\text{cov}(X(t_1), X(t_2))$.

2.32 Assume that the price of an asset follows the Geometric Brownian process with an expected rate of return of 10% per annum and a volatility of 20% per annum. Suppose the asset price at present is \$100, find the expected value and variance of the asset price half a year from now and its 90% confidence limits.

2.33 Let $Z(t)$ denote the standard Brownian process. Show that

(a) $dZ(t)^{2+n} = 0$, for any positive integer n,

(b) $$\int_{t_0}^{t_1} Z(t)^n \, dZ(t) = \frac{1}{n+1}[Z(t_1)^{n+1} - Z(t_0)^{n+1}] - \frac{n}{2}\int_{t_0}^{t_1} Z(t)^{n-1} \, dt,$$

for any positive integer n,

(c) $E[Z^4(t)] = 3t^2$,

(d) $E[e^{\alpha Z(t)}] = e^{\alpha^2 t/2}$.

2.34 Let the stochastic process $X(t), t \geq 0$, be defined by

$$X(t) = \int_0^t e^{\alpha(t-u)} \, dZ(u),$$

where $Z(t)$ is the standard Brownian process. Show that

$$\text{cov}(X(s), X(t)) = \frac{e^{\alpha(s+t)} - e^{\alpha|s-t|}}{2\alpha}, \qquad s \geq 0, t \geq 0.$$

2.35 Let $Z(t), t \geq 0$, be the standard Brownian process, $f(t)$ and $g(t)$ be differentiable functions over $[a, b]$. Show that

$$E\left[\int_a^b f'(t)[Z(t) - Z(a)] \, dt \int_a^b g'(t)[Z(t) - Z(a)] \, dt\right] = \int_a^b [f(b) - f(t)][g(b) - g(t)] \, dt.$$

Hint: Interchange the order of expectation and integration, and observe

$$E[[Z(t) - Z(a)][Z(s) - Z(a)]] = \min(t, s) - a.$$

2.36 Let $Z(t), t \geq 0$, be the standard Brownian process. Show that

$$\sigma \int_t^T [Z(u) - Z(t)] \, du$$

has zero mean and variance $\sigma^2(T-t)^3/3$.

Hint: Consider

$$\begin{aligned}
&\mathrm{var}\left(\int_t^T [Z(u) - Z(t)]\,du\right) \\
&= E\left[\int_t^T \int_t^T [Z(u) - Z(t)][Z(v) - Z(t)]\,du dv\right] \\
&= \int_t^T \int_t^T E[\{Z(u) - Z(t)\}\{Z(v) - Z(t)\}]\,du dv \\
&= \int_t^T \int_t^T [\min(u, v) - t]\,du dv.
\end{aligned}$$

2.37 Show that

$$N_2(a, b; \rho) + N_2(a, -b; -\rho) = N(a).$$

Also, show that

$$N_2(a, b; \rho) = \int_{-\infty}^{a} n(x) N\left(\frac{b - \rho x}{\sqrt{1 - \rho^2}}\right) dx.$$

2.38 Suppose the stochastic state variables S_1 and S_2 follow the Geometric Brownian processes where

$$\frac{dS_i}{S_i} = \mu_i\,dt + \sigma_i\,dZ_i, \qquad i = 1, 2.$$

Let ρ_{12} denote the correlation coefficient between the Brownian processes dZ_1 and dZ_2. Let $f = S_1 S_2$, show that f also follows the Geometric Brownian process of the form

$$\begin{aligned}
\frac{df}{f} &= (\mu_1 + \mu_2 + \rho_{12}\sigma_1\sigma_2)\,dt + \sigma_1\,dZ_1 + \sigma_2\,dZ_2 \\
&= \mu\,dt + \sigma\,dZ_f,
\end{aligned}$$

where $\mu = \mu_1 + \mu_2 + \rho_{12}\sigma_1\sigma_2$ and $\sigma^2 = \sigma_1^2 + \sigma_2^2 + 2\rho_{12}\sigma_1\sigma_2$. Similarly, let $g = \dfrac{S_1}{S_2}$, show that

$$\begin{aligned}
\frac{dg}{g} &= (\mu_1 - \mu_2 - \rho_{12}\sigma_1\sigma_2 + \sigma_2^2)\,dt + \sigma_1\,dZ_1 - \sigma_2\,dZ_2 \\
&= \widetilde{\mu}\,dt + \widetilde{\sigma}\,dZ_g,
\end{aligned}$$

where $\widetilde{\mu} = \mu_1 - \mu_2 - \rho_{12}\sigma_1\sigma_2 + \sigma_2^2$ and $\widetilde{\sigma}^2 = \sigma_1^2 + \sigma_2^2 - 2\rho_{12}\sigma_1\sigma_2$.
Hint: Note that

$$\frac{d\left(\frac{1}{S_2}\right)}{\frac{1}{S_2}} = -\mu_2\,dt + \sigma_2^2\,dt - \sigma_2\,dZ_2.$$

Treat S_1/S_2 as the product of S_1 and $1/S_2$ and use the result obtained for the product of Geometric Brownian processes.

2.39 Suppose the function $F(x,t)$ satisfies

$$\frac{\partial F}{\partial t} + \mu(x,t)\frac{\partial F}{\partial x} + \frac{\sigma^2(x,t)}{2}\frac{\partial^2 F}{\partial x^2} - rF = 0$$

with terminal condition: $F(X(T),T) = h(X(T))$. Show that

$$F(x,t) = e^{-r(T-t)} E_t[h(X(T))|X(t) = x], \quad t < T,$$

where $X(t)$ follows the Ito process

$$dX(t) = \mu(X(t),t)\,dt + \sigma(X(t),t)\,dZ(t).$$

2.40 Define the discrete random variable X by

$$X(\omega) = \begin{cases} 2 & \text{if } \omega = \omega_1 \\ 3 & \text{if } \omega = \omega_2 \\ 4 & \text{if } \omega = \omega_3, \end{cases}$$

where the sample space $\Omega = \{\omega_1, \omega_2, \omega_3\}$, $P(\omega_1) = P(\omega_2) = P(\omega_3) = 1/3$. Find a new probability measure $\widetilde{P}$ such that the mean becomes $E_{\widetilde{P}}[X] = 3.5$ while the variance remains unchanged. Is $\widetilde{P}$ unique?

2.41 Given that the process S_t is a Geometric Brownian process, it follows that

$$\frac{dS_t}{S_t} = \mu\,dt + \sigma\,dZ_t,$$

where Z_t is a P-Brownian process. Find another measure $\widetilde{P}$ by specifying the Radon–Nikodym derivative $\frac{d\widetilde{P}}{dP}$ such that S_t is governed by

$$\frac{dS_t}{S_t} = \mu'\,dt + \sigma\,d\widetilde{Z}_t$$

under the measure $\widetilde{P}$, where $\widetilde{Z}_t$ is a $\widetilde{P}$-Brownian process and μ' is the new drift rate.

2.42 Let $u^\mu(x,t)$ denote the solution to the partial differential equation

$$\frac{\partial u}{\partial t} = -\mu\frac{\partial u}{\partial x} + \frac{1}{2}\frac{\partial^2 u}{\partial x^2}, \qquad -\infty < x < \infty, \quad t > 0,$$

with $u(x,0^+) = \delta(x)$. From (2.3.12), it is seen that

$$u^\mu(x,t) = \frac{1}{\sqrt{2\pi t}} e^{-(x-\mu t)^2/2t}.$$

By applying the change of variable: $x = y + \mu t$, show that the above equation becomes

$$\frac{\partial u}{\partial t} = \frac{1}{2}\frac{\partial^2 u}{\partial y^2}, \qquad -\infty < y < \infty, \quad t > 0.$$

With the initial condition: $u(y, 0) = \delta(y)$. The new solution is given by

$$u^0(y, t) = \frac{1}{\sqrt{2\pi t}}\, e^{-y^2/2t}.$$

We observe the following relation between $u^0(y, t)$ and $u^0(x, t)$:

$$u^0(x, t) = \frac{1}{\sqrt{2\pi t}}\, e^{-(y+\mu t)^2/2t} = e^{-\mu y - \frac{\mu^2}{2}t} u^0(y, t).$$

Relate the above result to the Girsanov Theorem.

2.43 Let P and Q be two probability measures on the same measurable space $(\Omega, \mathcal{F})$ and let $f = \frac{dQ}{dP}$ denote the Radon–Nikodym derivative of Q with respect to P. Show that

$$E_Q[X|\mathcal{G}] = \frac{E_P[Xf|\mathcal{G}]}{E_P[f|\mathcal{G}]},$$

where $\mathcal{G}$ is a sub-sigma-algebra of $\mathcal{F}$ and X is a measurable random variable. This formula is considered as a generalization of the *Bayes Rule*.

3

Option Pricing Models: Black–Scholes–Merton Formulation and Martingale Pricing Theory

The revolution in trading and pricing derivative securities began in the early 1970's. In 1973, the Chicago Board of Options Exchange started the trading of options in exchanges, though options had been regularly traded by financial institutions in the *over-the-counter markets* in earlier years. In the same year, Black and Scholes (1973) and Merton (1973) published their seminal papers on the theory of option pricing. Since then the field of financial engineering has grown phenomenally. The Black–Scholes–Merton risk neutrality formulation of the option pricing theory is attractive because the pricing formula of a derivative deduced from their model is a function of several directly *observable* parameters (except one, which is the volatility parameter). The derivative can be priced as if the market price of the underlying asset's risk is zero. When judged by its ability to explain the empirical data, the option pricing theory is widely acclaimed to be the most successful theory not only in finance, but in all areas of economics. In recognition of their pioneering and fundamental contributions to the pricing theory of derivatives, Scholes and Merton were awarded the 1997 Nobel Prize in Economics.

In the first section, we first show how Black and Scholes applied the riskless hedging principle to derive the differential equation that governs the price of a derivative security. We also discuss Merton's approach of dynamically replicating an option by a portfolio of the riskless asset in the form of a money market account and the risky underlying asset. The cost of constructing the replicating portfolio gives the fair price of the option. Furthermore, we present an alternative perspective of the risk neutral valuation approach by showing that tradeable securities should have the same market price of risk if they are hedgeable with each other.

In Sect. 3.2, we discuss the renowned martingale pricing theory of options, which gives rise to the risk neutral valuation approach for pricing contingent claims. The price of a derivative is given by the expectation of the discounted terminal payoff under the risk neutral measure, in accordance with the property that discounted security prices are martingales under this measure. The choice of the money market account as the numeraire (accounting unit) is not unique. If a contingent claim is attainable under the numeraire-measure pair of money market account and risk neutral measure, then it is also attainable under an alternative numeraire-measure pair. We also discuss the versatile change of numeraire technique and examine how it can be used to effect efficient option pricing calculations.

In Sect. 3.3, we solve the Black–Scholes pricing equation for several types of European vanilla options. The contractual specifications are translated into an appropriate set of auxiliary conditions of the corresponding option pricing models. The most popular option price formulas are those for the European vanilla call and put options where the underlying asset price follows the Geometric Brownian process with constant drift rate and variance rate. The comparative statics of these price formulas with respect to different parameters in the option model are derived and their properties are discussed.

The generalization of the option pricing methodologies to other European-style derivative securities, like futures options, chooser options, compound options, exchange options and quanto options are considered in Sect. 3.4. We also consider extensions of the Black–Scholes–Merton formulation, which include the effects of dividends, time-dependent interest rate and volatility, etc. In addition, we illustrate how to apply the contingent claims approach to analyze the credit spread of a defaultable bond.

Practitioners using the Black–Scholes model are aware that it is less than perfect. The major criticisms are the assumptions of constant volatility, continuity of asset price process without jump and zero transaction costs on trading securities. When we try to equate the Black–Scholes option prices with actual quoted market prices of European calls and puts, we have to use different volatility values (called implied volatilities) for options with different maturities and strike prices. In Sect. 3.5, we consider the phenomena of volatility smiles exhibited in implied volatilities. We derive the Dupire equation, which may be considered as the forward version of the option pricing equation. From the Dupire equation, we can compute the local volatility function that gives the theoretical Black–Scholes option prices which agree with the market option prices. We also consider option pricing models that allow jumps in the underlying price process and include the effects of transaction costs in the trading of the underlying asset. Jumps in asset price occur when there are sudden arrivals of information about the firm or the economy as a whole. Transaction costs represent market frictions in the trading of assets. We examine how asset price jumps and transaction costs can be incorporated into the option pricing models.

Though most practitioners are aware of the limitations of the Black–Scholes model, why is it still so popular on the trading floor? One simple reason: the model involves only one parameter that is not directly observable in the market, namely, volatility. That gives an option trader the straight and simple insight: sell when volatility is high and buy when it is low. For a simple pricing model like the Black–Scholes model, traders can understand the underlying assumptions and limitations, and make appropriate adjustments if necessary. Also, pricing methodologies associated with the Black–Scholes model are relatively simple. For many European-style derivatives, closed form pricing formulas are readily available. For more complicated options, though pricing formulas do not exist, there exist an arsenal of efficient numerical schemes to calculate the option values and their comparative statics.

3.1 Black–Scholes–Merton Formulation

Black and Scholes (1973) revolutionalized the pricing theory of options by showing how to hedge continuously the exposure on the short position of an option. Consider the writer of a European call option on a risky asset. He or she is exposed to the risk of unlimited liability if the asset price rises above the strike price. To protect the writer's short position in the call option, he or she should consider purchasing a certain amount of the underlying asset so that the loss in the short position in the call option is offset by the long position in the asset. In this way, the writer is adopting the *hedging procedure*. A hedged position combines an option with its underlying asset so as to achieve the goal that either the asset compensates the option against loss or otherwise. Practitioners commonly use this risk-monitoring strategy in financial markets. By adjusting the proportion of the underlying asset and option continuously in a portfolio, Black and Scholes demonstrated that investors can create *a riskless hedging portfolio* where the risk exposure associated with the stochastic asset price is eliminated. In an efficient market with no riskless arbitrage opportunity, a riskless portfolio must earn an expected rate of return equal to the riskless interest rate.

Readers may be interested in Black's article (1989) which tells the story of how Black and Scholes came up with the idea of a riskless hedging portfolio.

3.1.1 Riskless Hedging Principle

We illustrate how to use the riskless hedging principle to derive the governing partial differential equation for the price of a European call option. In their seminal paper (1973), Black and Scholes made the following assumptions on the financial market.

(i) Trading takes place continuously in time.
(ii) The riskless interest rate r is known and constant over time.
(iii) The asset pays no dividend.
(iv) There are no transaction costs in buying or selling the asset or the option, and no taxes.
(v) The assets are perfectly divisible.
(vi) There are no penalties to short selling and the full use of proceeds is permitted.
(vii) There are no riskless arbitrage opportunities.

The stochastic process of the asset price S_t is assumed to follow the Geometric Brownian motion

$$\frac{dS_t}{S_t} = \mu\, dt + \sigma\, dZ_t, \tag{3.1.1}$$

where μ is the expected rate of return, σ is the volatility and Z_t is the standard Brownian process. Both μ and σ are assumed to be constant. Consider a portfolio that involves short selling of one unit of a European call option and long holding of Δ_t units of the underlying asset. The portfolio value $\Pi(S_t, t)$ at time t is given by

$$\Pi = -c + \Delta_t S_t,$$

where $c = c(S_t, t)$ denotes the call price. Note that Δ_t changes with time t, reflecting the dynamic nature of hedging. Since c is a stochastic function of S_t, we apply the Ito lemma to compute its differential as follows:

$$dc = \frac{\partial c}{\partial t}\,dt + \frac{\partial c}{\partial S_t}\,dS_t + \frac{\sigma^2}{2}S_t^2\frac{\partial^2 c}{\partial S_t^2}\,dt$$

so that

$$\begin{aligned} &-dc + \Delta_t\,dS_t \\ &= \left(-\frac{\partial c}{\partial t} - \frac{\sigma^2}{2}S_t^2\frac{\partial^2 c}{\partial S_t^2}\right)dt + \left(\Delta_t - \frac{\partial c}{\partial S_t}\right)dS_t \\ &= \left[-\frac{\partial c}{\partial t} - \frac{\sigma^2}{2}S_t^2\frac{\partial^2 c}{\partial S_t^2} + \left(\Delta_t - \frac{\partial c}{\partial S_t}\right)\mu S_t\right]dt + \left(\Delta_t - \frac{\partial c}{\partial S_t}\right)\sigma S_t\,dZ_t. \end{aligned}$$

The cumulative financial gain on the portfolio at time t is given by

$$\begin{aligned} G(\Pi(S_t,t)) &= \int_0^t -dc + \int_0^t \Delta_u\,dS_u \\ &= \int_0^t \left[-\frac{\partial c}{\partial u} - \frac{\sigma^2}{2}S_u^2\frac{\partial^2 c}{\partial S_u^2} + \left(\Delta_u - \frac{\partial c}{\partial S_u}\right)\mu S_u\right]du \\ &\quad + \int_0^t \left(\Delta_u - \frac{\partial c}{\partial S_u}\right)\sigma S_u\,dZ_u. \end{aligned} \tag{3.1.2}$$

The stochastic component of the portfolio gain stems from the last term: $\int_0^t (\Delta_u - \frac{\partial c}{\partial S_u})\sigma S_u\,dZ_u$. Suppose we adopt the dynamic hedging strategy by choosing $\Delta_u = \frac{\partial c}{\partial S_u}$ at all times $u < t$, then the financial gain becomes deterministic at all times. By virtue of no arbitrage, the financial gain should be the same as the gain from investing on the risk free asset with dynamic position whose value equals $-c + S_u\frac{\partial c}{\partial S_u}$. The deterministic gain from this dynamic position of the riskless asset is given by

$$M_t = \int_0^t r\left(-c + S_u\frac{\partial c}{\partial S_u}\right)du. \tag{3.1.3}$$

By equating these two deterministic gains: $G(\Pi(S_t,t))$ and M_t, we then have

$$-\frac{\partial c}{\partial u} - \frac{\sigma^2}{2}S_u^2\frac{\partial^2 c}{\partial S_u^2} = r\left(-c + S_u\frac{\partial c}{\partial S_u}\right), \quad 0 < u < t,$$

which is satisfied for any asset price S if $c(S,t)$ satisfies the equation

$$\frac{\partial c}{\partial t} + \frac{\sigma^2}{2}S^2\frac{\partial^2 c}{\partial S^2} + rS\frac{\partial c}{\partial S} - rc = 0. \tag{3.1.4}$$

The above parabolic partial differential equation is called the *Black–Scholes equation*. Note that the parameter μ, which is the expected rate of return of the asset, does not appear in the equation.

To complete the formulation of the option pricing model, we need to prescribe the auxiliary condition. The terminal payoff at time T of the European call with strike price X is translated into the following terminal condition:

$$c(S, T) = \max(S - X, 0) \tag{3.1.5}$$

for the differential equation.

Since both the equation and the auxiliary condition do not contain ρ, one can conclude that the call price does not depend on the actual expected rate of return of the asset price. The option pricing model involves five parameters: S, T, X, r and σ. Except for the volatility σ, all others are directly observable parameters. The independence of the pricing model on μ is related to the concept of *risk neutrality*. In a risk neutral world, investors do not demand extra returns above the riskless interest rate for bearing risks. This is in contrast to usual risk averse investors who would demand extra returns above r for risks borne in their investment portfolios. Apparently, the option is priced as if the rates of return on the underlying asset and the option are both equal to the riskless interest rate. This risk neutral valuation approach is viable if the risks from holding the underlying asset and option are hedgeable, as evidenced from Black–Scholes' riskless hedging argument. The concept of risk neutrality will be revisited using different arguments in later sections.

The governing equation for a European put option can be derived similarly and the same Black–Scholes equation is obtained. Let $V(S, t)$ denote the price of a derivative security with dependence on S and t, it can be shown that V is governed by

$$\frac{\partial V}{\partial t} + \frac{\sigma^2}{2} S^2 \frac{\partial^2 V}{\partial S^2} + rS\frac{\partial V}{\partial S} - rV = 0.$$

The price of a particular derivative security is obtained by solving the Black–Scholes equation subject to an appropriate set of auxiliary conditions that model the corresponding contractual specifications in the derivative security.

Remarks.

1. The original derivation of the governing partial differential equation by Black and Scholes (1973) focuses on the financial notion of riskless hedging but misses the precise analysis of the dynamic change in the value of the hedged portfolio. The inconsistencies in their derivation stem from the assumption of keeping the number of units of the underlying asset in the hedged portfolio to be instantaneously constant. They take the differential change of portfolio value Π to be

$$d\Pi = -dc + \Delta_t \, dS_t,$$

which misses the effect arising from the differential change in Δ_t. Recall that the product rule in calculus gives $d(\Delta_t S_t) = \Delta_t \, dS_t + S_t \, d\Delta_t$. Here, the notion of financial gain on the hedged portfolio is used to remedy the deficiencies in the original Black–Scholes' derivation (Carr and Bandyopadhyay, 2000). Interestingly, the "pragmatic" approach of the Black–Scholes derivation leads to the same differential equation for the option price function. Indeed, $-dc + \Delta_t \, dS_t$ is seen to be the differential financial gain on the portfolio over dt.

2. The ability to construct a perfectly hedged portfolio relies on the assumption of continuous trading and continuous asset price path. These two and other assumptions in the Black–Scholes pricing model have been critically examined by later works in derivative pricing theory. It has been commonly agreed that the assumed Geometric Brownian process of the asset price may not truly reflect the actual behavior of the asset price process. The asset price may exhibit jumps upon the arrival of a sudden news in the financial market.
3. The interest rate is widely recognized to be fluctuating over time in an irregular manner rather than being constant. For an option on a risky asset, the interest rate appears only in the discount factor so that the assumption of constant/deterministic interest rate is quite acceptable for a short-lived option.
4. The Black–Scholes pricing approach assumes continuous hedging at all times. In the real world of trading with transaction costs, this would lead to infinite transaction costs in the hedging procedure.

Even with all these limitations, the Black–Scholes model is still considered to be the most fundamental in derivative pricing theory. Various modifications to this basic model have been proposed to accommodate the above shortcomings. Some of these enhanced pricing models are addressed in Sect. 3.5.

3.1.2 Dynamic Replication Strategy

As an alternative to the riskless hedging approach, Merton (1973, Chap. 1) derived the option pricing equation via the construction of a self-financing and dynamically hedged portfolio containing the risky asset, option and riskless asset (in the form of money market account). Let $Q_S(t)$ and $Q_V(t)$ denote the number of units of asset and option in the portfolio, respectively, and $M_S(t)$ and $M_V(t)$ denote the dollar value of $Q_S(t)$ units of asset and $Q_V(t)$ units of option, respectively. The *self-financing portfolio* is set up with zero initial net investment cost and no additional funds are added or withdrawn afterwards. The additional units acquired for one security in the portfolio is completely financed by the sale of another security in the same portfolio. The portfolio is said to be *dynamic* since its composition is allowed to change over time. For notational convenience, we drop the subscript t for the asset price process S_t, the option value process V_t and the standard Brownian process Z_t. The portfolio value at time t can be expressed as

$$\begin{aligned}\Pi(t) &= M_S(t) + M_V(t) + M(t) \\ &= Q_S(t)S + Q_V(t)V + M(t),\end{aligned} \tag{3.1.6}$$

where $M(t)$ is the dollar value of the riskless asset invested in a riskless money market account. Suppose the asset price process is governed by the stochastic differential equation (3.1.1), we apply the Ito lemma to obtain the differential of the option value V as follows:

$$
\begin{aligned}
dV &= \frac{\partial V}{\partial t}dt + \frac{\partial V}{\partial S}dS + \frac{\sigma^2}{2}S^2\frac{\partial^2 V}{\partial S^2}dt \\
&= \left(\frac{\partial V}{\partial t} + \mu S\frac{\partial V}{\partial S} + \frac{\sigma^2}{2}S^2\frac{\partial^2 V}{\partial S^2}\right)dt + \sigma S\frac{\partial V}{\partial S}dZ.
\end{aligned}
$$

Suppose we formally write the stochastic dynamics of V as

$$
\frac{dV}{V} = \mu_V\, dt + \sigma_V\, dZ, \tag{3.1.7}
$$

then μ_V and σ_V are given by

$$
\mu_V = \frac{\frac{\partial V}{\partial t} + \rho S\frac{\partial V}{\partial S} + \frac{\sigma^2}{2}S^2\frac{\partial^2 V}{\partial S^2}}{V} \tag{3.1.8a}
$$

$$
\sigma_V = \frac{\sigma S\frac{\partial V}{\partial S}}{V}. \tag{3.1.8b}
$$

The instantaneous dollar return $d\Pi(t)$ of the above portfolio is attributed to the differential price changes of asset and option and interest accrued, and the differential changes in the amount of asset, option and money market account held. The differential of $\Pi(t)$ is computed as follows:

$$
\begin{aligned}
d\Pi(t) = {} & [Q_S(t)\, dS + Q_V(t)\, dV + rM(t)\, dt] \\
& + [S\, dQ_S(t) + V\, dQ_V(t) + dM(t)],
\end{aligned}
$$

where $rM(t)\, dt$ gives the interest amount earned from the money market account over dt and $dM(t)$ represents the change in the money market account held due to net dollar gained/lost from the sale of the underlying asset and option in the portfolio. Since the portfolio is self-financing, the sum of the last three terms in the above equation is zero. The instantaneous portfolio return $d\Pi(t)$ can then be expressed as

$$
\begin{aligned}
d\Pi(t) &= Q_S(t)\, dS + Q_V(t)\, dV + rM(t)\, dt \\
&= M_S(t)\frac{dS}{S} + M_V(t)\frac{dV}{V} + rM(t)\, dt.
\end{aligned} \tag{3.1.9}
$$

Eliminating $M(t)$ between (3.1.6) and (3.1.9) and expressing dS/S and dV/V in terms of their stochastic dynamics, we obtain

$$
\begin{aligned}
d\Pi(t) = {} & [(\mu - r)M_S(t) + (\mu_V - r)M_V(t)]dt \\
& + [\sigma M_S(t) + \sigma_V M_V(t)]dZ.
\end{aligned} \tag{3.1.10}
$$

How can we make the above self-financing portfolio instantaneously riskless so that its return is nonstochastic? This can be achieved by choosing an appropriate proportion of asset and option according to

$$
\sigma M_S(t) + \sigma_V M_V(t) = \sigma S Q_S(t) + \sigma S\frac{\partial V}{\partial S}Q_V(t) = 0,
$$

that is, the number of units of asset and option in the self-financing portfolio must be in the ratio

$$\frac{Q_S(t)}{Q_V(t)} = -\frac{\partial V}{\partial S} \tag{3.1.11}$$

at all times. The above ratio is time dependent, so continuous readjustment of the portfolio is necessary. We now have a dynamic replicating portfolio that is riskless and requires zero initial net investment, so the nonstochastic portfolio return $d\Pi(t)$ must be zero. Equation (3.1.10) now becomes

$$0 = [(\mu - r)M_S(t) + (\mu_V - r)M_V(t)]dt.$$

Substituting relation (3.1.11) into the above equation, we obtain

$$(\mu - r)S\frac{\partial V}{\partial S} = (\mu_V - r)V. \tag{3.1.12}$$

Finally, substituting μ_V from (3.1.8a) into the above equation, this results the same Black–Scholes equation for V:

$$\frac{\partial V}{\partial t} + \frac{\sigma^2}{2}S^2\frac{\partial^2 V}{\partial S^2} + rS\frac{\partial V}{\partial S} - rV = 0.$$

Suppose we take $Q_V(t) = -1$ in the above dynamically hedged self-financing portfolio, that is, the portfolio always shorts one unit of the option. By (3.1.11), the number of units of risky asset held is always kept at the level of $\frac{\partial V}{\partial S}$ units, which is changing continuously over time. To maintain a self-financing hedged portfolio that constantly keeps shorting one unit of the option, we need to have both the underlying asset and the riskfree asset (money market account) in the portfolio. The net cash flow resulted in the buying/selling of the risky asset in the dynamic procedure of maintaining $\frac{\partial V}{\partial S}$ units of the risky asset is siphoned to the money market account.

Replicating Portfolio
With the choice of $Q_V(t) = -1$ and knowing that

$$0 = \Pi(t) = -V + \Delta S + M(t),$$

the value of the option is found to be

$$V = \Delta S + M(t), \quad \text{with} \quad \Delta = \frac{\partial V}{\partial S}. \tag{3.1.13}$$

The above equation implies that the position of the option can be replicated by a self-financing dynamic trading strategy using the risky asset and the riskless asset (money market account) with hedge ratio Δ equals $\frac{\partial V}{\partial S}$.

3.1.3 Risk Neutrality Argument

We would like to present an alternative perspective by which the argument of *risk neutrality* can be explained in relation to the concept of market price of risk (Cox and Ross, 1976). Suppose we write the stochastic process followed by the option price $V(S,t)$ formally as in (3.1.7), then μ_V and σ_V are given by (3.1.8a,b). By rearranging the terms in (3.1.8a), we obtain the following form of the governing equation for $V(S,t)$:

$$\frac{\partial V}{\partial t} + \mu S \frac{\partial V}{\partial S} + \frac{\sigma^2}{2} S^2 \frac{\partial^2 V}{\partial S^2} - \mu_V V = 0. \tag{3.1.14}$$

Unlike the Black–Scholes equation, the above equation contains the parameters μ and μ_V. To solve for the option price, we need to determine μ and μ_V, or find some other means to avoid such a nuisance. The clue lies in the formation of a riskless hedged portfolio.

By forming the riskless dynamically hedged portfolio, we have observed that μ and μ_V are governed by (3.1.12). By combining (3.1.8b) and (3.1.12), we obtain

$$\frac{\mu_V - r}{\sigma_V} = \frac{\mu - r}{\sigma}. \tag{3.1.15}$$

The above equation has the following financial interpretation. The quantities $\mu_V - r$ and $\mu - r$ represent the extra rates of return over the riskless interest rate r on the option and the asset, respectively. When each is divided by its respective volatility (a measure of the risk of the associated security), the corresponding ratio is called the *market price of risk*. The market price of risk is interpreted as the extra rate of return above the riskless interest rate per unit of risk. Equation (3.1.15) reveals that the two hedgeable securities, option and its underlying asset, should have the same market price of risk. Substituting (3.1.8a,b) into (3.1.15) and rearranging the terms, we obtain

$$\frac{\partial V}{\partial t} + \frac{\sigma^2}{2} S^2 \frac{\partial^2 V}{\partial S^2} + rS \frac{\partial V}{\partial S} - rV = 0.$$

This is precisely the Black–Scholes equation, which is identical to the equation obtained by setting $\mu_V = \mu = r$ in (3.1.14). The relation: $\mu_V = \mu = r$ implies zero market price of risk of holding the underlying asset or option. In the world of zero market price of risk, investors are said to be risk neutral since they do not demand extra returns on holding the risky assets. We have shown that the risk neutrality argument holds when the risk of the option can be hedged by that of the underlying asset and the riskless hedged portfolio earns the risk free interest rate as its expected rate of return. Consequently, the market prices of risk of the option and the underlying asset do not enter into the governing equation. The Black–Scholes equation demonstrates that option valuation can be performed in the risk neutral world by *artificially* taking the expected rate of return of both the underlying asset and option to be the riskless interest rate. Readers should be cautioned that the actual rate of return of the underlying asset would affect the asset price and thus indirectly affects the *absolute* derivative price. We simply use the convenience of risk neutrality to arrive at the derivative price relative to that of the underlying asset. The mathematical relationship between the prices of the derivative and asset is invariant to the risk preferences of investors. However, the apparent assumption of risk neutrality of the investors is not necessary.

3.2 Martingale Pricing Theory

Under the discrete multiperiod securities models (see Sect. 2.2.4), the existence of equivalent martingale measures is equivalent to the absence of arbitrage. The market is said to be complete if all contingent claims can be replicated. Under market completeness, if equivalent martingale measures exist, then they are unique. In an arbitrage free complete market, the arbitrage price of a contingent claim is given by the expectation of the discounted terminal payoff under the equivalent martingale measure. Completeness of market can be interpreted in the sense that all sources of financial risk are priced uniquely and all future states of the world can be replicated by a dynamically rebalanced portfolio of traded assets.

In this section, we extend the discrete securities models to their continuous counterparts and examine the pricing theory of derivatives under the framework of martingale pricing measures (Harrison and Pliska, 1983). It is relatively straightforward to show that the existence of an equivalent martingale measure implies the absence of arbitrage. However, the converse statement is not true under continuous models. The restriction on trading strategies based on "no arbitrage" is not sufficient for the existence of an equivalent martingale measure. Additional technical conditions on trading strategies are required in order to establish the existence of an equivalent martingale measure (Duffie and Huang, 1985; Bingham and Kiesel, 2004, Chap. 2).

A numeraire defines the units in which security prices are measured. For example, if the traded security S is used as the numeraire, then the price of other securities $S_j, j = 1, 2, \cdots, n$, relative to the numeraire S is given by S_j/S. Be cautious that we adopt the usual convention where "numeraire" refers to both the physical instrument and the price of the instrument. Suppose the money market account is used as the numeraire, then the price of a security relative to the money market account is simply the discounted security price. A continuous time financial market consisting of traded securities and trading strategies is said to be arbitrage free and complete if for every choice of numeraire there exists a unique equivalent martingale measure such that all security prices relative to that numeraire are martingales under that measure. When we compute the price of an attainable contingent claim, the prices obtained using different martingale measures coincide. That is, the derivative price is invariant with respect to the choice of martingale pricing measure. Depending on the nature of the pricing model of the derivative, a numeraire is carefully chosen for a given pricing problem in order to achieve efficiency in the analytic valuation procedures.

In Sect. 3.2.1, we discuss the notion of absence of arbitrage and equivalent martingale measure, and present the risk neutral valuation formula for an attainable contingent claim in a continuous time securities model. Within the continuous securities models, absence of arbitrage is implied by the existence of by an equivalent martingale measure under which security prices normalized by the numeraire are martingales. We derive the risk neutral valuation principle from the observation that the discounted value process of the portfolio that replicates the contingent claim is a martingale under the risk neutral measure. We then consider the versatile numeraire invariant theorem on valuation of contingent claims. We discuss the associated change of numeraire technique and demonstrate how to compute the Radon–Nikodym derivative that effects the change of measure.

In Sect. 3.2.2, the Black–Scholes model will be revisited. We show that the martingale pricing theory gives the price of a European option as the expectation of the discounted terminal payoff under the equivalent martingale measure. Provided that the option price function satisfies the Black–Scholes equation, the risk neutral valuation formula is seen to be consistent with the Feynman–Kac representation formula (see Sect. 2.4.2).

3.2.1 Equivalent Martingale Measure and Risk Neutral Valuation

Under the continuous time framework, the investors are allowed to trade continuously in the financial market up to finite time T. Many of the tools and results in the continuous time securities models can be extended from those in discrete multiperiod models. Uncertainty in the financial market is modeled by the filtered probability space $(\Omega, \mathcal{F}, (\mathcal{F}_t)_{0\le t\le T}, P)$, where Ω is a sample space, $\mathcal{F}$ is a σ-algebra on Ω, P is a probability measure on $(\Omega, \mathcal{F})$, $\mathcal{F}_t$ is the filtration and $\mathcal{F}_T = \mathcal{F}$. In the securities model, there are $M+1$ securities whose price processes are modeled by the adapted stochastic processes $S_m(t)$, $m = 0, 1, \cdots, M$. Also, we define $h_m(t)$ to be the number of units of the mth security held in the portfolio. The trading strategy $\mathbf{h}(t)$ is the vector stochastic process $(h_0(t) \quad h_1(t) \cdots h_M(t))^T$, where $\mathbf{h}(t)$ is a $(M+1)$-dimensional predictable process since the portfolio composition is determined by the investor based on the information available before time t.

The value process associated with a trading strategy $\mathbf{h}(t)$ is defined by

$$V(t) = \sum_{m=0}^{M} h_m(t) S_m(t), \quad 0 \le t \le T, \tag{3.2.1}$$

and the gain process $G(t)$ is given by

$$G(t) = \sum_{m=0}^{M} \int_0^t h_m(u)\, dS_m(u), \quad 0 \le t \le T. \tag{3.2.2}$$

Likewise as in the discrete setting, $\mathbf{h}(t)$ is self-financing if and only if

$$V(t) = V(0) + G(t). \tag{3.2.3}$$

The above equation indicates that the change in portfolio value associated with a self-financing strategy comes only from the changes in the security prices since there are no additional cash inflows or outflows occur after the initial date $t = 0$.

We use $S_0(t)$ to denote the money market account process that grows at the riskless interest rate $r(t)$, that is,

$$dS_0(t) = r(t) S_0(t)\, dt. \tag{3.2.4}$$

The discounted security price process $S_m^*(t)$ is defined by

$$S_m^*(t) = S_m(t)/S_0(t), \quad m = 1, 2, \cdots, M. \tag{3.2.5}$$

The discounted value process $V^*(t)$ is obtained by dividing $V(t)$ by $S_0(t)$. From financial intuition, we see that self-financing portfolios remain self-financing after a numeraire change (see Problem 3.5). Using $S_0(t)$ as the numeraire and by virtue of (3.2.3), we deduce that the discounted gain process $G^*(t)$ of a self-financing strategy is given by

$$G^*(t) = V^*(t) - V^*(0). \tag{3.2.6}$$

A self-financing trading strategy is said to be Q-admissible if the discounted gain process $G^*(t)$ is a Q-martingale, where Q is a risk neutral measure. The corresponding discounted portfolio value $V^*(t)$ is a Q-martingale.

No Arbitrage Principle and Equivalent Martingale Measure

A self-financing trading strategy $\mathbf{h}$ represents an arbitrage opportunity if and only if (i) $G^*(T) \geq 0$ and (ii) $E_P[G^*(T)] > 0$, where P is the actual probability measure of the states of occurrence associated with the securities model. A probability measure Q on the space $(\Omega, \mathcal{F})$ is said to be an equivalent martingale measure if it satisfies

(i) Q is equivalent to P, that is, both P and Q have the same null set;
(ii) the discounted security price processes $S_m^*(t)$, $m = 1, 2, \cdots, M$, are martingales under Q, that is,

$$E_Q[S_m^*(u)|\mathcal{F}_t] = S_m^*(t), \quad \text{for all } 0 \leq t \leq u \leq T. \tag{3.2.7}$$

By following a similar argument as in the discrete multiperiod securities model, the absence of arbitrage is implied by the existence of an equivalent martingale measure. To show the claim, suppose an equivalent martingale measure exists and $\mathbf{h}$ is a self-financing strategy under P, and also under Q. The time-t discounted value $V^*(t)$ of the portfolio generated by $\mathbf{h}$ is a Q-martingale so that $V^*(0) = E_Q[V^*(T)]$. Now, we start with $V(0) = V^*(0) = 0$, and suppose we claim that $V^*(T) \geq 0$. Since $Q(\omega) > 0$ and $E_Q[V^*(T)] = V^*(0) = 0$ should be observed, we can only have $V^*(T) = 0$. In other words, starting with $V^*(0) = 0$, it is impossible to have $V^*(T) \geq 0$ and $V^*(T)$ is strictly positive under some states of the world. Hence, there cannot exist any arbitrage opportunities.

Contingent claims are modeled as $\mathcal{F}_T$-measurable random variables. A contingent claim Y is said to be *attainable* if there exists at least an admissible trading strategy $\mathbf{h}$ such that the time-T portfolio value $V(T)$ equals Y. We then say that Y is generated by $\mathbf{h}$. The arbitrage price of Y can be obtained by the *risk neutral valuation approach* as stated in Theorem 3.1.

Theorem 3.1. *Assume that an equivalent martingale measure Q exists. Let Y be an attainable contingent claim generated by a Q-admissible self-financing trading strategy* $\mathbf{h}$. *For each time t, $0 \leq t \leq T$, the arbitrage price of Y is given by*

$$V(t; \mathbf{h}) = S_0(t) E_Q\left[\frac{Y}{S_0(T)}\bigg|\mathcal{F}_t\right]. \tag{3.2.8}$$

The validity of Theorem 3.1 is readily seen since Y is generated by a Q-admissible self-financing trading strategy $\mathbf{h}$ so that the discounted portfolio value process $V^*(t;\mathbf{h})$ is a martingale under Q. This leads to

$$V(t;\mathbf{h}) = S_0(t)V^*(t;\mathbf{h}) = S_0(t)E_Q[V^*(T;\mathbf{h})|\mathcal{F}_t].$$

Furthermore, by observing that $V^*(T;\mathbf{h}) = Y/S_0(T)$, the risk neutral valuation formula (3.2.8) follows.

Though there may be two replicating portfolios that generate Y, the above risk neutral valuation formula shows that the arbitrage price is uniquely determined by the expectation of discounted terminal payoff, independent of the choice of the replicating portfolio. This is a consequence of the law of one price.

Change of Numeraire

The risk neutral valuation formula (3.2.8) uses the riskless asset $S_0(t)$ (money market account) as the numeraire. Geman, El Karoui and Rochet (1995) showed that the choice of $S_0(t)$ as the numeraire is not necessary to be unique in order that the risk neutral valuation formula holds. It will be demonstrated in Sect. 8.1. that the choice of another numeraire, like the stochastic bond price, may be more convenient for analytic pricing calculations of option models under stochastic interest rates. The following discussion summarizes the powerful tool developed by Geman, El Karoui and Rochet (1995) on the change of numeraire.

Let $N(t)$ be a numeraire and we assume the existence of an equivalent probability measure Q_N such that all security prices normalized with respect to $N(t)$ are Q_N-martingales. In addition, if a contingent claim Y is attainable under the numeraire-measure pair $(S_0(t), Q)$, then it is also attainable under an alternative pair $(N(t), Q_N)$. The arbitrage price of any security given by the risk neutral valuation formula under both measures should agree. We then have

$$S_0(t)E_Q\left[\frac{Y}{S_0(T)}\bigg|\mathcal{F}_t\right] = N(t)E_{Q_N}\left[\frac{Y}{N(T)}\bigg|\mathcal{F}_t\right]. \tag{3.2.9}$$

To derive the Radon–Nikodym derivative

$$L(t) = \frac{dQ_N}{dQ}\bigg|_{\mathcal{F}_t}, \quad t \in [0,T],$$

that effects the change of measure from Q_N to Q, we apply the Bayes rule (see Problem 2.43) and obtain

$$E_{Q_N}\left[\frac{Y}{N(T)}\bigg|\mathcal{F}_t\right] = \frac{1}{E_Q[L(T)|\mathcal{F}_t]}E_Q\left[\frac{Y}{N(T)}L(T)\bigg|\mathcal{F}_t\right]. \tag{3.2.10a}$$

From (3.2.9), we have

$$E_{Q_N}\left[\frac{Y}{N(T)}\bigg|\mathcal{F}_t\right] = \frac{N(0)S_0(t)}{N(t)}E_Q\left[\frac{Y}{N(0)S_0(T)}\bigg|\mathcal{F}_t\right]. \tag{3.2.10b}$$

Recall the relation: $E_Q[L(T)|\mathcal{F}_t] = L(t)$. Now, if we define the Radon–Nikodym derivative $L(t)$ by

$$L(t) = \left.\frac{dQ_N}{dQ}\right|_{\mathcal{F}_t} = \frac{N(t)}{N(0)S_0(t)}, \quad t \in [0, T], \tag{3.2.11}$$

then both (3.2.10a,b) become consistent.

The change of numeraire technique is seen to be the one of the most powerful tools for analytic pricing of financial derivatives. In general, the Radon–Nikodym derivative $L_{N,M}(t)$ that effects the change of measure from the numeraire-measure pair $(N(t), Q_N)$ to the other pair $(M(t), Q_M)$ is given by

$$L_{N,M}(t) = \left.\frac{dQ_N}{dQ_M}\right|_{\mathcal{F}_t} = \frac{N(t)}{N(0)} \Big/ \frac{M(t)}{M(0)}, \quad t \in [0, T]. \tag{3.2.12}$$

3.2.2 Black–Scholes Model Revisited

The Black–Scholes option model is revisited under the martingale pricing framework. We assume the existence of a risk neutral measure Q under which all discounted price processes are Q-martingales. The securities model has two basic tradeable securities, the underlying risky asset and riskless asset in the form of a money market account. The price processes of the risky asset and riskless asset under the actual probability measure P are governed by

$$\frac{dS_t}{S_t} = \mu\, dt + \sigma\, dZ_t \tag{3.2.13a}$$

$$dM_t = rM_t\, dt, \tag{3.2.13b}$$

respectively, where Z_t is P-Brownian. Suppose we take the money market account as the numeraire, and define the price of the discounted risky asset by $S_t^* = S_t/M_t$. By Ito's lemma, the price process S_t^* becomes

$$\frac{dS_t^*}{S_t^*} = (\mu - r)dt + \sigma\, dZ_t.$$

We would like to find the equivalent martingale measure Q under which the discounted asset price S_t^* is a Q-martingale. By the Girsanov Theorem, suppose we choose $\gamma(t)$ in the Radon–Nikodym derivative [see (2.4.32a,b)] such that

$$\gamma(t) = \frac{\mu - r}{\sigma},$$

then $\widetilde{Z}_t$ is a Brownian process under the probability measure Q and

$$d\widetilde{Z}_t = dZ_t + \frac{\mu - r}{\sigma}dt. \tag{3.2.14}$$

Under the Q-measure, the process of S_t^* now becomes

$$\frac{dS_t^*}{S_t^*} = \sigma\, d\widetilde{Z}_t. \tag{3.2.15a}$$

Since $\widetilde{Z}_t$ is Q-Brownian, so S_t^* is a Q-martingale. Substituting (3.2.14) into (3.2.13a), the asset price S_t under the Q-measure is governed by

$$\frac{dS_t}{S_t} = r\, dt + \sigma\, d\widetilde{Z}_t, \tag{3.2.15b}$$

where the drift rate equals the riskless interest rate r. When the money market account is used as the numeraire, the corresponding equivalent martingale measure is commonly called the *risk neutral measure* and the drift rate of S_t under the Q-measure is called the *risk neutral drift rate*.

By virtue of the risk neutral valuation formula (3.2.8), the arbitrage price of a derivative is given by

$$V(S,t) = E_Q^{t,S}\left[\frac{M_T}{M_t} h(S_T)\right] = e^{-r(T-t)} E_Q^{t,S}[h(S_T)], \tag{3.2.16}$$

where $E_Q^{t,S}$ is the expectation under the risk neutral measure Q conditional on the filtration $\mathcal{F}_t$ with $S_t = S$. Under the assumption of constant interest rate r, the discount factor $M_T/M_t = e^{-r(T-t)}$ is constant so that it can be taken out from the expectation term. In our future discussion, when there is no ambiguity, we choose to write E_Q instead of the full notation $E_Q^{t,S}$. The terminal payoff of the derivative is some function h of the terminal asset price S_T. Suppose $V(S,t)$ is governed by the Black–Scholes equation:

$$\frac{\partial V}{\partial t} + \frac{\sigma^2}{2} S^2 \frac{\partial^2 V}{\partial S^2} + rS\frac{\partial V}{\partial S} - rV = 0,$$

by virtue of the Feynman–Kac representation formula, the price function $V(S,t)$ admits the expectation representation as defined by (3.2.16). This illustrates the consistency between the risk neutral valuation principle and the Black–Scholes–Merton pricing formulation.

As an example, consider the European call option whose terminal payoff is $\max(S_T - X, 0)$. Using (3.2.16), the call price $c(S,t)$ is given by

$$\begin{aligned} c(S,t) &= e^{-r(T-t)} E_Q[\max(S_T - X, 0)] \\ &= e^{-r(T-t)}\{E_Q[S_T \mathbf{1}_{\{S_T \geq X\}}] - X E_Q[\mathbf{1}_{\{S_T \geq X\}}]\}. \end{aligned} \tag{3.2.17}$$

In the next section, we show how to derive the call price formula by computing the above expectations [see (3.3.12a,b)].

Exchange Rate Process under Domestic Risk Neutral Measure

Consider a foreign currency option whose terminal payoff function depends on the exchange rate F, which is defined as the domestic currency price of one unit of the foreign currency. Let M_d and M_f denote the respective money market account process in the domestic market and foreign market. Suppose the processes of $M_d(t)$, $M_f(t)$ and $F(t)$ under the actual probability measure are governed by

$$\begin{aligned} dM_d(t) &= rM_d(t)\,dt, \\ dM_f(t) &= r_f M_f(t)\,dt, \\ \frac{dF(t)}{F(t)} &= \mu_F\,dt + \sigma\,dZ_F(t), \end{aligned} \tag{3.2.18}$$

where r and r_f denote the constant riskless domestic and foreign interest rate, respectively. We would like to find the risk neutral drift rate of the exchange rate process $F(t)$ under the domestic risk neutral measure Q_d.

We may treat the domestic money market account and the foreign money market account in domestic dollars (whose value is given by FM_f) as traded securities in the domestic currency world. Under the domestic risk neutral measure Q_d, $M_d(t)$ is used as the numeraire. By Ito's lemma, the relative price process $X(t) = F(t)M_f(t)/M_d(t)$ is governed by

$$\frac{dX(t)}{X(t)} = (r_f - r + \mu_F)\,dt + \sigma\,dZ_F(t). \tag{3.2.19a}$$

Here, $X(t)$ can be considered as the discounted price process of a domestic asset in the domestic currency world so that $X(t)$ should be a martingale under Q_d. Taking $\gamma = (r_f - r + \mu_F)/\sigma$ in the Girsanov Theorem, we define

$$dZ_d(t) = dZ_F(t) + \gamma\,dt,$$

where $Z_d(t)$ is Q_d-Brownian. Now, under the domestic risk neutral measure Q_d, the process $X(t)$ satisfies

$$\frac{dX(t)}{X(t)} = \sigma\,dZ_d(t). \tag{3.2.19}$$

Since $Z_d(t)$ is Q_d-Brownian, so $X(t)$ is a Q_d-martingale. The exchange rate process $F(t)$ under Q_d is then given by

$$\frac{dF(t)}{F(t)} = (r - r_f)dt + \sigma\,dZ_d(t). \tag{3.2.20}$$

We deduce that the risk neutral drift rate of the exchange rate process $F(t)$ under the domestic risk neutral measure Q_d is found to be $r - r_f$.

3.3 Black–Scholes Pricing Formulas and Their Properties

In this section, we first derive the Black–Scholes price formula for a European call option by solving directly the Black–Scholes equation augmented with appropriate auxiliary conditions. The European put price formula can be obtained easily from the put-call parity relation once the corresponding European call price formula is known. We also derive the call price function using the risk neutral valuation formula by computing the discounted expectation of the terminal payoff. By substituting the expectation representation of the call price function into the Black–Scholes equation, we deduce the backward Fokker–Planck equation for the transition density function of the asset price under the risk neutral measure.

For hedging and other trading purposes, it is important to estimate the rate of change of option price with respect to the price of the underlying asset and other option parameters, like the strike price, volatility etc. The formulas for these comparative statics (commonly called the *greeks* of the option formulas) are derived and their analytic properties are analyzed.

3.3.1 Pricing Formulas for European Options

Recall that the Black–Scholes equation for a European vanilla call option takes the form

$$\frac{\partial c}{\partial \tau} = \frac{\sigma^2}{2} S^2 \frac{\partial^2 c}{\partial S^2} + rS\frac{\partial c}{\partial S} - rc, \qquad 0 < S < \infty,\ \tau > 0, \tag{3.3.1}$$

where $c = c(S, \tau)$ is the European call value, S and $\tau = T - t$ are the asset price and time to expiry, respectively. We use the time to expiry τ instead of the calendar time t as the time variable so that the Black–Scholes equation becomes the usual parabolic type partial differential equation. The auxiliary conditions of the option pricing model are prescribed as follows:

Initial condition (payoff at expiry)

$$c(S, 0) = \max(S - X, 0), \qquad X \text{ is the strike price.} \tag{3.3.2a}$$

Solution behaviors at the boundaries

(i) When the asset value hits zero for some $t < T$, it will stay at zero at all subsequent times so that the call option is sure to expire out-of-the-money. As a consequence, the call option has zero value, that is,

$$c(0, \tau) = 0. \tag{3.3.2b}$$

(ii) When S is sufficiently large, it becomes almost certain that the call will be exercised. Since the present value of the strike price is $Xe^{-r\tau}$, we have

$$c(S, \tau) \sim S - Xe^{-r\tau} \quad \text{as} \quad S \to \infty. \tag{3.3.2c}$$

We illustrate how to apply the Green function technique in partial differential equation theory to determine the solution to $c(S, \tau)$. Using the transformation: $y = \ln S$ and $c(y, \tau) = e^{-r\tau} w(y, \tau)$, the Black–Scholes equation is transformed into the following constant-coefficient parabolic equation

$$\frac{\partial w}{\partial \tau} = \frac{\sigma^2}{2}\frac{\partial^2 w}{\partial y^2} + \left(r - \frac{\sigma^2}{2}\right)\frac{\partial w}{\partial y}, \qquad -\infty < y < \infty,\ \tau > 0. \tag{3.3.3a}$$

The initial condition (3.3.2a) for the model now becomes

$$w(y, 0) = \max(e^y - X, 0). \tag{3.3.3b}$$

Since the domain of the pricing model is infinite, the differential equation together with the initial condition are sufficient to determine the call price function. Once the price function is obtained, we check whether the solution values at the boundaries agree with those stated in (3.3.2b,c).

Green Function Approach

Recall that the probability density function of the Brownian process $X(t)$ with drift rate μ, variance rate σ^2 and $X(0) = 0$ is given by

$$u(x, t) = \frac{1}{\sqrt{2\pi\sigma^2 t}} \exp\left(-\frac{(x - \mu t)^2}{2\sigma^2 t}\right),$$

where $u(x, t)$ satisfies (2.3.13). We then deduce that the infinite domain Green function of (3.3.3a) is given by

$$\phi(y, \tau; \xi) = \frac{1}{\sqrt{2\pi\sigma^2\tau}} \exp\left(-\frac{[y + (r - \frac{\sigma^2}{2})\tau - \xi]^2}{2\sigma^2\tau}\right). \tag{3.3.4}$$

Here, $\phi(y, \tau; \xi)$ satisfies the initial condition:

$$\lim_{\tau \to 0^+} \phi(y, \tau; \xi) = \delta(y - \xi),$$

where $\delta(y - \xi)$ is the Dirac function representing a unit impulse at the position ξ. The Green function $\phi(y, \tau; \xi)$ can be considered as the response in the position y and at time τ due to a unit impulse placed at the position ξ initially. On the other hand, from the property of the Dirac function, the initial condition can be expressed as

$$w(y, 0) = \int_{-\infty}^{\infty} w(\xi, 0)\delta(y - \xi)\, d\xi,$$

so that $w(y, 0)$ can be considered as the superposition of impulses with varying magnitude $w(\xi, 0)$ ranging from $\xi \to -\infty$ to $\xi \to \infty$. Since (3.3.3a) is linear, the response in position y and at time τ due to an impulse of magnitude $w(\xi, 0)$ in position ξ at $\tau = 0$ is given by $w(\xi, 0)\phi(y, \tau; \xi)$. From the principle of superposition for a linear differential equation, the solution to the initial value problem posed in (3.3.3a,b) is obtained by summing up the responses due to these impulses. This amounts to integration from $\xi \to -\infty$ to $\xi \to \infty$. Hence, the solution to $c(y, \tau)$ is given by

$$\begin{aligned} c(y, \tau) &= e^{-r\tau} w(y, \tau) \\ &= e^{-r\tau} \int_{-\infty}^{\infty} w(\xi, 0)\, \phi(y, \tau; \xi)\, d\xi \end{aligned}$$

$$= e^{-r\tau} \int_{\ln X}^{\infty} (e^{\xi} - X) \frac{1}{\sigma\sqrt{2\pi\tau}} \exp\left(-\frac{[y + (r - \frac{\sigma^2}{2})\tau - \xi]^2}{2\sigma^2\tau}\right) d\xi. \tag{3.3.5}$$

Here, $\phi(y, \tau; \xi)$ can be interpreted as the kernel of the integral transform that transforms the initial value $w(\xi, 0)$ to the solution $w(y, \tau)$ at time τ.

The integral in (3.3.5) can be evaluated in closed form as follows. Consider the following integral, by completing square in the exponential expression, we obtain

$$\begin{aligned}
&\int_{\ln X}^{\infty} e^{\xi} \frac{1}{\sigma\sqrt{2\pi\tau}} \exp\left(-\frac{[y + (r - \frac{\sigma^2}{2})\tau - \xi]^2}{2\sigma^2\tau}\right) d\xi \\
&= \exp(y + r\tau) \int_{\ln X}^{\infty} \frac{1}{\sigma\sqrt{2\pi\tau}} \exp\left(-\frac{\left[y + \left(r + \frac{\sigma^2}{2}\right)\tau - \xi\right]^2}{2\sigma^2\tau}\right) d\xi \\
&= e^{r\tau} S N\left(\frac{\ln \frac{S}{X} + (r + \frac{\sigma^2}{2})\tau}{\sigma\sqrt{\tau}}\right), \quad y = \ln S.
\end{aligned}$$

The other integral can be expressed as:

$$\begin{aligned}
&\int_{\ln X}^{\infty} \frac{1}{\sigma\sqrt{2\pi\tau}} \exp\left(-\frac{[y + (r - \frac{\sigma^2}{2})\tau - \xi]^2}{2\sigma^2\tau}\right) d\xi \\
&= N\left(\frac{y + (r - \frac{\sigma^2}{2})\tau - \ln X}{\sigma\sqrt{\tau}}\right) = N\left(\frac{\ln \frac{S}{X} + (r - \frac{\sigma^2}{2})\tau}{\sigma\sqrt{\tau}}\right), \quad y = \ln S.
\end{aligned}$$

Hence, the European call price formula is found to be

$$c(S, \tau) = SN(d_1) - Xe^{-r\tau} N(d_2), \tag{3.3.6}$$

where

$$d_1 = \frac{\ln \frac{S}{X} + (r + \frac{\sigma^2}{2})\tau}{\sigma\sqrt{\tau}}, \quad d_2 = d_1 - \sigma\sqrt{\tau}.$$

The initial condition is seen to be satisfied by observing that the limits of both d_1 and d_2 tend to either one or zero depending on $S > X$ or $S < X$, respectively, as $\tau \to 0^+$. By observing

$$\lim_{S\to\infty} N(d_1) = \lim_{S\to\infty} N(d_2) = 1,$$

and

$$\lim_{S\to 0^+} N(d_1) = \lim_{S\to 0^+} N(d_2) = 0,$$

one can easily check that boundary conditions (3.3.2b,c) are satisfied by the analytic call price formula (3.3.6). The European call price formula also gives the price of an American call on a nondividend paying asset since the early exercise privilege of this American call is rendered useless.

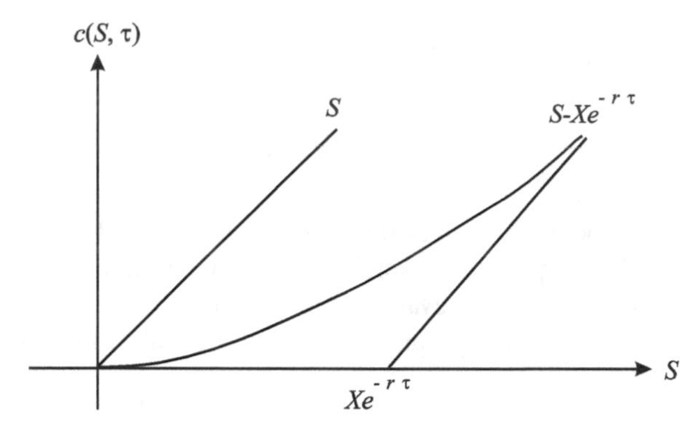

Fig. 3.1. A plot of $c(S, \tau)$ against S at a given τ. The European call price is bounded between S and $\max(S - Xe^{-r\tau}, 0)$.

The call price can be shown to lie within the bounds

$$\max(S - Xe^{-r\tau}, 0) \leq c(S, \tau) \leq S, \quad S \geq 0, \tau \geq 0, \tag{3.3.7}$$

which agrees with the distribution free results on the bounds of the call price function [see (1.2.12)]. Also, the call price function $c(S, \tau)$ can be shown to be an increasing convex function of S [see (3.3.28)]. A plot of $c(S, \tau)$ against S is shown in Fig. 3.1.

Risk Neutral Valuation Approach

Using the risk neutral valuation approach, the European call price can be obtained by computing the expectation of the discounted terminal payoff under the risk neutral measure. Let $\psi(S_T, T; S, t)$ denote the transition density function under the risk neutral measure of the terminal asset price S_T at time T, given asset price S at an earlier time t. According to (3.2.17), the call price $c(S, t)$ can be written as

$$\begin{aligned} c(S, t) &= e^{-r(T-t)} E_Q[(S_T - X)\mathbf{1}_{\{S_T \geq X\}}] \\ &= e^{-r(T-t)} \int_0^\infty \max(S_T - X, 0)\psi(S_T, T; S, t)\, dS_T. \end{aligned} \tag{3.3.8}$$

Under the risk neutral measure Q, the asset price follows the Geometric Brownian process with drift rate r and variance rate σ^2. By applying the results in (2.4.16)–(2.4.17), we deduce that

$$\ln \frac{S_T}{S} = \left(r - \frac{\sigma^2}{2}\right)\tau + \sigma \widetilde{Z}(\tau), \quad \tau = T - t, \tag{3.3.9}$$

where $\widetilde{Z}(\tau)$ is Q-Brownian. We observe that $\ln \frac{S_T}{S}$ is normally distributed with mean $(r - \frac{\sigma^2}{2})\tau$ and variance $\sigma^2\tau$. As deduced from the density function of a normal random variable [see $f_X(x, t)$ defined in (2.3.12)], the transition density function is given by

$$\psi(S_T, T; S, t) = \frac{1}{S_T \sigma \sqrt{2\pi\tau}} \exp\left(-\frac{\left[\ln \frac{S_T}{S} - \left(r - \frac{\sigma^2}{2}\right)\tau\right]^2}{2\sigma^2\tau}\right). \tag{3.3.10}$$

We set $\xi = \ln S_T$ and $y = \ln S$ so that $\ln \frac{S_T}{S} = \xi - y$ and $d\xi = \frac{dS_T}{S_T}$. Substituting the above transition density function into (3.3.8), the European call price can be expressed as

$$c(S, \tau) = e^{-r\tau} \int_{-\infty}^{\infty} \max(e^{\xi} - X, 0) \\ \frac{1}{\sigma\sqrt{2\pi\tau}} \exp\left(-\frac{\left[y + \left(r - \frac{\sigma^2}{2}\right)\tau - \xi\right]^2}{2\sigma^2\tau}\right) d\xi, \tag{3.3.11}$$

which is consistent with the result shown in (3.3.5).

If we compare the call price formula (3.3.6) with the expectation representation in (3.2.17), we deduce that

$$N(d_2) = E_Q[\mathbf{1}_{\{S_T > X\}}] = Q[S_T > X] \tag{3.3.12a}$$

$$e^{r\tau} S N(d_1) = E_Q[S_T \mathbf{1}_{\{S_T > X\}}]. \tag{3.3.12b}$$

Hence, $N(d_2)$ is recognized as the probability under the risk neutral measure Q that the call expires in-the-money, so $Xe^{-r\tau}N(d_2)$ represents the present value of the risk neutral expectation of the payment of strike made by the option holder at expiry. Also, $SN(d_1)$ is the risk neutral expectation of the discounted terminal asset price conditional on the call being in-the-money at expiry. An alternative approach to derive (3.3.12b) using the change of measure formula (3.2.11) is illustrated in Problem 3.10.

Fokker–Planck Equations

If we substitute the integral in (3.3.8) into the Black–Scholes equation, we obtain

$$0 = e^{-r(T-t)} \int_0^{\infty} \max(S_T - X, 0) \left(\frac{\partial \psi}{\partial t} + \frac{\sigma^2}{2} S^2 \frac{\partial^2 \psi}{\partial S^2} + rS \frac{\partial \psi}{\partial S}\right) dS_T.$$

The integrand function must vanish and thus leads to the following governing equation for $\psi(S_T, T; S, t)$:

$$\frac{\partial \psi}{\partial t} + \frac{\sigma^2}{2} S^2 \frac{\partial^2 \psi}{\partial S^2} + rS \frac{\partial \psi}{\partial S} = 0. \tag{3.3.13}$$

This is called the *backward Fokker–Planck equation* since the dependent variables S and t are backward variables. In terms of the forward variables S_T and T, one can show that $\psi(S_T, T; S, t)$ satisfies the following *forward Fokker–Planck equation* (see Problem 3.9):

$$\frac{\partial \psi}{\partial T} - \frac{\sigma^2}{2} S_T^2 \frac{\partial^2 \psi}{\partial S_T^2} + rS_T \frac{\partial \psi}{\partial S_T} = 0. \tag{3.3.14}$$

The forward equation reduces to the same form as that in (2.3.13) if we set $x = \ln S_T$, $\mu = r - \frac{\sigma^2}{2}$ and visualize the calendar time variable t in (2.3.13) as the forward time variable T in (3.3.14).

Put Price Function

Using the put-call parity relation [see (1.2.17)], the price of a European put option is given by

$$\begin{aligned} p(S, \tau) &= c(S, \tau) + Xe^{-r\tau} - S \\ &= S[N(d_1) - 1] + Xe^{-r\tau}[1 - N(d_2)] \\ &= Xe^{-r\tau}N(-d_2) - SN(-d_1). \end{aligned} \tag{3.3.15}$$

At a sufficiently low asset value, we see that $N(-d_2) \to 1$ and $SN(-d_1) \to 0$ so that

$$p(S, \tau) \sim Xe^{-r\tau} \quad \text{as} \quad S \to 0^+. \tag{3.3.16}$$

The put value can be below its intrinsic value, $X - S$, when S is sufficiently low in value.

On the other hand, though $SN(-d_1)$ is of the indeterminate form $\infty \cdot 0$ as $S \to \infty$, one can show that $SN(-d_1) \to 0$ as $S \to \infty$. Hence, we obtain

$$\lim_{S \to \infty} p(S, \tau) = 0. \tag{3.3.17}$$

This is not surprising since the European put is certain to expire out-of-the-money as $S \to \infty$, so it has zero value. The put price function is a decreasing convex function of S and it is bounded above by the strike price X. A plot of the put price $p(S, \tau)$ against S is shown in Fig. 3.2.

For a perpetual European put option with infinite time to expiry. This is expected since both $N(-d_1) \to 0$ and $N(-d_2) \to 0$ as $\tau \to \infty$, we obtain

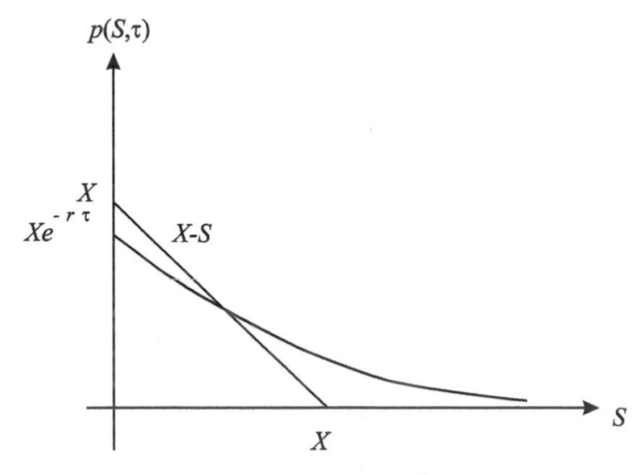

Fig. 3.2. A plot of $p(S, \tau)$ against S at a given τ. The European put price may be below the intrinsic value $X - S$ at a sufficiently low asset value S.

$$\lim_{\tau \to \infty} p(S, \tau) = 0. \tag{3.3.18}$$

The value of a perpetual European put is zero since the present value of the strike price becomes zero if it is received at infinite time from now.

3.3.2 Comparative Statics

The option price formulas are price functions of five parameters: S, τ, X, r and σ. To understand better the pricing behavior of the European vanilla options, we analyze the comparative statics. We examine the rate of change of the option price with respect to each of these parameters. We commonly use different Greek letters to denote different types of comparative statics, so these rates of change are also called the greeks of an option price function.

Delta—Derivative with Respect to Asset Price

The *delta* Δ of the value of a derivative security is defined to be $\frac{\partial V}{\partial S}$, where V is the value of the derivative security and S is the asset price. Delta plays a crucial role in the hedging of portfolios. Recall that in the implementation of the Black–Scholes riskless hedging procedure, a covered call position is maintained by creating a riskless portfolio where the writer of a call sells one unit of call and holds Δ units of asset. From the call price formula (3.3.6), the delta of the price of a European call option is found to be

$$\begin{aligned}
\Delta_c = \frac{\partial c}{\partial S} &= N(d_1) + S\,\frac{1}{\sqrt{2\pi}}\,e^{-\frac{d_1^2}{2}}\,\frac{\partial d_1}{\partial S} - Xe^{-r\tau}\,\frac{1}{\sqrt{2\pi}}\,e^{-\frac{d_2^2}{2}}\,\frac{\partial d_2}{\partial S} \\
&= N(d_1) + \frac{1}{\sigma\sqrt{2\pi\tau}}[e^{-\frac{d_1^2}{2}} - e^{-(r\tau + \ln\frac{S}{X})}e^{-\frac{d_2^2}{2}}] \\
&= N(d_1) > 0.
\end{aligned} \tag{3.3.19}$$

Interestingly, Δ_c is finally reduced to $N(d_1)$. This is not surprising since a European call can be replicated by Δ_c units of the risky asset plus a negative amount of the money market account. The multiplicative factor $N(d_1)$ in front of S in the call price formula thus gives the hedge ratio Δ_c. From the put-call parity relation, the delta of the price of a European vanilla put option is

$$\Delta_p = \frac{\partial p}{\partial S} = \Delta_c - 1 = N(d_1) - 1 = -N(-d_1) < 0. \tag{3.3.20}$$

The delta of the price of a call is always positive since an increase in the asset price will increase the probability of a positive terminal payoff resulting in a higher call price. The reverse argument is used to explain why the delta of the put price is always negative. The negativity of Δ_p means that a long position in a put option should be hedged by a continuously rebalancing long position in the underlying asset.

Both the call and put deltas are functions of S and τ. Note that Δ_c is an increasing function of S since $N(d_1)$ is always an increasing function of S. Also, the value of

Δ_c is bounded between 0 and 1. The curve of Δ_c against S changes concavity at the critical value

$$S_c = X \exp\left(-\left(r + \frac{3\sigma^2}{2}\right)\tau\right).$$

The curve is then concave upward for $0 \leq S < S_c$ and concave downward for $S_c < S < \infty$. To estimate the limiting value of Δ_c at $\tau \to \infty$ and $\tau \to 0^+$, we apply the following properties of the normal distribution function $N(x)$:

$$\lim_{x\to\infty} N(x) = 1, \quad \lim_{x\to-\infty} N(x) = 0 \quad \text{and} \quad \lim_{x\to 0} N(x) = \frac{1}{2}.$$

Note that $d_1 \to \infty$ when $\tau \to \infty$ for all values of S. When $\tau \to 0^+$, we have

(i) $d_1 \to \infty$ if $S > X$,
(ii) $d_1 \to 0$ if $S = X$, and
(iii) $d_1 \to -\infty$ if $S < X$.

Hence, we can deduce that

$$\lim_{\tau\to\infty} \frac{\partial c}{\partial S} = 1 \quad \text{for all values of } S, \tag{3.3.21a}$$

while

$$\lim_{\tau\to 0^+} \frac{\partial c}{\partial S} = \begin{cases} 1 & \text{if } S > X \\ \frac{1}{2} & \text{if } S = X \\ 0 & \text{if } S < X \end{cases}. \tag{3.3.21b}$$

The variation of the delta of the call price with respect to asset price S and time to expiry τ are shown in Figs. 3.3 and 3.4, respectively.

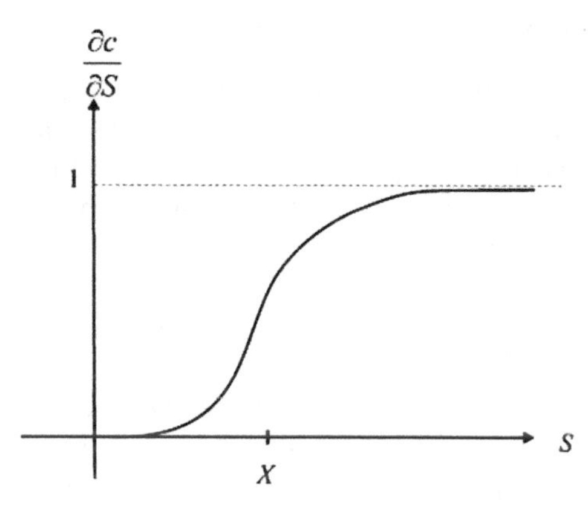

Fig. 3.3. Variation of the delta of the European call price with respect to the asset price S. The curve changes concavity at $S = Xe^{-(r+\frac{3\sigma^2}{2})\tau}$.

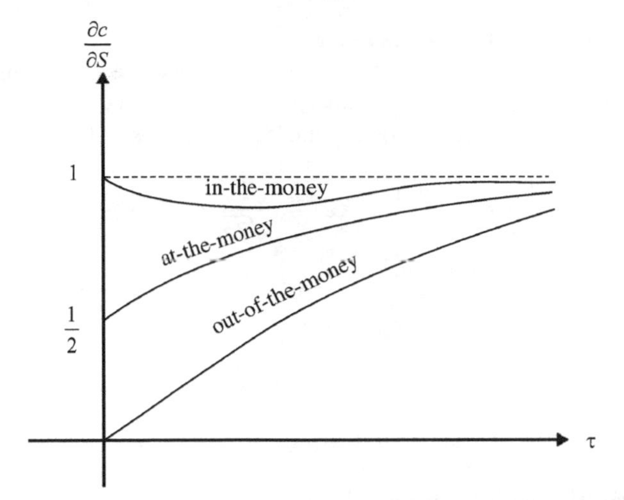

Fig. 3.4. Variation of the delta of the European call price with respect to time to expiry τ. The delta value always tends to one from below when $\tau \to \infty$. The delta value tends to different asymptotic limits as time comes close to expiry, depending on the moneyness of the option.

Elasticity with Respect to Asset Price

We define the *elasticity* of the call price with respect to the asset price as $\left(\frac{\partial c}{\partial S}\right)\left(\frac{S}{c}\right)$. The elasticity parameter gives the measure of the percentage change in call price for a unit percentage change in the asset price. For a European call on a nondividend paying asset, the elasticity e_c is found to be

$$e_c = \left(\frac{\partial c}{\partial S}\right)\left(\frac{S}{c}\right) = \frac{SN(d_1)}{SN(d_1) - Xe^{-r\tau}N(d_2)} > 1. \tag{3.3.22}$$

As $e_c > 1$, a call option is riskier in percentage change than the underlying asset. It can be shown that the elasticity of a call has a very high value when the asset price is small (out-of-the-money) and it decreases monotonically with an increase of the asset price. At sufficiently high value of S, the elasticity tends asymptotically to one since $c \sim S$ as $S \to \infty$.

The elasticity of the European put price is defined similarly by

$$e_p = \left(\frac{\partial p}{\partial S}\right)\left(\frac{S}{p}\right). \tag{3.3.23}$$

It can be shown that the put's elasticity is always negative, but its absolute value can be less than or greater than one (see Problem 3.13). Therefore, a European put option may or may not be riskier than the underlying asset.

For both put and call options, their elasticity increases in absolute value when the corresponding options become more out-of-the-money and the time comes closer to expiry.

Derivative with Respect to Strike Price

In Sect. 1.2, we argued that the European call (put) price is a decreasing (increasing) function of the strike price. These properties can be verified for the European call and put prices by computing the corresponding derivatives as follows:

$$\begin{aligned}\frac{\partial c}{\partial X} &= S\,\frac{1}{\sqrt{2\pi}}\,e^{-\frac{d_1^2}{2}}\,\frac{\partial d_1}{\partial X} - Xe^{-r\tau}\,\frac{1}{\sqrt{2\pi}}\,e^{-\frac{d_2^2}{2}}\,\frac{\partial d_2}{\partial X} - e^{-r\tau}N(d_2)\\ &= -e^{-r\tau}N(d_2) < 0, \end{aligned} \tag{3.3.24}$$

and

$$\begin{aligned}\frac{\partial p}{\partial X} &= \frac{\partial c}{\partial X} + e^{-r\tau} \qquad \text{(from the put-call parity relation)}\\ &= e^{-r\tau}[1 - N(d_2)] = e^{-r\tau}N(-d_2) > 0. \end{aligned} \tag{3.3.25}$$

Theta—Derivative with Respect to Time

The *theta* Θ of the value of a derivative security V is defined as $\frac{\partial V}{\partial t}$, where t is the calendar time. The theta of the European vanilla call and put prices are found, respectively, to be

$$\begin{aligned}\Theta_c &= \frac{\partial c}{\partial t} = -\frac{\partial c}{\partial \tau}\\ &= -\left[S\frac{1}{\sqrt{2\pi}}e^{-\frac{d_1^2}{2}}\,\frac{\partial d_1}{\partial \tau} + rXe^{-r\tau}N(d_2) - Xe^{-r\tau}\frac{1}{\sqrt{2\pi}}e^{-\frac{d_2^2}{2}}\,\frac{\partial d_2}{\partial \tau}\right]\\ &= -\frac{1}{\sqrt{2\pi}}\,\frac{Se^{-\frac{d_1^2}{2}}\sigma}{2\sqrt{\tau}} - rXe^{-r\tau}N(d_2) < 0, \end{aligned} \tag{3.3.26}$$

and

$$\begin{aligned}\Theta_p &= \frac{\partial p}{\partial t} = -\frac{\partial p}{\partial \tau} = -\frac{\partial c}{\partial \tau} + rXe^{-r\tau} \qquad \text{(from the put-call parity relation)}\\ &= -\frac{1}{\sqrt{2\pi}}\,\frac{Se^{-\frac{d_1^2}{2}}\sigma}{2\sqrt{\tau}} + rXe^{-r\tau}N(-d_2). \end{aligned} \tag{3.3.27}$$

In Sect. 1.2, we deduced that the longer-lived American options are worth more than their shorter-lived counterparts. Since an American call option on a nondividend paying asset will not be exercised early, the above property also holds for a European call option on a nondividend paying asset. The negativity of $\frac{\partial c}{\partial t}$ confirms the above observation. The theta has its greatest absolute value when the call option is at-the-money since the option may become in-the-money or out-of-the-money an instant later. Also, the theta has a small absolute value when the option is sufficiently out-of-the-money since it will be quite unlikely for the option to become in-the-money a later time. Further, the theta tends asymptotically to $-rXe^{-r\tau}$ at a sufficiently high value of S. The variation of the theta of the European call price with respect to the asset price is sketched in Fig. 3.5.

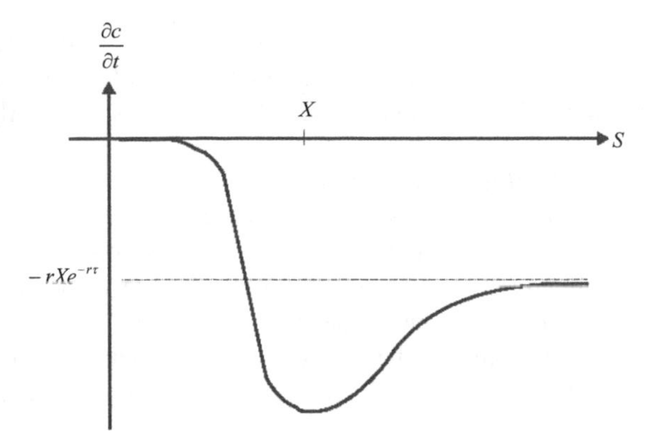

Fig. 3.5. Variation of the theta of the price of a European call option with respect to asset price S. The theta value tends asymptotically to $-rXe^{-r\tau}$ from below when the asset price is sufficiently high.

The sign of the theta of the price of a European put option may be positive or negative depending on the relative magnitude of the two terms with opposing signs in (3.3.27). When the put is deep in-the-money, S assumes a small value so that $N(-d_2)$ tends to one. In this case, the theta is positive since the second term is dominant over the first term. The positivity of theta is consistent with the observation that the European put price can be below the intrinsic value $X - S$ when S is sufficiently small, which then grows to $X - S$ at expiry. On the other hand, when the option is at-the-money or out-of-the-money, there will be a higher chance of a positive payoff for the put option as the time to expiry is lengthened. The European put price then becomes a decreasing function of time and so $\frac{\partial p}{\partial t} < 0$. For an American put option, the corresponding theta is always negative. This is because the longer-lived American put is always worth more than its shorter-lived counterpart.

Actually, the details of the sign behavior of the theta of a European put option can be quite complicated. Its full analysis is relegated to Problem 3.15.

Gamma—Second-Order Derivative with Respect to Asset Price

The *gamma* Γ of the value of a derivative security V is defined as the rate of change of the delta with respect to the asset price S, that is, $\Gamma = \frac{\partial^2 V}{\partial S^2}$. The gammas of the European call and put options are the same since their deltas differ by a constant. The gamma of the price of a European put/call option is given by

$$\Gamma_p = \Gamma_c = \frac{e^{-\frac{d_1^2}{2}}}{S\sigma\sqrt{2\pi\tau}} > 0. \tag{3.3.28}$$

The curve of Γ_c against S resembles a slightly skewed belt-shaped curve centered at $S = X$ above the S-axis. Since the gamma is always positive for any European call or put option, this explains why the option price curves plotted against asset price are always concave upward. From calculus, we know that gamma assumes a small value when the curvature of the option value curve is small. A small value of gamma implies that the delta changes slowly with the asset price and so portfolio rebalancing required to keep the portfolio delta neutral can be made less frequently.

Vega—Derivative with Respect to Volatility

In the Black–Scholes model, we assume the volatility of the underlying asset price process to be constant. In reality, the asset price volatility changes over time. Sometimes, we may be interested to see how the option price responds to changes in volatility value. The *vega* $\wedge$ of the value of a derivative security is defined to be the rate of change of the value of the derivative security with respect to asset price volatility. For the European vanilla call and put options, their vegas are found to be

$$\wedge_c = \frac{\partial c}{\partial \sigma} = SN'(d_1)\,\frac{\partial d_1}{\partial \sigma} - Xe^{-r\tau}N'(d_2)\,\frac{\partial d_2}{\partial \sigma} = \frac{S\sqrt{\tau}e^{-\frac{d_1^2}{2}}}{\sqrt{2\pi}} > 0, \tag{3.3.29}$$

and

$$\wedge_p = \frac{\partial p}{\partial \sigma} = \frac{\partial c}{\partial \sigma} + \frac{\partial}{\partial \sigma}(Xe^{-r\tau} - S) = \wedge_c. \tag{3.3.30}$$

The above results indicate that both the European call and put prices increase with increasing volatility. Since an increase in the asset price volatility will lead to a wider spread of the terminal asset price, there is a higher chance that the option may end up either deeper in-the-money or deeper out-of-the-money. However, there is no increase in penalty for the option to be deeper out-of-the-money but the payoff increases when the option expires deeper in-the-money. Due to this non-symmetry in the payoff pattern, the vega of any option is always positive.

Rho—Derivative with Respect to Interest Rate

A higher interest rate lowers the present value of the cost of exercising the European call option at expiration (the effect is similar to the lowering of the strike price), in turn this increases the call price. Reverse effect holds for the put price. The rho ρ of the value of a derivative security is defined to be the rate of change of the value of the derivative security with respect to the interest rate. The rhos of the European call and put prices are found to be

$$\begin{aligned}\rho_c = \frac{\partial c}{\partial r} &= SN'(d_1)\,\frac{\partial d_1}{\partial r} + \tau Xe^{-r\tau}N(d_2) - Xe^{-r\tau}N'(d_2)\,\frac{\partial d_2}{\partial r} \\ &= \tau Xe^{-r\tau}N(d_2) > 0 \quad \text{for } r > 0 \text{ and } X > 0,\end{aligned} \tag{3.3.31}$$

and

$$\begin{aligned}\rho_p = \frac{\partial p}{\partial r} &= \frac{\partial c}{\partial r} - \tau Xe^{-r\tau} \qquad \text{(from the put-call parity relation)} \\ &= -\tau Xe^{-r\tau}N(-d_2) < 0 \quad \text{for } r > 0 \text{ and } X > 0.\end{aligned} \tag{3.3.32}$$

The signs of ρ_c and ρ_p confirm the above claims on the impact of changing interest rate on the call and put prices.

3.4 Extended Option Pricing Models

In this section, we first show how to extend the original Black–Scholes formulation by relaxing some of the model's assumptions. It is a common practice that assets pay *dividends* either as discrete payments or continuous yield. We examine the modification of the governing differential equation and the price formulas with the inclusion of continuous dividend yield/discrete dividends. We also present the analytic techniques of solving the option pricing models with time-dependent parameters. We derive the price formulas of *futures options* where the underlying asset is a futures contract. We also consider the valuation of the *chooser option*, where the holder can choose whether the option is a call or a put after a specified period of time has lapsed from the starting date of the option contract. Other valuation models analyzed in this section include pricing models of the compound options and quanto options. A compound option is an option on an option while a quanto option is an option on a foreign asset, but the option's payoff is denominated in the domestic currency. In addition, we consider the structural approach of analyzing the credit risk structure of a risky debt using option pricing theory. The payment to the debt holders at maturity is contingent on the terminal value of the firm value process.

3.4.1 Options on a Dividend-Paying Asset

The dividends received by holding an underlying asset may be stochastic or deterministic. The modeling of stochastic dividends is more complicated since we have to assume the dividend to be another random state variable in addition to the asset price. Here, we assume dividends to be deterministic, possibly quite an acceptable assumption. How about the impact of dividends on the asset price? Using the principle of no arbitrage, the asset price falls right after an ex-dividend date by the same amount as the dividend payment (see Sect. 1.2.1).

Continuous Dividend Yield Models

First, we consider the effect of continuous dividend yield on the price of a European call option. Let q denote the constant continuous dividend yield paid by the underlying asset. That is, the holder receives a dividend of dollar amount $q S_t \, dt$ within a time interval dt, where S_t is the asset price. The asset price dynamics is assumed to follow the Geometric Brownian process

$$\frac{dS_t}{S_t} = \mu \, dt + \sigma \, dZ_t, \tag{3.4.1}$$

where μ and σ^2 are the expected rate of return and variance rate of the asset price, respectively. To derive the governing differential equation of the price function $c(S_t, t)$ of a European call option, we form a riskless hedging portfolio by short selling one unit of the European call and long holding Δ_t units of the underlying asset. The financial gain on the portfolio at time t is given by

$$\int_0^t -dc + \int_0^t \Delta_u\, dS_u + q \int_0^t \Delta_u S_u\, du$$
$$= \int_0^t \left\{ -\frac{\partial c}{\partial u} - \frac{\sigma^2}{2} S_u^2 \frac{\partial^2 c}{\partial S_u^2} + \left[\left(\Delta_u - \frac{\partial c}{\partial S_u} \right) \mu + q \Delta_u \right] S_u \right\} du$$
$$+ \int_0^t \left(\Delta_u - \frac{\partial c}{\partial S_u} \right) \sigma S_u\, dZ_u.$$

The last term $q \int_0^t \Delta_u S_u\, du$ represents the gain added to the portfolio due to the dividend payment received. By choosing $\Delta_u = \frac{\partial c}{\partial S_u}$, the stochastic term vanishes so that the financial gain becomes deterministic at all times. On the other hand, the deterministic gain from a riskless asset with a dynamic position of $-c + \Delta_u S_u = -c + \frac{\partial c}{\partial S_u} S_u$ is given by

$$\int_0^t r \left(-c + S_u \frac{\partial c}{\partial S_u} \right) du.$$

By invoking the no arbitrage argument, these two gains are equal. Equating the above two gains, we obtain

$$-\frac{\partial c}{\partial u} - \frac{\sigma^2}{2} S_u^2 \frac{\partial^2 c}{\partial S_u^2} + q S_u \frac{\partial c}{\partial S_u} = r \left(-c + S_u \frac{\partial c}{\partial S_u} \right), \quad 0 < u < t,$$

which is satisfied for any asset price S_u if $c(S, t)$ satisfies the following modified version of the Black–Scholes equation

$$\frac{\partial c}{\partial t} + \frac{\sigma^2}{2} S^2 \frac{\partial^2 c}{\partial S^2} + (r - q) S \frac{\partial c}{\partial S} - rc = 0. \tag{3.4.2}$$

The terminal payoff of the European call option on a continuous dividend paying asset is identical to that of the nondividend paying counterpart.

Risk Neutral Drift Rate

From the modified Black–Scholes equation (3.4.2), we deduce that the risk neutral drift rate of the price process of an asset paying dividend yield q is $r - q$. One can also show this result via the martingale pricing approach. Suppose all the dividend yields received are used to purchase additional units of the underlying asset, then the wealth process of holding one unit of the underlying asset initially is given by

$$\widehat{S}_t = e^{qt} S_t,$$

where e^{qt} represents the growth factor in the number of units. Suppose S_t follows the price dynamics as defined in (3.4.1), then the wealth process $\widehat{S}_t$ follows

$$\frac{d\widehat{S}_t}{\widehat{S}_t} = (\mu + q)\, dt + \sigma\, dZ_t.$$

The discounted wealth process $\widehat{S}_t^* = \widehat{S}_t / S_0(t)$ is a martingale under the equivalent risk neutral measure Q. It amounts to finding Q under which the discounted wealth process $\widehat{S}_t^*$ is a Q-martingale. We choose $\gamma(t)$ in the Radon–Nikodym derivative to be

$$\gamma(t) = \frac{\mu + q - r}{\sigma}$$

so that $\widehat{Z}_t$ is a Brownian process under Q and

$$d\widehat{Z}_t = dZ_t + \frac{\mu + q - r}{\sigma}\, dt.$$

Now, $\widehat{S}_t^*$ becomes a Q-martingale since

$$\frac{d\widehat{S}_t^*}{\widehat{S}_t^*} = \sigma\, d\widehat{Z}_t.$$

The asset price S_t under the equivalent risk neutral measure Q becomes

$$\frac{dS_t}{S_t} = (r - q)dt + \sigma\, d\widehat{Z}_t. \tag{3.4.3}$$

Hence, the risk neutral drift rate of S_t is deduced to be $r - q$.

Analogy with Foreign Currency Options

The continuous yield model is also applicable to *options on foreign currencies* where the continuous dividend yield can be considered as the yield due to the interest earned by the foreign currency at the foreign interest rate r_f. In the pricing model for a foreign currency call option, we can simply set $q = r_f$ in the modified Black–Scholes equation [see (3.4.2)]. This is consistent with the observation that the risk neutral drift rate of the exchange rate process under the domestic equivalent martingale measure Q_d is $r - r_f$ [see (3.2.20)].

Call and Put Price Formulas

If we set $\widehat{S} = Se^{-q\tau}$ in the modified Black–Scholes equation (3.4.2), $\tau = T - t$, then the equation becomes

$$\frac{\partial c}{\partial \tau} = \frac{\sigma^2}{2}\widehat{S}^2\frac{\partial^2 c}{\partial \widehat{S}^2} + r\widehat{S}\frac{\partial c}{\partial \widehat{S}} - rc.$$

The terminal payoff of a European call option in terms of $\widehat{S}$ is given by $\max(\widehat{S} - X, 0)$. Hence the price of a European call option on a continuous dividend paying asset can be obtained by a simple modification of the Black–Scholes call price formula (3.3.6) as follows: changing S to $Se^{-q\tau}$ in the price formula. Now, the European call price formula with continuous dividend yield q is found to be

$$c(S, \tau) = Se^{-q\tau}N(\widehat{d}_1) - Xe^{-r\tau}N(\widehat{d}_2), \tag{3.4.4}$$

where

$$\widehat{d}_1 = \frac{\ln\frac{S}{X} + (r - q + \frac{\sigma^2}{2})\tau}{\sigma\sqrt{\tau}}, \quad \widehat{d}_2 = \widehat{d}_1 - \sigma\sqrt{\tau}.$$

Similarly, the European put formula with continuous dividend yield q can be deduced from the Black–Scholes put price formula to be

$$p(S, \tau) = Xe^{-r\tau}N(-\widehat{d_2}) - Se^{-q\tau}N(-\widehat{d_1}). \tag{3.4.5}$$

Put-Call Parity and Symmetry Relations

Note that the new put and call prices satisfy the put-call parity relation [see (1.2.24)]

$$p(S, \tau) = c(S, \tau) - Se^{-q\tau} + Xe^{-r\tau}. \tag{3.4.6}$$

Furthermore, the following *put-call symmetry relation* can also be deduced from the above call and put price formulas

$$c(S, \tau; X, r, q) = p(X, \tau; S, q, r), \tag{3.4.7}$$

that is, the put price can be obtained from the corresponding call price by interchanging S with X and r with q in the formula.

To provide an intuitive argument behind the put-call symmetry relation, we recall that a call option entitles its holder the right to exchange the riskless asset for the risky asset, and vice versa for a put option. The dividend yield earned from the risky asset is q while that from the riskless asset is r. If we interchange the roles of the riskless asset and risky asset in a call option, the call becomes a put, thus giving the justification for the put-call symmetry relation.

The call and put price formulas of the foreign currency options mimic the above price formulas, where the dividend yield q is replaced by the foreign interest rate r_f (Garman and Kohlhagen, 1983). Accordingly, the asset price process S_t is replaced by the exchange rate process F_t, where F_t represents the time-t domestic currency price of one unit of the foreign currency.

Time-Dependent Parameters

So far, we have assumed constant value for the dividend yield, interest rate and volatility. When these parameters become deterministic functions of time, the Black–Scholes equation has to be modified as follows

$$\frac{\partial V}{\partial \tau} = \frac{\sigma^2(\tau)}{2}S^2\frac{\partial^2 V}{\partial S^2} + [r(\tau) - q(\tau)]\,S\,\frac{\partial V}{\partial S} - r(\tau)V, \quad 0 < S < \infty, \quad \tau > 0, \tag{3.4.8}$$

where V is the price of the derivative security. When we apply the following transformations:

$$y = \ln S \text{ and } w = e^{\int_0^\tau r(u)\,du}V,$$

(3.4.8) then becomes

$$\frac{\partial w}{\partial \tau} = \frac{\sigma^2(\tau)}{2}\frac{\partial^2 w}{\partial y^2} + \left[r(\tau) - q(\tau) - \frac{\sigma^2(\tau)}{2}\right]\frac{\partial w}{\partial y}, \quad -\infty < y < \infty, \quad \tau > 0. \tag{3.4.9}$$

Consider the following analytic fundamental solution

$$f(y,\tau) = \frac{1}{\sqrt{2\pi s(\tau)}} \exp\left(-\frac{[y+e(\tau)]^2}{2s(\tau)}\right), \tag{3.4.10}$$

it can be shown that $f(y,\tau)$ satisfies the following differential equation

$$\frac{\partial f}{\partial \tau} = \frac{1}{2}s'(\tau)\frac{\partial^2 f}{\partial y^2} + e'(\tau)\frac{\partial f}{\partial y}. \tag{3.4.11}$$

Suppose we let

$$s(\tau) = \int_0^\tau \sigma^2(u)\,du$$
$$e(\tau) = \int_0^\tau [r(u) - q(u)]\,du - \frac{s(\tau)}{2},$$

and comparing (3.4.9), (3.4.11), one can deduce that the fundamental solution of (3.4.9) is given by

$$\phi(y,\tau) = \frac{1}{\sqrt{2\pi\int_0^\tau \sigma^2(u)\,du}} \exp\left(-\frac{\{y + \int_0^\tau [r(u) - q(u) - \frac{\sigma^2(u)}{2}]\,du\}^2}{2\int_0^\tau \sigma^2(u)\,du}\right). \tag{3.4.12}$$

Given the initial condition, $w(y,0)$, the solution to (3.4.9) can be expressed as

$$w(y,\tau) = \int_{-\infty}^{\infty} w(\xi,0)\,\phi(y-\xi,\tau)\,d\xi. \tag{3.4.13}$$

Note that the time dependency of the coefficients $r(\tau)$, $q(\tau)$ and $\sigma^2(\tau)$ will not affect the spatial integration with respect to ξ. The result of integration will be similar in analytic form to that obtained for the constant coefficient models, except that we have to make the following respective substitution in the option price formulas

$$r \text{ is replaced by } \frac{1}{\tau}\int_0^\tau r(u)\,du$$
$$q \text{ is replaced by } \frac{1}{\tau}\int_0^\tau q(u)\,du$$
$$\sigma^2 \text{ is replaced by } \frac{1}{\tau}\int_0^\tau \sigma^2(u)\,du.$$

For example, the European call price formula is modified as follows:

$$c = Se^{-\int_0^\tau q(u)\,du} N(\widetilde{d_1}) - Xe^{-\int_0^\tau r(u)\,du} N(\widetilde{d_2}), \tag{3.4.14}$$

where

$$\widetilde{d}_1 = \frac{\ln \frac{S}{X} + \int_0^\tau [r(u) - q(u) + \frac{\sigma^2(u)}{2}]\, du}{\sqrt{\int_0^\tau \sigma^2(u)\, du}}, \qquad \widetilde{d}_2 = \widetilde{d}_1 - \sqrt{\int_0^\tau \sigma^2(u)\, du},$$

when the option model has time-dependent parameters. The European put price formula can be deduced in a similar manner. In conclusion, the Black–Scholes call and put formulas are also applicable to models with time-dependent parameters except that the interest rate r, the dividend yield q and the variance rate σ^2 in the Black–Scholes formulas are replaced by the corresponding average value of the instantaneous interest rate, dividend yield and variance rate over the remaining life of the option.

Discrete Dividends

Suppose the underlying asset pays N discrete dividends at known payment times $t_1, t_2, \cdots, t_N$ of dollar amount $D_1, D_2, \cdots, D_N$, respectively. Taking the usual assumption for valuation of options with known discrete dividends, the asset price is taken to consist of two components: a riskless component that will be used to pay the known dividends during the remaining life of the option and a risky component which follows the Geometric Brownian process. The riskless component at a given time is taken to be the present value of all future dividends discounted from the ex-dividend dates to the present at the riskless interest rate. One can then apply the Black–Scholes formulas by setting the asset price equal the risky component and letting the volatility parameter be the volatility of the stochastic process followed by the risky component (which differs slightly from that followed by the whole asset price). The value of the risky component $\widetilde{S}_t$ is taken to be

$$\begin{aligned} \widetilde{S}_t &= S_t - D_1 e^{-r\tau_1} - D_2 e^{-r\tau_2} - \cdots - D_N e^{-r\tau_N} && \text{for } t < t_1 \\ \widetilde{S}_t &= S_t - D_2 e^{-r\tau_2} - \cdots - D_N e^{-r\tau_N} && \text{for } t_1 < t < t_2 \\ \vdots\ & \ \ \vdots \\ \widetilde{S}_t &= S_t && \text{for } t > t_N, \end{aligned} \tag{3.4.15}$$

where S_t is the current asset price, $\tau_i = t_i - t,\ i = 1, 2, \cdots, N$. It is customary to take the volatility of the risky component to be approximately given by the volatility of the whole asset price multiplied by the factor $\frac{S_t}{S_t - D}$, where D is the present value of the lumped future discrete dividends.

Note that the asset price may not fall by the same amount as the whole dividend due to tax and other considerations. In the above discussion, the "dividend" may be broadly interpreted as the decline in the asset price on the ex-dividend date caused by the dividend, rather than the actual amount of dividend payment.

3.4.2 Futures Options

The underlying asset in a futures option is a futures contract. When a futures call option is exercised, the holder acquires from the option writer a long position in the underlying futures contract plus a cash inflow equal to the excess of the spot futures price over the strike price. Since the newly opened futures contract has zero value, the value of the futures option upon exercise is equal to the above cash inflow. For example, suppose the strike price of an October futures call option on 10,000 ounces of gold is \$340 per ounce. On the expiration date of the option (say, August 15), the spot gold futures price is \$350 per ounce. The holder of the futures call option then receives \$100,000 = 10,000 × (\$350 − \$340), plus a long position in a futures contract to buy 10,000 ounces of gold on the October delivery date. The position of the futures contract can be immediately closed out at no cost, if the option holder chooses. The maturity dates of the futures option and the underlying futures may or may not coincide. Note that the maturity date of the futures should not be earlier than that of the option.

The trading of futures options is more popular than the trading of options on the underlying asset since futures contracts are more liquid and easier to trade than the underlying asset. Futures and futures options are often traded in the same exchange. Most futures options are settled in cash without the delivery of the underlying futures. For most commodities and bonds, the futures price is readily available from trading in the futures exchange whereas the spot price of the commodity or bond may have to be obtained through a dealer.

We would like to derive the governing differential equation for the value of a futures option based on the Black–Scholes–Merton formulation. The interest rate is assumed to be *constant* and the price dynamics of the underlying asset is assumed to follow the Geometric Brownian process. Under constant interest rate, the futures price is given by a deterministic time function times the asset price, so the volatility of the futures price should be the same as that of the underlying asset price. We write the dynamics of the futures price f_t as

$$\frac{df_t}{f_t} = \mu_f \, dt + \sigma \, dZ_t, \tag{3.4.16}$$

where μ_f is the expected rate of return of the futures and σ is the constant volatility of the asset price process. Let $V(f_t, t)$ denote the value of the futures option. Now, we consider a portfolio that contains α_t units of the futures in the long position and one unit of the futures option in the short position, where α_t is adjusted dynamically so as to create an instantaneously hedged (riskless) position of the portfolio at all times. The value of the portfolio Π is given by

$$\Pi(f_t, t) = -V(f_t, t),$$

since there is no cost incurred to enter into a futures contract. Be cautious that the portfolio also gains in value from the long position of the futures of net amount as given by $\int_0^t \alpha_u \, df_u$. Using Ito's lemma, we obtain

$$dV(f_t, t) = \left(\frac{\partial V}{\partial t} + \frac{\sigma^2}{2} f_t^2 \frac{\partial^2 V}{\partial f_t^2} + \mu_f f_t \frac{\partial V}{\partial f_t} \right) dt + \sigma f_t \frac{\partial V}{\partial f_t} \, dZ_t.$$

The financial gain on the portfolio at time t is given by

$$\int_0^t -dV(f_u, u)\,du + \int_0^t \alpha_u\,df_u$$
$$= \int_0^t \left[-\frac{\partial V}{\partial u} - \frac{\sigma^2}{2} f_u^2 \frac{\partial^2 V}{\partial f_u^2} + \left(\alpha_u - \frac{\partial V}{\partial f_u}\right)\mu_f f_u\right] du + \left(\alpha_u - \frac{\partial V}{\partial f_u}\right)\sigma f_u\,dZ_u.$$

The number of units of futures held at any time u is dynamically rebalanced so that the financial gain from the portfolio becomes deterministic at all times. This is achieved by the following judicious choice

$$\alpha_u = \frac{\partial V}{\partial f_u}.$$

In this case, the deterministic financial gain from this dynamically hedged portfolio becomes

$$\int_0^t \left(-\frac{\partial V}{\partial u} - \frac{\sigma^2}{2} f_u^2 \frac{\partial^2 V}{\partial f_u^2}\right) du.$$

To avoid arbitrage, the above deterministic gain should be the same as the gain from a risk free asset with a dynamic position of value equals $-V$. These two deterministic gains are equal at all times provided that $V(f, t)$ satisfies

$$\frac{\partial V}{\partial t} + \frac{\sigma^2}{2} f^2 \frac{\partial^2 V}{\partial f^2} - rV = 0. \tag{3.4.17}$$

When we compare (3.4.17) with the corresponding governing equation for the value of a European option on an asset that pays continuous dividend yield at the rate q, (3.4.17) can be obtained by setting $q = r$ in (3.4.2). Recall that the expected rate of growth of the continuous dividend paying asset under the risk neutral measure is $r - q$, so "$q = r$" apparently implies zero drift rate of the futures price process. Under the risk neutral measure Q, f_t is a martingale. Alternatively, using the relation: $f_t = e^{r(T-t)} S_t$ and knowing that the risk neutral drift rate of S_t is r, one can show using Ito's lemma that the risk neutral drift rate of f_t is zero.

The above observation enables us to obtain the prices of European futures call and put options by simply substituting $q = r$ in the price formulas of the corresponding call and put options on a continuous dividend paying asset. It then follows that the prices of European futures call option and put option are, respectively, given by (Black, 1976)

$$c(f, \tau; X) = e^{-r\tau}\,[fN(\widetilde{d_1}) - XN(\widetilde{d_2})] \tag{3.4.18}$$

and

$$p(f, \tau; X) = e^{-r\tau}\,[XN(-\widetilde{d_2}) - fN(-\widetilde{d_1})], \tag{3.4.19}$$

where

$$\widetilde{d_1} = \frac{\ln \frac{f}{X} + \frac{\sigma^2}{2}\tau}{\sigma\sqrt{\tau}}, \qquad \widetilde{d_2} = \widetilde{d_1} - \sigma\sqrt{\tau}, \quad \tau = T - t,$$

f and X are the current futures price and option's strike price, respectively, and τ is the time to expiry. The corresponding put-call parity relation is

$$p(f,\tau;X) + fe^{-r\tau} = c(f,\tau;X) + Xe^{-r\tau}. \tag{3.4.20}$$

Since the futures price of any asset is the same as its spot price at maturity of the futures, a European futures option must be worth the same as the corresponding European option on the underlying asset if the option and the futures contract are set to have the same date of maturity. Recall that τ is the time to expiry of the futures option. When the futures option and its underlying futures contract are set to have the same maturity, the futures price is equal to $f = Se^{r\tau}$. If we substitute $f = Se^{r\tau}$ into (3.4.18)–(3.4.19), then the resulting price formulas become the usual Black–Scholes price formulas for European vanilla options.

Under the assumption of constant interest rate, the price formulas of a futures option and its forward option counterpart are identical since the futures price and forward price are equal. When the interest rates become stochastic, the forward price and futures price differ due to intermediate payments under the mark-to-market mechanism of a futures contract. The price dynamics of futures and forward and pricing models of derivatives on futures and forward in a stochastic interest rate environment are examined in detail in Sect. 8.1.

3.4.3 Chooser Options

A standard *chooser option* (or called *as-you-like-it option*) entitles the holder to choose, at a predetermined time T_c in the future, whether the T-maturity option is a standard European call or put with a common strike price X for the remaining time to expiration $T - T_c$. The payoff of the chooser option on the date of choice T_c is

$$V(S_{T_c}, T_c) = \max(c(S_{T_c}, T - T_c; X), p(S_{T_c}, T - T_c; X)), \tag{3.4.21}$$

where $T - T_c$ is the time to expiry in both call and put price formulas above, and S_{T_c} is the asset price at time T_c. For notational convenience, we take the current time to be zero. Suppose the underlying asset pays a continuous dividend yield at the rate q. By the put-call parity relation, the above payoff function can be expressed as

$$\begin{aligned} V(S_{T_c}, T_c) &= \max(c, c + Xe^{-r(T-T_c)} - S_{T_c}e^{-q(T-T_c)}) \\ &= c + e^{-q(T-T_c)}\max(0, Xe^{-(r-q)(T-T_c)} - S_{T_c}). \end{aligned} \tag{3.4.22}$$

Hence, the chooser option is equivalent to the combination of one call with exercise price X and time to expiration T and $e^{-q(T-T_c)}$ units of put with strike price $Xe^{-(r-q)(T-T_c)}$ and time to expiration T_c. Applying the Black–Scholes–Merton pricing approach, the value of the standard chooser option at the current time is found to be (Rubinstein, 1992)

$$\begin{aligned} V(S,0) &= Se^{-qT}N(x) - Xe^{-rT}N(x - \sigma\sqrt{T}) + e^{-q(T-T_c)} \\ &\quad [Xe^{-(r-q)(T-T_c)}e^{-rT_c}N(-y + \sigma\sqrt{T_c}) - Se^{-qT_c}N(-y)] \\ &= Se^{-qT}N(x) - Xe^{-rT}N(x - \sigma\sqrt{T}) \\ &\quad + Xe^{-rT}N(-y + \sigma\sqrt{T_c}) - Se^{-qT}N(-y), \end{aligned} \tag{3.4.23}$$

where S is the current asset price and

$$x = \frac{\ln \frac{S}{X} + (r - q + \frac{\sigma^2}{2})T}{\sigma\sqrt{T}}, \quad y = \frac{\ln \frac{S}{X} + (r - q)T + \frac{\sigma^2}{2}T_c}{\sigma\sqrt{T_c}}.$$

The pricing models of more exotic payoff structures of chooser options are considered in Problems 3.26–3.27.

3.4.4 Compound Options

A compound option is simply *an option on an option*. There are four main types of compound options, namely, a call on a call, a call on a put, a put on a call and a put on a put. A compound option has two strike prices and two expiration dates. As an illustration, we consider a call on a call where both calls are European-style. On the first expiration date T_1, the holder of the compound option has the right to buy the underlying call option by paying the first strike price X_1. The underlying call option again gives the right to the holder to buy the underlying asset by paying the second strike price X_2 on a later expiration date T_2. Let $c(S, t)$ denote the value of the compound call-on-a-call option, where S is the asset price at current time t. The value of the underlying call option at the first expiration time T_1 is denoted by $\widetilde{c}(S_{T_1}, T_1)$, where S_{T_1} is the asset price at time T_1. Note that the compound option will be exercised at T_1 only when $\widetilde{c}(S_{T_1}, T_1) > X_1$.

Assume the usual Black–Scholes–Merton pricing framework, we would like to derive the analytic price formula of a European call-on-a-call option. First, the value of the underlying call option at time T_1 is given by the Black–Scholes call formula

$$\widetilde{c}(S_{T_1}, T_1) = S_{T_1} N(d_1) - X_2 e^{-r(T_2 - T_1)} N(d_2), \tag{3.4.24}$$

where

$$d_1 = \frac{\ln \frac{S_{T_1}}{X_2} + \left(r + \frac{\sigma^2}{2}\right)(T_2 - T_1)}{\sigma\sqrt{T_2 - T_1}}, \quad d_2 = d_1 - \sigma\sqrt{T_2 - T_1}.$$

Let $\widetilde{S}_{T_1}$ denote the critical value for S_{T_1}, above which the compound option should be exercised at T_1. The value of $\widetilde{S}_{T_1}$ is obtained by solving the following nonlinear algebraic equation

$$\widetilde{c}(\widetilde{S}_{T_1}, T_1) = X_1. \tag{3.4.25}$$

The payoff function of the compound call-on-a-call option at time T_1 is

$$c(S_{T_1}, T_1) = \max(\widetilde{c}(S_{T_1}, T_1) - X_1, 0). \tag{3.4.26}$$

The value of the compound option for $t < T_1$ is given by the following risk neutral valuation calculation, where

$$\begin{aligned} c(S, t) &= e^{-r(T_1 - t)} E_Q[\max(\widetilde{c}(S_{T_1}, T_1) - X_1, 0] \\ &= e^{-r(T_1 - t)} \int_0^\infty \max(\widetilde{c}(S_{T_1}, T_1) - X_1, 0)\, \psi(S_{T_1}; S)\, dS_{T_1} \end{aligned}$$

$$= e^{-r(T_1-t)} \int_{\widetilde{S}_{T_1}}^{\infty} [S_{T_1} N(d_1) - X_2 e^{-r(T_2-T_1)} N(d_2) - X_1] \psi(S_{T_1}; S)\, dS_{T_1}. \tag{3.4.27}$$

Here, Q denotes the risk neutral measure and the transition density function $\psi(S_{T_1}; S)$ is given by (3.3.10). The last term in (3.4.27) is easily recognized as

$$\text{3rd term} = -X_1 e^{-r(T_1-t)} E_Q[\mathbf{1}_{\{S_{T_1} \geq \widetilde{S}_{T_1}\}}] = -X_1 e^{-r(T_1-t)} N(a_2),$$

where

$$a_2 = \frac{\ln \frac{S}{\widetilde{S}_{T_1}} + (r - \frac{\sigma^2}{2})(T_1 - t)}{\sigma\sqrt{T_1 - t}}.$$

Here, $X_1 e^{-r(T_1-t)} N(a_2)$ represents the present value of the expected payment at time T_1 conditional on the first call being exercised.

If we define the random variables Y_1 and Y_2 to be the logarithm of the price ratios $\frac{S_{T_1}}{S}$ and $\frac{S_{T_2}}{S}$, respectively, then Y_1 and Y_2 are Brownian increments over the overlapping intervals $[t, T_1]$ and $[t, T_2]$. The second term in (3.4.27) can be expressed as

$$\begin{aligned} \text{2nd term} &= -X_2 e^{-r(T_2-t)} E_Q[\mathbf{1}_{\{S_{T_1} \geq \widetilde{S}_{T_1}\}} \mathbf{1}_{\{S_{T_2} \geq X_2\}}] \\ &= -X_2 e^{-r(T_2-t)} Q\left(Y_1 \geq \ln \frac{\widetilde{S}_{T_1}}{S}, Y_2 \geq \ln \frac{X_2}{S}\right). \end{aligned}$$

To evaluate the above probability, it is necessary to find the joint density function of Y_1 and Y_2. The correlation coefficient between Y_1 and Y_2 is found to be [see (2.3.16)]

$$\rho = \sqrt{\frac{T_1 - t}{T_2 - t}}. \tag{3.4.28}$$

The Brownian increments, Y_1 and Y_2, are bivariate normally distributed. Their respective mean are $(r - \frac{\sigma^2}{2})(T_1 - t)$ and $(r - \frac{\sigma^2}{2})(T_2 - t)$, the respective variance are $\sigma^2(T_1 - t)$ and $\sigma^2(T_2 - t)$, while the correlation coefficient ρ is given by (3.4.28). Suppose we define the standard normal random variables Y_1' and Y_2' by

$$Y_1' = \frac{Y_1 - \left(r - \frac{\sigma^2}{2}\right)(T_1 - t)}{\sigma\sqrt{T_1 - t}} \quad \text{and} \quad Y_2' = \frac{Y_2 - \left(r - \frac{\sigma^2}{2}\right)(T_2 - t)}{\sigma\sqrt{T_2 - t}},$$

and let

$$b_2 = \frac{\ln \frac{S}{X_2} + \left(r - \frac{\sigma^2}{2}\right)(T_2 - t)}{\sigma\sqrt{T_2 - t}},$$

then we can express the second term in the following form

$$\begin{aligned}
\text{2nd term} &= -X_2 e^{-r(T_2-t)} \int_{-a_2}^{\infty}\int_{-b_2}^{\infty} \frac{1}{2\pi}\frac{1}{\sqrt{1-\rho^2}} \\
&\qquad \exp\left(-\frac{y_1'^2 - 2\rho y_1' y_2' + y_2'^2}{2(1-\rho^2)}\right) dy_2' dy_1' \\
&= -X_2 e^{-r(T_2-t)} N_2(a_2, b_2; \rho),
\end{aligned}$$

where $N_2(a_2, b_2; \rho)$ is the standard bivariate normal distribution function with correlation coefficient ρ. Note that $N_2(a_2, b_2; \rho)$ can be interpreted as the probability that $S_{T_1} > \widetilde{S}_{T_1}$ at time T_1 and $S_{T_2} > X_2$ at time T_2, given that the asset price at time t equals S. Hence, $X_2 e^{-r(T_2-t)} N_2(a_2, b_2; \rho)$ represents the present value of expected payment made at time T_2 conditional on both calls being exercised at their respective expiration dates.

Consider the first term in (3.4.27):

$$\text{1st term} = e^{-r(T_1-t)} \int_{\widetilde{S}_{T_1}}^{\infty} S_{T_1} N(d_1) \psi(S_{T_1}; S)\, dS_{T_1},$$

by following the analytic procedures outlined in Problem 3.28, we can show that

$$\text{1st term} = S N_2(a_1, b_1; \rho),$$

where

$$a_1 = a_2 + \sigma\sqrt{T_1 - t} \quad \text{and} \quad b_1 = b_2 + \sigma\sqrt{T_2 - t}.$$

Combining the above results, the price of a European call-on-a-call compound option is found to be

$$c(S,t) = S N_2(a_1, b_1; \rho) - X_2 e^{-r(T_2-t)} N_2(a_2, b_2; \rho) - X_1 e^{-r(T_1-t)} N(a_2), \tag{3.4.29}$$

where the critical asset value $\widetilde{S}_{T_1}$ contained in a_2 is obtained by solving (3.4.25).

In a similar manner, the price of the European put-on-a-put is given by

$$\begin{aligned}
p(S,t) &= e^{-r(T_1-t)} \int_0^{\widetilde{S}_{T_1}} \{X_1 - [X_2 e^{-r(T_2-T_1)} N(-d_2) - S_{T_1} N(-d_1)]\} \\
&\qquad \psi(S_{T_1}; S)\, dS_{T_1} \\
&= X_1 e^{-r(T_1-t)} N(-a_2) - X_2 e^{-r(T_2-t)} N_2(-a_2, -b_2; \rho) \\
&\qquad + S N_2(-a_1, -b_1; \rho).
\end{aligned} \tag{3.4.30}$$

Here, $\widetilde{S}_{T_1}$ is the critical value for S_{T_1} below which the first put is exercised at T_1.

The compound option models were first used by Geske (1979) to find the value of an option on a firm's stock, where the firm is assumed to be defaultable. The firm's liabilities consist of claims to future cash flows by the bondholders and the stock holders. When the firm defaults, the holders of the common stock have the right but not the obligation to sell the entire firm to the bondholders for a strike price equal to the par value of the bond. Therefore, an option on a share of the common stock can be considered as a compound option since the stock received upon exercising the option can be visualized as an option on the firm value.

3.4.5 Merton's Model of Risky Debts

In their seminar papers, Black and Scholes (1973) and Merton (1974) introduce the contingent claims approach to valuing risky corporate debt using option pricing theory. In their approach, default is assumed to occur when the market value of the issuer's firm asset has fallen to a low level such that the issuer cannot meet the par payment at maturity. The issuer is essentially granted an option to default on its debt. When the value of the firm asset is less than the total debt, the debt holders can receive only the value of the firm. In the literature, the approach that uses the firm value as the fundamental state variable determining default is termed the *structural approach* or *firm value approach*. To analyze the credit risk structure of a risky debt using the structural approach, it is necessary to characterize the issuer's firm value process together with the information on the capital structure of the firm.

The Merton model of risky debts starts from the assumption that the value of the assets owned by the debt issuer's firm A_t evolves according to the Geometric Brownian process:

$$\frac{dA_t}{A_t} = \mu_A\, dt + \sigma\, dZ_t, \tag{3.4.31}$$

where μ_A is the instantaneous expected rate of return and σ is the volatility of the firm asset value process. We assume a simple capital structure of the firm, where the liabilities of the firm consist only of a single debt with face value F. The debt has zero coupon and no embedded option features. Merton views the debt as a contingent claim on the assets of the firm. At maturity of the debt, the payment to the debt holders will be the minimum of the face value F and the firm value at maturity A_T. Default can be triggered only at maturity and this occurs when $A_T < F$, that is, the firm asset value cannot meet its debt claim. It is then assumed that the firm is liquidated at zero cost and all the proceeds from liquidation are transferred to the debt holders. The terminal payoff to the debt holders can be expressed as

$$\min(A_T, F) = F - \max(F - A_T, 0), \tag{3.4.32}$$

where the last term can be visualized as a put payoff. The debt holders have essentially sold a put option to the issuer since the issuer has the right to put the firm asset at the price of the par F.

Let A denote the firm asset value at current time, $\tau = T - t$ is the time to expiry and we view the value of the risky debt $V(A, \tau)$ as a contingent claim on the firm asset value. By invoking the standard assumption of continuous time no arbitrage pricing framework (continuous trading and short selling of the firm assets, perfectly divisible assets, no borrowing-lending spread, etc.), we obtain the usual Black–Scholes pricing equation:

$$\frac{\partial V}{\partial \tau} = \frac{\sigma^2}{2} A^2 \frac{\partial^2 V}{\partial A^2} + rA\frac{\partial V}{\partial A} - rA, \tag{3.4.33}$$

where r is the riskless interest rate. The terminal payoff defined in (3.4.32) becomes the "initial" condition at $\tau = 0$:

$$V(A,0) = F - \max(F - A, 0).$$

By linearity of the Black–Scholes equation, $V(A,\tau)$ can be decomposed into

$$V(A,\tau) = Fe^{-r\tau} - p(A,\tau), \quad \tau = T - t, \tag{3.4.34}$$

where $p(A,\tau)$ is the price function of a European put option. The put price function takes the form

$$p(A,\tau) = Fe^{-r\tau}N(-d_2) - AN(-d_1), \tag{3.4.35}$$

$$d_1 = \frac{\ln\frac{A}{F} + \left(r + \frac{\sigma^2}{2}\right)\tau}{\sigma\sqrt{\tau}}, \quad d_2 = d_1 - \sigma\sqrt{\tau}.$$

The value of the risky debt $V(A,\tau)$ is seen to be the value of the default free debt $Fe^{-r\tau}$ less the present value of expected loss to the debt holders. In this model, the present value of the expected loss is simply the value of the put option granted to the issuer.

The equity value $E(A,\tau)$ (or shareholders' stake) is the firm value less the debt liability. By virtue of the put-call parity relation, we have

$$\begin{aligned} E(A,\tau) &= A - V(A,\tau) \\ &= A - [Fe^{-r\tau} - p(A,\tau)] = c(A,\tau), \end{aligned} \tag{3.4.36}$$

where $c(A,\tau)$ is the price function of the European call. This is not surprising since the shareholders have the call payoff at maturity equals $\max(A_T - F, 0)$.

Term Structure of Credit Spreads

The yield to maturity $Y(\tau)$ of the risky debt is defined as the rate of return of the debt, where

$$V(A,\tau) = Fe^{-Y(\tau)\tau}.$$

Rearranging the terms, we have

$$Y(\tau) = -\frac{1}{\tau}\ln\frac{V(A,\tau)}{F}. \tag{3.4.37}$$

The credit spread is the difference between the yields of risky and default free zero-coupon debts. This represents the risk premium demanded by the debt holders to compensate for the potential risk of default. Under the assumption of constant risk free interest rate, the credit spread is found to be

$$Y(\tau) - r = -\frac{1}{\tau}\ln\left(N(d_2) + \frac{1}{d}N(d_1)\right), \tag{3.4.38}$$

where

$$d = \frac{Fe^{-r\tau}}{A}, \quad d_1 = \frac{\ln d}{\sigma\sqrt{\tau}} - \frac{\sigma\sqrt{\tau}}{2} \quad \text{and} \quad d_2 = -\frac{\ln d}{\sigma\sqrt{\tau}} - \frac{\sigma\sqrt{\tau}}{2}.$$

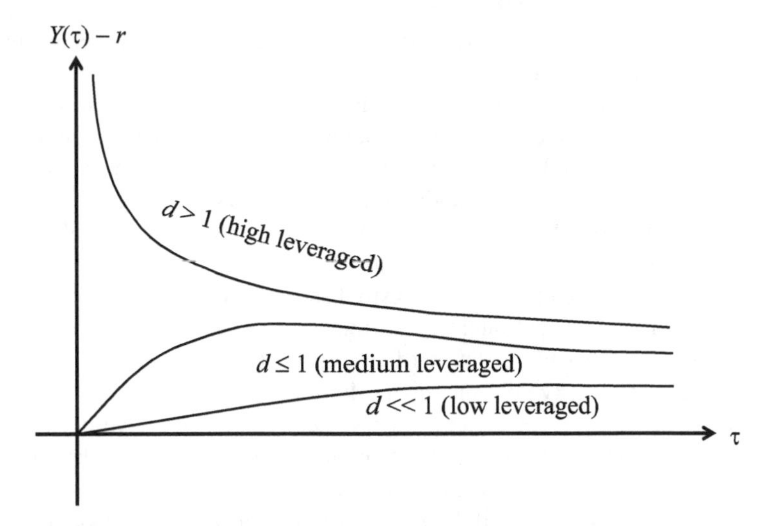

Fig. 3.6. The term structure of the credit spread as predicted by the Merton risky debt model. As the time approaches maturity, the credit spread always tends to zero when $d \leq 1$ but tends toward infinity when $d > 1$.

The quantity d is the ratio of the default free debt $Fe^{-r\tau}$ to the firm asset value A, thus it is coined the term "quasi" debt-to-firm ratio. The adjective "quasi" is added since all valuations are performed under the risk neutral measure instead of the "physical" measure.

The term structure of the credit spread, $Y(\tau)-r$, is seen to be a function of d and $\sigma^2\tau$. It can be seen readily that $Y(\tau)-r$ is an increasing function of σ^2. However, the time-dependent behavior of the credit spread depends on whether $d > 1$ or $d \leq 1$ (see Fig. 3.6). When $d > 1$, $Y(\tau) - r$ tends to infinity as $\tau \to 0^+$, which is a manifestation of the "sure" event of default. On the other hand, when $d \leq 1$, the credit spread always tends to zero as time approaches maturity. For low leveraged firms (corresponds to $d \ll 1$), the credit spread increases monotonically with τ; while for medium leveraged firms ($d \leq 1$ but not too small), the credit spread curve exhibits the humped shape. At times far from maturity, the credit spread appears to be small for highly or medium leveraged firms since a sufficient amount of time is allowed for the firm asset value to have a higher chance to grow beyond the promised claim F at maturity. On the other hand, the credit spread for a low leveraged (highly rated) firm increases monotonically with τ (though at a low rate) due to the potential downside move of the firm value below F given the life of the debt to be sufficiently long.

A strong criticism of Merton's risky debt model points to the fact that default can never occur by surprise under the model. This is because the firm value is assumed to follow a diffusion process and it takes finite time for the firm value to migrate to a level below the defaulting threshold. In order to capture the short maturity credit spreads observed in the market even for high quality bonds, Kijima and Suzuki (2001) introduced jump effects into the firm value process. Their jump-diffusion models reflect the more realistic scenario where a default can also occur due to an unexpected sudden drop in the firm value.

Though the option approach of analyzing default risk of debts has an elegant theoretical appeal, empirical studies show that the actual spreads are larger than those predicted by Merton's model even a high firm value volatility has been chosen. This reflects the fact that the conditions under which default will be triggered are far more complex than those conditions assumed by Merton's model. A more realistic default model should include (i) inter-temporal default where financial distress can occur throughout the whole life of the debt; (ii) stochastic interest rates and the correlation between credit migration and interest rate uncertainty; and (iii) multiple classes of seniority claims and possible violation of strict priority rule.

A wide variety of risky debt models with more refined modeling of the default mechanisms, recovery processes and interest rate fluctuations have been proposed in the literature. Among them, the most popular structural risky debt models are the Black–Cox (1976) model that examine the effects of bond indenture provisions (under constant interest rate), and the Longstaff–Schwartz (1995) model that allows for stochastic interest rate, inter-temporal default and flexibility of settlement rules upon default.

3.4.6 Exchange Options

An exchange option is an option that gives the holder the right but not the obligation to exchange one risky asset for another. Let X_t and Y_t be the price processes of the two assets. The terminal payoff of a European exchange option at maturity T of exchanging Y_T for X_T is given by $\max(X_T - Y_T, 0)$. Assuming that both X_t and Y_t follow the Geometric Brownian processes, the analytic price formula of an exchange option can be derived as a variant of the Black–Scholes formula (Margrabe, 1978). We would like to derive the price function of an exchange option using the change of numeraire technique (see Problem 3.34 for the alternative partial differential equation approach of deriving the price formula). Under the risk neutral measure Q, let X_t and Y_t be governed by

$$\frac{dX_t}{X_t} = r\,dt + \sigma_X\,dZ_t^X \quad \text{and} \quad \frac{dY_t}{Y_t} = r\,dt + \sigma_Y dZ_t^Y, \tag{3.4.39}$$

where r is the constant riskless interest rate, σ_X and σ_Y are the constant volatility of X_t and Y_t, respectively. Also, $dZ_t^X\,dZ_t^Y = \rho\,dt$, where ρ is correlation coefficient. Here, the two risky assets are assumed to be nondividend paying. Suppose X_t is used as the numeraire, we define the equivalent probability measure Q_X on $\mathcal{F}_T$ by [see (3.2.11)]

$$\frac{dQ_X}{dQ} = e^{-rT}\frac{X_T}{X_0}. \tag{3.4.40}$$

By the risk neutral valuation principle, the price function $V(X, Y, \tau)$ of the exchange option conditional on $X_t = X$ and $Y_t = Y$ is given by

$$V(X, Y, \tau) = e^{-r\tau} E_Q[\max(X_T - Y_T, 0)]$$
$$= e^{-r\tau} E_Q\left[X_T\left(1 - \frac{Y_T}{X_T}\right)\mathbf{1}_{\{Y_T/X_T<1\}}\right], \quad \tau = T - t. \quad (3.4.41)$$

Writing $W_t = Y_t/X_t$ and taking X_t as the numeraire, then by virtue of (3.2.9), we obtain

$$V(X, Y, \tau) = X E_{Q_X}\left[(1 - W_T)\mathbf{1}_{\{W_T<1\}}\right]. \quad (3.4.42)$$

The above expectation representation resembles that of a put option on the underlying asset $W_t = Y_t/X_t$ and with unit strike.

In the next step, we would like to derive the dynamics of $W_t = Y_t/X_t$ under Q_X. Note that (see Problem 2.38)

$$\frac{dW_t}{W_t} = \frac{d(Y_t/X_t)}{Y_t/X_t} = (-\rho\sigma_X\sigma_Y + \sigma_X^2)\,dt + \sigma_Y\,dZ_t^Y - \sigma_X\,dZ_t^X.$$

By the Girsanov Theorem, $\widetilde{Z}_t^X$ and $\widetilde{Z}_t^Y$ as defined by

$$d\widetilde{Z}_t^X = dZ_t^X - \sigma_X\,dt \quad \text{and} \quad d\widetilde{Z}_t^Y = dZ_t^Y - \rho\sigma_X\,dt$$

are Brownian processes under Q_X (see Problem 3.10). Combining the above relations, we obtain

$$\frac{dW_t}{W_t} = \sigma_Y\,d\widetilde{Z}_t^Y - \sigma_X\,d\widetilde{Z}_t^X.$$

Under Q_X, W_t is seen to be a Geometric Brownian motion with zero drift rate and whose volatility $\sigma_{Y/X}$ is given by

$$\sigma_{Y/X}^2 = \sigma_X^2 - 2\rho\sigma_X\sigma_Y + \sigma_Y^2. \quad (3.4.43)$$

Using the put price formula, the price of the exchange option is then given by

$$V(X, Y, \tau) = XN(d_X) - YN(d_Y), \quad (3.4.44)$$

where

$$d_X = \frac{\ln\frac{X}{Y} + \frac{\sigma_{Y/X}^2\tau}{2}}{\sigma_{Y/X}\sqrt{\tau}}, \quad d_Y = d_X - \sigma_{Y/X}\sqrt{\tau}, \quad \tau = T - t.$$

The extension of the exchange option price formula to dividend yield paying assets and time-dependent parameter functions are quite straightforward. Under the risk neutral measure Q, suppose X_t and Y_t are governed by

$$\frac{dX_t}{X_t} = \left[r(t) - q_X(t)\right]dt + \sigma_X(t)\,dZ_t^X \quad (3.4.45a)$$
$$\frac{dY_t}{Y_t} = \left[r(t) - q_Y(t)\right]dt + \sigma_Y(t)\,dZ_t^Y, \quad (3.4.45b)$$

where $q_X(t)$ and $q_Y(t)$ are the time-dependent dividend yield of X_t and Y_t, respectively. Writing $\delta_X(t)$ and $\delta_Y(t)$ as the drift rate of X_t and Y_t, respectively, the exchange option price formula becomes

$$V(X,Y,t) = e^{-\int_t^T r(u)\,du}\left[e^{\int_t^T \delta_X(u)\,du} XN(\widehat{d}_X) - e^{\int_t^T \delta_Y(u)\,du} YN(\widehat{d}_Y)\right]$$
$$= e^{-\int_t^T q_X(u)\,du} XN(\widehat{d}_X) - e^{-\int_t^T q_Y(u)\,du} YN(\widehat{d}_Y), \tag{3.4.46}$$

where

$$\widehat{d}_X = \frac{\ln \frac{X}{Y} + \int_t^T [q_Y(u) - q_X(u)]\,du + \widehat{\sigma}^2_{Y/X}(t,T)(T-t)/2}{\widehat{\sigma}_{Y/X}(t,T)\sqrt{T-t}},$$
$$\widehat{d}_Y = \widehat{d}_X - \widehat{\sigma}_{Y/X}(t,T)\sqrt{T-t},$$
$$\widehat{\sigma}^2_{Y/X}(t,T) = \frac{1}{T-t}\int_t^T [\sigma_X^2(u) - 2\rho\sigma_X(u)\sigma_Y(u) + \sigma_Y^2(u)]\,du.$$

3.4.7 Equity Options with Exchange Rate Risk Exposure

A quanto option is an option on a foreign currency donominated asset but the payoff is in domestic currency. The holder of a quanto option is exposed to both exchange rate risk and equity risk. There are many different forms of payoff that can be structured. Some examples of quanto call options are listed below:

1. Foreign equity call struck in foreign currency

$$c_1(S_T, F_T, T) = F_T \max(S_T - X_f, 0).$$

   Here, F_T is the terminal exchange rate, S_T is the terminal price of the underlying foreign currency denominated asset and X_f is the strike price in foreign currency.
2. Foreign equity call struck in domestic currency

$$c_2(S_T, T) = \max(F_T S_T - X_d, 0).$$

   Here, X_d is the strike price in domestic currency.
3. Fixed exchange rate foreign equity call

$$c_3(S_T, T) = F_0 \max(S_T - X_f, 0).$$

   Here, F_0 is some predetermined fixed exchange rate.
4. Equity-linked foreign exchange call

$$c_4(S_T, T) = S_T \max(F_T - X_F, 0).$$

   Here, X_F is the strike price on the exchange rate. The holder plans to purchase the foreign asset any way but wishes to place a floor value X_F on the exchange rate.

Quanto Prewashing Technique

Let S_t and F_t denote the price process of the foreign asset and the exchange rate, respectively. Define $\widehat{S}_t = F_t S_t$, which is the foreign asset price in domestic currency.

Let r_d and r_f denote the constant domestic and foreign interest rate, respectively, and let q denote the dividend yield of the foreign asset. We assume that both S_t and F_t are Geometric Brownian processes. Since $\widehat{S}_t$ and F_t can be considered as price processes in domestic currency, so under the domestic risk neutral measure Q_d, the drift rate of $\widehat{S}_t$ and F_t are

$$\delta^d_{\widehat{S}} = r_d - q \quad \text{and} \quad \delta^d_F = r_d - r_f, \tag{3.4.47a}$$

respectively. The reciprocal of F_t can be considered as the foreign currency price of one unit of the domestic currency. The drift rate of S_t and $1/F_t$ under the foreign risk neutral measure Q_f are given by

$$\delta^f_S = r_f - q \quad \text{and} \quad \delta^f_{1/F} = r_f - r_d, \tag{3.4.47b}$$

respectively. The quanto prewashing technique means finding δ^d_S, that is, the drift rate of the price process of the foreign currency denominated asset S_t under the domestic risk neutral measure Q_d. Let the dynamics of S_t and F_t under Q_d be governed by

$$\frac{dS_t}{S_t} = \delta^d_S\,dt + \sigma_S\,dZ^d_S \tag{3.4.48a}$$

$$\frac{dF_t}{F_t} = \delta^d_F\,dt + \sigma_F\,dZ^d_F, \tag{3.4.48b}$$

where the Q_d-Brownian processes Z^d_S and Z^d_F observe $dZ^d_S\,dZ^d_F = \rho\,dt$, σ_S and σ_F are the volatility of S_t and F_t, respectively. Since $\widehat{S}_t = F_t S_t$, we then have (see Problem 2.38)

$$\delta^d_{\widehat{S}} = \delta^d_{FS} = \delta^d_F + \delta^d_S + \rho\sigma_F\sigma_S.$$

Using the results in (3.4.47a,b), we obtain

$$\delta^d_S = \delta^d_{\widehat{S}} - \delta^d_F - \rho\sigma_F\sigma_S = r_f - q - \rho\sigma_F\sigma_S. \tag{3.4.49}$$

Comparing δ^d_S with δ^f_S, there is an extra term $-\rho\sigma_F\sigma_S$. That is, the risk neutral drift rate of the price process of the foreign asset is adjusted by the amount $-\rho\sigma_F\sigma_S$ when the risk neutral measure is changed from the foreign currency world to the domestic currency world.

From (3.4.48b) and Problem 2.38, we obtain

$$\frac{d(1/F_t)}{1/F_t} = (r_f - r_d + \sigma_F^2)\,dt - \sigma_F\,dZ^d_F.$$

Thus the risk neutral drift rate of $1/F_t$ under Q_d is seen to be $r_f - r_d + \sigma_F^2$. This result may puzzle some people (known as Siegel's paradox) since they would expect the risk neutral drift rate of $1/F_t$ to be $r_f - r_d$. Actually, the drift rate of $1/F_t$ under Q_d can be deduced easily from the quanto-prewashing formula (3.4.49). Observing that $\sigma_F = \sigma_{1/F}$ and $\rho_{F,1/F} = -1$, an additional quanto prewashing term

$-\rho_{F,1/F}\sigma_{1/F}\sigma_F = \sigma_F^2$ has to be added to $\delta_{1/F}^f = r_f - r_d$ [see (3.4.47b)], thus giving

$$\delta_{1/F}^d = r_f - r_d + \sigma_F^2. \tag{3.4.50}$$

Price Formulas of Various Quanto Options

Using the quanto-prewashing technique, we derive the price formulas of quanto options with various forms of terminal payoff. Though these quanto options involve two-state variables, S_t and F_t, we manage to reduce them to one-state option models.

1. Foreign equity call struck in foreign currency
   Let $c_1^f(S, \tau)$ denote the usual vanilla call option on the foreign currency asset in the foreign currency world, where S is the current asset price and $\tau = T - t$ is time to expiry. The terminal payoff is

   $$c_1^f(S, 0) = \max(S - X_f, 0).$$

   We write F as the current exchange rate. It is readily seen that

   $$c_1(S, F, \tau) = Fc_1^f(S, \tau) = F\big[Se^{-q\tau}N(d_1^{(1)}) - X_f e^{-r_f\tau}N(d_2^{(1)})\big], \tag{3.4.51}$$

   where

   $$d_1^{(1)} = \frac{\ln \frac{S}{X_f} + \left(\delta_S^f + \frac{\sigma_S^2}{2}\right)\tau}{\sigma_S\sqrt{\tau}}, \quad d_2^{(1)} = d_1^{(1)} - \sigma_S\sqrt{\tau}.$$

   Interestingly, the exchange rate risk does not enter into the price formula. At any time, the value of the option is converted from foreign currency into domestic currency using the prevailing exchange rate. The reduction to a one-state model is achieved through performing valuation in an appropriate currency world.
2. Foreign equity call struck in domestic currency
   The terminal payoff at $\tau = 0$ in domestic currency is

   $$c_2(S, F, 0) = \max(\widehat{S} - X_d, 0),$$

   where $\widehat{S} = FS$ is a domestic currency denominated asset. The drift rate and volatility of $\widehat{S}$ under the domestic risk neutral measure Q_d are

   $$\delta_{\widehat{S}}^d = r_d - q \quad \text{and} \quad \sigma_{\widehat{S}}^2 = \sigma_S^2 + 2\rho\sigma_S\sigma_F + \sigma_F^2, \tag{3.4.52}$$

   respectively. The price formula of the foreign equity call is then given by

   $$c_2(S, F, \tau) = \widehat{S}e^{-q\tau}N(d_1^{(2)}) - X_d e^{-r_d\tau}N(d_2^{(2)}), \tag{3.4.53}$$

   where

   $$d_1^{(2)} = \frac{\ln \frac{\widehat{S}}{X_d} + \left(\delta_{\widehat{S}}^d + \frac{\sigma_{\widehat{S}}^2}{2}\right)\tau}{\sigma_{\widehat{S}}\sqrt{\tau}}, \quad d_2^{(2)} = d_1^{(2)} - \sigma_{\widehat{S}}\sqrt{\tau}.$$

3. Fixed exchange rate foreign equity call
The terminal payoff is denominated in domestic currency, so the drift rate δ_S^d of the foreign asset in Q_d should be used. The price function of the fixed exchange rate foreign equity call is given by

$$c_3(S,\tau) = F_0 e^{-r_d \tau}\left[S e^{\delta_S^d \tau} N(d_1^{(3)}) - X_f N(d_2^{(3)})\right], \tag{3.4.54}$$

where

$$d_1^{(3)} = \frac{\ln \frac{S}{X_f} + \left(\delta_S^d + \frac{\sigma_S^2}{2}\right)\tau}{\sigma_S \sqrt{\tau}}, \quad d_2^{(3)} = d_1^{(3)} - \sigma_S \sqrt{\tau}.$$

The price formula does not depend on the exchange rate F since the exchange rate has been chosen to be the fixed value F_0. The currency exposure of the call is embedded in the quanto-prewashing term $-\rho\sigma_S\sigma_F$ in δ_S^d, exhibiting dependence on the exchange rate volatility σ_F and the correlation coefficient ρ.

4. Equity-linked foreign exchange call
We may write the terminal payoff as an exchange option, where

$$c_4(S, F, 0) = \max(\widehat{S} - XS, 0).$$

Taking the two assets that are exchanged to be $\widehat{S}$ and XS, the ratio of the two assets is $\frac{\widehat{S}}{XS} = \frac{F}{X}$ and the difference of the drift rates under Q_d is $\delta_{\widehat{S}}^d - \delta_S^d = r_d - r_f + \rho\sigma_F\sigma_S$. The value of the equity-linked foreign exchange call is given by [see (3.4.46)]

$$\begin{aligned} c_4(S,\tau) &= e^{-r_d\tau}\left[\widehat{S} e^{\delta_{\widehat{S}}^d \tau} N(d_1^{(4)}) - XS e^{\delta_S^d \tau} N(d_2^{(4)})\right] \\ &= S e^{-q\tau}\left[F N(d_1^{(4)}) - X e^{(r_f - r_d - \rho\sigma_F\sigma_S)\tau} N(d_2^{(4)})\right], \end{aligned} \tag{3.4.55}$$

where

$$d_1^{(4)} = \frac{\ln \frac{F}{X} + \left(r_d - r_f + \rho\sigma_F\sigma_S + \frac{\sigma_F^2}{2}\right)\tau}{\sigma_F \sqrt{\tau}}, \quad d_2^{(4)} = d_1^{(4)} - \sigma_F \sqrt{\tau}.$$

3.5 Beyond the Black–Scholes Pricing Framework

In the Black–Scholes–Merton option pricing framework, it is assumed that the portfolio composition changes continuously according to a dynamic hedging strategy at zero transaction costs. In the presence of transaction costs associated with buying and selling of the asset, the continuous portfolio adjustment required by the Black–Scholes–Merton model will lead to an infinite number of transactions and so infinite total transaction costs. As a hedger, one has to strike the balance between transaction costs required for rebalancing the portfolio and the implied costs of hedging errors. The presence of transaction costs implies that absence of arbitrage no longer leads to a single option price but rather a range of feasible prices. An option can be overpriced or underpriced up to the extent where the profit obtained by an arbitrageur is offset by the transaction costs. The transaction costs model will be presented in Sect. 3.5.1.

Another assumption in the Black–Scholes–Merton framework is the continuity of the asset price path. Numerous empirical studies have revealed that the asset price may jump discontinuously, say, due to the arrival of sudden news. In Sect. 3.5.2, we consider the jump-diffusion model proposed by Merton (1976). Under the assumption that the jump components are uncorrelated with the market (jump risks can be diversified away), we can derive the governing equation for the derivative price. Also, when the random jump arrivals follow a Poisson process and the logarithm of the jump ratio is normally distributed, it is possible to obtain closed form option price formulas under the jump-diffusion model.

The Black–Scholes–Merton model gives the option price as a function of volatility and quantifies the randomness in the asset price dynamics through a constant volatility parameter. Instead of computing the option price given the volatility value using the Black–Scholes price formula, as an inverse problem, we solve for the volatility from the observed market option price. The volatility value implied by an observed option price is called the *implied volatility*. If the Black–Scholes option pricing model were perfect, the implied volatility would be the same for all option market prices. However, empirical studies have revealed that the implied volatilities depend on the strike price and the maturity of the options. Such phenomena are called the volatility smiles. We consider the relaxation of the constant volatility assumption and attempt to model volatility either as a deterministic volatility function (local volatility) or as a mean-reverting stochastic process. By following the local volatility approach, we derive the Dupire equation that governs option prices with maturity and strike price as independent variables. Various issues on implied volatilities and local volatilities are discussed in Sect. 3.5.3.

The local volatility model is in general too restrictive to describe the behavior of the volatility variations. In the literature, there have been extensive research efforts to develop different types of volatility models. In the class of stochastic volatility models, volatility itself is modeled as a mean reverting Ito process (Hull and White, 1987). The mean reverting characteristics of volatility agree with our intuition that the level of volatility should revert to the mean level of its long-run distribution. The pricing of options under the stochastic volatility assumption is quite challenging. When the asset price process and the volatility process are uncorrelated, it can be shown that the price of a European option is the Black–Scholes price integrated over the probability distribution of the average variance rate for the remaining life of the option. In the general case where the two processes are correlated, the analytic solution can be obtained via the Fourier transform method (Heston, 1993). Fouque, Papanicolaou and Sircar (2000) provided a comprehensive review of different aspects of stochastic volatility. Another class of volatility models that have gained popularity in recent years are the family of GARCH (generalized autoregressive conditional heteroscedasticity) models (Duan, 1995). In the GARCH models, the variance rate at the current time step is a weighted average of a constant long-run average variance rate, the variance rate at the previous time steps and the most recent information about the variance rate. Continued research efforts are directed to explore better volatility models to explain the volatility smile and extract useful market information from the smile itself.

3.5.1 Transaction Costs Models

How can we construct the hedging strategy that best replicates the payoff of a derivative security in the presence of transaction costs? Recall that one can create a portfolio containing Δ units of the underlying asset and a money market account that replicates the payoff of the option. By the portfolio replication argument, the value of an option is equal to the initial cost of setting up the replicating portfolio which mimics the payoff of the option. Leland (1985) proposed a modification to the Black–Scholes model where the portfolio is adjusted at regular time intervals. His model assumes proportional transaction costs where the costs in buying and selling the asset are proportional to the monetary value of the transaction. Let k denote the round trip transaction cost per unit dollar of the transaction. Suppose α units of assets are bought ($\alpha > 0$) or sold ($\alpha < 0$) at the price S, then the transaction cost is given by $\frac{k}{2}\ |\alpha|S$ in either buying or selling.

In the following proportional transaction costs option pricing model (Leland, 1985; Whalley and Wilmott, 1993), the asset price dynamics is assumed to follow the Geometric Brownian process where the volatility is taken to be constant. Also, the underlying asset is assumed to pay no dividends during the life of the option. We consider a hedged portfolio of the writer of the option, where he or she is shorting one unit of option and long holding Δ units of the underlying asset. For notational convenience, we drop the subscript t in the asset price process S_t in later exposition. The value of this hedged portfolio at time t is given by

$$\Pi(t) = -V(S,t) + \Delta S, \tag{3.5.1}$$

where $V(S,t)$ is the value of the option and S is the asset price at time t. Let δt denote the fixed and finite time interval between successive rebalancing of the portfolio. After the time interval δt, the change in value of the portfolio is

$$\delta\Pi = -\delta V + \Delta\,\delta S - \frac{k}{2}|\delta\Delta|S, \tag{3.5.2}$$

where δS is the change in asset price and $\delta\Delta$ is the change in the number of units of asset held in the portfolio. A cautious reader may doubt why the proportional transaction cost term $-\frac{k}{2}|\delta\Delta|S$ appears in $\delta\Pi$ while the term $S\,\delta\Delta$ is missing. The transaction cost term represents the single trip transaction cost paid due to rebalancing of the position in the underlying asset. On the other hand, by following the "pragmatic" approach used by Black and Scholes (1973), the number of units Δ is treated to be instantaneously constant (see the remarks at the end of Sect. 3.1.1). By Ito's lemma, the change in option value in time δt is given by

$$\delta V - \frac{\partial V}{\partial S}\,\delta S + \left(\frac{\partial V}{\partial t} + \frac{\sigma^2}{2}S^2\frac{\partial^2 V}{\partial S^2}\right)\delta t. \tag{3.5.3}$$

In order to cancel the stochastic terms in (3.5.2)–(3.5.3), one chooses $\Delta = \frac{\partial V}{\partial S}$ so as to hedge against the risk due to the asset price fluctuation. The change in the number of units of asset in time δt is given by

$$\delta \Delta = \frac{\partial V}{\partial S}(S + \delta S, t + \delta t) - \frac{\partial V}{\partial S}(S, t). \tag{3.5.4}$$

By Ito's lemma, the leading order of $|\delta \Delta|$ is found to be

$$|\delta \Delta| \approx \sigma S \left| \frac{\partial^2 V}{\partial S^2} \right| |\delta Z|. \tag{3.5.5}$$

Formally, we may treat δZ as $\widetilde{x}\sqrt{\delta t}$, where $\widetilde{x}$ is the standard normal variable. It can be shown that the expectation of the reflected Brownian process $|\delta Z|$ is given by

$$E(|\delta Z|) = \sqrt{\frac{2}{\pi}}\sqrt{\delta t} \tag{3.5.6}$$

(see Problem 2.29). This hedged portfolio should be expected to earn a return same as that of a riskless asset. This gives

$$E[\delta \Pi] = r\left(-V + \frac{\partial V}{\partial S} S\right)\delta t. \tag{3.5.7}$$

By putting all the above results together, (3.5.7) can be rewritten as

$$\left(-\frac{\partial V}{\partial t} - \frac{\sigma^2}{2} S^2 \frac{\partial^2 V}{\partial S^2} - \frac{k}{2}\sigma S^2 \sqrt{\frac{2}{\pi \delta t}} \left| \frac{\partial^2 V}{\partial S^2} \right| \right)\delta t = r\left(-V + \frac{\partial V}{\partial S} S\right)\delta t.$$

If we define the Leland number to be $Le = \sqrt{\frac{2}{\pi}}\left(\frac{k}{\sigma\sqrt{\delta t}}\right)$, we obtain

$$\frac{\partial V}{\partial t} + \frac{\sigma^2}{2} S^2 \frac{\partial^2 V}{\partial S^2} + \frac{\sigma^2}{2} Le\, S^2 \left| \frac{\partial^2 V}{\partial S^2} \right| + rS \frac{\partial V}{\partial S} - rV = 0. \tag{3.5.8}$$

In the *proportional transaction costs model*, the term $\frac{\sigma^2}{2} Le\, S^2 \left|\frac{\partial^2 V}{\partial S^2}\right|$ is in general nonlinear, except when the comparative static $\Gamma = \frac{\partial^2 V}{\partial S^2}$ does not change sign for al S. The transaction cost term is dependent on Γ, where Γ measures the degree of mishedging of the portfolio. One may rewrite (3.5.8) in a form that resembles the Black–Scholes equation

$$\frac{\partial V}{\partial t} + \frac{\widetilde{\sigma}^2}{2} S^2 \frac{\partial^2 V}{\partial S^2} + rS \frac{\partial V}{\partial S} - rV = 0, \tag{3.5.9}$$

where the modified volatility is given by

$$\widetilde{\sigma}^2 = \sigma^2[1 + Le\ \mathrm{sign}(\Gamma)]. \tag{3.5.10}$$

Equation (3.5.9) becomes mathematically ill-posed when $\widetilde{\sigma}^2$ becomes negative. This occurs when $\Gamma < 0$ and $Le > 1$. However, it is known that Γ is always positive for the vanilla European call and put options in the absence of transaction costs. If we postulate the same sign behavior for Γ in the presence of transaction costs, then $\widetilde{\sigma}^2 = \sigma^2(1 + Le) > \sigma^2$. Equation (3.5.9) then becomes linear under the above assumption so that the Black–Scholes formulas become applicable except that the modified volatility $\widetilde{\sigma}$ is now used as the volatility parameter. We can deduce $V(S,t)$ to be an increasing function of Le since we expect a higher option value for a high value of modified volatility. Financially speaking, the more frequent the rebalancing (smaller δt) the higher the transaction costs and so the writer of an option should charge higher for the price of the option. Let $V(S,t;\widetilde{\sigma})$ and $V(S,t;\sigma)$ denote the option values obtained from the Black–Scholes formula with volatility values $\widetilde{\sigma}$ and σ, respectively. The total transaction costs associated with the replicating strategy is then given by

$$\mathcal{T} = V(S,t;\widetilde{\sigma}) - V(S,t;\sigma). \tag{3.5.11}$$

When Le is small, $\mathcal{T}$ can be approximated by

$$\mathcal{T} \approx \frac{\partial V}{\partial \sigma}(\widetilde{\sigma} - \sigma), \tag{3.5.12}$$

where $\widetilde{\sigma} - \sigma \approx \frac{k}{\sqrt{2\pi\,\delta t}}$. Note that $\frac{\partial V}{\partial \sigma}$ is the same for both call and put options and the vega value is given by (3.3.29)–(3.3.30). For $Le \ll 1$, the total transaction costs for either a call or a put is approximately given by

$$\mathcal{T} \approx \frac{kSe^{-\frac{d_1^2}{2}}}{2\pi}\sqrt{\frac{T-t}{\delta t}}, \tag{3.5.13}$$

where d_1 is defined in (3.3.6).

Rehedging at regular time intervals is one of the many possible hedging strategies. The natural question is: How would we characterize the optimality condition of a given hedging strategy? The usual approach is to define an appropriate utility function, which is used as the reference for which optimization is being taken. For the discussion of utility-based hedging strategies in the presence of transaction costs, one may refer to the papers by Hodges and Neuberger (1989), and by Davis, Panas and Zariphopoulou (1993). In their models, they attempted to find the set of optimal portfolio policies that maximize the expected utility over an infinite horizon. Neuberger (1994) showed that it is possible to use arbitrage strategies to set tight and preference-free bounds on option prices in the presence of transaction costs when the underlying asset follows a pure jump process. Other aspects of option pricing models with transaction costs were discussed by Bensaid et al. (1992) and Grannan and Swindle (1996).

3.5.2 Jump-Diffusion Models

In the Black–Scholes option pricing model, we assume that trading takes place continuously in time and the asset price process has a continuous sample path. There have been numerous empirical studies on asset price dynamics that show occasional jumps in asset price. Such jumps may reflect the arrival of new important information on the firm or its industry or economy as a whole.

Merton (1976) initiated the modeling of the asset price process S_t by a combination of normal fluctuation and abnormal jumps. The normal fluctuation is modeled by the Geometric Brownian process and the associated sample paths are continuous. The jumps are modeled by Poisson distributed events where their arrivals are assumed to be independent and identically distributed with intensity λ. That is, the probability that a jump event occurs over the time interval $(t, t+dt)$ is equal to $\lambda\, dt$. We may define the Poisson process dq_t by

$$dq_t = \begin{cases} 0 & \text{with probability } 1 - \lambda\, dt \\ 1 & \text{with probability } \lambda\, dt \end{cases}. \tag{3.5.14}$$

Here, λ is interpreted as the mean number of arrivals per unit time.

Let J denote the jump ratio of the asset price upon the arrival of a jump event, that is, S_t jumps immediately to JS_t when $dq_t = 1$. The jump ratio itself is a random variable with density function f_J. For example, suppose we assume $\ln J$ to be a Gaussian distribution with mean μ_J and variance σ_J^2, then

$$E[J-1] = \exp\left(\mu_J + \frac{\sigma_J^2}{2}\right) - 1. \tag{3.5.15}$$

Assume that the asset price dynamics is a combination of the Geometric Brownian diffusion process and the Poisson jump process, the asset price process S_t is then governed by

$$\frac{dS_t}{S_t} = \mu\, dt + \sigma\, dZ_t + (J-1)dq_t, \tag{3.5.16}$$

where μ and σ are the drift rate and volatility of the Geometric Brownian process, respectively. The change in asset price upon the arrival of a jump event is $(J-1)S_t$.

Imagine that a writer of an option follows the Black–Scholes hedging strategy, where he or she is long holding Δ units of the underlying asset and shorting one unit of the option. Let $V(S,t)$ denote the price function of the option. For convenience, we drop the subscript t in the asset price process S_t and Poisson process dq_t. Again, we adopt the "pragmatic" Black–Scholes approach of keeping Δ to be instantaneously constant. The portfolio value Π and its differential $d\Pi$ are given by

$$\Pi = \Delta S - V(S,t),$$

and

$$\begin{aligned} d\Pi = & -\left(\frac{\partial V}{\partial t} + \frac{\sigma^2}{2}S^2\frac{\partial^2 V}{\partial S^2}\right)dt + \left(\Delta - \frac{\partial V}{\partial S}\right)(\mu S\, dt + \sigma S\, dZ_t) \\ & + \{\Delta(J-1)S - [V(JS,t) - V(S,t)]\}\, dq_t. \end{aligned} \tag{3.5.17}$$

The two sources of risk come from the diffusion component dZ_t and the jump component dq_t. To hedge the diffusion risk, we may choose $\Delta = \frac{\partial V}{\partial S}$, like the usual

Black–Scholes hedge ratio. How about the jump risk? Merton (1976) argued that if the jump component is firm specific and uncorrelated with the market (nonsystematic risk), then the jump risk should not be priced into the option. In this case, the beta (from the Capital Asset Pricing Model) of the portfolio is zero. Since the expected return on all zero-beta securities is equal to the riskless interest rate, we then have

$$E_J[d\Pi] = r\Pi\, dt, \tag{3.5.18}$$

where the expectation E_J is taken over the jump ratio J. Note that

$$\begin{aligned}&E_J\big[\{\Delta(J-1)S - \big[V(JS,t) - V(S,t)\big]\}dq_t\big]\\ &= \lambda\left(E_J[J-1]S\frac{\partial V}{\partial S} - E_J\big[V(JS,t) - V(S,t)\big]\right)dt.\end{aligned}$$

Combining all the results together, we obtain the following governing differential equation of the option price function $V(S,\tau)$ under the jump-diffusion asset price process:

$$\begin{aligned}\frac{\partial V}{\partial \tau} &= \frac{\sigma^2}{2}S^2\frac{\partial^2 V}{\partial S^2} + \big(r - \lambda E_J[J-1]\big)S\frac{\partial V}{\partial S} - rV\\ &\quad + \lambda E_J[V(JS,\tau) - V(S,\tau)], \quad \tau = T - t.\end{aligned} \tag{3.5.19}$$

To solve for $V(S,\tau)$, one has to specify the distribution for J. Let $k = E_J[J-1]$ and define the random variable X_n which has the same distribution as the product of n independent and identically distributed random variables, each is identically distributed to J (with $X_0 = 1$). We write $V_{BS}(S,\tau)$ as the Black–Scholes price function of the same option contract. The representation of the solution to (3.5.19) in terms of expectations is given by (Merton, 1976)

$$V(S,\tau) = \sum_{n=0}^{\infty}\frac{e^{-\lambda\tau}(\lambda\tau)^n}{n!}\{E_{X_n}[V_{BS}(SX_ne^{-\lambda k\tau},\tau)]\}, \quad \tau = T - t, \tag{3.5.20}$$

where E_{X_n} is the expectation over the distribution of X_n. Hints to the proof of the above representation are given in Problem 3.39.

In general, it is not easy to obtain closed form price formulas for options under the jump-diffusion models, except for a few exceptions. When the jump ratio J follows the lognormal distribution, it is possible to obtain a closed form price formula for a European call option (see Problem 3.40). Also, Das and Foresi (1996) obtained closed form price formulas for bonds and options when the interest rate follows the jump-diffusion models.

3.5.3 Implied and Local Volatilities

The option prices obtained from the Black–Scholes pricing framework are functions of five parameters: asset price S, strike price X, riskless interest rate r, time to expiry τ and volatility σ. Except for the volatility parameter, the other four parameters are observable quantities. The difficulties of setting volatility value in the price formulas lie in the fact that the input value should be the forecast volatility value over the remaining life of the option rather than an estimated volatility value (*historical volatility*) from the past market data of the asset price. Suppose we treat the option price function $V(\sigma)$ as a function of the volatility σ and let V_{market} denote the option price observed in the market. The implied volatility σ_{imp} is defined by

$$V(\sigma_{imp}) = V_{market}. \tag{3.5.21}$$

The volatility value implied by an observed market option price (*implied volatility*) indicates a consensual view about the volatility level determined by the market. In particular, several implied volatility values obtained simultaneously from different options with varying maturities and strike prices on the same underlying asset provide an extensive market view about the volatility at varying strikes and maturities. Such information may be useful for a trader to set the volatility value for the underlying asset of an option that he or she is interested in. In financial markets, it becomes a common practice for traders to quote an option's market price in terms of implied volatility σ_{imp}. This provides the direct comparison on the implied volatility values based on market information of the option prices.

Since σ cannot be solved explicitly in terms of S, X, r, τ and option price V from the pricing formulas, the determination of the implied volatility must be accomplished by an iterative algorithm as commonly performed for the root-finding procedure for a nonlinear equation. Manaster and Koehler (1982) proposed an iterative algorithm based on the well-known Newton–Raphson iterative method. The iteration exhibits the quadratic rate of convergence and the sequence of iterates $\{\sigma_1, \sigma_2, \cdots\}$ converge monotonically to σ_{imp}. By quadratic rate of convergence, we mean

$$\sigma_{n+1} - \sigma_{imp} = K(\sigma_n - \sigma_{imp})^2 \tag{3.5.22}$$

for some K independent of n.

Numerical Calculations of Implied Volatilities

When applied to the implied volatility calculations, the Newton–Raphson iterative scheme is given by

$$\sigma_{n+1} = \sigma_n - \frac{V(\sigma_n) - V_{market}}{V'(\sigma_n)}, \tag{3.5.23}$$

where σ_n denotes the nth iterate of σ_{imp}. Provided that the first iterate σ_1 is properly chosen, the limit of the sequence $\{\sigma_n\}$ converges to the unique solution σ_{imp}. The above iterative scheme may be rewritten in the following form

$$\frac{\sigma_{n+1} - \sigma_{imp}}{\sigma_n - \sigma_{imp}} = 1 - \frac{V(\sigma_n) - V(\sigma_{imp})}{\sigma_n - \sigma_{imp}} \frac{1}{V'(\sigma_n)} = 1 - \frac{V'(\sigma_n^*)}{V'(\sigma_n)}. \tag{3.5.24}$$

One can show that σ_n^* lies between σ_n and σ_{imp}, by virtue of the Mean Value Theorem in calculus. Manaster and Koehler (1982) proposed choosing the first iterate σ_1 such that $V'(\sigma)$ is maximized by $\sigma = \sigma_1$. As explained below, this choice of the starting iterate would guarantee monotonic convergence of the sequence of iterates to σ_{imp}. Recall from (3.3.29) that

$$V'(\sigma) = \frac{S\sqrt{\tau}\, e^{-\frac{d_1^2}{2}}}{\sqrt{2\pi}} > 0 \quad \text{for all } \sigma,$$

and so

$$V''(\sigma) = \frac{S\sqrt{\tau} d_1 d_2 e^{-\frac{d_1^2}{2}}}{\sqrt{2\pi}\sigma} = \frac{V'(\sigma) d_1 d_2}{\sigma},$$

where d_1 and d_2 are defined in (3.3.6). Therefore, the critical points of the function $V'(\sigma)$ are given by $d_1 = 0$ and $d_2 = 0$, which lead respectively to

$$\sigma^2 = -2\frac{\ln \frac{S}{X} + r\tau}{\tau} \quad \text{and} \quad \sigma^2 = 2\frac{\ln \frac{S}{X} + r\tau}{\tau}.$$

The above two values of σ^2 both give $V'''(\sigma) < 0$. Hence, we can choose the first iterate σ_1 to be

$$\sigma_1 = \sqrt{\left|\frac{2}{\tau}\left(\ln \frac{S}{X} + r\tau\right)\right|}. \tag{3.5.25}$$

With this choice of σ_1, $V'(\sigma)$ is maximized at $\sigma = \sigma_1$. Setting $n = 1$ in (3.5.24) and observing $V'(\sigma_1^*) < V'(\sigma_1)$ [note that $V'(\sigma)$ is maximized at $\sigma = \sigma_1$], we obtain

$$0 < \frac{\sigma_2 - \sigma_{imp}}{\sigma_1 - \sigma_{imp}} < 1.$$

In general, suppose we can establish (see Problem 3.42)

$$0 < \frac{\sigma_{n+1} - \sigma_{imp}}{\sigma_n - \sigma_{imp}} < 1, \qquad n \geq 1, \tag{3.5.26}$$

then the sequence $\{\sigma_n\}$ is monotonic and bounded, so $\{\sigma_n\}$ converges to the unique solution σ_{imp}. In conclusion, if we start with the first iterate σ_1 given by (3.5.25), then the sequence $\{\sigma_n\}$ generated by (3.5.23) converges to σ_{imp} monotonically with a quadratic rate of convergence.

Volatility Smiles

The Black–Scholes model assumes a lognormal probability distribution of the asset price at all future times. Since volatility is the only unobservable parameter in the Black–Scholes model, the model gives the option price as a function of volatility. It would be interesting to examine the dependence of volatility on the option's strike price.

If we plot the implied volatility of the exchange-traded options, like index options, against their strike price for a fixed maturity, the curve is typically convex

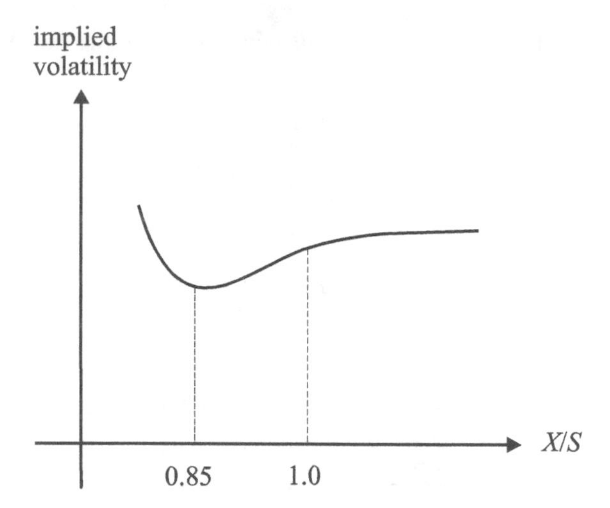

Fig. 3.7a. A typical pattern of pre-crash smile. The implied volatility curve is convex with a dip.

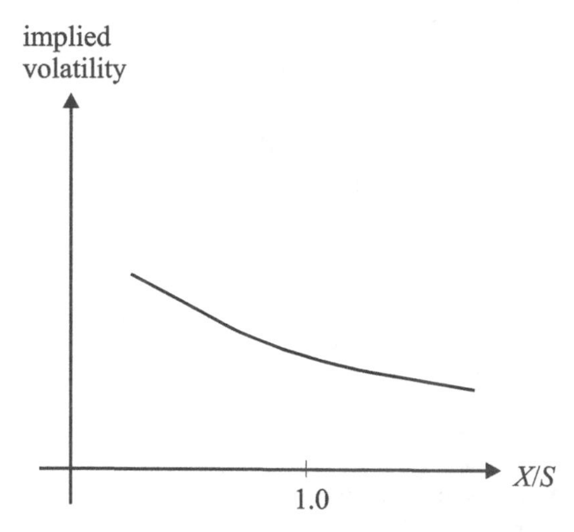

Fig. 3.7b. A typical pattern of post-crash smile. The implied volatility drops against X/S, indicating that out-of-the-money puts ($X/S < 1$) are traded at higher implied volatility than out-of-the-money calls ($X/S > 1$).

in shape, rather than a straight horizontal line as suggested by the simple Black–Scholes model. This phenomenon is commonly called the *volatility smile* by market practitioners. These smiles exhibit widely differing properties, depending on whether the market data were taken before or after the October, 1987 market crash. Figures 3.7a,b show the shapes of typical pre-crash smile and post-crash smile of the exchange-traded European index options. The implied volatility values are obtained by averaging options of different maturities.

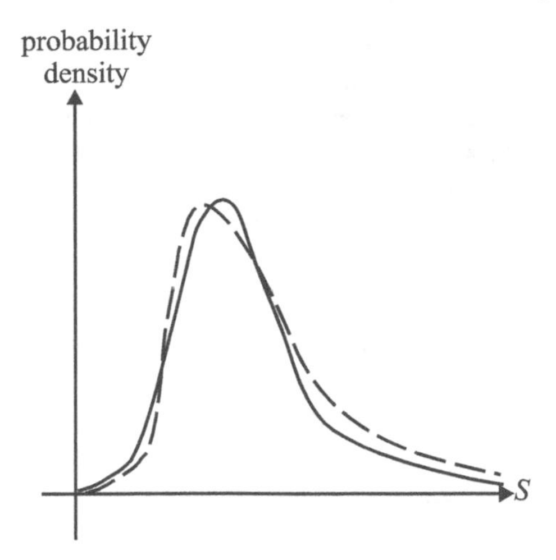

Fig. 3.8. Comparison of the true probability density of asset price (solid curve) implied from market data and the lognormal distribution (dotted curve). The true probability density is thicker at the left tail and thinner at the right tail.

In real market situation, it is a common occurrence that when the asset price is high, volatility tends to decrease, making it less probable for a higher asset price to be realized. On the other hand, when the asset price is low, volatility tends to increase, that is, it is more probable that the asset price plummets further down. Suppose we plot the true probability distribution of the asset price and compare with the lognormal distribution, one observes that the left-hand tail of the true distribution is thicker than that of the lognormal one, while the reverse situation occurs at the right-hand tail (see Fig. 3.8).

As reflected from the implied probabilities calculated from the market data of option prices, this market behavior of higher probability of large decline in stock index is better known to market practitioners after the October, 1987 market crash. In other words, the market price of the out-of-the-money calls (puts) became cheaper (more expensive) than the Black–Scholes price after the 1987 crash because of the thickening (thinning) of the left- (right-) hand tail of the true probability distribution. In common market situations, the out-of-the-money stock index puts are traded at higher implied volatilities than the out-of-the-money stock index calls.

Local Volatility

Instead of introducing stochastic volatility, which requires assumptions about investor's risk preferences, one may choose to stay within the framework of one-factor diffusion process but allow the volatility to be time or state dependent or both. If the volatility function is assumed to be time dependent, it can be shown easily that $\sigma(T)$ can be deduced from the known information of implied volatility $\sigma_{imp}(t, T)$ that is available for all $T > t$ (see Problem 3.43). Now, suppose European option prices at all strikes and maturities are available so that $\sigma_{imp}(t, T; X)$ can be computed. Can we find a state-time dependent volatility function $\sigma(S_t, t)$ that gives the theoretical

Black–Scholes option prices that are consistent with the market option prices? In the literature, $\sigma(S_t, t)$ is called the *local volatility function*.

Assuming that all market European option prices are available, Breeden and Litzenberger (1979) showed that the risk neutral probability distribution of the asset price can be recovered. Let $\psi(S_T, T; S_t, t)$ denote the transition density function of the asset price. The time-t price of a European call with maturity date T and strike price X is given by

$$c(S_t, t; X, T) = e^{-r(T-t)} \int_X^{\infty} (S_T - X)\psi(S_T, T; S_t, t)\, dS_T. \tag{3.5.27}$$

If we differentiate c with respect to X, we obtain

$$\frac{\partial c}{\partial X} = -e^{-r(T-t)} \int_X^{\infty} \psi(S_T, T; S_t, t)\, dS_T; \tag{3.5.28}$$

and differentiate once more, we have

$$\psi(X, T; S_t, t) = e^{r(T-t)} \frac{\partial^2 c}{\partial X^2}. \tag{3.5.29}$$

The above equation indicates that the transition density function can be inferred completely from the market prices of optons with the same maturity and different strikes, without knowing the volatility function.

The Black–Scholes equation that governs the European call price can be considered as a backward equation since it involves the backward state and time variables. Can we find the forward version of the option pricing equation that involves the forward state variables? Such an equation does exist, and it is commonly known as the Dupire equation (Dupire, 1994).

Assume that the asset price dynamics under the risk neutral measure is governed by

$$\frac{dS_t}{S_t} = (r - q)dt + \sigma(S_t, t)\, dZ_t, \tag{3.5.30}$$

where the volatility has both state and time dependence. Suppose we write the call price function in the form of $c = c(X, T)$, the Dupire equation takes the form

$$\frac{\partial c}{\partial T} = -qc - (r - q)X\frac{\partial c}{\partial X} + \frac{\sigma^2(X, T)}{2}X^2\frac{\partial^2 c}{\partial X^2}. \tag{3.5.31}$$

The Black–Scholes equation and Dupire equation somewhat resemble the pair of backward and forward Fokker–Planck equations.

To derive the Dupire equation, we start with the differentiation with respect to T of (3.5.29) to obtain

$$\frac{\partial \psi}{\partial T} = e^{r(T-t)}\left(r\frac{\partial^2 c}{\partial X^2} + \frac{\partial^2}{\partial X^2}\frac{\partial c}{\partial T}\right). \tag{3.5.32}$$

Recall that $\psi(X, T; S, t)$ satisfies the forward Fokker–Planck equation, where

$$\frac{\partial \psi}{\partial T} = \frac{\partial^2}{\partial X^2}\left[\frac{\sigma^2(X,T)}{2}X^2\psi\right] - \frac{\partial}{\partial X}[(r-q)X\psi]. \tag{3.5.33}$$

Combining (3.5.29), (3.5.32), (3.5.33) and eliminating the common factor $e^{r(T-t)}$, we have

$$\begin{aligned} & r\frac{\partial^2 c}{\partial X^2} + \frac{\partial^2}{\partial X^2}\frac{\partial c}{\partial T} \\ &= \frac{\partial^2}{\partial X^2}\left[\frac{\sigma^2(X,T)}{2}X^2\frac{\partial^2 c}{\partial X^2}\right] - (r-q)\frac{\partial^2}{\partial X^2}\left(X\frac{\partial c}{\partial X} - c\right). \end{aligned} \tag{3.5.34}$$

Integrating the above equation with respect to X twice, we obtain

$$\begin{aligned} & \frac{\partial c}{\partial T} + rc + (r-q)\left(X\frac{\partial c}{\partial X} - c\right) \\ &= \frac{\sigma^2(X,T)}{2}X^2\frac{\partial^2 c}{\partial X^2} + \alpha(T)X + \beta(T), \end{aligned} \tag{3.5.35}$$

where $\alpha(T)$ and $\beta(T)$ are arbitrary functions of T. Since all functions involving c in the above equation vanish as X tends to infinity, hence $\alpha(T)$ and $\beta(T)$ must be zero. Grouping the remaining terms in the equation, we obtain the Dupire equation.

From the Dupire equation, we may express the local volatility $\sigma(X,T)$ explicitly in terms of the call price function and its derivatives, where

$$\sigma^2(X,T) = \frac{2\left[\frac{\partial c}{\partial T} + qc + (r-q)X\frac{\partial c}{\partial X}\right]}{X^2\frac{\partial^2 c}{\partial X^2}}. \tag{3.5.36}$$

Suppose a sufficiently large number of market option prices are available at many maturities and strikes, we can estimate the local volatility from the above equation by approximating the derivatives of c with respect to X and T using the market data. However, in real market conditions, market prices of options are available only at a limited of number of maturities and strikes. Given a finite number of market option prices, how can we construct a discrete binomial tree that simulates the asset price movement based on the one-factor local volatility assumption? Unlike the constant volatility binomial tree, the implied binomial tree will be distorted in shape. The upward and downward moves and their associated probabilities are determined by an induction procedure such that the implied tree gives the numerical estimated option prices that agree with the observed option prices. In other words, the tree structure is *implied* by the market data. Unfortunately, the number of nodes in the binomial tree is in general far more than the number of available option prices. This would cause numerical implementation of the implied binomial tree highly unstable. For a discussion of the theory of local volatility and implied tree techniques, one may read Derman and Kani (1998).

3.5.4 Stochastic Volatility Models

The daily fluctuations of the return of asset prices typically exhibit volatility clustering where large moves follow large moves and small moves follow small moves.

Also, the distribution of asset price returns is highly peaked and fat-tailed, indicating mixtures of distribution with different variances. It is natural to model volatility as a random variable. The volatility clustering feature reflects the mean reversion characteristic of volatility. The modeling of the stochastic behavior of volatility is more difficult because volatility is a *hidden* process. Although volatility is driving asset prices, it is not directly observable. In this section, we describe the stochastic volatility model (Heston, 1993) which takes the price variance v as a mean reversion process that is correlated with the asset price process. Using the riskless hedging principle, we derive the governing differential equation of the price of an option on an underlying asset whose price volatility follows a mean reversion stochastic process. We then show how to use the Fourier transform method to solve for the value of a European futures call option.

Differential Equation Formulation

Heston (1993) assumed the asset price S_t and the variance of asset price v_t to follow the joint stochastic processes

$$dS_t = \mu S_t\,dt + \sqrt{v_t}S_t\,dZ_S \tag{3.5.37a}$$

$$dv_t = k(\overline{v} - v_t)\,dt + \eta\sqrt{v_t}\,dZ_v, \tag{3.5.37b}$$

where the Brownian processes are correlated with $dZ_S\,dZ_v = \rho\,dt$. The variance process is seen to have a mean reversion level $\overline{v}$ and reversion speed k, and η is the volatility of variance. The asset price process has the drift rate μ under the physical measure. All model parameters are assumed to be constant. For convenience of notation, we drop the subscript t in S_t and v_t in later exposition. The price of an option on the underlying asset should be a function of S, v, t. Let $V(S, v, t; T)$ denote the price of an option with maturity date T. Applying Ito's lemma, the differential dV is given by

$$dV = \left(\frac{\partial V}{dt} + \frac{v}{2}S^2\frac{\partial^2 V}{\partial S^2} + \rho\eta v S\frac{\partial^2 V}{\partial S\partial v} + \frac{\eta^2 v}{2}\frac{\partial^2 V}{\partial v^2}\right)dt + \frac{\partial V}{\partial S}\,dS + \frac{\partial V}{\partial v}\,dv. \tag{3.5.38}$$

Since the price variance v is not a traded security, it is necessary to include options of different maturity dates T_1 and T_2 and the underlying asset in order to construct a riskless hedged portfolio. Let the portfolio contain Δ_1 units of the option with maturity date T_1, Δ_2 units of the option with maturity date T_2 and Δ_S units of the underlying asset. The value of the portfolio is given by

$$\Pi = \Delta_1 V(S, v, t; T_1) + \Delta_2 V(S, v, t; T_2) + \Delta_S S.$$

Henceforth, we suppress the dependence of S, v and t when options of different maturities are referred. Suppose we write formally

$$\frac{dV(T_i)}{V(T_i)} = \mu_i\,dt + \sigma_i^S\,dZ_S + \sigma_i^v\,dZ_v, \quad i = 1, 2, \tag{3.5.39}$$

and using the result in (3.5.38), we obtain

$$\mu_i = \frac{1}{V(T_i)}\left[\frac{\partial V(T_i)}{\partial t} + \frac{v}{2}S^2\frac{\partial^2 V(T_i)}{\partial S^2} + \rho\eta v S\frac{\partial^2 V(T_i)}{\partial S\partial v} + \frac{\eta^2 v}{2}\frac{\partial V(T_i)}{\partial v^2} + \mu S\frac{\partial V(T_i)}{\partial S} + k(\overline{v} - v)\frac{\partial V(T_i)}{\partial v}\right],$$

$$\sigma_i^S = \frac{1}{V(T_i)}\sqrt{v}S\frac{\partial V(T_i)}{\partial S}, \quad \sigma_i^v = \frac{1}{V(T_i)}\eta\sqrt{v}\frac{\partial V(T_i)}{\partial v}, \quad i = 1, 2.$$

Since there are only two risk factors (as modeled by the two Brownian processes Z_S and Z_v) and three traded securities are available in the portfolio, it is always possible to form an instantaneously riskless portfolio. Assume the trading strategy is self-financing so that the change in portfolio value arises only from changes in the prices of the traded securities. By following the "pragmatic" Black–Scholes approach of taking the units of securities held to be instantaneously constant, the differential change in the portfolio value is then given by

$$\begin{aligned} d\Pi &= \Delta_1\, dV(T_1) + \Delta_2\, dV(T_2) + \Delta_S\, dS \\ &= [\Delta_1\mu_1 V(T_1) + \Delta_2\mu_2 V(T_2) + \Delta_S\mu S]\, dt \\ &\quad + [\Delta_1\sigma_1^S V(T_1) + \Delta_2\sigma_2^S V(T_2) + \Delta_S\sqrt{v}S]\, dZ_S \\ &\quad + [\Delta_1\sigma_1^v V(T_1) + \Delta_2\sigma_2^v V(T_2)]\, dZ_v. \end{aligned}$$

In order to cancel the stochastic terms in $d\Pi$, we must choose Δ_1, Δ_2 and Δ_S such that they satisfy the following pair of equations

$$\begin{aligned} &\Delta_1\sigma_1^S V(T_1) + \Delta_2\sigma_2^S V(T_2) + \Delta_S\sqrt{v}S = 0 \\ &\Delta_1\sigma_1^v V(T_1) + \Delta_2\sigma_2^v V(T_2) = 0. \end{aligned}$$

The portfolio now becomes instantaneously riskless. Using the no arbitrage principle, the instantaneously riskless portfolio must earn the riskless interest rate r, that is,

$$\begin{aligned} d\Pi &= [\Delta_1\mu_1 V(T_1) + \Delta_2\mu_2 V(T_2) + \Delta_S\mu S]\, dt \\ &= r[\Delta_1 V(T_1) + \Delta_2 V(T_2) + \Delta_S S]\, dt \end{aligned}$$

giving the third equation for Δ_1, Δ_2 and Δ_S:

$$\Delta_1(\mu_1 - r)V(T_1) + \Delta_2(\mu_2 - r)V(T_2) + \Delta_S(\mu - r)S = 0.$$

We put the three linear equations for Δ_1, Δ_2 and Δ_S in the following matrix form:

$$\begin{pmatrix} (\mu_1 - r)V(T_1) & (\mu_2 - r)V(T_2) & (\mu - r)S \\ \sigma_1^S V(T_1) & \sigma_2^S V(T_2) & \sqrt{v}S \\ \sigma_1^v V(T_1) & \sigma_2^v V(T_2) & 0 \end{pmatrix}\begin{pmatrix} \Delta_1 \\ \Delta_2 \\ \Delta_S \end{pmatrix} = \begin{pmatrix} 0 \\ 0 \\ 0 \end{pmatrix}. \qquad (3.5.40)$$

The second and third rows are seen to be independent. Nontrivial solutions to Δ_1, Δ_2 and Δ_S exist for the above homogeneous system of equations only if the first row in the above coefficient matrix can be expressed as a linear combination of the second and third rows. In this case, the coefficient matrix becomes singular. This is equivalent to the existence of a pair of multipliers $\lambda_S(S, v, t)$ and $\lambda_v(S, v, t)$ such that

$$\mu_i - r = \lambda_S \sigma_i^S + \lambda_v \sigma_i^v, \quad i = 1, 2, \quad \text{and} \quad \mu - r = \lambda_S \sqrt{v}. \tag{3.5.41}$$

In other words, we set the first row to be the sum of λ_S times the second row and λ_v times the third row. The multipliers λ_S and λ_v can be interpreted as the market price of risk of the asset price and variance, respectively. In general, they are functions of S, v and t. Substituting the expression for μ_i, σ_i^S and σ_i^v [see (3.5.39)] into (3.5.41), we obtain (dropping the subscript "i")

$$\begin{aligned} &\frac{\partial V}{\partial t} + \frac{v}{2} S^2 \frac{\partial^2 V}{\partial S^2} + \rho \eta v S \frac{\partial^2 V}{\partial S \partial v} + \frac{\eta^2 v}{2} \frac{\partial V}{\partial v^2} + rS \frac{\partial V}{\partial S} \\ &\quad + [k(\overline{v} - v) - \lambda_v \eta \sqrt{v}] \frac{\partial V}{\partial v} - rV = 0. \end{aligned} \tag{3.5.42}$$

Interestingly, only the market price of variance risk λ_v appears in the governing equation while the market price of asset price risk λ_S is eliminated by the relation: $\mu - r = \lambda_S \sqrt{v}$. This is because the underlying asset is a tradeable security while the price variance is not directly tradeable, though options whose values dependent on the price variance are tradeable.

Heston (1993) made the assumption that $\lambda_v(S, v, t)$ is a constant multiple of $\sqrt{v}$ so that the coefficient of $\frac{\partial V}{\partial v}$ in (3.5.42) becomes a linear function of v. Without loss of generality, we may express the drift term in the form $k'(\overline{v}' - v)$ for some constants k' and $\overline{v}'$, where k' and $\overline{v}'$ can be treated as the risk adjusted parameters for the drift of v.

Price Function of a European Call Option

We would like to find the price function of a European call with strike price X and expiration date T on the underlying asset whose price dynamics is governed by (3.5.37a,b). It may be more convenient to work with the futures call option. Let f_t denote the time-t price of the futures on the underlying asset with expiration date T and define $x_t = \ln \frac{f_t}{X}$. Let $c(x, v, \tau; X)$ denote the futures call price function, $\tau = T - t$, whose governing equation is given by

$$\frac{\partial c}{\partial \tau} = \frac{v}{2} \frac{\partial^2 c}{\partial x^2} - \frac{v}{2} \frac{\partial c}{\partial x} + \frac{\eta^2 v}{2} \frac{\partial^2 c}{\partial v^2} + \rho \eta v \frac{\partial^2 c}{\partial x \partial v} + k'(\overline{v}' - v) \frac{\partial c}{\partial v} \tag{3.5.43}$$

with initial condition:

$$c(x, v, 0) = \max(e^x - 1, 0).$$

The futures call price function takes the form:

$$c(x, v, \tau) = e^x G_1(x, v, \tau) - G_0(x, v, \tau), \tag{3.5.44}$$

where $G_0(x, v, \tau)$ is the risk neutral probability that the futures call option is in-the-money at expiration and $G_1(x, v, \tau)$ is related to the risk neutral expectation of the terminal futures price given that the option expires in-the-money. The two functions $G_j(x, v, \tau)$, $j = 0, 1$, satisfy the following differential equations:

$$\frac{\partial G_j}{\partial \tau} = \frac{v}{2}\frac{\partial^2 G_j}{\partial x^2} - \left(\frac{1}{2} - j\right) v \frac{\partial G_j}{\partial x} + \frac{\eta^2 v}{2}\frac{\partial^2 G_j}{\partial v^2} + \rho\eta v \frac{\partial^2 G_j}{\partial x \partial v} + k'(\overline{v}' - v)\frac{\partial G_j}{\partial v}, \quad j = 0, 1, \tag{3.5.45}$$

with initial condition:

$$G_j(x, v, 0) = \mathbf{1}_{\{x \geq 0\}}.$$

Heston (1993) illustrated the use of the Fourier transform method to solve the above differential equation. Let $\widehat{G}_j(m, v, \tau)$ denote the Fourier transform of $G_j(x, v, \tau)$, where

$$\widehat{G}_j(m, v, \tau) = \int_{-\infty}^{\infty} e^{-imx} G_j(x, v, \tau)\, dx, \quad j = 0, 1.$$

The Fourier transform of the initial condition is

$$\begin{aligned}\widehat{G}_j(m, v, 0) &= \int_{-\infty}^{\infty} e^{-imx} G_j(x, v, 0)\, dx \\ &= \int_0^{\infty} e^{-imx}\, dx = \frac{1}{im}, \quad j = 0, 1.\end{aligned}$$

Taking the Fourier transform of the differential equation (3.5.45), we obtain

$$\begin{aligned}\frac{\partial \widehat{G}_j}{\partial \tau} &= -\frac{m^2}{2} v \widehat{G}_j - imv\left(\frac{1}{2} - j\right)\widehat{G}_j \\ &\quad + \frac{\eta^2}{2} v \frac{\partial^2 \widehat{G}_j}{\partial v^2} + im\rho\eta v \frac{\partial \widehat{G}_j}{\partial v} + k'(\overline{v}' - v)\frac{\partial \widehat{G}_j}{\partial v} \\ &= v\left(\alpha \widehat{G}_j + \beta \frac{\partial \widehat{G}_j}{\partial v} + \gamma \frac{\partial^2 \widehat{G}_j}{\partial v^2}\right) + \delta \frac{\partial \widehat{G}_j}{\partial v}, \quad j = 0, 1,\end{aligned} \tag{3.5.46}$$

where

$$\alpha = -\frac{m^2}{2} - im\left(\frac{1}{2} - j\right), \quad \beta = im\rho\eta - k',$$

$$\gamma = \frac{\eta^2}{2}, \quad \delta = k'\overline{v}'.$$

We seek solution of the affine form for $\widehat{G}_j$ such that

$$\widehat{G}_j(m, v, \tau) = \exp(A(m, \tau) + B(m, \tau)v) G_j(m, v, 0).$$

By substituting the above assumed form into (3.5.46), we obtain

$$\begin{aligned}\frac{\partial B}{\partial \tau} &= \alpha + \beta B + \gamma B^2 = \gamma(B - \rho_+)(B - \rho_-) \\ \frac{\partial A}{\partial \tau} &= \delta B\end{aligned}$$

with $B(m, 0) = 0$ and $A(m, 0) = 0$. Here, $\rho_\pm = \frac{-\beta \pm \sqrt{\beta^2 - 4\alpha\gamma}}{2\gamma}$. Writing

$$\rho = \rho_-/\rho_+ \quad \text{and} \quad \xi = \sqrt{\beta^2 - 4\alpha\gamma},$$

the solutions to $B(m, \tau)$ and $A(m, \tau)$ are found to be

$$B(m, \tau) = \rho_- \frac{1 - e^{-\xi\tau}}{1 - \rho e^{-\xi\tau}}$$

$$A(m, \tau) = \delta\left(\rho_-\tau - \frac{2}{\eta^2} \ln \frac{1 - \rho e^{-\xi\tau}}{1 - \rho}\right).$$

Finally, the solution to $G_j(x, v, \tau)$ is obtained by taking the Fourier inversion of $\widehat{G}_j(m, v, \tau)$, giving

$$G_j(x, v, \tau) = \frac{1}{2} + \frac{1}{\pi} \int_0^\infty \mathrm{Re}\left(\frac{\exp(imx + A(m, \tau) + B(m, \tau)v)}{im}\right) dm, \quad j = 0, 1. \quad (3.5.47)$$

3.6 Problems

3.1 Consider a forward contract on an underlying commodity, find the portfolio consisting of the underlying commodity and a bond (bond's maturity coincides with forward's maturity) that replicates the forward contract.

(a) Show that the hedge ratio Δ is always equal to one. Give the financial argument to justify $\Delta = 1$.

(b) Let $B(t, T)$ denote the time-t price of the unit-par zero-coupon bond maturing at time T and let S denote the price of the commodity at time t. Show that the forward price $F(S, \tau)$ is given by

$$F(S, \tau) = S/B(t, T), \quad \tau = T - t.$$

3.2 Consider a portfolio containing Δ_t units of the risky asset and M_t dollars of the riskless asset in the form of a money market account. The portfolio is dynamically adjusted so as to replicate an option. Let S_t and $V(S_t, t)$ denote the price process of the underlying asset and the option value, respectively. Let r denote the riskless interest rate and Π_t denote the value of the *self-financing* replicating portfolio. When the self-financing trading strategy is adopted, we obtain

$$\Pi_t = \Delta_t S_t + M_t \quad \text{and} \quad d\Pi_t = \Delta_t \, dS_t + rM_t \, dt,$$

where r is the riskless interest rate. Explain why the differential term $S_t \, d\Delta_t$ does not appear in $d\Pi_t$. The asset price dynamics is assumed to follow the Geometric Brownian process:

$$\frac{dS_t}{S_t} = \mu\, dt + \sigma\, dZ_t.$$

In order that the option value and the value of the replicating portfolio match at all times, show that the number of units of asset held is given by

$$\Delta_t = \frac{\partial V}{\partial S_t}.$$

How should we proceed in order to obtain the Black–Scholes equation for V?

3.3 The following statement is quoted from Black (1989):
"...the expected return on a warrant (call option) should depend on the risk of the warrant in the same way that a common stock's expected return depends on its risk ...".
Explain the meaning of the above statement in relation to the concept of market price of risk and risk neutrality.

3.4 Consider a *self-financing* portfolio that contains α_t units of the underlying risky asset whose price process is S_t and β_t dollars of the money market account with riskless interest rate r. Suppose the initial portfolio contains α_0 units of the risky asset and β_0 dollars of money market account. Show that the time-t value of the portfolio value V_t is given by

$$\begin{aligned} V_t &= \alpha_t S_t + \beta_t e^{rt} \\ &= \alpha_0 S_0 + \beta_0 + \int_0^t \alpha_u\, dS_u + \int_0^t r\beta_u e^{ru}\, du. \end{aligned}$$

3.5 Show that

$$V(t) = V(0) + \sum_{m=0}^{M} \int_0^t h_m(u)\, dS_m(u)$$

if and only if

$$V^*(t) = V^*(0) + \sum_{m=0}^{M} \int_0^t h_m(u)\, dS_m^*(u),$$

where $h_m(t)$ is the asset holding of the mth security at time t, $V^*(t) = V(t)/S_0(t)$ and $S_m^*(t) = S_m(t)/S_0(t)$, $m = 1, 2, \cdots, M$. Deduce that a self-financing portfolio remains self-financing after a numeraire change.

3.6 Suppose the cost of carry of a commodity is b. Show that the governing differential equation for the price of the option on the commodity under the Black–Scholes formulation is given by

$$\frac{\partial V}{\partial t} + \frac{\sigma^2}{2} S^2 \frac{\partial^2 V}{\partial S^2} + bS\, \frac{\partial V}{\partial S} - rV = 0,$$

where $V(S,t)$ is the price of the option, σ and r are the constant volatility and riskless interest rate, respectively. Find the put-call parity relation for the price functions of the European put and call options on the commodity.

3.7 Suppose the price process of an asset follows the diffusion process

$$dS_t = \mu(S_t,t)\,dt + \sigma(S_t,t)\,dZ_t.$$

Show that the corresponding governing equation for the price of a derivative security V contingent on the above asset takes the form

$$\frac{\partial V}{\partial t} + \frac{1}{2}\sigma^2(S,t)\frac{\partial^2 V}{\partial S^2} + rS\frac{\partial V}{\partial S} - rV = 0,$$

where r is the riskless interest rate. Again, the derivative price V does not depend on the instantaneous mean $\mu(S_t,t)$ of the diffusion process.

3.8 Let the dynamics of the stochastic state variable S_t be governed by the Ito process

$$dS_t = \mu(S_t,t)\,dt + \sigma(S_t,t)\,dZ_t.$$

For a twice differentiable function $f(S_t)$, the differential of $f(S_t)$ is given by

$$df = \left[\mu(S_t,t)\frac{\partial f}{\partial S_t} + \frac{\sigma^2(S_t,t)}{2}\frac{\partial^2 f}{\partial S_t^2}\right]dt + \sigma(S_t,t)\frac{\partial f}{\partial S_t}\,dZ_t.$$

We let $\psi(S_t,t;S_0,t_0)$ denote the transition density function of S_t at the future time t, conditional on the value S_0 at an earlier time t_0. By considering the time-derivative of the expected value of $f(S_t)$ and equating $\frac{d}{dt}E[f(S_t)]$ and $E[\frac{df(S_t)}{dt}]$, where

$$\frac{d}{dt}E[f(S_t)] = \int_{-\infty}^{\infty} f(\xi)\frac{\partial \psi}{\partial t}(\xi,t;S_0,t_0)\,d\xi \tag{i}$$

$$E\left[\frac{df(S_t)}{dt}\right] = \int_{-\infty}^{\infty}\left[\mu(\xi,t)\frac{\partial f}{\partial \xi} + \frac{\sigma^2(\xi,t)}{2}\frac{\partial^2 f}{\partial \xi^2}\right]\psi(\xi,t;S_0,t_0)\,d\xi, \tag{ii}$$

show that $\psi(S_t,t;S_0,t_0)$ is governed by the following forward Fokker–Planck equation:

$$\frac{\partial \psi}{\partial t} + \frac{\partial}{\partial S_t}[\mu(S_t,t)\psi] - \frac{\partial^2}{\partial S_t^2}\left[\frac{\sigma^2(S_t,t)}{2}\psi\right] = 0.$$

Hint: Perform parts integration of the integral in (ii).

3.9 To derive the backward Fokker–Planck equation, we consider

$$\psi(S_t,t;S_0,t_0) = \int_{-\infty}^{\infty}\psi(S_t,t;\xi,u)\psi(\xi,u;S_0,t_0)\,d\xi,$$

where u is some intermediate time satisfying $t_0 < u < t$. Differentiating with respect to u on both sides, we obtain

$$0 = \int_{-\infty}^{\infty} \frac{\partial \psi}{\partial u}(S_t, t; \xi, u)\psi(\xi, u; S_0, t_0)\, d\xi + \int_{-\infty}^{\infty} \psi(S_t, t; \xi, u)\frac{\partial \psi}{\partial u}(\xi, u; S_0, t_0)\, d\xi.$$

From the forward Fokker–Planck equation derived in Problem 3.8, we obtain

$$\int_{-\infty}^{\infty} \frac{\partial \psi}{\partial u}(S_t, t; \xi, u)\psi(\xi, u; S_0, t_0)\, d\xi = -\int_{-\infty}^{\infty}\left\{-\frac{\partial}{\partial \xi}[\mu(\xi, u)\psi(\xi, u; S_0, t_0)] + \frac{\partial^2}{\partial \xi^2}\left[\frac{\sigma^2(\xi, u)}{2}\psi(\xi, u; S_0, t_0)\right]\right\}\psi(S_t, t; \xi, u)\, d\xi.$$

By performing parts integration of the last integral and taking the limit $u \to t_0$, show that $\psi(S_t, t; S_0, t_0)$ satisfies

$$\frac{\partial \psi}{\partial t_0} + \mu(S_0, t_0)\frac{\partial \psi}{\partial S_0} + \frac{\sigma^2(S_0, t_0)}{2}\frac{\partial^2 \psi}{\partial S_0^2} = 0.$$

Hint: $\psi(\xi, u; S_0, t_0) \to \delta(\xi - S_0)$ as $u \to t_0$.

3.10 Let Q be the martingale measure with the money market account as the numeraire and Q^* denote the equivalent martingale measure where the asset price S_t is used as the numeraire. Suppose S_t follows the Geometric Brownian process with drift rate r and volatility σ under Q, where r is the riskless interest rate. By using (3.2.11), show that

$$\left.\frac{dQ^*}{dQ}\right|_{\mathcal{F}_T} = \frac{S_T}{S_0}e^{-rT} = e^{-\frac{\sigma^2}{2}T + \sigma Z_T},$$

where Z_T is Q-Brownian. Using the Girsanov Theorems, show that

$$Z_T^* = Z_T - \sigma T$$

is Q^*-Brownian. Explain why

$$E_{Q^*}[\mathbf{1}_{\{S_T > X\}}] = N\left(\frac{\ln \frac{S_0}{X} + \left(r + \frac{\sigma^2}{2}\right)T}{\sigma\sqrt{T}}\right),$$

then deduce that [see (3.3.12b)]

$$E_Q[S_T \mathbf{1}_{\{S_T > X\}}] = e^{rT} S_0 N\left(\frac{\ln \frac{S_0}{X} + \left(r + \frac{\sigma^2}{2}\right)T}{\sigma\sqrt{T}}\right).$$

Let U_t be another asset whose price dynamics under Q is governed by

$$\frac{dU_t}{U_t} = r\,dt + \sigma_U\,dZ_t^U,$$

where $dZ_t^U\,dZ_t = \rho\,dt$ and ρ is the correlation coefficient. Show that

$$Z_t^{*U} = Z_t^U - \rho\sigma_U T$$

is a Q^*-Brownian process.

Hint: Since dZ_t^U and dZ_t are correlated with correlation coefficient ρ, we may write

$$dZ_t^U = \rho\,dZ_t + \sqrt{1-\rho^2}\,dZ_t^{\perp},$$

where $Z_t^{\perp}$ is uncorrelated with Z_t.

3.11 From the Black–Scholes price function $c(S,\tau)$ for a European vanilla call, show that the limiting values of the call price at vanishing volatility and infinite volatility are the lower and upper bounds of the European call price respectively, namely,

$$\lim_{\sigma\to 0^+} c(S,\tau) = \max(S - Xe^{-r\tau}, 0),$$

and

$$\lim_{\sigma\to\infty} c(S,\tau) = S.$$

Give an appropriate financial interpretation of the above results. Apparently, X does not appear in $c(S,\tau)$ when $\sigma \to \infty$. Is it justifiable based on financial intuition?

3.12 Show that when a European option is currently out-of-the-money, then higher volatility of the asset price or longer time to expiry makes it more likely for the option to expire in-the-money. What would be the impact on the value of delta? Do we have the same effect or opposite effect when the option is currently in-the-money? Also, give the financial interpretation of the asymptotic behavior of the delta curves in Fig. 3.4 at the respective limit $\tau \to 0^+$ and $\tau \to \infty$.

3.13 Show that when the European call price is a convex function of the asset price, the elasticity of the call price is always greater than or equal to one. Give the financial argument to explain why the elasticity of the price of a European option increases in absolute value when the option becomes more out-of-the-money and closer to expiry. Can you think of a situation where the European put's elasticity has absolute value less than one, that is, the European put option is less risky than the underlying asset?

3.14 Suppose the greeks of the value of a derivative security are defined by

$$\Theta = \frac{\partial f}{\partial t}, \quad \Delta = \frac{\partial f}{\partial S}, \quad \Gamma = \frac{\partial^2 f}{\partial S^2}.$$

(a) Find the relation between Θ and Γ for a delta-neutral portfolio where $\Delta = 0$.
(b) Show that the theta may become positive for an in-the-money European call option on a continuous dividend paying asset when the dividend yield is sufficiently high.
(c) Explain by financial argument why the theta value tends asymptotically to $-rXe^{-r\tau}$ from below when the asset value is sufficiently high.

3.15 Let $P_\alpha(\tau)$ denote the European put price normalized by the asset price, that is,

$$P_\alpha(\tau) = p(S,\tau)/S = \alpha e^{-r\tau}N(-d_-) - N(-d_+),$$

where

$$\gamma_- = \frac{r - \frac{\sigma^2}{2}}{\sigma}, \quad \gamma_+ = \frac{r + \frac{\sigma^2}{2}}{\sigma}, \quad \alpha = \frac{X}{S}, \quad \beta = \frac{\ln\frac{1}{\alpha}}{\sigma},$$
$$d_- = \gamma_-\sqrt{\tau} + \frac{\beta}{\sqrt{\tau}}, \quad d_+ = \gamma_+\sqrt{\tau} + \frac{\beta}{\sqrt{\tau}}.$$

We would like to explore the behavior of the temporal rate of change of the European put price. The derivative of $P_\alpha(\tau)$ with respect to τ is found to be

$$P'_\alpha(\tau) = \alpha e^{-r\tau}\left[-rN(-d_-) + n(-d_-)\frac{\sigma}{2\sqrt{\tau}}\right].$$

Define $f(\tau)$ by the relation $P'_\alpha(\tau) = \alpha e^{-r\tau}f(\tau)$, and the quadratic polynomial $p_2(\tau)$ by

$$p_2(\tau) = \gamma_-\gamma_+\tau^2 - [\beta(\gamma_- + \gamma_+) + 1]\tau + \beta^2.$$

Let τ_1 and τ_2 denote the two real roots of $p_2(\tau)$, where $\tau_1 < \tau_2$, and let $\tau_0 = \frac{\beta^2\sigma}{2r\beta+\sigma}$. The sign behavior of $P'_\alpha(\tau)$ exhibits the following properties (Dai and Kwok, 2005c).

1. When $r \le 0$, $P'_\alpha(\tau) > 0$ for all $\tau \ge 0$.
2. When $r > 0$ and $\beta \ge 0$ (equivalent to $S \ge X$), there exists unique $\tau^* > 0$ at which $P'_\alpha(\tau)$ changes sign, and that $P'_\alpha(\tau) > 0$ for $\tau < \tau^*$ and $P'_\alpha(\tau) < 0$ for $\tau > \tau^*$.
3. When $r > 0$ and $\beta < 0$ (equivalent to $S < X$), there are two possibilities:
   (a) There may exist a time interval (τ_1^*, τ_2^*) such that $P'_\alpha(\tau) > 0$ when $\tau \in (\tau_1^*, \tau_2^*)$ and $P'_\alpha(\tau) \le 0$ if otherwise. This occurs only when either one of the following conditions is satisfied.
      (i) $\gamma_- < 0$ and $f(\tau_2) > 0$;
      (ii) $\gamma_- > 0$, $\beta(\gamma_- + \gamma_+) + 1 > 0$, $\Delta = \beta^2\sigma^2 + 1 + \dfrac{4\beta r}{\sigma} > 0$ and $f(\tau_1) > 0$;
      (iii) $\gamma_- = 0$, $\beta(\gamma_- + \gamma_+) + 1 > 0$ and $f(\tau_0) > 0$.
   (b) When none of the above conditions (i)–(iii) hold, then $P'_\alpha(\tau) \le 0$ for all $\tau \ge 0$.

3.16 Show that the value of a European call option satisfies

$$c(S,\tau;X) = S\frac{\partial c}{\partial S}(S,\tau;X) + X\frac{\partial c}{\partial X}(S,\tau;X).$$

Hint: The call price function is a linear homogeneous function of S and X, that is,

$$c(\lambda S,\tau;\lambda X) = \lambda c(S,\tau;X).$$

3.17 Consider a European capped call option whose terminal payoff function is given by

$$c_M(S,0;X,M) = \min(\max(S-X,0),M),$$

where X is the strike price and M is the cap. Show that the value of the European capped call is given by

$$c_M(S,\tau;X,M) = c(S,\tau;X) - c(S,\tau;X+M),$$

where $c(S,\tau;X+M)$ is the value of a European vanilla call with strike price $X+M$.

3.18 Consider the value of a European call option written by an issuer whose only asset is α (< 1) units of the underlying asset. At expiration, the terminal payoff of this call is then given by

$$\begin{array}{rl} S_T - X & \text{if} \quad \alpha S_T \geq S_T - X \geq 0 \\ \alpha S_T & \text{if} \quad S_T - X > \alpha S_T \end{array}$$

and zero otherwise. Show that the value of this European call option is given by (Johnson and Stulz, 1987)

$$c_L(S,\tau;X,\alpha) = c(S,\tau;X) - (1-\alpha)c\left(S,\tau;\frac{X}{1-\alpha}\right), \quad \alpha < 1,$$

where $c(S,\tau;\frac{X}{1-\alpha})$ is the value of a European vanilla call with strike price $\frac{X}{1-\alpha}$.

3.19 Consider the price functions of European call and put options on an underlying asset which pays a dividend yield at the rate q, show that their deltas and thetas are given by

$$\begin{aligned}
\frac{\partial c}{\partial S} &= e^{-q\tau}\,N(\widehat{d_1}) \\
\frac{\partial p}{\partial S} &= e^{-q\tau}\,[N(\widehat{d_1}) - 1] \\
\frac{\partial c}{\partial t} &= -\frac{Se^{-q\tau}\sigma N'(\widehat{d_1})}{2\sqrt{\tau}} + qSe^{-q\tau}N(\widehat{d_1}) - rXe^{-r\tau}N(\widehat{d_2}) \\
\frac{\partial p}{\partial t} &= -\frac{Se^{-q\tau}\sigma N'(\widehat{d_1})}{2\sqrt{\tau}} - qSe^{-q\tau}N(-\widehat{d_1}) + rXe^{-r\tau}N(-\widehat{d_2}),
\end{aligned}$$

where $\widehat{d_1}$ and $\widehat{d_1}$ are defined in (3.4.4). Deduce the expressions for the gammas, vegas and rhos of the above call and put option prices.

3.20 Deduce the corresponding put-call parity relation when the parameters in the European option models are time dependent, namely, volatility of the asset price is $\sigma(t)$, dividend yield is $q(t)$ and riskless interest rate is $r(t)$.

3.21 Explain why the option price should be continuous across a dividend date though the asset price experiences a jump. Using no arbitrage principle, deduce the following jump condition:

$$V(S(t_d^+), t_d^+) = V(S(t_d^-), t_d^-),$$

where V denotes option price, t_d^- and t_d^+ denote the time just before and after the dividend date t_d.

3.22 Suppose the dividends and interest incomes are taxed at the rate R but capital gains taxes are zero. Find the price formulas of the European put and call on an asset which pays a continuous dividend yield at the constant rate q, assuming that the riskless interest rate r is also constant.
Hint: Explain why the riskless interest rate r and dividend yield q should be replaced by $r(1-R)$ and $q(1-R)$, respectively, in the Black–Scholes formulas.

3.23 Consider futures on an underlying asset that pays N discrete dividends between t and T and let D_i denote the amount of the ith dividend paid on the ex-dividend date t_i. Show that the futures price is given by

$$F(S,t) = Se^{r(T-t)} - \sum_{i=1}^{N} D_i e^{r(T-t_i)},$$

where S is the current asset price and r is the riskless interest rate. Consider a European call option on the above futures. Show that the governing differential equation for the price of the call, $c_F(F,t)$, is given by (Brenner, Courtadon and Subrahmanyan, 1985)

$$\frac{\partial c_F}{\partial t} + \frac{\sigma^2}{2}\left[F + \sum_{i=1}^{N} D_i e^{r(T-t_i)}\right]^2 \frac{\partial^2 c_F}{\partial F^2} - rc_F = 0.$$

3.24 A *forward start* option is an option that comes into existence at some future time T_1 and expires at T_2 ($T_2 > T_1$). The strike price is set equal the asset price at T_1 such that the option is at-the-money at the future option's initiation time T_1. Consider a forward start call option whose underlying asset has value S at current time t and constant dividend yield q, show that the value of the forward start call is given by

$$e^{-qT_1}c(S, T_2 - T_1; S),$$

where $c(S, T_2 - T_1; S)$ is the value of an at-the-money call (strike price same as asset price) with time to expiry $T_2 - T_1$.

Hint: The value of an at-the-money call option is proportional to the asset price.

3.25 Show that the payoff function of a chooser option on the date of choice T_c can be alternatively decomposed into the following form:

$$\begin{aligned} V(S_{T_c}, T_c) &= \max(p + S_{T_c}e^{-q(T-T_c)} - Xe^{-r(T-T_c)}, p) \\ &= p + e^{-q(T-T_c)} \max(S_{T_c} - Xe^{-(r-q)(T-T_c)}, 0) \end{aligned}$$

[see (3.4.22)]. Find the alternative representation of the price formula of the chooser option based on the above decomposition. Show that your new formula agrees with that given by (3.4.23).

3.26 Suppose the holder of the chooser option can make the choice of either a call or a put at any time between now and a later cutoff date T_c. Is it optimal for the holder to make the choice at some time before T_c?

Hint: Apparently, the price function of the chooser option depends on T_c [see (3.4.23)]. Check whether the price function is an increasing or decreasing function of T_c. Consider the extreme case where T_c coincides with the expiration date of the option.

3.27 Consider a chooser option that entitles the holder to choose, on the choice date T_c periods from now, whether the option is a European call with exercise price X_1 and time to expiration $T_1 - T_c$ or a European put with exercise price X_2 and time to expiration $T_2 - T_c$. Show that the price of the chooser option at the current time (taken to be time zero) is given by (Rubinstein, 1992)

$$Se^{-qT_1}N_2(x, y_1; \rho_1) - X_1e^{-rT_1}N_2(x - \sigma\sqrt{T_c}, y_1 - \sigma\sqrt{T_1}; \rho_1)$$

$$-Se^{-qT_2}N_2(-x, -y_2; \rho_2) + X_2e^{-rT_2}N_2(-x + \sigma\sqrt{T_c}, -y_2 + \sigma\sqrt{T_2}; \rho_2),$$

where q is the continuous dividend yield of the underlying asset. The parameters are defined by

$$x = \frac{\ln\frac{S}{X} + (r - q + \frac{\sigma^2}{2})T_c}{\sigma\sqrt{T_c}}, \quad \rho_1 = \sqrt{\frac{T_c}{T_1}}, \quad \rho_2 = \sqrt{\frac{T_c}{T_2}},$$

$$y_1 = \frac{\ln\frac{S}{X_1} + (r - q + \frac{\sigma^2}{2})T_1}{\sigma\sqrt{T_1}}, \quad y_2 = \frac{\ln\frac{S}{X_2} + (r - q + \frac{\sigma^2}{2})T_2}{\sigma\sqrt{T_2}}.$$

Here, X solves the following nonlinear algebraic equation

$$Xe^{-q(T_1-T_c)}N(z_1) - X_1e^{-r(T_1-T_c)}N(z_1 - \sigma\sqrt{T_1 - T_c})$$
$$+ Xe^{-q(T_2-T_c)}N(-z_2) - X_2e^{-r(T_2-T_c)}N(-z_2 + \sigma\sqrt{T_2 - T_c}) = 0,$$

where

$$z_1 = \frac{\ln \frac{X}{X_1} + (r - q + \frac{\sigma^2}{2})\,(T_1 - T_c)}{\sigma\sqrt{T_1 - T_c}},$$
$$z_2 = \frac{\ln \frac{X}{X_2} + (r - q + \frac{\sigma^2}{2})\,(T_2 - T_c)}{\sigma\sqrt{T_2 - T_c}}.$$

Hint: The two overlapping standard Brownian increments $Z(T_c)$ and $Z(T_1)$ have the joint normal distribution with zero means, unit variances and correlation coefficient $\sqrt{\frac{T_c}{T_1}}$, $T_c < T_1$.

3.28 Show that the first term in the last integral in (3.4.27) can be expressed as

$$e^{-r(T_1-t)} \int_{\widetilde{S}_{T_1}}^{\infty} S_{T_1} N_1(d_1) \psi(S_{T_1}; S)\, dS_{T_1}$$
$$= e^{-r(T_1-t)} \int_{\ln \widetilde{S}_{T_1}}^{\infty} \int_{\ln X_2}^{\infty} \frac{1}{2\pi} \frac{1}{\sigma\sqrt{T_1 - t}} \frac{1}{\sigma\sqrt{T_2 - T_1}} Se^{r(T_1-t)}$$
$$\exp\left(-\frac{\{y - [\ln S + (r + \frac{\sigma^2}{2})(T_1 - t)]\}^2}{2\sigma^2(T_1 - t)}\right)$$
$$\exp\left(-\frac{\{x - [y + (r + \frac{\sigma^2}{2})(T_2 - T_1)]\}^2}{2\sigma^2(T_2 - T_1)}\right) dxdy,$$

where $x = \ln S_{T_2}$ and $y = \ln S_{T_1}$. Through comparison with the second term in (3.4.27), show that the above integral reduces to the first term given in (3.4.29).

Hint: The second term in (3.4.27) can be expressed as

$$-X_2e^{-r(T_2-t)} \int_{\ln \widetilde{S}_{T_1}}^{\infty} \int_{\ln X_2}^{\infty} \frac{1}{2\pi} \frac{1}{\sigma\sqrt{T_1 - t}} \frac{1}{\sigma\sqrt{T_2 - T_1}}$$
$$\exp\left(-\frac{\{y - [\ln S + (r - \frac{\sigma^2}{2})(T_1 - t)]\}^2}{2\sigma^2(T_1 - t)}\right)$$
$$\exp\left(-\frac{\{x - [y + (r - \frac{\sigma^2}{2})(T_2 - T_1)]\}^2}{2\sigma^2(T_2 - T_1)}\right) dxdy.$$

3.29 Explain why the sum of prices of the call-on-a-call and call-on-a-put is equal to the price of the call with expiration T_2. Show that the price of a European call-on-a-put is given by

$$c(S,t) = X_2 e^{-r(T_2-t)} N_2(a_2, -b_2; -\rho) - S N_2(a_1, -b_1; -\rho) \\ - e^{-r(T_1-t)} X_1 N(a_2),$$

where a_1, b_1, a_2 and b_2 are defined in Sect. 3.4.4.
Hint: Use the relation

$$N_2(a, b; \rho) + N_2(a, -b; -\rho) = N(a).$$

3.30 Find the price formulas for the following European compound options:
(a) put-on-a-call option when the underlying asset pays a continuous dividend yield q;
(b) call-on-a-put option when the underlying asset is a futures;
(c) put-on-a-put option when the underlying asset has a constant cost of carry b.

3.31 Consider a contingent claim whose value at maturity T is given by

$$\min(S_{T_0}, S_T),$$

where T_0 is some intermediate time before maturity, $T_0 < T$, and S_T and S_{T_0} are the asset price at T and T_0, respectively. Assuming the usual Geometric Brownian process for the price of the underlying asset that pays no dividend, show that the value of the contingent claim at time t is given by

$$V = S[1 - N(d_1) + e^{-r(T-T_0)} N(d_2)],$$

where S is the asset price at time t and

$$d_1 = \frac{r(T-T_0) + \frac{\sigma^2}{2}(T-T_0)}{\sigma\sqrt{T-T_0}}, \quad d_2 = d_1 - \sigma\sqrt{T-T_0}.$$

3.32 In the Merton model of risky debt, suppose we define

$$\sigma_V(\tau; d) = \frac{\sigma A}{V}\frac{\partial V}{\partial A},$$

which gives the volatility of the value of the risky debt. Also, we denote the credit spread by $s(\tau; d)$, where $s(\tau; d) = Y(\tau) - r$. Show that (Merton, 1974)
(a) $\dfrac{\partial s}{\partial d} = \dfrac{1}{\sigma\tau d}\sigma_V(\tau; d) > 0$;
(b) $\dfrac{\partial s}{\partial \sigma^2} = \dfrac{1}{2\sqrt{\tau}}\dfrac{N'(d_1)}{N(d_1)}\sigma_V(\tau; d) > 0$, where $d_1 = \dfrac{\ln d}{\sigma\sqrt{\tau}} - \dfrac{\sigma\sqrt{\tau}}{2}$;
(c) $\dfrac{\partial s}{\partial r} = -\dfrac{\sigma_V}{\sigma}(\tau; d) < 0$.
Give the financial interpretation to each of the above results.

3.33 A firm is an entity that consists of its assets and let A_t denote the market value of the firm's assets. Assume that the total asset value follows a stochastic process modeled by

$$\frac{dA_t}{A_t} = \mu\, dt + \sigma\, dZ_t,$$

where μ and σ^2 (assumed to be constant) are the instantaneous mean and variance, respectively, of the rate of return on A_t. Let C and D denote the market value of the current liabilities and market value of debt, respectively. Let T be the maturity date of the debt with face value D_T. Suppose the current liabilities of amount C_T are also payable at time T, and it constitutes a claim senior to the debt. Also, let F denote the present value of total amount of interest and dividends paid over the term T. For simplicity, F is assumed to be prepaid at time $t = 0$.

The debt is in default if A_T is less than the total amount payable at maturity date T, that is,

$$A_T < D_T + C_T.$$

(a) Show that the probability of default is given by

$$p = N\left(\frac{\ln \frac{D_T+C_T}{A_0-F} - \mu T + \frac{\sigma^2 T}{2}}{\sigma\sqrt{T}}\right).$$

(b) Explain why the expected loan loss L on the debt is given by

$$EL = \int_{C_T}^{D_T+C_T} (D_T + C_T - a) f(a)\, da + \int_0^{C_T} D_T f(a)\, da,$$

where f is the density function of A_T. Give the financial interpretation to each of the above integrals.

3.34 Consider the exchange option that gives the holder the right but not the obligation to exchange risky asset S_2 for another risky asset S_1. Let the price dynamics of S_1 and S_2 under the risk neutral measure be governed by

$$\frac{dS_i}{S_i} = (r - q_i)\, dt + \sigma_i\, dZ_i, \quad i = 1, 2,$$

where $dZ_1\, dZ_2 = \rho\, dt$. Let $V(S_1, S_2, \tau)$ denote the price function of the exchange option, whose terminal payoff takes the form

$$V(S_1, S_2, 0) = \max(S_1 - S_2, 0).$$

Show that the governing equation for $V(S_1, S_2, \tau)$ is given by

$$\frac{\partial V}{\partial \tau} = \frac{\sigma_1^2}{2} S_1^2 \frac{\partial^2 V}{\partial S_1^2} + \rho\sigma_1\sigma_2 S_1 S_2 \frac{\partial^2 V}{\partial S_1 \partial S_2} + \frac{\sigma_2^2}{2} S_2^2 \frac{\partial^2 V}{\partial S_2^2}$$
$$+ (r - q_1) S_1 \frac{\partial V}{\partial S_1} + (r - q_2) S_2 \frac{\partial V}{\partial S_2} - rV.$$

By taking S_2 as the numeraire and defining the similarity variables:

$$x = \frac{S_1}{S_2} \quad \text{and} \quad W(x, \tau) = \frac{V(S_1, S_2, \tau)}{S_2},$$

show that the governing equation for $W(x, \tau)$ becomes

$$\frac{\partial W}{\partial \tau} = \frac{\sigma^2}{2} x^2 \frac{\partial^2 W}{\partial x^2} + (q_1 - q_2) x \frac{\partial W}{\partial x} - q_2 W.$$

Verify that the solution to $W(x, \tau)$ is given by

$$W(x, \tau) = e^{-q_1 \tau} x N(d_1) - e^{-q_2 \tau} N(-d_2),$$

where

$$d_1 = \frac{\ln \frac{S_1}{S_2} + \left(q_2 - q_1 + \frac{\sigma^2}{2}\right)\tau}{\sigma\sqrt{\tau}},$$

$$d_2 = \frac{\ln \frac{S_2}{S_1} + (q_1 - q_2 + \frac{\sigma^2}{2})\tau}{\sigma\sqrt{\tau}},$$

$$\sigma^2 = \sigma_1^2 - 2\rho\sigma_1\sigma_2 + \sigma_2^2.$$

Referring to the original option price function $V(S_1, S_2, \tau)$, we obtain

$$V(S_1, S_2, \tau) = e^{-q_1 \tau} S_1 N(d_1) - e^{-q_2 \tau} S_2 N(-d_2).$$

3.35 Suppose the terminal payoff of an exchange rate option is $F_T \mathbf{1}_{\{F_T > X\}}$. Let $V_d(F, t)$ denote the value of the option in domestic currency, show that

$$\begin{aligned} V_d(F, t) &= e^{-r_f(T-t)} F E_{Q_f}\left[\mathbf{1}_{\{F_T > X\}} | F_t = F\right] \\ &= e^{-r_f \tau} F N(d) = e^{-r_d \tau} e^{\delta_F^d \tau} F N(d), \end{aligned}$$

where $\delta_F^d = r_d - r_f$ and

$$d = \frac{\ln \frac{F}{K} + \left(r_d - r_f + \frac{\sigma_F^2}{2}\right)\tau}{\sigma_F \sqrt{\tau}}, \quad \tau = T - t.$$

3.36 Let $F_{S\backslash U}$ denote the Singaporean currency price of one unit of U.S. currency and $F_{H\backslash S}$ denote the Hong Kong currency price of one unit of Singaporean currency. We may interpret $F_{S\backslash U}$ as the price process of a tradeable asset in Singaporean currency. Assume $F_{S\backslash U}$ to be governed by the following dynamics under the risk neutral measure Q_S in the Singaporean currency world:

$$\frac{dF_{S\backslash U}}{F_{S\backslash U}} = (r_{SGD} - r_{USD})\, dt + \sigma_{F_{S\backslash U}}\, dZ^S_{F_{S\backslash U}},$$

where r_{SGD} and r_{USD} are the Singaporean and U.S. riskless interest rates, respectively. The digital quanto option pays one Hong Kong dollar if $F_{S\backslash U}$ is above the strike level X. Find the value of the digital quanto option in terms of the riskless interest rates of the different currency worlds and volatility values $\sigma_{F_{S\backslash U}}$ and $\sigma_{F_{H\backslash S}}$. Recalculate the digital quanto option if it pays one Hong Kong dollar when $F_{S\backslash U} > \alpha F_{H\backslash U}$, where α is a fixed constant and $F_{H\backslash U}$ is the Hong Kong currency price of one unit of U.S. currency.
Hint: $F_{H\backslash U} = F_{H\backslash S} F_{S\backslash U}$.

3.37 Show that the total transaction costs in Leland's model (Leland, 1985) increases (decreases) with the strike price X when $X < X^*$ $(X > X^*)$, where

$$X^* = Se^{(r+\frac{\sigma^2}{2})(T-t)}.$$

Hint: Use the result $\frac{\partial}{\partial X}(\frac{\partial V}{\partial \sigma}) = \frac{S}{\sqrt{2\pi(T-t)}\,\sigma} d_1 \exp(-\frac{d_1^2}{2})$.

3.38 Suppose the transaction costs are proportional to the number of units of asset traded rather than the dollar value of the asset traded as in Leland's original model. Find the corresponding governing equation for the price of a derivative based on this new transaction costs assumption.

3.39 By writing $P_n(\tau) = e^{-\lambda\tau}\frac{(\lambda\tau)^n}{n!}$ and $\widetilde{S}_n = SX_n e^{-\lambda k\tau}$, and considering

$$V(S,\tau) = \sum_{n=0}^{\infty} P_n(\tau) E_{X_n}[V_{BS}(\widetilde{S}_n, \tau)]$$

[see (3.5.20)], show that

$$\frac{\partial V}{\partial \tau} = -\lambda V - \lambda k S \frac{\partial V}{\partial S} + \sum_{n=0}^{\infty} P_n(\tau) E_{X_n}\left[\frac{\partial V_{BS}}{\partial \tau}(\widetilde{S}_n, \tau)\right]$$
$$+ \lambda \sum_{m=0}^{\infty} P_m(\tau) E_{X_{m+1}}[V_{BS}(\widetilde{S}_{m+1}, \tau)].$$

Furthermore, by observing that

$$E_J[V(JS,\tau)] = \sum_{n=0}^{\infty} P_n(\tau) E_{X_{n+1}}[V_{BS}(\widetilde{S}_{n+1}, \tau)],$$

show that $V(S,\tau)$ satisfies the governing equation (3.5.19). Also, show that $V(S,\tau)$ and $V_{BS}(S,\tau)$ satisfy the same terminal payoff condition.

3.40 Suppose $\ln J$ is normally distributed with standard deviation σ_J. Show that the price of a European vanilla option under the jump-diffusion model can be expressed as (Merton, 1976)

$$V(S,\tau) = \sum_{n=0}^{\infty} \frac{e^{-\lambda'\tau}(\lambda'\tau)^n}{n!} V_{BS}(S,\tau;\sigma_n, r_n),$$

where

$$\lambda' = \lambda(1+k),\ \sigma_n^2 = \sigma^2 + \frac{n\sigma_J^2}{\tau} \quad \text{and} \quad r_n = r - \lambda k + \frac{n\ln(1+k)}{\tau}.$$

3.41 Consider the expression for $d\Pi$ given in (3.5.17). Show that the variance of $d\Pi$ is given by

$$\begin{aligned}\text{var}(d\Pi) = &\left(\Delta - \frac{\partial V}{\partial S}\right)^2 \sigma^2 S^2\, dt \\ &+ \lambda E_J[\{\Delta(J-1)S - [V(JS,t) - V(S,t)]\}^2]\, dt.\end{aligned}$$

Suppose we try to hedge the diffusion and jump risks as much as possible by minimizing var$(d\Pi)$. Show that this can be achieved by choosing Δ such that

$$\Delta = \frac{\lambda E_J[(J-1)\{V(JS,t) - V(S,t)\}] + \sigma^2 S \frac{\partial V}{\partial S}}{\lambda S E_J[(J-1)^2] + \sigma^2 S}.$$

With this choice of Δ, find the corresponding governing equation for the option price function under the jump-diffusion asset price dynamics.

3.42 Suppose $V(\sigma)$ is the option price function with dependence on volatility σ. Show that

$$V''(\sigma) = \frac{V'(\sigma)\tau}{4\sigma^3}(\sigma_1^4 - \sigma^4) \quad \text{for all } \sigma,$$

where σ_1 is given by (3.5.25). Hence, deduce that $V'' > 0$ if $\sigma_1 > \sigma_{imp}$ and $V'' < 0$ if $\sigma_1 < \sigma_{imp}$, where σ_{imp} is the implied volatility. Explain why $V(\sigma)$ is strictly convex if $\sigma_1 > \sigma_{imp}$ and strictly concave if $\sigma_1 < \sigma_{imp}$, and deduce that

$$0 < \frac{\sigma_{n+1} - \sigma_{imp}}{\sigma_n - \sigma_{imp}} < 1$$

for both cases (Manaster and Koehler, 1982).

3.43 Assume that the time dependent volatility function $\sigma(t)$ is deterministic. Suppose we write $\sigma_{imp}(t,T)$ as the implied volatility obtained from the time-t price of a European option with maturity T, for $T > t$. Show that

$$\sigma(T) = \sqrt{\sigma_{imp}^2(t,T) + 2(T-t)\sigma_{imp}(t,T)\frac{\partial}{\partial T}\sigma_{imp}(t,T)}.$$

In real situations, we may have the implied volatility available only at discrete times $T_i, i = 1, 2, \cdots, N$. Assuming the volatility $\sigma(T)$ to be piecewise constant over each time interval $[T_{i-1}, T_i], i = 1, 2, \cdots, N$, show that

$$\sigma(u) = \sqrt{\frac{(T_i - t)\sigma_{imp}^2(t, T_i) - (T_{i-1} - t)\sigma_{imp}^2(t, T_{i-1})}{T_i - T_{i-1}}}$$
$$\text{for} \quad T_{i-1} < u < T_i.$$

Hint: The implied volatility $\sigma_{imp}(t, T)$ and the time dependent volatility function $\sigma(t)$ are related by

$$\sigma_{imp}^2(t, T)(T - t) = \int_t^T \sigma^2(u)\, du.$$

3.44 We would like to compute $d(S_T - X)^+$, where S_t follows the Geometric Brownian process

$$\frac{dS_t}{S_t} = (r - q)\, dt + \sigma(S_t, t)\, dZ_t.$$

The function $(S_T - X)^+$ has a discontinuity at $S_T = X$. Rossi (2002) proposed to approximate $(S_T - X)^+$ by the following function $f(S_T)$ whose first derivative is continuous, where

$$f(S_T) = \begin{cases} 0 & \text{if } S_T < X \\ \frac{(S_T - X)^2}{2\epsilon} & \text{if } X \le S_T \le X + \epsilon \\ S_T - X - \frac{\epsilon}{2} & \text{if } S_T > X + \epsilon \end{cases}.$$

Here, ϵ is a small positive quantity. By applying Ito's lemma, show that

$$f(S_T) = f(S_0) + \int_0^T f''(S_t)\, dS_t + \frac{1}{2}\int_0^T \sigma^2(S_t, t) S_t^2 f''(S_t)\, dt.$$

By taking the limit $\epsilon \to 0$, explain why

$$\int_0^T f'(S_t)\, dS_t \quad \to \quad \int_0^T \mathbf{1}_{\{S_t > X\}}\, dS_t$$
$$\int_0^T \sigma^2(S_t, t) S_t^2 f''(S_t)\, dt \quad \to \quad \int_0^T \sigma^2(S_t, t) S_t^2 \delta(S_t - X)\, dt.$$

Finally, show that

$$d(S_T - X)^+ = \mathbf{1}_{\{S_T > X\}} dS_T + \frac{\sigma^2(S_T, T)}{2} S_T^2 \delta(S_T - X)\, dT.$$

3.45 Under the risk neutral measure Q, the stochastic process of the logarithm of the asset price $x_t = \ln S_t$ and its instantaneous volatility σ_t are assumed to be governed by

$$dx_t = \left(r - \frac{\sigma_t^2}{2}\right) dt + \sigma_t \, dZ_x$$
$$d\sigma_t = k(\theta - \sigma_t) \, dt + \eta \, dZ_\sigma,$$

where $dZ_x \, dZ_\sigma = \rho \, dt$. All model parameters are taken to be constant. The price function of a European call option with strike price X and maturity date T takes the form

$$c(S_t, \sigma_t, t; T) = S_t F_1 - e^{-r(T-t)} X F_0,$$

where

$$F_j = \frac{1}{2} + \frac{1}{\pi} \int_0^\infty \mathrm{Re}\left(\frac{f_j(\phi) e^{-i\phi X}}{i\phi} \right) d\phi, \quad j = 0, 1,$$
$$f_0(\phi) = E_Q^t[\exp(i\phi x_T)],$$
$$f_1(\phi) = E_Q^t[\exp(-r(T-t) - x_t + (1 + i\phi) x_T)].$$

Solve for $f_0(\phi)$ and $f_1(\phi)$ (Schöbel and Zhu, 1999).

4

Path Dependent Options

The quest for new, innovative derivative products pushes financial institutions to design and develop more exotic forms of structured products, many of which are aimed toward the specific needs of the customers. Recently, there has been a growing popularity for path dependent options, so named since their payoff structures are related to the underlying asset price path history during the whole or part of the life of the option. The *barrier option* is the most popular path dependent option that is either nullified, activated or exercised when the underlying asset price breaches a *barrier* during the life of the option. The capped stock-index options, based on the Standard and Poor's (S&P) 100 and 500 Indexes, are well-known examples of barrier options traded in option exchanges (they were launched by the Chicago Board of Exchange in 1991). These capped options will be exercised automatically when the index value exceeds the cap at the close of the day. The payoff of a *lookback option* depends on the minimum or maximum price of the underlying asset attained during a certain period of the life of the option, while the payoff of an *average option* (usually called an *Asian option*) depends on the average asset price over some period within the life of the option. An interesting example is the *Russian option*, which is in fact a perpetual American lookback option. The owner of a Russian option on an asset receives the historical maximum value of the asset price when the option is exercised and the option has no preset expiration date.

In this chapter, we discuss the product nature of barrier options, lookback options and Asian options, and present analytic procedures for their valuation. Due to the path dependent nature of these options, the asset price process is monitored over the life of the option contract either for breaching of a barrier level, observation of a new extremum value or sampling of asset prices for computing the average value. In actual implementation, these monitoring procedures can be only performed at discrete times rather than continuously at all times. However, most pricing models of path dependent options assume continuous monitoring of the asset price in order to achieve good analytic tractability. We derive analytic price formulas for the most common types of continuously monitored barrier and lookback options, and geometric averaged Asian options. For the arithmetic averaged Asian options, we manage to obtain analytic approximation price formulas. Under the Black–Scholes pricing paradigm, we assume that the uncertainty in the financial market over the time horizon $[0, T]$ is modeled by a filtered probability space $(\Omega, \mathcal{F}, \{\mathcal{F}_t\}_{t\in[0,T]}, Q)$, where Q is the risk neutral (equivalent martingale) probability measure and the filtration $\mathcal{F}_t$ is generated by the standard Brownian process $\{Z(u) : 0 \leq u \leq t\}$. All discounted prices of securities are Q-martingales. Under Q, the asset price process S_t follows the Geometric Brownian process with riskless interest rate r as the drift rate and constant volatility σ.

When the asset price path is monitored at discrete time instants, the analytic forms of the price formulas become quite daunting because they involve multidimensional cumulative normal distribution functions and whose dimension is equal to the number of monitoring instants. We discuss briefly some effective analytic approximation techniques for estimating the prices of discretely monitored path dependent options.

4.1 Barrier Options

Options with the barrier feature are considered to be the simplest types of path dependent options. Barrier option's distinctive feature is that the payoff depends not only on the final price of the underlying asset, but also on whether the asset price has breached (one-touch) some barrier level during the life of the option. An *out-barrier option* (or knock-out option) is one where the option is nullified prior to expiration if the underlying asset price touches the barrier. The holder of the option may be compensated by a rebate payment for the cancellation of the option. An *in-barrier option* (or knock-in option) is one where the option only comes in existence if the asset price crosses the in-barrier, though the holder has already paid the option premium up front. When the barrier is upstream with respect to the asset price, the barrier option is called an *up-option*; otherwise, it is called a *down-option*. One can identify eight types of European barrier options, such as down-and-out calls, up-and-out calls, down-and-in puts, down-and-out puts, etc. Also, we may have two-sided barrier options that have both upside and downside barriers. Nullification or activation of the contract occurs either when one of the barriers is touched or only when the two barriers are breached in a prespecified sequential order. The latter type of options are called *sequential barrier options*. Suppose the knock-in or knock-out feature is activated only when the asset price breaches the barrier for a prespecified length of time (rather than one touch of the barrier), we call this special type of barrier options as *Parisian options* (Chesney, Jeanblanc-Picqué and Yor, 1997).

Why are barrier options popular? From the perspective of the buyer of an option contract, he or she can achieve *option premium reduction* through the barrier provision by not paying a premium to cover scenarios he or she views as unlikely. For example, the buyer of a down-and-out call believes that the asset price would never fall below some floor value, so he or she can reduce the option premium by allowing the option to be nullified when the asset price does fall below the perceived floor value. As another example, consider the up-and-out call. How can both buyer and writer benefit from the barrier structure? With an appropriate rebate being paid upon breaching the upstream barrier, this type of barrier options provide upside exposure for the option buyer but at a lower cost. On the other hand, the option writer is not exposed to unlimited liabilities when the asset price rises acutely. In general, barrier options are attractive because they give investors more flexibility to express their view on the asset price movement in the option contract design.

The very nature of discontinuity at the barrier (circuit breaker effect upon knock-out) creates hedging problems with the barrier options. It is extremely difficult for option writers to hedge barrier options when the asset price is around the barrier level. Pitched battles often erupt around popular knock-out barriers in currency barrier options and these add much unwanted volatility to the markets. George Soros once said "knock-out options relate to ordinary options the way crack relates to cocaine." More details on the discussion of the hedging problems of barrier options can be found in Linetsky (1999). Also, Hsu (1997) discussed the difficulties in market implementation of different criteria for determining barrier events. In order to avoid the unpleasantness of being knocked out, the criteria used should be impartial, objective and consistent.

Consider a portfolio of one European in-option and one European out-option: both have the same barrier, strike price and date of expiration. The sum of their values is simply the same as that of a corresponding European option with the same strike price and date of expiration. This is obvious since only one of the two barrier options survives at expiry and either payoff is the same as that of the European option. Hence, provided there is no rebate payment upon knock-out, we have

$$c_{\text{ordinary}} = c_{\text{down-and-out}} + c_{\text{down-and-in}} \tag{4.1.1a}$$

$$p_{\text{ordinary}} = p_{\text{up-and-out}} + p_{\text{up-and-in}}, \tag{4.1.1b}$$

where c and p denote call and put values, respectively. Therefore, the value of an out-option can be found easily once the value of the corresponding in-option is available, or vice versa.

In this section, we derive analytic price formulas for European options with either one-sided barrier or two-sided barriers based on continuous monitoring of the asset price process. Under the Black–Scholes pricing paradigm, we can solve the pricing models using both the partial differential equation approach and the martingale pricing approach. We derive the Green function (fundamental solution) of the governing Black–Scholes equation in a restricted domain using the *method of images* in partial differential equation theory. When the martingale approach is used, we obtain the transition density function using the *reflection principle* in the Brownian process literature. To compute the expected present value of the rebate payment, we derive the density function of the first passage time to the barrier. We also extend our pricing methodologies to options with double barriers. We discuss the effects of discrete monitoring of the barrier on option prices at the end of the section.

4.1.1 European Down-and-Out Call Options

The down-and-out call options have been available in the U.S. market since 1967 and the analytic price formula first appears in Merton (1973, Chap. 1). A down-

and-out call has features similar to an ordinary call option, except that it becomes nullified when the asset price S_t falls below the downstream knock-out level B.

Partial Differential Equation Formulation

Let B denote the constant down-and-out barrier. The domain of definition for the barrier option model now becomes $[B, \infty) \times [0, T]$ in the S-τ plane. Let $R(\tau)$ denote the time-dependent rebate paid to the holder when the barrier is hit. Taking the usual Black–Scholes assumptions (frictionless market, continuous trading, etc.), the partial differential equation formulation of the down-and-out barrier call option model is given by

$$\frac{\partial c}{\partial \tau} = \frac{\sigma^2}{2} S^2 \frac{\partial^2 c}{\partial S^2} + rS\frac{\partial c}{\partial S} - rc, \quad S > B \text{ and } \tau \in (0, T], \tag{4.1.2}$$

subject to

$$\begin{aligned} &\text{knock-out condition:} \quad && c(B, \tau) = R(\tau) \\ &\text{terminal payoff:} \quad && c(S, 0) = \max(S - X, 0), \end{aligned}$$

where $c = c(S, \tau)$ is the barrier option value, r and σ are the constant riskless interest rate and volatility, respectively. The down-barrier is normally set below the strike price X, otherwise the down-and-out call may be knocked out even if it expires in-the-money. The partial differential equation formulation implies that knock-out occurs when the barrier is breached at any time during the life of the option.

Suppose we apply the transformation of the independent variable, $y = \ln S$, the barrier becomes the line $y = \ln B$. Now, the Black–Scholes equation (4.1.2) is reduced to the following constant coefficient equation for $c(y, \tau)$

$$\frac{\partial c}{\partial \tau} = \frac{\sigma^2}{2}\frac{\partial^2 c}{\partial y^2} + \left(r - \frac{\sigma^2}{2}\right)\frac{\partial c}{\partial y} - rc \tag{4.1.3a}$$

defined in the semi-infinite domain: $y > \ln B$ and $\tau \in (0, T]$. The auxiliary conditions become

$$c(\ln B, \tau) = R(\tau) \text{ and } c(y, 0) = \max(e^y - X, 0). \tag{4.1.3b}$$

The Green function of (4.1.3a) in the infinite domain: $-\infty < y < \infty$ is given by [see (3.3.4)]

$$G_0(y, \tau; \xi) = \frac{e^{-r\tau}}{\sigma\sqrt{2\pi\tau}} \exp\left(-\frac{(y + \mu\tau - \xi)^2}{2\sigma^2\tau}\right), \tag{4.1.4}$$

where $\mu = r - \frac{\sigma^2}{2}$ and $G_0(y, \tau; \xi)$ satisfies the initial condition:

$$\lim_{\tau \to 0^+} G_0(y, \tau; \xi) = \delta(y - \xi).$$

Method of Images
We would like to solve for the restricted Green function in the semi-infinite domain: $\ln B < y < \infty$ with zero Dirichlet boundary condition at $y = \ln B$. As a judicious guess, assuming that the Green function takes the form

$$G(y, \tau; \xi) = G_0(y, \tau; \xi) - H(\xi)G_0(y, \tau; \eta), \tag{4.1.5}$$

we are required to determine $H(\xi)$ and η in terms of ξ such that the zero Dirichlet boundary condition $G(\ln B, \tau; \xi) = 0$ is satisfied. Note that $G(y, \tau; \xi)$ satisfies (4.1.3a) since both $G_0(y, \tau; \xi)$ and $H(\xi)G_0(y, \tau; \eta)$ satisfy the differential equation. Also, provided that $\eta \notin (\ln B, \infty)$, then $\lim_{\tau\to 0^+} G_0(y, \tau; \eta) = 0$ for all $y > \ln B$. Hence, the initial condition is satisfied. By imposing the boundary condition along the barrier, $H(\xi)$ has to satisfy

$$H(\xi) = \frac{G_0(\ln B, \tau; \xi)}{G_0(\ln B, \tau; \eta)} = \exp\left(\frac{(\xi - \eta)[2(\ln B + \mu\tau) - (\xi + \eta)]}{2\sigma^2\tau}\right). \tag{4.1.6}$$

The assumed form of $G(y, \tau; \xi)$ is feasible only if the right-hand side of (4.1.6) becomes a function of ξ only. This can be achieved by the judicious choice of

$$\eta = 2\ln B - \xi, \tag{4.1.7}$$

so that

$$H(\xi) = \exp\left(\frac{2\mu}{\sigma^2}(\xi - \ln B)\right). \tag{4.1.8}$$

As a remark, this method works only if μ/σ^2 is a constant, independent of τ. The parameter η can be visualized as the *mirror image* of ξ with respect to the barrier $y = \ln B$. This is how the name of this method is derived (see Fig. 4.1). By grouping the terms involving exponentials, the second term in (4.1.5) can be expressed as

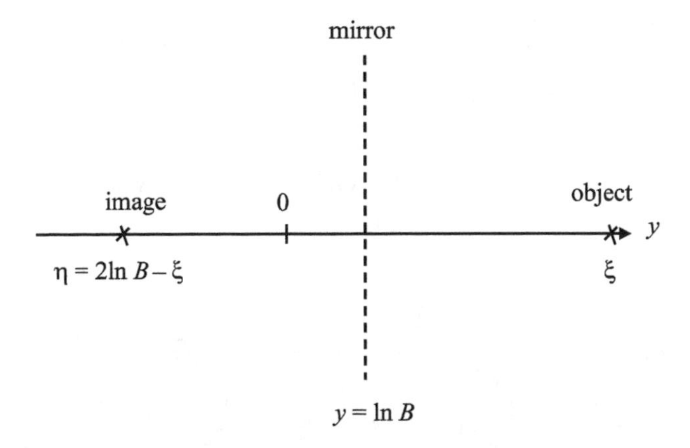

Fig. 4.1. A graphical representation of the method of image. The mirror is placed at $y = \ln B$.

$$
\begin{aligned}
&H(\xi)G_0(y,\tau;\eta)\\
&= \exp\left(\frac{2\mu}{\sigma^2}(\xi-\ln B)\right)\frac{e^{-r\tau}}{\sigma\sqrt{2\pi\tau}}\exp\left(-\frac{[y+\mu\tau-(2\ln B-\xi)]^2}{2\sigma^2\tau}\right)\\
&= \left(\frac{B}{S}\right)^{2\mu/\sigma^2}\frac{e^{-r\tau}}{\sigma\sqrt{2\pi\tau}}\exp\left(-\frac{[(y-\xi)+\mu\tau-2(y-\ln B)]^2}{2\sigma^2\tau}\right).
\end{aligned}
$$

Collecting all the terms together, the Green function in the specified semi-infinite domain: $\ln B < y < \infty$ becomes

$$
\begin{aligned}
G(y,\tau;\xi) = \frac{e^{-r\tau}}{\sigma\sqrt{2\pi\tau}}&\left\{\exp\left(-\frac{(u-\mu\tau)^2}{2\sigma^2\tau}\right)\right.\\
&\left.-\left(\frac{B}{S}\right)^{2\mu/\sigma^2}\exp\left(-\frac{(u-2\beta-\mu\tau)^2}{2\sigma^2\tau}\right)\right\}, \qquad (4.1.9)
\end{aligned}
$$

where $u=\xi-y$ and $\beta=\ln B-y=\ln\frac{B}{S}$.

We consider the barrier option with zero rebate, where $R(\tau)=0$, and let $K=\max(B,X)$. The price of the zero-rebate European down-and-out call can be expressed as

$$
\begin{aligned}
c_{do}(y,\tau) &= \int_{\ln B}^{\infty}\max(e^{\xi}-X,0)G(y,\tau;\xi)\,d\xi\\
&= \int_{\ln K}^{\infty}(e^{\xi}-X)G(y,\tau;\xi)\,d\xi\\
&= \frac{e^{-r\tau}}{\sigma\sqrt{2\pi\tau}}\int_{\ln K/S}^{\infty}(Se^{u}-X)\left[\exp\left(-\frac{(u-\mu\tau)^2}{2\sigma^2\tau}\right)\right.\\
&\qquad\left.-\left(\frac{B}{S}\right)^{2\mu/\sigma^2}\exp\left(-\frac{(u-2\beta-\mu\tau)^2}{2\sigma^2\tau}\right)\right]du.
\end{aligned}
$$

The direct evaluation of the integral gives

$$
\begin{aligned}
c_{do}(S,\tau) = S&\left[N(d_1)-\left(\frac{B}{S}\right)^{\delta+1}N(d_3)\right]\\
&-Xe^{-r\tau}\left[N(d_2)-\left(\frac{B}{S}\right)^{\delta-1}N(d_4)\right], \qquad (4.1.10)
\end{aligned}
$$

where

$$
\begin{aligned}
d_1 &= \frac{\ln\frac{S}{K}+\left(r+\frac{\sigma^2}{2}\right)\tau}{\sigma\sqrt{\tau}}, \quad d_2=d_1-\sigma\sqrt{\tau},\\
d_3 &= d_1+\frac{2}{\sigma\sqrt{\tau}}\ln\frac{B}{S}, \quad d_4=d_2+\frac{2}{\sigma\sqrt{\tau}}\ln\frac{B}{S}, \quad \delta=\frac{2r}{\sigma^2}.
\end{aligned}
$$

Suppose we define

$$\widetilde{c}_E(S, \tau; X, K) = SN(d_1) - Xe^{-r\tau}N(d_2),$$

then $c_{do}(S, \tau; X, B)$ can be expressed in the following succinct form

$$c_{do}(S, \tau; X, B) = \widetilde{c}_E(S, \tau; X, K) - \left(\frac{B}{S}\right)^{\delta-1} \widetilde{c}_E\left(\frac{B^2}{S}, \tau; X, K\right). \tag{4.1.11}$$

One can show by direct calculation that the function $(\frac{B}{S})^{\delta-1}\widetilde{c}_E(\frac{B^2}{S}, \tau)$ satisfies the Black–Scholes equation identically (see Problem 4.1). The above form allows us to observe readily the satisfaction of the boundary condition: $c_{do}(B, \tau) = 0$, and the terminal payoff condition.

The barrier option price formula (4.1.11) indicates that $c_{do}(S, \tau; X, B) < \widetilde{c}_E(S, \tau; X, K) < c_E(S, \tau; X)$, that is, a down-and-out call is less expensive than the corresponding vanilla European call. This is obvious since the inherent knock-out property lowers the option premium.

Remarks.

1. Closed form analytic price formulas for barrier options with exponential time dependent barrier, $B(\tau) = Be^{-\gamma\tau}$, can also be derived [see Problem 4.2]. However, when the barrier level has an arbitrary time dependence, the search for an analytic price formula for the barrier option fails. Roberts and Shortland (1997) showed how to derive the analytic approximation formula by estimating the boundary hitting times of the asset price process via the Brownian bridge technique.
2. Closed form price formulas for barrier options can also be obtained for other types of diffusion processes followed by the underlying asset price. Lo, Yuen and Hui (2002) derived the barrier option price formulas under the square root constant elasticity of variance process, while Sepp (2004) used the Laplace transform method to obtain price formulas under the double exponential jump diffusion process.
3. By the relation (4.1.1a) and assuming $B < X$, the price of a down-and-in call option can be deduced to be

$$c_{di}(S, \tau; X, B) = \left(\frac{B}{S}\right)^{\delta-1} c_E\left(\frac{B^2}{S}, \tau; X\right). \tag{4.1.12}$$

4. The barrier option with rebate payment will be considered later when the density function of the first passage time is available. Alternatively, one can use the known solution for the diffusion equation in the semi-infinite domain with time dependent boundary condition to derive the additional option premium due to the rebate (see Problem 4.3).
5. The method of images can be extended to derive the density functions of restricted multi-state diffusion processes where a barrier is placed on only one of the state variables (Kwok, Wu and Yu, 1998). The pricing of a two-asset down-and-out call is illustrated in Problem 4.6.
6. The monitoring period for breaching of the barrier may be limited to only part of the life of the option. The price formulas for these *partial barrier options* were obtained by Heynen and Kat (1994a) (see Problem 4.7).

4.1.2 Transition Density Function and First Passage Time Density

We may formulate the pricing models of barrier options using the martingale pricing approach and derive the corresponding price formulas by computing the expectation of the discounted terminal payoff (subject to knock-out or knock-in provision) under the risk neutral measure Q. Let the time interval $[0, T]$ denote the life of the barrier option, that is, the option is initiated at time zero and will expire at time T. The realized maximum and minimum value of the asset price from time zero to time t (under continuous monitoring) are defined by

$$\begin{aligned} m_0^t &= \min_{0 \le u \le t} S_u \\ M_0^t &= \max_{0 \le u \le t} S_u, \end{aligned} \tag{4.1.13}$$

respectively. The terminal payoffs of various types of barrier options can be expressed in terms of m_0^T and M_0^T. For example, consider the down-and-out call and up-and-out put, their respective terminal payoff can be expressed as

$$\begin{aligned} c_{do}(S_T, T; X, B) &= \max(S_T - X, 0)\mathbf{1}_{\{m_0^T > B\}} \\ p_{uo}(S_T, T; X, B) &= \max(X - S_T, 0)\mathbf{1}_{\{M_0^T < B\}}, \end{aligned} \tag{4.1.14}$$

respectively. Suppose B is the down-barrier, we define τ_B to be the stopping time at which the underlying asset price crosses the barrier for the first time:

$$\tau_B = \inf\{t | S_t \le B\}, \quad S_0 = S. \tag{4.1.15a}$$

Assume $S > B$ and due to path continuity, we may express τ_B (commonly called the *first passage time*) as

$$\tau_B = \inf\{t | S_t = B\}. \tag{4.1.15b}$$

In a similar manner, if B is the up-barrier and $S < B$, we have

$$\tau_B = \inf\{t | S_t \ge B\} = \inf\{t | S_t = B\}. \tag{4.1.15c}$$

It is easily seen that $\{\tau_B > T\}$ and $\{m_0^T > B\}$ are equivalent events if B is a down-barrier. By virtue of the risk neutral valuation principle, the price of a down-and-out call at time zero is given by

$$\begin{aligned} c_{do}(S, 0; X, B) &= e^{-rT} E_Q[\max(S_T - X, 0)\mathbf{1}_{\{m_0^T > B\}}] \\ &= e^{-rT} E_Q[(S_T - X)\mathbf{1}_{\{S_T > \max(X,B)\}}\mathbf{1}_{\{\tau_B > T\}}]. \end{aligned} \tag{4.1.16}$$

Here, E_Q denotes the expectation under the risk neutral measure Q conditional on $S_t = S$ (the same notation is used throughout this chapter). The determination of the price function $c_{do}(S, 0; X, B)$ requires the determination of the joint distribution function of S_T and m_0^T.

Reflection Principle

We illustrate how the reflection principle is applied to derive the joint law of the minimum value over $[0, T]$ and terminal value of a Brownian motion. Let W_t^0 (W_t^μ) denote the Brownian motion that starts at zero, with constant volatility σ and zero drift rate (constant drift rate μ). We would like to find $P(m_0^T < m, W_T^\mu > x)$, where $x \geq m$ and $m \leq 0$. First, we consider the zero-drift Brownian motion W_t^0. Given that the minimum value m_0^T falls below m, then there exists some time instant $\xi, 0 < \xi < T$, such that ξ is the first time that W_ξ^0 equals m. As Brownian paths are continuous, there exist some times during which $W_t^0 < m$. In other words, W_t^0 decreases at least below m and then increases at least up to level x (higher than m) at time T. Suppose we define a random process

$$\widetilde{W}_t^0 = \begin{cases} W_t^0 & \text{for } t < \xi \\ 2m - W_t^0 & \text{for } \xi \leq t \leq T, \end{cases} \tag{4.1.17}$$

that is, $\widetilde{W}_t^0$ is the mirror reflection of W_t^0 at the level m within the time interval between ξ and T (see Fig. 4.2). It is then obvious that $\{W_T^0 > x\}$ is equivalent to $\{\widetilde{W}_T^0 < 2m - x\}$. Also, the reflection of the Brownian path dictates that

$$\widetilde{W}_{\xi+u}^0 - \widetilde{W}_\xi^0 = -(W_{\xi+u}^0 - W_\xi^0), \quad u > 0. \tag{4.1.18}$$

The stopping time ξ depends only on the path history $\{W_t^0 : 0 \leq t \leq \xi\}$ and it will not affect the Brownian motion at later times. By the strong Markov property of Brownian motions, we argue that the two Brownian increments in (4.1.18) have the same distribution. The distribution has zero mean and variance $\sigma^2 u$. For every Brownian path that starts at 0, travels at least m units (downward, $m \leq 0$) before T and later travels at least $x - m$ units (upward, $x \geq m$), there is an equally likely path

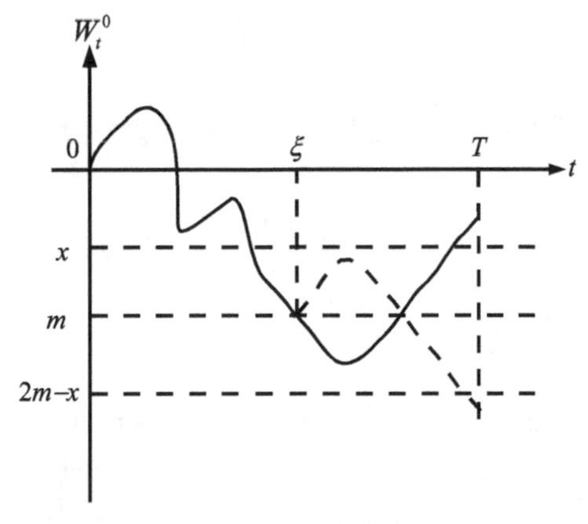

Fig. 4.2. A graphical representation of the reflection principle of the Brownian motion W_t^0. The dotted path after time ξ is the mirror reflection of the Brownian path at the level m. Suppose W_T^0 ends up at a value higher than x, then the reflected path at time T has a value lower than $2m - x$.

that starts at 0, travels m units (downward, $m \leq 0$) some time before T and travels at least $m - x$ units (further downward, $m \leq x$). Suppose $W_T^0 > x$, then $\widetilde{W}_T^0 < 2m - x$, and together with relation (4.1.18), we obtain the joint distribution function for the zero-drift case as follows:

$$P(W_T^0 > x, m_0^T < m) = P(\widetilde{W}_T^0 < 2m - x) = P(W_T^0 < 2m - x)$$
$$= N\left(\frac{2m - x}{\sigma\sqrt{T}}\right), \quad m \leq \min(x, 0). \tag{4.1.19}$$

Next, we apply the Girsanov Theorem to effect the change of measure for finding the above joint distribution when the Brownian motion has nonzero drift. Suppose under the measure Q, W_t^μ is a Brownian motion with drift rate μ. We change the measure from Q to $\widetilde{Q}$ such that W_t^μ becomes a Brownian process with zero drift under $\widetilde{Q}$. Consider the following joint distribution

$$\begin{aligned} &P(W_T^\mu > x, m_0^T < m) \\ &= E_Q[\mathbf{1}_{\{W_T^\mu > x\}} \mathbf{1}_{\{m_0^T < m\}}] \\ &= E_{\widetilde{Q}}\left[\mathbf{1}_{\{W_T^\mu > x\}} \mathbf{1}_{\{m_0^T < m\}} \exp\left(\frac{\mu W_T^\mu}{\sigma^2} - \frac{\mu^2 T}{2\sigma^2}\right)\right], \end{aligned}$$

where the Radon–Nikodym derivative: $\exp\left(\frac{\mu W_T^\mu}{\sigma^2} - \frac{\mu^2 T}{2\sigma^2}\right)$ is appended due to the Girsanov Theorem [see (2.4.30)]. Next, by applying the reflection principle and observing that W_T^μ is a zero-drift Brownian motion under $\widetilde{Q}$, we obtain

$$\begin{aligned} &P(W_T^\mu > x, m_0^T < m) \\ &= E_{\widetilde{Q}}\left[\mathbf{1}_{\{2m - W_T^\mu > x\}} \exp\left(\frac{\mu}{\sigma^2}(2m - W_T^\mu) - \frac{\mu^2 T}{2\sigma^2}\right)\right] \\ &= e^{\frac{2\mu m}{\sigma^2}} E_{\widetilde{Q}}\left[\mathbf{1}_{\{W_T^\mu < 2m - x\}} \exp\left(-\frac{\mu}{\sigma^2} W_T^\mu - \frac{\mu^2 T}{2\sigma^2}\right)\right] \\ &= e^{\frac{2\mu m}{\sigma^2}} \int_{-\infty}^{2m-x} \frac{1}{\sqrt{2\pi\sigma^2 T}} e^{-\frac{z^2}{2\sigma^2 T}} e^{-\frac{\mu z}{\sigma^2} - \frac{\mu^2 T}{2\sigma^2}} \, dz \\ &= e^{\frac{2\mu m}{\sigma^2}} \int_{-\infty}^{2m-x} \frac{1}{\sqrt{2\pi\sigma^2 T}} \exp\left(-\frac{(z + \mu T)^2}{2\sigma^2 T}\right) dz \\ &= e^{\frac{2\mu m}{\sigma^2}} N\left(\frac{2m - x + \mu T}{\sigma\sqrt{T}}\right), \quad m \leq \min(x, 0). \end{aligned} \tag{4.1.20}$$

As a remark, when μ and σ^2 have arbitrary time dependence, the Radon–Nikodym derivative would involve stochastic integral terms. In this case, the reflection principle cannot be applied.

Suppose the Brownian motion W_t^μ has a downstream barrier m over the period $[0, T]$ so that $m_0^T > m$, we would like to derive the joint distribution

$$P(W_T^\mu > x, m_0^T > m), \quad \text{where} \quad m \leq \min(x, 0).$$

By applying the law of total probabilities, we obtain

$$
\begin{aligned}
&P(W_T^\mu > x, m_0^T > m) \\
&= P(W_T^\mu > x) - P(W_T^\mu > x, m_0^T < m) \\
&= N\left(\frac{-x+\mu T}{\sigma\sqrt{T}}\right) - e^{\frac{2\mu m}{\sigma^2}} N\left(\frac{2m - x + \mu T}{\sigma\sqrt{T}}\right), \quad m \le \min(x, 0). \qquad (4.1.21)
\end{aligned}
$$

Under the special case $m = x$, since $W_T^\mu > m$ is implicitly implied from $m_0^T > m$, we have

$$
P(m_0^T > m) = N\left(\frac{-m+\mu T}{\sigma\sqrt{T}}\right) - e^{\frac{2\mu m}{\sigma^2}} N\left(\frac{m + \mu T}{\sigma\sqrt{T}}\right). \qquad (4.1.22)
$$

Extension to the Upstream Barrier

When the Brownian motion W_t^μ has an upstream barrier M over the period $[0, T]$ so that $M_0^T < M$, the joint distribution function of W_T^μ and M_0^T can be deduced using the following relation between M_0^T and m_0^T:

$$
M_0^T = \max_{0\le t\le T} (\sigma Z_t + \mu t) = - \min_{0\le t\le T} (-\sigma Z_t - \mu t),
$$

where Z_t is the standard Brownian motion. Since $-Z_t$ has the same distribution as Z_t, the distribution of the maximum value of W_t^μ is the same as that of the negative of the minimum value of $W_t^{-\mu}$. By swapping $-\mu$ for μ, $-M$ for m and $-y$ for x in (4.1.20), we obtain

$$
P(W_T^\mu < y, M_0^T > M) = e^{\frac{2\mu M}{\sigma^2}} N\left(\frac{y - 2M - \mu T}{\sigma\sqrt{T}}\right), M \ge \max(y, 0). \qquad (4.1.23)
$$

In a similar manner, we obtain

$$
\begin{aligned}
&P(W_T^\mu < y, M_0^T < M) \\
&= P(W_T^\mu < y) - P(W_T^\mu < y, M_0^T > M) \\
&= N\left(\frac{y-\mu T}{\sigma\sqrt{T}}\right) - e^{\frac{2\mu M}{\sigma^2}} N\left(\frac{y - 2M - \mu T}{\sigma\sqrt{T}}\right), \quad M \ge \max(y, 0), \qquad (4.1.24)
\end{aligned}
$$

and by setting $y = M$, we obtain

$$
P(M_0^T < M) = N\left(\frac{M-\mu T}{\sigma\sqrt{T}}\right) - e^{\frac{2\mu M}{\sigma^2}} N\left(-\frac{M + \mu T}{\sigma\sqrt{T}}\right). \qquad (4.1.25)
$$

Density Functions of Restricted Brownian Processes

We define $f_{down}(x, m, T)$ to be the density function of W_T^μ with the downstream barrier m, where $m \le \min(x, 0)$, that is,

$$f_{down}(x, m, T)\,dx = P(W_T^\mu \in dx, m_0^T > m).$$

By differentiating (4.1.21) with respect to x and swapping the sign, we obtain

$$\begin{aligned} & f_{down}(x, m, T) \\ &= \frac{1}{\sigma\sqrt{T}}\left[n\left(\frac{x-\mu T}{\sigma\sqrt{T}}\right) - e^{\frac{2\mu m}{\sigma^2}} n\left(\frac{x-2m-\mu T}{\sigma\sqrt{T}}\right)\right] \\ & \mathbf{1}_{\{m\le\min(x,0)\}}. \end{aligned} \tag{4.1.26}$$

Similarly, we define $f_{up}(x, M, T)$ to be the density function of W_T^μ with the upstream barrier M, where $M > \max(y, 0)$, then

$$\begin{aligned} & P(W_T^\mu \in dy, M_0^T < M) \\ &= f_{up}(y, M, T)\,dy \\ &= \frac{1}{\sigma\sqrt{T}}\left[n\left(\frac{y-\mu T}{\sigma\sqrt{T}}\right) - e^{\frac{2\mu M}{\sigma^2}} n\left(\frac{y-2M-\mu T}{\sigma\sqrt{T}}\right)\right] dy \\ & \mathbf{1}_{\{M>\max(y,0)\}}. \end{aligned} \tag{4.1.27}$$

Suppose the asset price S_t follows the Geometric Brownian process under the risk neutral measure such that $\ln \frac{S_t}{S} = W_t^\mu$, where S is the asset price at time zero and the drift rate $\mu = r - \frac{\sigma^2}{2}$. Let $\psi(S_T; S, B)$ denote the transition density of the asset price S_T at time T conditional on $S_t > B$ for $0 \le t \le T$. Here, B is the downstream barrier. By (4.1.26), we deduce that $\psi(S_T; S, B)$ is given by

$$\begin{aligned} \psi(S_T; S, B) = & \frac{1}{\sigma\sqrt{T}S_T}\left[n\left(\frac{\ln\frac{S_T}{S} - \left(r-\frac{\sigma^2}{2}\right)T}{\sigma\sqrt{T}}\right)\right. \\ & \left. - \left(\frac{B}{S}\right)^{\frac{2r}{\sigma^2}-1} n\left(\frac{\ln\frac{S_T}{S} - 2\ln\frac{B}{S} - \left(r-\frac{\sigma^2}{2}\right)T}{\sigma\sqrt{T}}\right)\right]. \end{aligned} \tag{4.1.28}$$

First Passage Time Density Functions

Let $Q(u; m)$ denote the density function of the first passage time at which the downstream barrier m is first hit by the Brownian path W_t^μ, that is, $Q(u; m)\,du = P(\tau_m \in du)$. First, we determine the distribution function $P(\tau_m > u)$ by observing that $\{\tau_m > u\}$ and $\{m_0^u > m\}$ are equivalent events. By (4.1.22), we obtain

$$\begin{aligned} P(\tau_m > u) &= P(m_0^u > m) \\ &= N\left(\frac{-m+\mu u}{\sigma\sqrt{u}}\right) - e^{\frac{2\mu m}{\sigma^2}} N\left(\frac{m+\mu u}{\sigma\sqrt{u}}\right). \end{aligned} \tag{4.1.29}$$

The density function $Q(u; m)$ is then given by

$$\begin{aligned}Q(u;m)\,du &= P(\tau_m \in du)\\ &= -\frac{\partial}{\partial u}\left[N\left(\frac{-m+\mu u}{\sigma\sqrt{u}}\right) - e^{\frac{2\mu m}{\sigma^2}} N\left(\frac{m+\mu u}{\sigma\sqrt{u}}\right)\right] du\,\mathbf{1}_{\{m<0\}}\\ &= \frac{-m}{\sqrt{2\pi\sigma^2 u^3}}\exp\left(-\frac{(m-\mu u)^2}{2\sigma^2 u}\right) du\,\mathbf{1}_{\{m<0\}}. \qquad (4.1.30a)\end{aligned}$$

In a similar manner, let $Q(u; M)$ denote the first passage time density associated with the upstream barrier M. Using the result in (4.1.25), we have

$$\begin{aligned}Q(u;M) &= -\frac{\partial}{\partial u}\left[N\left(\frac{M-\mu u}{\sigma\sqrt{u}}\right) - e^{\frac{2\mu M}{\sigma^2}} N\left(-\frac{M+\mu u}{\sigma\sqrt{u}}\right)\right]\mathbf{1}_{\{M>0\}}\\ &= \frac{M}{\sqrt{2\pi\sigma^2 u^3}}\exp\left(-\frac{(M-\mu u)^2}{2\sigma^2 u}\right)\mathbf{1}_{\{M>0\}}. \qquad (4.1.30b)\end{aligned}$$

We write B as the barrier level, either upstream or downstream. When the barrier is downstream (upstream), we have $\ln\frac{B}{S} < 0$ ($\ln\frac{B}{S} > 0$). We may combine (4.1.30a,b) into one equation as follows:

$$Q(u;B) = \frac{\left|\ln\frac{B}{S}\right|}{\sqrt{2\pi\sigma^2 u^3}}\exp\left(-\frac{\left[\ln\frac{B}{S} - \left(r - \frac{\sigma^2}{2}\right)u\right]^2}{2\sigma^2 u}\right). \qquad (4.1.31)$$

Suppose a rebate $R(t)$ is paid to the option holder upon breaching the barrier at level B by the asset price path at time t, $0 < t < T$. Since the expected rebate payment over the time interval $[u, u+du]$ is given by $R(u)Q(u; B)\,du$, the expected present value of the rebate is given by

$$\text{rebate value} = \int_0^T e^{-ru} R(u) Q(u;B)\,du. \qquad (4.1.32)$$

When $R(t) = R_0$, a constant value, the direct integration of the above integral gives

$$\begin{aligned}\text{rebate value} = R_0\Bigg[&\left(\frac{B}{S}\right)^{\alpha_+} N\left(\delta\frac{\ln\frac{B}{S} + \beta T}{\sigma\sqrt{T}}\right)\\ &+ \left(\frac{B}{S}\right)^{\alpha_-} N\left(\delta\frac{\ln\frac{B}{S} - \beta T}{\sigma\sqrt{T}}\right)\Bigg], \qquad (4.1.33)\end{aligned}$$

where

$$\beta = \sqrt{\left(r - \frac{\sigma^2}{2}\right)^2 + 2r\sigma^2}, \quad \alpha_\pm = \frac{r - \frac{\sigma^2}{2} \pm \beta}{\sigma^2},$$

$$\delta = \operatorname{sign}\left(\ln\frac{S}{B}\right).$$

Here, δ is a binary variable indicating whether the barrier is downstream ($\delta = 1$) or upstream ($\delta = -1$).

Transition Density Function

We would like to find the partial differential equation formulation of the transition density function $\psi_B(x, t; x_0, t_0)$ for the restricted Brownian process with upstream absorbing barrier B. The absorbing condition resembles the knock-out feature in barrier options. The appropriate boundary condition for an absorbing barrier is given by (Cox and Miller, 1995)

$$\psi_B(x, t; x_0, t_0)\Big|_{x=B} = 0. \tag{4.1.34}$$

The forward Fokker–Planck equation that governs ψ_B is known to be [see (2.3.11)]

$$\frac{\partial \psi_B}{\partial t} = -\mu \frac{\partial \psi_B}{\partial x} + \frac{\sigma^2}{2} \frac{\partial^2 \psi_B}{\partial x^2}, \quad -\infty < x < B, t > t_0, \tag{4.1.35}$$

with boundary condition: $\psi_B(B, t) = 0$. Since $x \to x_0$ as $t \to t_0$ so that

$$\lim_{t \to t_0} \psi_B(x, t; x_0, t_0) = \delta(x - x_0). \tag{4.1.36}$$

As deduced from the density function in (4.1.27), ψ_B is found to be

$$\begin{aligned}\psi_B(x, t; x_0, t_0) = \frac{1}{\sigma\sqrt{t - t_0}}\Bigg[& n\left(\frac{x - x_0 - \mu(t - t_0)}{\sigma\sqrt{t - t_0}}\right) \\ & - e^{\frac{2\mu(B - x_0)}{\sigma^2}} n\left(\frac{(x - x_0) - 2(B - x_0) - \mu(t - t_0)}{\sigma\sqrt{t - t_0}}\right)\Bigg], \\ & x < B, t > t_0, x_0 < B.\end{aligned} \tag{4.1.37}$$

The probability that W_t^μ never crosses the barrier over $[t_0, t]$ is given by

$$\begin{aligned}P(\tau_B > t) &= P\left(W_t^\mu \le B, M_{t_0}^t \le B \,\middle|\, W_{t_0}^\mu = x_0\right) \\ &= \int_{-\infty}^{B} \psi_B(x, t; x_0, t_0)\, dx.\end{aligned} \tag{4.1.38}$$

Price Formula of European Up-and-Out Call

Consider a European up-and-out call with strike X and upstream barrier B. Since the writer uses the knock-out feature to cap the upside liability, the payoff structure makes sense only if we choose $X < B$. As the option is always in-the-money upon knock-out, some form of rebate should be paid upon breaching the barrier.

The nonrebate portion of the value of the up-and-out call is computed as follows. By the risk neutral valuation principle, the nonrebate call value is given by

$$\begin{aligned}&e^{-rT} E_Q\left[(S_T - X)\mathbf{1}_{\{X < S_T < B\}}\mathbf{1}_{\{M_0^T < B\}}\right] \\ &= e^{-rT} \int_{\ln X/S}^{\ln B/S} (Se^y - X) f_{up}(y, B, T)\, dy.\end{aligned} \tag{4.1.39}$$

Recall that the value of the down-and-out call ($B < X$) is given by

$$e^{-rT}\int_{\ln X/S}^{\infty}(Se^{x}-X)f_{down}(x,B,T)\,dx$$
$$= c_E(S,T;X)-\left(\frac{B}{S}\right)^{\delta-1}c_E\left(\frac{B^2}{S},T;X\right), \tag{4.1.40}$$

where $c_E(S,\tau;X)$ is the price function of a European vanilla call option with time to expiry τ. Since $f_{down}(x,B,T)$ and $f_{up}(y,B,T)$ have the same analytic form, the integral in (4.1.39) can be related to the integral in (4.1.40). We obtain

$$\begin{aligned}&\text{up-and-out call value}\\&=\left[c_E(S,T;X)-\left(\frac{B}{S}\right)^{\delta-1}c_E\left(\frac{B^2}{S},T;X\right)\right]\\&\quad-\left[c_E(S,T;B)-\left(\frac{B}{S}\right)^{\delta-1}c_E\left(\frac{B^2}{S},T;B\right)\right].\end{aligned} \tag{4.1.41}$$

The price of the corresponding up-and-in call option is given by

$$\begin{aligned}\text{up-and-in call value}&=\left(\frac{B}{S}\right)^{\delta-1}c_E\left(\frac{B^2}{S},T;X\right)+c_E(S,T;B)\\&\quad-\left(\frac{B}{S}\right)^{\delta-1}c_E\left(\frac{B^2}{S},T;B\right).\end{aligned} \tag{4.1.42}$$

Other price formulas of the European out/in put options with either an upstream or downstream barrier can be derived in a similar manner (see Problem 4.5). A comprehensive list of price formulas of European options with an one-sided barrier can be found in Rich (1994).

4.1.3 Options with Double Barriers

A double barrier option has two barriers, upstream barrier at U and downstream barrier at L. In its simplest form, when the asset price reaches one of the two barriers, either nullification (knock-out) or activation (knock-in) of the option contract is triggered. A more complicated payoff structure may include a rebate payment upon hitting either one of the knock-out barriers as a compensation to the option holder. In sequential barrier options, knock-out is triggered only when the two barriers are hit at a prespecified sequential order (see Problem 4.12). Luo (2001) and Kolkiewicz (2002) presented comprehensive discussions on various types of double barrier options that can be structured.

With the presence of two barriers, the following first passage times of the asset price process S_t can be defined

$$\tau_U=\inf\{t|S_t=U\}\quad\text{and}\quad\tau_L=\inf\{t|S_t=L\}. \tag{4.1.43}$$

During the life of the option $[0, T]$, we can distinguish the following three mutually exclusive events: (i) the upper barrier is first reached, (ii) the lower barrier is first reached, (iii) neither of the two barriers is reached.

We take the usual assumption that S_t is a Geometric Brownian process with volatility σ under the risk neutral measure Q. Let $X_t = \ln \frac{S_t}{S}$ so that $X_0 = 0$ and X_t is a Brownian process with drift rate $r - \frac{\sigma^2}{2}$ and variance rate σ^2. We are interested to find the following density functions:

$$g(x, T)\, dx = P(X_T \in dx, \min(\tau_L, \tau_U) > T) \tag{4.1.44a}$$
$$g^+(x, T)\, dx = P(X_T \in dx, \min(\tau_L, \tau_U) \leq T, \tau_U < \tau_L) \tag{4.1.44b}$$
$$g^-(x, T)\, dx = P(X_T \in dx, \min(\tau_L, \tau_U) \leq T, \tau_L < \tau_U). \tag{4.1.44c}$$

Most double barrier options can be priced using the above density functions. Some examples are listed below.

1. Double knock-out call option (call payoff will be received at maturity if none of the barriers is breached)
$$\begin{aligned} c_{LU}^o &= e^{-rT} E_Q\big[(S_T - X)\mathbf{1}_{\{S_T > X\}}\mathbf{1}_{\{\min(\tau_L,\tau_U) > T\}}\big] \\ &= e^{-rT}\int_{\ln X/S}^{\ln U/S} (Se^x - X) g(x, T)\, dx, \quad X \in (L, U). \end{aligned}$$
2. Upper-barrier knock-in call option (a vanilla call comes into being if the upper barrier is breached before the lower barrier is breached during the option life, that is, $\tau_U < \tau_L$ and $\tau_U \leq T$)
$$\begin{aligned} c_U^i &= e^{-rT} E_Q\big[(S_T - X)\mathbf{1}_{\{S_T > X\}}\mathbf{1}_{\{\tau_U < \tau_L\}}\mathbf{1}_{\{\min(\tau_L,\tau_U) \leq T\}}\big] \\ &= e^{-rT}\int_{\ln X/S}^{\ln U/S} (Se^x - X) g^+(x, T)\, dx, \quad X \in (L, U). \end{aligned}$$
The analytic form for $g^+(x, T)$ can be found in Problem 4.11.
3. Lower-barrier knock-out call option (the call payoff will be received at maturity if the lower barrier is never breached or the upper barrier is breached before the lower barrier, that is, $\tau_L > T$ or $\tau_U < \tau_L$).
Since the sum of a lower-barrier knock-out call and a lower-barrier knock-in call equals a vanilla call, we have
$$\begin{aligned} c_L^o &= c_E - c_L^i \\ &= c_E - e^{-rT}\int_{\ln X/S}^{\ln U/S} (Se^x - X) g^-(x, T)\, dx, \quad X \in (L, U), \end{aligned}$$
where c_E is the price of a European vanilla call option.

Density Functions of Brownian processes with Two-Sided Barriers

The density functions defined in (4.1.44a,b,c) satisfy the forward Fokker–Planck equation. Their full partial differential equation formulations require the prescription of appropriate auxiliary conditions.

We take the initial position $X_0 = 0$. Let $g(x, t; \ell, u)$ denote the density function of the restricted Brownian process X_t with two-sided absorbing barriers at $x = \ell$ and $x = u$, where the barriers are positioned such that $\ell < 0 < u$. Recall that $X_t = \ln \frac{S_t}{S}$, and if L and U are the absorbing barriers of the asset price process S_t, respectively, then $\ell = \ln \frac{L}{S}$ and $u = \ln \frac{U}{S}$. The partial differential equation formulation for $g(x, t; \ell, u)$ is given by

$$\frac{\partial g}{\partial t} = -\mu \frac{\partial g}{\partial x} + \frac{\sigma^2}{2} \frac{\partial^2 g}{\partial x^2}, \quad \ell < x < u, \quad t > 0, \tag{4.1.45}$$

with auxiliary conditions:

$$g(\ell, t) = g(u, t) = 0 \quad \text{and} \quad g(x, 0^+) = \delta(x).$$

Defining the transformation

$$g(x, t) = e^{\frac{\mu x}{\sigma^2} - \frac{\mu^2 t}{2\sigma^2}} \widehat{g}(x, t),$$

we observe that $\widehat{g}(x, t)$ satisfies the forward Fokker–Planck equation with zero drift:

$$\frac{\partial \widehat{g}}{\partial t}(x, t) = \frac{\sigma^2}{2} \frac{\partial^2 \widehat{g}}{\partial x^2}(x, t). \tag{4.1.46}$$

The auxiliary conditions for $\widehat{g}(x, t)$ are seen to remain the same as those for $g(x, t)$. Without the barriers, the infinite-domain fundamental solution to (4.1.46) is known to be

$$\phi(x, t) = \frac{1}{\sqrt{2\pi\sigma^2 t}} \exp\left(-\frac{x^2}{2\sigma^2 t}\right). \tag{4.1.47}$$

Like the one-sided barrier case, we try to add extra terms to the above solution such that the homogeneous boundary conditions at $x = \ell$ and $x = u$ are satisfied. The following procedure is an extension of the method of images to two-sided barriers. First, we attempt to add the pair of negative terms $-\phi(x - 2\ell, t)$ and $-\phi(x - 2u, t)$ whereby

$$[\phi(x, t) - \phi(x - 2\ell, t)]\Big|_{x=\ell} = 0,$$

and

$$[\phi(x, t) - \phi(x - 2u, t)]\Big|_{x=u} = 0.$$

Note that $\phi(x - 2\ell, t)$ and $\phi(x - 2u, t)$ correspond to the fundamental solution with initial condition: $\delta(x - 2\ell)$ and $\delta(x - 2u)$, respectively. Writing the above partial sum with three terms as

$$\widehat{g}_3(x, t) = \phi(x, t) - \phi(x - 2\ell, t) - \phi(x - 2u, t),$$

we observe that the homogeneous boundary conditions are not yet satisfied since

$$\widehat{g}_3(\ell, t) = -\phi(x - 2u, t)\Big|_{x=\ell} \neq 0$$

$$\widehat{g}_3(u, t) = -\phi(2 - 2\ell, t)\Big|_{x=u} \neq 0.$$

To nullify the nonzero value of $-\phi(x - 2u, t)|_{x=\ell}$ and $-\phi(x - 2\ell, t)|_{x=u}$, we add a new pair of positive terms $\phi(x - 2(u - \ell), t)$ and $\phi(x + 2(u - \ell), t)$. Similarly, we write the partial sum with five terms as

$$\widehat{g}_5(x, t) = \widehat{g}_3(x, t) + \phi(x - 2(u - \ell), t) + \phi(x + 2(u - \ell), t),$$

and observe that

$$\widehat{g}_5(\ell, t) = \phi(x - 2(u - \ell), t)\Big|_{x=\ell} \neq 0$$

$$\widehat{g}_5(u, t) = \phi(x + 2(u - \ell), t)\Big|_{x=u} \neq 0.$$

Whenever a new pair of positive terms or negative terms are added, the value of the partial sum at $x = \ell$ and $x = u$ becomes closer to zero. In a recursive manner, we add successive pairs of positive and negative terms so as to come closer to the satisfaction of the homogeneous boundary conditions at $x = \ell$ and $x = u$. Apparently, the two absorbing barriers may be visualized as a pair of mirrors with the object placed at the origin (see Fig. 4.3). The source at the origin generates a sink at $x = 2\ell$ due to the mirror at $x = \ell$ and another sink at $x = 2u$ due to the mirror at $x = u$. To continue, the sink at $x = 2\ell$ $(x = 2u)$ generates a source at $x = 2(u - \ell)$ $[x = 2(\ell - u)]$ due to the mirror at $x = u$ $(x = \ell)$. As the procedure continues, this leads to the sum of an infinite number of positive and negative terms. The solution to $g(x, t)$ is deduced to be

$$\begin{aligned} g(x, t) &= e^{\frac{\mu x}{\sigma^2} - \frac{\mu^2 t}{2\sigma^2}} \widehat{g}(x, t) \\ &= e^{\frac{\mu x}{\sigma^2} - \frac{\mu^2 t}{2\sigma^2}} \sum_{n=-\infty}^{\infty} [\phi(x - 2n(u - \ell, t), t) - \phi(x - 2\ell - 2n(u - \ell), t)] \end{aligned}$$

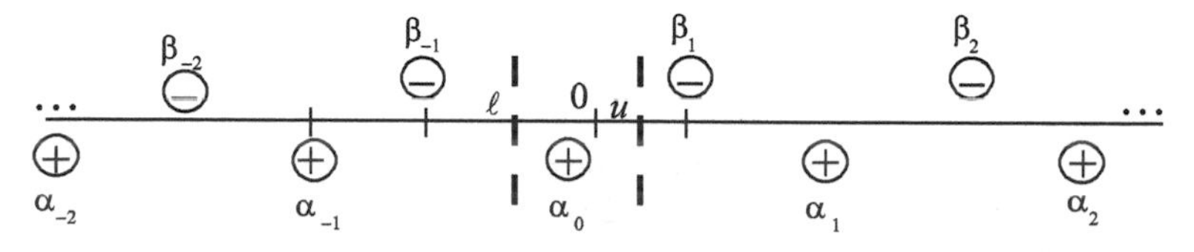

Fig. 4.3. A graphical representation of the infinite number of sources and sinks due to a pair of absorbing barriers (mirrors) with the object placed at the origin. The positions of the sources and sinks are $\alpha_j = 2(u - \ell)j$, $j = 0, \pm 1, \pm 2, \cdots$; $\beta_j = 2u + 2(u - \ell)(j - 1)$ if $j > 0$ and $\beta_j = 2\ell + 2(u - \ell)(j - 1)$ if $j < 0$.

$$= \frac{e^{\frac{\mu x}{\sigma^2} - \frac{\mu^2 t}{2\sigma^2}}}{\sqrt{2\pi\sigma^2 t}} \sum_{n=-\infty}^{\infty} \left[\exp\left(-\frac{[x - 2n(u - \ell)]^2}{2\sigma^2 t} \right) - \exp\left(-\frac{[(x - 2\ell) - 2n(u - \ell)]^2}{2\sigma^2 t} \right) \right]. \quad (4.1.48)$$

The double-mirror analogy provides the intuitive argument showing why $g(x, t)$ involves an infinite number of terms. Once $g(x, t)$ is known, it becomes quite straightforward to derive the price formula of c^o_{LU} (see Problem 4.8).

Next, we would like to derive the density function of the first passage time to either barrier, which is defined by

$$q(t; \ell, u)\, dt = P(\min(\tau_\ell, \tau_u) \in dt), \quad (4.1.49)$$

where $\tau_\ell = \inf\{t | X_t = \ell\}$ and $\tau_u = \inf\{t | X_t = u\}$. First, we consider the corresponding distribution function

$$\begin{aligned} P(\min(\tau_\ell, \tau_u) \le t) &= 1 - P(\min(\tau_\ell, \tau_u) > t) \\ &= 1 - \int_\ell^u g(x, t)\, dx \end{aligned}$$

so that

$$q(t; \ell, u) = -\frac{\partial}{\partial t} \int_\ell^u g(x, t)\, dx.$$

We manage to obtain

$$\begin{aligned} q(t; \ell, u) = \frac{1}{\sqrt{2\pi\sigma^2 t^3}} & \\ \sum_{n=-\infty}^{\infty} & [2n(u - \ell) - \ell] \exp\left(\frac{\mu\ell}{\sigma^2} - \frac{\mu^2 t}{2\sigma^2} \right) \exp\left(-\frac{(2n(u - \ell) - \ell]^2}{2\sigma^2 t} \right) \\ & + [2n(u - \ell) + u] \exp\left(\frac{\mu u}{\sigma^2} - \frac{\mu^2 t}{2\sigma^2} \right) \exp\left(-\frac{[2n(u - \ell) + u]^2}{2\sigma^2 t} \right). \end{aligned} \quad (4.1.50)$$

We may also be interested to find the exit time to a particular barrier. The density function of the exit time to the lower and upper barriers are defined by

$$q^-(t; \ell, u)\, dt = P(\tau_\ell \in dt, \tau_\ell < \tau_u) \quad (4.1.51a)$$
$$q^+(t; \ell, u)\, dt = P(\tau_u \in dt, \tau_u < \tau_\ell). \quad (4.1.51b)$$

Since $\{\tau_\ell \in dt, \tau_\ell < \tau_u\} \cup \{\tau_u \in dt, \tau_u < \tau_\ell\} = \{\min(\tau_\ell, \tau_u) \in dt\}$, we deduce that

$$q(t; \ell, u) = q^-(t; \ell, u) + q^+(t; \ell, u). \quad (4.1.52)$$

A judicious decomposition of $q(t; \ell, u)$ in (4.1.50) into its two components would suggest (Karatzas and Shreve, 1991; see also Problem 4.10)

$$q^-(t;\ell,u) = \frac{1}{\sqrt{2\pi\sigma^2 t^3}} \sum_{n=-\infty}^{\infty} [2n(u-\ell)-\ell] \exp\left(\frac{\mu\ell}{\sigma^2} - \frac{\mu^2 t}{2\sigma^2}\right) \exp\left(-\frac{[2n(u-\ell)-\ell]^2}{2\sigma^2 t}\right) \tag{4.1.53a}$$

$$q^+(t;\ell,u) = \frac{1}{\sqrt{2\pi\sigma^2 t^3}} \sum_{n=-\infty}^{\infty} [2n(u-\ell)+u] \exp\left(\frac{\mu u}{\sigma^2} - \frac{\mu^2 t}{2\sigma^2}\right) \exp\left(-\frac{[2n(u-\ell)+u]^2}{2\sigma^2 t}\right). \tag{4.1.53b}$$

To show the claim, we define the probability flow by

$$J(x,t) = \mu g(x,t) - \frac{\sigma^2}{2}\frac{\partial g}{\partial x}(x,t)$$

and observe that

$$q(t;\ell,u) = -\frac{\partial}{\partial t}\int_\ell^u g(x,t)\,dx = \int_\ell^u -\frac{\partial g}{\partial t}\,dx.$$

Since g satisfies the forward Fokker–Planck equation, we have

$$q(t;\ell,u) = \int_\ell^u \left(\mu\frac{\partial g}{\partial x} - \frac{\sigma^2}{2}\frac{\partial^2 g}{\partial x^2}\right) dx = J(u,t) - J(\ell,t). \tag{4.1.54}$$

One may visualize the probability flow across $x = \ell$ and $x = u$ as

$$\begin{aligned} -J(\ell,t) &= P(\tau_\ell \in dt, \tau_\ell < \tau_u) \\ J(u,t) &= P(\tau_u \in dt, \tau_u < \tau_\ell). \end{aligned}$$

The exit time densities $q^-(t;\ell,u)$ and $q^+(t;\ell,u)$ are seen to satisfy

$$q^-(t;\ell,u) = -J(\ell,t) = -\left[\mu g(x,t) - \frac{\sigma^2}{2}\frac{\partial g}{\partial x}(x,t)\right]\Bigg|_{x=\ell} \tag{4.1.55a}$$

$$q^+(t;\ell,u) = J(u,t) = \mu g(x,t) - \frac{\sigma^2}{2}\frac{\partial g}{\partial x}(x,t)\Bigg|_{x=u}. \tag{4.1.55b}$$

An alternative proof to (4.1.55a,b) was given by Kolkiewicz (2002). Suppose rebate $R^-(t)\,[R^+(t)]$ was paid when the lower (upper) barrier is first breached during the life of the option, then the value of the rebate portion of the double-barrier option is given by

$$\text{rebate value} = \int_0^T e^{-r\xi}[R^-(\xi)q^-(\xi;\ell,u) + R^+(\xi)q^+(\xi;\ell,u)]\,d\xi. \tag{4.1.56}$$

4.1.4 Discretely Monitored Barrier Options

The barrier option pricing models have good analytic tractability when the barrier is monitored continuously (knock-in or knock-out is presumed to occur if the barrier is breached at any instant). In financial markets, practitioners necessarily have to specify a discrete monitoring frequency. Kat and Verdonk (1995) showed that the price differences between the discrete and continuous barrier options can be quite substantial, even under daily monitoring of the barrier. We would expect that discrete monitoring would lower the cost of knock-in options but raise the cost of knock-out options, when compared to their counterparts with continuous monitoring. The analytic price formulas of discrete monitoring barrier options can be derived, but their numerical valuation would be very tedious. The analytic representation involves the multi-variate normal distribution functions (see Problem 4.13).

Correction Formula for Discretely Monitored Barrier Options
Broadie, Glasserman and Kou (1997) obtained an approximation formula of the discretely monitored barrier options, which requires only a simple continuity correction to the continuous barrier option formulas. Let δt denote the uniform time interval between monitoring instants and there are m monitoring instants prior to expiration. Let $V_d(B)$ be the price of a discretely monitored knock-in or knock-out down call or up put option with constant barrier B and $V(B)$ be the price of the corresponding continuously monitored barrier option. Their analytic approximation formula is given by

$$V_d(B) = V(Be^{\pm\beta\sigma\sqrt{\delta t}}) + o\left(\frac{1}{\sqrt{m}}\right), \tag{4.1.57}$$

where $\beta = -\xi(\frac{1}{2})/\sqrt{2\pi} \approx 0.5826$, ξ is the Riemann zeta function, σ is the volatility. The "+" sign is chosen when $B > S$, while the "−" sign is chosen when $B < S$. One observes that the correction term shifts the barrier away from the current underlying asset price by a factor of $e^{\beta\sigma\sqrt{\delta t}}$. The extensive numerical experiments performed by Broadie, Glasserman and Kou (1997) reveal the remarkably good accuracy of the above approximation formula.

4.2 Lookback Options

Lookback options are path dependent options whose payoffs depend on the maximum or the minimum of the underlying asset price attained over a certain period of time (called the *lookback period*). We first consider lookback options where the lookback period is taken to be the whole life of the option. Let T denote the time of expiration of the option and $[T_0, T]$ be the lookback period. We denote the minimum value and maximum value of the asset price realized from T_0 to the current time t $(T_0 \leq t \leq T)$ by

$$m_{T_0}^t = \min_{T_0 \leq \xi \leq t} S_\xi \tag{4.2.1a}$$

and

$$M_{T_0}^t = \max_{T_0 \le \xi \le t} S_\xi, \tag{4.2.1b}$$

respectively. The above formulas implicitly imply continuous monitoring of the asset price, though discrete monitoring for the extremum value is normally adopted in practical implementation. Lookback options can be classified into two types: *fixed strike* and *floating strike*. A floating strike lookback call gives the holder the right to buy at the lowest realized price while a floating strike lookback put allows the holder to sell at the highest realized price over the lookback period. Since $S_T \ge m_{T_0}^T$ and $M_{T_0}^T \ge S_T$ so that the holder of a floating strike lookback option always exercises the option. Hence, the respective terminal payoff of the lookback call and put are given by $S_T - m_{T_0}^T$ and $M_{T_0}^T - S_T$. In a sense, floating strike lookback options are not options. A fixed strike lookback call (put) is a call (put) option on the maximum (minimum) realized price. The respective terminal payoff of the fixed strike lookback call and put are $\max(M_{T_0}^T - X, 0)$ and $\max(X - m_{T_0}^T, 0)$, where X is the strike price. Lookback options guarantee a "no-regret" outcome for the holders, and thus the holders are relieved from making difficult decisions on the optimal timing for entry into or exit from the market. Generally speaking, lookback options are most desirable for investors who have confidence in the view of the range of the asset price movement over a certain period. One would expect that the prices of lookback options are more sensitive to volatility. Also, the writer would charge a much higher premium in view of the favorable payoff to the holder.

Lookback option models can be shown to be closely related to the dynamic investment fund protection. The basic form of fund protection is a guarantee that the sponsor instantaneously provides extra capital into the fund when the fund value falls below a threshold; thus the upgraded fund value never falls below the protection level. It is seen that the upgraded fund value is related to the minimum value realized by the original fund over the protection period. A discussion of pricing the dynamic investment fund protection with a constant protection level can be found in Problem 4.24. Chu and Kwok (2004) presented a comprehensive analysis of the dynamic fund protection with a stochastic guaranteed level (say, the level is benchmarked against a stock index) and withdrawal right.

We would like to derive the price formulas of various types of lookback options, including those with exotic forms of lookback payoff structures. The analytic formulas are limited to options which are European style. Also, *continuous monitoring* of the asset price for the extremum value is assumed. We adopt the usual Black–Scholes pricing framework and the underlying asset price is assumed to follow the Geometric Brownian process. Under the risk neutral measure, the process for the stochastic variable $U_\xi = \ln \frac{S_\xi}{S}$ is a Brownian process with drift rate $\mu = r - \frac{\sigma^2}{2}$ and variance rate σ^2, where r is the riskless interest rate and S is the asset price at current time t (dropping the subscript t for brevity).

We define the following stochastic variables

$$y_T = \ln \frac{m_t^T}{S} = \min\{U_\xi, \xi \in [t, T]\} \tag{4.2.2a}$$

$$Y_T = \ln \frac{M_t^T}{S} = \max\{U_\xi, \xi \in [t, T]\}, \tag{4.2.2b}$$

and write $\tau = T - t$. We can deduce the following joint distribution function of U_T and y_T from the transition density function of the Brownian process with the presence of a downstream barrier [see (4.1.21)]:

$$P(U_T \geq u, y_T \geq y) = N\left(\frac{-u + \mu\tau}{\sigma\sqrt{\tau}}\right) - e^{\frac{2\mu y}{\sigma^2}} N\left(\frac{-u + 2y + \mu\tau}{\sigma\sqrt{\tau}}\right), \tag{4.2.3a}$$

where $y \leq 0$ and $y \leq u$. The corresponding joint distribution function of U_T and Y_T is given by [see (4.1.24)]

$$P(U_T \leq u, Y_T \leq y) = N\left(\frac{u - \mu\tau}{\sigma\sqrt{\tau}}\right) - e^{\frac{2\mu y}{\sigma^2}} N\left(\frac{u - 2y - \mu\tau}{\sigma\sqrt{\tau}}\right), \tag{4.2.3b}$$

where $y \geq 0$ and $y \geq u$. By taking $y = u$ in the above two joint distribution functions, we obtain the following distribution functions for y_T and Y_T

$$P(y_T \geq y) = N\left(\frac{-y + \mu\tau}{\sigma\sqrt{\tau}}\right) - e^{\frac{2\mu y}{\sigma^2}} N\left(\frac{y + \mu\tau}{\sigma\sqrt{\tau}}\right), \quad y \leq 0, \tag{4.2.4a}$$

$$P(Y_T \leq y) = N\left(\frac{y - \mu\tau}{\sigma\sqrt{\tau}}\right) - e^{\frac{2\mu y}{\sigma^2}} N\left(\frac{-y - \mu\tau}{\sigma\sqrt{\tau}}\right), \quad y \geq 0. \tag{4.2.4b}$$

The density functions of y_T and Y_T can be obtained by differentiating the above distribution functions (see Problem 4.14).

4.2.1 European Fixed Strike Lookback Options

Consider a European fixed strike lookback call option whose terminal payoff is $\max(M_{T_0}^T - X, 0)$. The value of this lookback call option at the current time t is given by

$$c_{fix}(S, M, t) = e^{-r(T-t)} E_Q\left[\max(\max(M, M_t^T) - X, 0)\right], \tag{4.2.5}$$

where $S_t = S$, $M_{T_0}^t = M$ and Q is the risk neutral measure. The terminal payoff function can be simplified into the following forms, depending on $M \leq X$ or $M > X$:

(i) $M \leq X$

$$\max(\max(M, M_t^T) - X, 0) = \max(M_t^T - X, 0), \text{ and}$$

(ii) $M > X$

$$\max(\max(M, M_t^T) - X, 0) = (M - X) + \max(M_t^T - M, 0).$$

When $M > X$, the terminal payoff is guaranteed to have the floor value $M - X$. Apparently, the original strike price X is replaced by the "new" strike M. A higher terminal payoff above $M - X$ is resulted when M_t^T assumes a value larger than M.

Now, we define the function H by

$$H(S, t; K) = e^{-r(T-t)} E_Q[\max(M_t^T - K, 0)],$$

where K is a positive constant. Once $H(S, t; K)$ is determined, then

$$\begin{aligned} c_{fix}(S, M, t) &= \begin{cases} H(S, t; X) & \text{if } M \leq X \\ e^{-r(T-t)}(M - X) + H(S, t; M) & \text{if } M > X \end{cases} \\ &= e^{-r(T-t)} \max(M - X, 0) + H(S, t; \max(M, X)). \end{aligned} \quad (4.2.6)$$

Interestingly, $c_{fix}(S, M, t)$ is independent of M when $M \leq X$. This is obvious because the terminal payoff is independent of M when $M \leq X$. On the other hand, when $M > X$, the terminal payoff is guaranteed to have the floor value $M - X$. If we subtract the present value of this guaranteed floor value, then the remaining value of the fixed strike call option is equal to a new fixed strike call but with the strike being increased from X to M.

Since $\max(M_t^T - K, 0)$ is a nonnegative random variable, its expected value is given by the integral of the tail probabilities where

$$\begin{aligned} & e^{-r\tau} E_Q[\max(M_t^T - K, 0)] \\ &= e^{-r\tau} \int_0^\infty P(M_t^T - K \geq x)\, dx \\ &= e^{-r\tau} \int_K^\infty P\left(\ln \frac{M_t^T}{S} \geq \ln \frac{z}{S}\right) dz \qquad z = x + K \\ &= e^{-r\tau} \int_{\ln \frac{K}{S}}^\infty S e^y P(Y_T \geq y)\, dy \qquad y = \ln \frac{z}{S} \\ &= e^{-r\tau} \int_{\ln \frac{K}{S}}^\infty S e^y \left[N\left(\frac{-y + \mu\tau}{\sigma\sqrt{\tau}}\right) + e^{\frac{2\mu y}{\sigma^2}} N\left(\frac{-y - \mu\tau}{\sigma\sqrt{\tau}}\right)\right] dy, \end{aligned}$$

where $\tau = T - t$ and the last integral is obtained by using the distribution function in (4.2.4b). By performing straightforward integration, we obtain

$$\begin{aligned} H(S, \tau; K) &= SN(d) - e^{-r\tau} KN(d - \sigma\sqrt{\tau}) \\ &\quad + e^{-r\tau} \frac{\sigma^2}{2r} S\left[e^{r\tau} N(d) - \left(\frac{S}{K}\right)^{-\frac{2r}{\sigma^2}} N\left(d - \frac{2r}{\sigma}\sqrt{\tau}\right)\right], \end{aligned} \quad (4.2.7)$$

where

$$d = \frac{\ln \frac{S}{K} + \left(r + \frac{\sigma^2}{2}\right)\tau}{\sigma\sqrt{\tau}}.$$

The European fixed strike lookback put option with terminal payoff $\max(X - m_{T_0}^T, 0)$ can be priced in a similar manner. Write $m = m_{T_0}^t$ and define the function

$$h(S,t;K) = e^{-r(T-t)} E_Q[\max(K - m_t^T, 0)].$$

The value of this lookback put can be expressed as

$$p_{fix}(S,m,t) = e^{-r(T-t)} \max(X - m, 0) + h(S,t;\min(m,X)), \tag{4.2.8}$$

where

$$\begin{aligned}
h(S,\tau;K) &= e^{-r\tau} \int_0^\infty P(\max(K - m_t^T, 0) \geq x)\, dx \\
&= e^{-r\tau} \int_0^K P(K - m_t^T \geq x)\, dx \quad \text{since } 0 \leq \max(K - m_t^T, 0) \leq K \\
&= e^{-r\tau} \int_0^K P(m_t^T \leq z)\, dz \qquad z = K - x \\
&= e^{-r\tau} \int_0^{\ln \frac{K}{S}} Se^y P(y_T \leq y)\, dy \qquad y = \ln \frac{y}{S} \\
&= e^{-r\tau} \int_0^{\ln \frac{K}{S}} Se^y \left[N\left(\frac{y - \mu\tau}{\sigma\sqrt{\tau}} \right) + e^{\frac{2\mu y}{\sigma^2}} N\left(\frac{y + \mu\tau}{\sigma\sqrt{\tau}} \right) \right] dy \\
&= e^{-r\tau} KN(-d + \sigma\sqrt{\tau}) - SN(-d) + e^{-r\tau} \frac{\sigma^2}{2r} S \\
&\quad \left[\left(\frac{S}{K} \right)^{-2r/\sigma^2} N\left(-d + \frac{2r}{\sigma}\sqrt{\tau} \right) - e^{r\tau} N(-d) \right], \quad \tau = T - t.
\end{aligned}$$

4.2.2 European Floating Strike Lookback Options

By exploring the pricing relations between the fixed and the floating lookback options, we can deduce the price functions of the floating strike lookback options from those of the fixed strike options. Consider a European floating strike lookback call option whose terminal payoff is $S_T - m_{T_0}^T$, and write $\tau = T - t$, the present value of this call option is given by

$$\begin{aligned}
c_{f\ell}(S,m,\tau) &= e^{-r\tau} E_Q[S_T - \min(m, m_t^T)] \\
&= e^{-r\tau} E_Q[(S_T - m) + \max(m - m_t^T, 0)] \\
&= S - me^{-r\tau} + h(S,\tau;m) \\
&= SN(d_m) - e^{-r\tau} mN(d_m - \sigma\sqrt{\tau}) + e^{-r\tau} \frac{\sigma^2}{2r} S \\
&\quad \left[\left(\frac{S}{m} \right)^{-\frac{2r}{\sigma^2}} N\left(-d_m + \frac{2r}{\sigma}\sqrt{\tau} \right) - e^{r\tau} N(-d_m) \right],
\end{aligned} \tag{4.2.9}$$

where

$$d_m = \frac{\ln \frac{S}{m} + \left(r + \frac{\sigma^2}{2}\right)\tau}{\sigma\sqrt{\tau}}.$$

In a similar manner, consider a European floating strike lookback put option whose terminal payoff is $M_{T_0}^T - S_T$, the present value of this put option is given by

$$\begin{aligned}
p_{f\ell}(S, M, \tau) &= e^{-r\tau} E_Q[\max(M, M_t^T) - S_T] \\
&= e^{-r\tau} E_Q[\max(M_t^T - M, 0) - (S_T - M)] \\
&= H(S, \tau; M) - (S - Me^{-r\tau}) \\
&= e^{-r\tau} MN(-d_M + \sigma\sqrt{\tau}) - SN(-d_M) + e^{-r\tau}\frac{\sigma^2}{2r}S \\
&\quad \left[e^{r\tau}N(d_M) - \left(\frac{S}{M}\right)^{-\frac{2r}{\sigma^2}} N\left(d_M - \frac{2r}{\sigma}\sqrt{\tau}\right)\right], \qquad (4.2.10)
\end{aligned}$$

where

$$d_M = \frac{\ln\frac{S}{M} + \left(r + \frac{\sigma^2}{2}\right)\tau}{\sigma\sqrt{\tau}}.$$

Boundary Condition at $S = m$

What would happen when $S = m$, that is, the current asset price is at the minimum value realized so far? The probability that the current minimum value remains the realized minimum value at expiration is expected to be zero (see Problem 4.16). We then argue that the value of the floating strike lookback call should be insensitive to infinitesimal changes in m since the change in option value with respect to marginal changes in m is proportional to the probability that m will be the realized minimum at expiry (Goldman, Sosin and Gatto, 1979). Mathematically, this is represented by

$$\left.\frac{\partial c_{f\ell}}{\partial m}(S, m, \tau)\right|_{S=m} = 0. \qquad (4.2.11)$$

The above property can be verified by direct differentiation of the call price formula (4.2.9).

Rollover Strategy and Strike Bonus Premium

The sum of the first two terms in $c_{f\ell}$ can be seen as the price function of a European vanilla call with strike price m, while the third term can be interpreted as the strike bonus premium (Garman, 1992). To interpret the strike bonus premium, we consider the hedging of the floating strike lookback call by the following rollover strategy. At any time, we hold a European vanilla call with the strike price set at the current realized minimum asset value. In order to replicate the payoff of the floating strike lookback call at expiry, whenever a new realized minimum value of the asset price is established at a later time, one should sell the original call option and buy a new call with the same expiration date but with the strike price set equal to the newly established minimum value. Since the call with a lower strike is always more expensive, an extra premium is required to adopt the rollover strategy. The present value of the sum of these expected costs of rollover is termed the strike bonus premium.

We would like to show how the strike bonus premium can be obtained by integrating a joint probability distribution function involving m_t^T and S_T. First, we observe that

$$\begin{aligned}\text{strike bonus premium} &= c_{fl}(S, m, \tau) - c_E(S, \tau; m) \\ &= h(S, \tau; m) + S - me^{-r\tau} - c_E(S, \tau; m) \\ &= h(S, \tau; m) - p_E(S, \tau; m), \qquad (4.2.12)\end{aligned}$$

where $c_E(S, \tau, m)$ and $p_E(S, \tau; m)$ are the price functions of the European vanilla call and put, respectively. The last result is due to the put-call parity relation. Recall that

$$h(S, \tau; m) = e^{-r\tau} \int_0^m P(m_t^T \leq \xi)\, d\xi$$

and in a similar manner

$$\begin{aligned}p_E(S, \tau; m) &= e^{-r\tau} \int_0^\infty P(\max(m - S_T, 0) \geq x)\, dx \\ &= e^{-r\tau} \int_0^m P(S_T \leq \xi)\, d\xi. \qquad (4.2.13)\end{aligned}$$

Since the two stochastic state variables satisfy: $0 \leq m_t^T \leq S_T$, we have

$$P(m_t^T \leq \xi) - P(S_T \leq \xi) = P(m_t^T \leq \xi < S_T)$$

so that (Wong and Kwok, 2003)

$$\text{strike bonus premium} = e^{-r\tau} \int_0^m P(m_t^T \leq \xi \leq S_T)\, d\xi. \qquad (4.2.14)$$

4.2.3 More Exotic Forms of European Lookback Options

The lookback options discussed above are the most basic types where the payoff functions at expiry are of standard forms and the lookback period spans the whole life of the option. How can we structure a lookback option whose payoff structure is similar to the above prototype lookback options but is less expensive? Some examples are: partial lookback call option with terminal payoff $\max(S_T - \lambda m_{T_0}^T, 0)$, $\lambda > 1$, and partial lookback put option with terminal payoff $\max(\lambda M_{T_0}^T - S_T, 0)$, $0 < \lambda < 1$ (see Problem 4.19). Also, the price of a lookback option will be lowered if the lookback period spans only a part of the life of the option. When an investor is faced with the problem of deciding the optimal timing for market entry (market exit), he or she may be interested in purchasing an option whose lookback period covers the early part (time period near expiration) of the life of the option. In what follows, we discuss the properties of a "limited period" floating strike lookback call option that is designed for optimal market entry. Discussion of the corresponding fixed strike lookback options for an optimal market exit is relegated to Problem 4.21.

Lookback Options for Market Entry

Suppose an investor thinks that the asset price will rise substantially in the next 12 months and he or she buys a call option on the asset with the strike price set equal to the current asset price. Suppose the asset price drops a few percent within a few weeks after the purchase, though it does rise up strongly at expiration. The investor should have a better return if he or she had bought the option a few weeks later. Timing the market for optimal entry is always difficult. The investor could have avoided difficulty by purchasing a "limited period" floating strike lookback call option whose lookback period covers only the early part of the option's life. It would cost the investor too much if a full period floating strike lookback call were purchased instead.

Let $[T_0, T_1]$ denote the lookback period where $T_1 < T$, T is the expiration time, and let the current time $t \in [T_0, T_1]$. The terminal payoff function of the "limited period" lookback call is $\max(S_T - m_{T_0}^{T_1}, 0)$. We write $S_t = S$, $m_{T_0}^t = m$ and $\tau = T - t$. The value of this lookback call is given by

$$\begin{aligned} c(S, m, \tau) &= e^{-r\tau} E_Q[\max(S_T - m_{T_0}^{T_1}, 0)] \\ &= e^{-r\tau} E_Q[\max(S_T - m, 0)\mathbf{1}_{\{m \le m_t^{T_1}\}}] \\ &\quad + e^{-r\tau} E_Q[\max(S_T - m_t^{T_1}, 0)\mathbf{1}_{\{m > m_t^{T_1}\}}] \\ &= e^{-r\tau} E_Q[S_T \mathbf{1}_{\{S_T > m, m \le m_t^{T_1}\}}] \\ &\quad - e^{-r\tau} m E_Q\big[\mathbf{1}_{\{S_T > m, m \le m_t^{T_1}\}}\big] \\ &\quad + e^{-r\tau} E_Q[S_T \mathbf{1}_{\{S_T > m_t^{T_1}, m > m_t^{T_1}\}}] \\ &\quad - e^{-r\tau} E_Q[m_t^{T_1} \mathbf{1}_{\{S_T > m_t^{T_1}, m > m_t^{T_1}\}}], \quad t < T_1, \end{aligned} \tag{4.2.15}$$

where the expectation is taken under the risk neutral measure Q. The solution for the call price requires the derivation of the appropriate distribution functions. For the first term, the expectation can be expressed as

$$E_Q[S_T \mathbf{1}_{\{S_T > m, m \le m_t^{T_1}\}}] = \int_{\ln \frac{m}{S}}^{\infty} \int_y^{\infty} \int_{\ln \frac{m}{S} - x}^{\infty} S e^{xz} k(z) h(x, y)\, dz\, dx\, dy,$$

where $k(z)$ is the density function for $z = \ln \frac{S_T}{S_{T_1}}$ and $h(x, y)$ is the bivariate density function for $x = \ln \frac{S_{T_1}}{S}$ and $y = \ln \frac{m_t^{T_1}}{S}$ [the corresponding distribution function is presented in (4.2.3a)]. Similarly, the third and fourth terms can be expressed as

$$E_Q[S_T \mathbf{1}_{\{S_T > m_t^{T_1}, m > m_t^{T_1}\}}] = \int_{-\infty}^{\ln \frac{m}{S}} \int_y^{\infty} \int_{y-x}^{\infty} S e^{xz} k(z) h(x, y)\, dz\, dx\, dy$$

and

$$E_Q[m_t^{T_1} \mathbf{1}_{\{S_T > m_t^{T_1}, m > m_t^{T_1}\}}] = \int_{-\infty}^{\ln \frac{m}{S}} \int_y^{\infty} \int_{y-x}^{\infty} S e^{y} k(z) h(x, y)\, dz\, dx\, dy.$$

After performing the tedious integration procedures in the above discounted expectation calculations, the price formula of the "limited-period" lookback call is found to be (Heynen and Kat, 1994b)

$$
\begin{aligned}
&c(S, m, \tau) \\
&= SN(d_1) - me^{-r\tau}N(d_2) + SN_2\left(-d_1, e_1; -\sqrt{\frac{T-T_1}{T-t}}\right) \\
&\quad + e^{-r\tau}mN_2\left(-f_2, d_2; -\sqrt{\frac{T_1-t}{T-t}}\right) \\
&\quad + e^{-r\tau}\frac{\sigma^2}{2r}S\left[\left(\frac{S}{m}\right)^{-\frac{2r}{\sigma^2}}N_2\left(-f_1 + \frac{2r}{\sigma}\sqrt{T_1-t}, -d_1 + \frac{2r}{\sigma}\sqrt{\tau}; \sqrt{\frac{T_1-t}{T-t}}\right)\right. \\
&\qquad\qquad \left. - e^{r\tau}N_2\left(-d_1, e_1; -\sqrt{\frac{T-T_1}{T-t}}\right)\right] \\
&\quad + e^{-r(T-T_1)}\left(1 + \frac{\sigma^2}{2r}\right)SN(e_2)N(-f_1), \quad t < T_1, \qquad (4.2.16)
\end{aligned}
$$

where

$$
\begin{aligned}
d_1 &= \frac{\ln\frac{S}{m} + \left(r + \frac{\sigma^2}{2}\right)\tau}{\sigma\sqrt{\tau}}, & d_2 &= d_1 - \sigma\sqrt{\tau}, \\
e_1 &= \frac{\left(r + \frac{\sigma^2}{2}\right)(T - T_1)}{\sigma\sqrt{T-T_1}}, & e_2 &= e_1 - \sigma\sqrt{T-T_1}, \\
f_1 &= \frac{\ln\frac{S}{m} + \left(r + \frac{\sigma^2}{2}\right)(T_1 - t)}{\sigma\sqrt{T_1-t}}, & f_2 &= f_1 - \sigma\sqrt{T_1-t}.
\end{aligned}
$$

One can check easily that when $T_1 = T$ (full lookback period), the above price formula reduces to price formula (4.2.9). Suppose the current time passes beyond the lookback period, $t > T_1$, the realized minimum value $m_{T_0}^{T_1}$ is now a known quantity. This "limited period" lookback call option then becomes a European vanilla call option with the known strike price $m_{T_0}^{T_1}$.

4.2.4 Differential Equation Formulation

Here we illustrate how to derive the governing partial differential equation and the associated auxiliary conditions for the European floating strike lookback put option. When we consider the partial differential equation formulation, it is convenient to drop the subscript t in the state variables. First, we define the quantity

$$
M_n = \left[\int_{T_0}^{t}(S_\xi)^n d\xi\right]^{1/n}, \quad t > T_0, \qquad (4.2.17)
$$

the derivative of which is given by

$$dM_n = \frac{1}{n}\frac{S^n}{(M_n)^{n-1}}\,dt \tag{4.2.18}$$

so that dM_n is deterministic. Taking the limit $n \to \infty$, we obtain

$$M = \lim_{n\to\infty} M_n = \max_{T_0 \le \xi \le t} S_\xi, \tag{4.2.19}$$

giving the realized maximum value of the asset price process over the lookback period $[T_0, t]$. We attempt to construct a hedged portfolio that contains one unit of a put option whose payoff depends on M_n and shorts Δ units of the underlying asset. Again, we choose Δ so that the stochastic components associated with the option and the underlying asset cancel. Let $p(S, M_n, t)$ denote the value of the lookback put option. We follow the "pragmatic" Black–Scholes derivation procedure by keeping Δ to be instantaneously "frozen." Writing the value Π of the above portfolio as

$$\Pi = p(S, M_n, t) - \Delta S,$$

the differential change of the portfolio value is given by

$$d\Pi = \frac{\partial p}{\partial t}dt + \frac{1}{n}\frac{S^n}{(M_n)^{n-1}}\frac{\partial p}{\partial M_n}dt + \frac{\partial p}{\partial S}dS + \frac{\sigma^2}{2}S^2\frac{\partial^2 p}{\partial S^2}dt - \Delta dS$$

by virtue of Ito's lemma. Again, we choose $\Delta = \frac{\partial p}{\partial S}$ so that the stochastic terms cancel. Using the usual no-arbitrage argument, the riskless hedged portfolio should earn its expected rate of return at the riskless interest rate so that

$$d\Pi = r\Pi\, dt,$$

where r is the riskless interest rate. Putting all the equations together, we obtain

$$\frac{\partial p}{\partial t} + \frac{1}{n}\frac{S^n}{(M_n)^{n-1}}\frac{\partial p}{\partial M_n} + \frac{\sigma^2}{2}S^2\frac{\partial^2 p}{\partial S^2} + rS\frac{\partial p}{\partial S} - rp = 0. \tag{4.2.20}$$

Now, we take the limit $n \to \infty$ and note that $S \le M$. When $S < M$, $\lim_{n\to\infty}\frac{1}{n}\frac{S^n}{(M_n)^{n-1}} = 0$; and when $S = M$, the lookback put value is insensitive to the current realized maximum value, so that $\frac{\partial p}{\partial M} = 0$ [see (4.2.11)]. Hence, the second term in (4.2.20) becomes zero as $n \to \infty$. We conclude that the governing equation for the floating strike lookback put value is given by

$$\frac{\partial p}{\partial t} + \frac{\sigma^2}{2}S^2\frac{\partial^2 p}{\partial S^2} + rS\frac{\partial p}{\partial S} - rp = 0, \quad 0 < S < M, \quad t > T_0, \tag{4.2.21}$$

which is identical to the usual Black–Scholes equation (Goldman, Sosin and Gatto, 1979) except that the domain of the pricing model has an upper bound M on S. It is interesting to observe that the variable M does not appear in the equation, though M appears as a parameter in the auxiliary conditions. The final condition is the terminal payoff function, namely,

$$p(S, M, T) = M - S. \tag{4.2.22a}$$

In this European floating strike lookback put option, the boundary conditions are applied at $S = 0$ and $S = M$. Once S becomes zero, it stays at the zero value at all subsequent times and the payoff at expiry is certain to be M. Discounting at the riskless interest rate, the lookback put value at the current time t is

$$p(0, M, t) = e^{-r(T-t)}M. \tag{4.2.22b}$$

The boundary condition at the other end $S = M$ is given by

$$\frac{\partial p}{\partial M} = 0 \quad \text{at} \quad S = M. \tag{4.2.22c}$$

Remarks.

1. One can show by direct differentiation that the put price formula given in (4.2.10) satisfies (4.2.21) and the auxiliary conditions (4.2.22a,b,c).
2. When the terminal payoff assumes the more general form $f(S_T, M_{T_0}^T)$, Xu and Kwok (2005) managed to derive an integral representation of the lookback option price formula using the partial differential equation approach.

4.2.5 Discretely Monitored Lookback Options

In actual implementation, practitioners necessarily specify a discrete monitoring frequency because continuous monitoring of the asset price movement is almost impractical. We would expect discrete monitoring causes the price of the lookback options to go lower since a new extremum value may be missed out in discrete monitoring. Heynen and Kat (1995) showed in their numerical experiments that monitoring the asset price discretely instead of continuously may have a significant effect on the values of lookback options.

The analytic price formulas for lookback options with discrete monitoring (Heynen and Kat, 1995) involve the n-variate normal distribution functions, where n is the number of monitoring instants within the remaining life of the option. Levy and Mantion (1997) proposed a simple but effective analytic approximation method to price discretely monitored lookback options. The method assumes a second-order Taylor expansion of the option value in powers of $\sqrt{\delta t}$, where δt is the time between successive monitoring instants (assumed to be at regular intervals). The two coefficients of the Taylor expansion are determined by fitting the approximate formula with option values corresponding to $\delta t = \tau$ and $\delta t = \tau/2$, where τ is the time to expiry. The construction and implementation of the method are quite straightforward, the details of which are presented in Problem 4.23. As demonstrated by their numerical experiments, the accuracy of this analytic approximation method is quite remarkable.

Similar to the discretely monitored barrier options, Broadie, Glasserman and Kou (1999) derived analytic approximation formulas of the price functions of discretely monitored lookback options in terms of the price functions of the continuously monitored counterparts.

4.3 Asian Options

Asian options are averaging options whose terminal payoff depends on some form of averaging of the price of the underlying asset over a part or the whole of the option's life. There are frequent market situations where traders may be interested to hedge against the average price of a commodity over a period rather than, say, the end-of-period price. For example, suppose a manufacturer expects to make a string of copper purchases for his factory over some fixed time horizon. The company would be interested in acquiring price protection that is linked to the average price over the period. The hedging of risk against the average price may be achieved through the purchase of an appropriate averaging option. Averaging options are particularly useful for business involving thinly traded commodities. The use of such financial instruments may avoid price manipulation near the end of the period.

Most Asian options are of European style since an Asian option with the American early exercise feature may be redeemed as early as the start of the averaging period and lose the intent of protection from averaging. There are two main classes of Asian options, the *fixed strike* (*average rate*) and the *floating strike* (*average strike*) *options*. The corresponding terminal call payoff are $\max(A_T - X, 0)$ and $\max(S_T - A_T, 0)$, respectively. Here, S_T is the asset price at expiry, X is the strike price and A_T denotes some form of average of the price of the underlying asset over the averaging period $[0, T]$. The value of A_T depends on the realization of the asset price path. The most common averaging procedures are the discrete arithmetic averaging defined by

$$A_T = \frac{1}{n} \sum_{i=1}^{n} S_{t_i} \tag{4.3.1a}$$

and the discrete geometric averaging defined by

$$A_T = \left[\prod_{i=1}^{n} S_{t_i} \right]^{1/n}. \tag{4.3.1b}$$

Here, S_{t_i} is the asset price at discrete time t_i, $i = 1, 2, \cdots, n$. In the limit $n \to \infty$, the discrete sampled averages become the continuous sampled averages. The continuous arithmetic average is given by

$$A_T = \frac{1}{T} \int_0^T S_t \, dt, \tag{4.3.2a}$$

while the continuous geometric average is defined to be

$$A_T = \exp\left(\frac{1}{T} \int_0^T \ln S_t \, dt \right). \tag{4.3.2b}$$

A wide variety of averaging options have been proposed. Good, comprehensive summaries of them can be found in Boyle (1993) and Zhang (1994). The most commonly used sampled average is the discrete arithmetic average. If the Geometric

Brownian process is assumed for the underlying asset price process, the pricing of this type of Asian option is in general analytically intractable since the sum of lognormal densities has no explicit representation. On the other hand, the analytic derivation of the price formula of a European Asian option with geometric averaging is feasible because the product of lognormal prices remains to be lognormal.

In this section, we first derive the general partial differential equation formulation for pricing Asian options. We then consider the pricing of continuously monitored Asian options with geometric or arithmetic averaging. We deduce put-call parity relations and fixed-floating symmetry relations between the prices of continuously monitored Asian options. For discretely monitored Asian options, we derive closed form price formulas for geometric averaging options and deduce analytic approximation formulas for arithmetic averaging options using the Edgeworth expansion technique.

4.3.1 Partial Differential Equation Formulation

In this section we derive the governing differential equation for the price of an Asian option using the Black–Scholes approach. The price function $V(S, A, \tau)$ is a function of time to expiry τ and the two state variables, asset price S and average asset value A. When we formulate an option pricing model using the partial differential equation approach, it is convenient to drop the subscript t in the state variables. Suppose we write the average of the asset price as

$$A = \int_0^t f(S, u)\, du, \tag{4.3.3}$$

where $f(S, t)$ is chosen according to the type of average adopted in the Asian option. For example, $f(S, t) = \frac{1}{t} S$ corresponds to continuous arithmetic average, $f(S, t) = \exp\left(\frac{1}{n}\sum_{i=1}^{n} \delta(t - t_i) \ln S\right)$ corresponds to discrete geometric average, etc. Suppose $f(S, t)$ is a continuous time function, then by the mean value theorem

$$dA = \lim_{\Delta t \to 0} \int_t^{t+\Delta t} f(S, u)\, du = \lim_{\Delta t \to 0} f(S, u^*)\, dt = f(S, t)\, dt,$$
$$t < u^* < t + \Delta t, \tag{4.3.4}$$

so dA is deterministic. Hence, a riskless hedge for the Asian option requires only eliminating the asset-induced risk, so the exposure on the Asian option can be hedged by holding an appropriate number of units of the underlying asset.

Consider a portfolio that contains one unit of the Asian option and $-\Delta$ units of the underlying asset. We then choose Δ such that the stochastic components associated with the option and the underlying asset cancel each other out. Assume the asset price dynamics to be given by

$$\frac{dS}{S} = \mu\, dt + \sigma\, dZ, \tag{4.3.5}$$

where Z is the standard Brownian process, q is the dividend yield on the asset, μ and σ are the expected rate of return and volatility of the asset price, respectively. Let $V(S, A, t)$ denote the value of the Asian option and let Π denote the value of the above portfolio. The portfolio value is given by

$$\Pi = V(S, A, t) - \Delta S,$$

and assuming Δ to be kept instantaneously "frozen," the differential of Π is found to be

$$d\Pi = \frac{\partial V}{\partial t}\,dt + f(S,t)\frac{\partial V}{\partial A}\,dt + \frac{\partial V}{\partial S}\,dS + \frac{\sigma^2}{2}S^2\frac{\partial^2 V}{\partial S^2}\,dt - \Delta\,dS - \Delta q S\,dt.$$

The last term in the above equation corresponds to the contribution of the dividend dollar amount from the asset to the portfolio value. As usual, we choose $\Delta = \frac{\partial V}{\partial S}$ so that the stochastic terms containing dS cancel. The absence of arbitrage dictates

$$d\Pi = r\Pi\,dt,$$

where r is the riskless interest rate. Putting the above results together, we obtain the following governing differential equation for $V(S, A, t)$

$$\frac{\partial V}{\partial t} + \frac{\sigma^2}{2}S^2\frac{\partial^2 V}{\partial S^2} + (r-q)S\frac{\partial V}{\partial S} + f(S,t)\frac{\partial V}{\partial A} - rV = 0. \tag{4.3.6}$$

The equation is a degenerate diffusion equation since it contains diffusion term corresponding to S only but not A. The auxiliary conditions in the pricing model depend on the specific details of the Asian option contract.

4.3.2 Continuously Monitored Geometric Averaging Options

Here we derive analytic price formulas for the European Asian options whose terminal payoff depends on the continuously monitored geometric averaging of the underlying asset price. We take time zero to be the initiation time of the averaging period, t is the current time and T denotes the expiration time. We define the continuously monitored geometric averaging of the asset price S_u over the time period $[0, t]$ by

$$G_t = \exp\left(\frac{1}{t}\int_0^t \ln S_u\,du\right). \tag{4.3.7}$$

The terminal payoff of the fixed strike call option and floating strike call option are, respectively,

$$\begin{aligned} c_{fix}(S_T, G_T, T; X) &= \max(G_T - X, 0) \\ c_{f\ell}(S_T, G_T, T) &= \max(S_T - G_T, 0), \end{aligned} \tag{4.3.8}$$

where X is the fixed strike price. We illustrate how to use the risk neutral valuation approach to derive the price formula of the European fixed strike Asian call option.

On the other hand, the partial differential equation method is used to derive the price formula of the floating strike counterpart.

European Fixed Strike Asian Call Option

We assume the existence of a risk neutral pricing measure Q under which discounted asset prices are martingales, implying the absence of arbitrage. Under the measure Q, the asset price dynamics follows

$$\frac{dS_t}{S_t} = (r - q)\,dt + \sigma\,dZ_t, \tag{4.3.9}$$

where Z_t is Q-Brownian and q is the constant dividend yield of the underlying asset. For $0 < t < T$, the solution of the above stochastic differential equation is given by [see (2.4.16)]

$$\ln S_u = \ln S_t + \left(r - q - \frac{\sigma^2}{2}\right)(u - t) + \sigma(Z_u - Z_t). \tag{4.3.13}$$

Substituting the above relation into (4.3.7) and performing the integration, we obtain

$$\begin{aligned}\ln G_T &= \frac{t}{T}\ln G_t + \frac{1}{T}\left[(T - t)\ln S_t + \left(r - q - \frac{\sigma^2}{2}\right)\frac{(T - t)^2}{2}\right] \\ &\quad + \frac{\sigma}{T}\int_t^T (Z_u - Z_t)\,du.\end{aligned} \tag{4.3.10}$$

The stochastic term $\frac{\sigma}{T}\int_t^T (Z_u - Z_t)\,du$ can be shown to be Gaussian with zero mean and variance $\frac{\sigma^2}{T^2}\frac{(T-t)^3}{3}$ (see Problem 2.36). By the risk neutral valuation principle, the value of the European fixed strike Asian call option is given by

$$c_{fix}(S_t, G_t, t) = e^{-r(T-t)}E_Q[\max(G_T - X, 0)], \tag{4.3.11}$$

where the expectation is taken under Q conditional on the filtration generated by the Q-Brownian process. We assume the current time t to be within the averaging period. By defining

$$\overline{\mu} = \left(r - q - \frac{\sigma^2}{2}\right)\frac{(T - t)^2}{2T} \quad \text{and} \quad \overline{\sigma} = \frac{\sigma}{T}\sqrt{\frac{(T - t)^3}{3}},$$

G_T can be written as

$$G_T = G_t^{t/T} S_t^{(T-t)/T} \exp(\overline{\mu} + \overline{\sigma}\widehat{Z}), \tag{4.3.12}$$

where $\widehat{Z}$ is the standard normal random variable. Recall from our usual expectation calculations with call payoff:

$$\begin{aligned}&E_Q[\max(F\exp(\overline{\mu} + \overline{\sigma}\widehat{Z}) - X, 0] \\ &= Fe^{\overline{\mu}+\overline{\sigma}^2/2}N\left(\frac{\ln\frac{F}{X} + \overline{\mu} + \overline{\sigma}^2}{\overline{\sigma}}\right) - XN\left(\frac{\ln\frac{F}{X} + \overline{\mu}}{\overline{\sigma}}\right),\end{aligned}$$

we then deduce that

$$c_{fix}(S_t, G_t, t) = e^{-r(T-t)}\left[G_t^{t/T} S_t^{(T-t)/T} e^{\overline{\mu}+\overline{\sigma}^2/2} N(d_1) - XN(d_2)\right], \quad (4.3.13)$$

where

$$d_2 = \left(\frac{t}{T}\ln G_t + \frac{T-t}{T}\ln S_t + \overline{\mu} - \ln X\right)\Big/\overline{\sigma},$$
$$d_1 = d_1 + \overline{\sigma}.$$

European Floating Strike Asian Call Option

Since the terminal payoff of the floating strike Asian call option involves S_T and G_T, pricing the Asian option by the risk neutral valuation approach would require the joint distribution of S_T and G_T. For floating strike Asian options, the partial differential equation method provides an alternative approach to derive the price formula for $c_{f\ell}(S, G, t)$. We show how the similarity reduction technique can be applied to reduce the dimension of the differential equation.

When the continuously monitored geometric averaging is adopted, the governing equation for $c_{f\ell}(S, G, t)$ can be expressed as

$$\frac{\partial c_{f\ell}}{\partial t} + \frac{\sigma^2}{2}S^2\frac{\partial^2 c_{f\ell}}{\partial S^2} + (r-q)S\frac{\partial c_{f\ell}}{\partial S} + \frac{G}{t}\ln\frac{S}{G}\frac{\partial c_{f\ell}}{\partial G} - rc_{f\ell} = 0, \quad 0 < t < T. \quad (4.3.14)$$

Next, we define the similarity variables:

$$y = t\ln\frac{G}{S} \quad \text{and} \quad W(y, t) = \frac{c_{f\ell}(S, G, t)}{S}. \quad (4.3.15)$$

This is equivalent to choosing S as the numeraire. In terms of the similarity variables, the governing equation for $c_{f\ell}(S, G, t)$ becomes

$$\frac{\partial W}{\partial t} + \frac{\sigma^2 t^2}{2}\frac{\partial^2 W}{\partial y^2} - \left(r - q + \frac{\sigma^2}{2}\right)t\frac{\partial W}{\partial y} - qW = 0, \quad 0 < t < T, \quad (4.3.16)$$

with terminal condition: $W(y, T) = \max(1 - e^{y/T}, 0)$.

We write $\tau = T - t$ and let $F(y, \tau; \eta)$ denote the Green function to the following parabolic equation with time dependent coefficients

$$\frac{\partial F}{\partial \tau} = \frac{\sigma^2(T-\tau)^2}{2}\frac{\partial^2 F}{\partial y^2} - \left(r - q + \frac{\sigma^2}{2}\right)(T-\tau)\frac{\partial F}{\partial y}, \quad \tau > 0,$$

with initial condition at $\tau = 0$ (corresponding to $t = T$) given as

$$F(y, 0; \eta) = \delta(y - \eta).$$

Though the differential equation has time dependent coefficients, the fundamental solution is readily found to be [see (3.4.10)]

$$F(y,\tau;\eta) = n\left(\frac{y-\eta-\left(r-q+\frac{\sigma^2}{2}\right)\int_0^\tau (T-u)\,du}{\sigma\sqrt{\int_0^\tau (T-u)^2\,du}}\right). \tag{4.3.17}$$

The solution to $W(y,\tau)$ is then given by

$$W(y,\tau) = e^{-q\tau}\int_{-\infty}^{\infty} \max(1-e^{\eta/T},0)F(y,\tau;\eta)\,d\eta. \tag{4.3.18}$$

The evaluation of the above integral gives (Wu, Kwok and Yu, 1999)

$$c_{f\ell}(S,G,t) = Se^{-q(T-t)}N(\widehat{d}_1) - G^{t/T}S^{(T-t)/T}e^{-q(T-t)}e^{-\widehat{Q}}N(\widehat{d}_2), \tag{4.3.19}$$

where

$$\widehat{d}_1 = \frac{t\ln\frac{S}{G} + \left(r-q+\frac{\sigma^2}{2}\right)\frac{T^2-t^2}{2}}{\sigma\sqrt{\frac{T^3-t^3}{3}}},$$

$$\widehat{d}_2 = \widehat{d}_1 - \frac{\sigma}{T}\sqrt{\frac{T^3-t^3}{3}},$$

$$\widehat{Q} = \frac{r-q+\frac{\sigma^2}{2}}{2}\frac{T^2-t^2}{T} - \frac{\sigma^2}{6}\frac{T^3-t^3}{T^2}.$$

4.3.3 Continuously Monitored Arithmetic Averaging Options

We consider a European fixed strike Asian call based on continuously monitored arithmetic averaging. The terminal payoff is defined by

$$c_{fix}(S_T, A_T, T; X) = \max(A_T - X, 0). \tag{4.3.20}$$

He and Takahashi (2000) proposed a variable reduction method that reduces the dimension of the governing differential equation by one. To motivate an appropriate choice of the transformation of variable, we consider the following expectation representation of the price of the Asian call at time t

$$\begin{aligned} c_{fix}(S_t, A_t, t) &= e^{-r(T-t)}E_Q\big[\max(A_T - X, 0)\big] \\ &= e^{-r(T-t)}E_Q\left[\max\left(\frac{1}{T}\int_0^t S_u\,du - X + \frac{1}{T}\int_t^T S_u\,du, 0\right)\right] \\ &= \frac{S_t}{T}e^{-r(T-t)}E_Q\left[\max\left(x_t + \int_t^T \frac{S_u}{S_t}\,du, 0\right)\right], \end{aligned} \tag{4.3.21}$$

where the state variable x_t is defined by

$$x_t = \frac{1}{S_t}(I_t - XT), \quad \text{where} \quad I_t = \int_0^t S_u\,du = tA_t. \tag{4.3.22}$$

In our subsequent discussion, it is more convenient to use I_t instead of A_t as the averaging state variable. By virtue of the Markovian property of the asset price process, the price ratio $S_u/S_t, u > t$, is independent of the history of the asset price up to time t. The conditional expectation in (4.3.21) is a function of x_t only. We then deduce that

$$c_{fix}(S_t, I_t, t) = S_t f(x_t, t) \tag{4.3.23}$$

for some function of f. It is seen that $f(x_t, t)$ is given by

$$f(x_t, t) = \frac{e^{-r(T-t)}}{T} E_Q\left[\max\left(x_t + \int_t^T \frac{S_u}{S_t}\, du, 0\right)\right]. \tag{4.3.24}$$

Recall that the governing equation for the price function $c_{fix}(S, I, t)$ of the fixed strike call is given by

$$\frac{\partial c_{fix}}{\partial t} + \frac{\sigma^2}{2} S^2 \frac{\partial^2 c_{fix}}{\partial S^2} + (r - q)S \frac{\partial c_{fix}}{\partial S} + S \frac{\partial c_{fix}}{\partial I} - r c_{fix} = 0. \tag{4.3.25}$$

The expectation representation of c_{fix} in (4.3.21) motivates us to define the following set of transformation of variables

$$x = \frac{1}{S}(I - XT) \quad \text{and} \quad f(x, t) = \frac{c_{fix}(S, I, t)}{S}.$$

The governing differential equation for $f(x, t)$ can then be shown to be

$$\frac{\partial f}{\partial t} + \frac{\sigma^2}{2} x^2 \frac{\partial^2 f}{\partial x^2} + [1 - (r - q)x] \frac{\partial f}{\partial x} - qf = 0, \quad -\infty < x < \infty, t > 0. \tag{4.3.26}$$

The terminal condition is given by

$$f(x, T) = \frac{1}{T} \max(x, 0). \tag{4.3.27}$$

As a remark, by finding a judicious trading strategy that replicates the average of asset prices, a similar form of the governing equation with one state variable can also be derived (see Problem 4.31).

When $x_t \geq 0$, which corresponds to $\frac{1}{T} \int_0^t S_u\, du \geq X$, it is possible to find a closed form analytic solution to $f(x, t)$. Since x_t is an increasing function of t so that $x_T \geq 0$, the terminal condition $f(x, T)$ reduces to x/T. In this case, $f(x, t)$ admits a solution of the form

$$f(x, t) = a(t)x + b(t).$$

By substituting the assumed form of solution into (4.3.26), we obtain the following pair of governing equations for $a(t)$ and $b(t)$

$$\frac{da(t)}{dt} - ra(t) = 0, \qquad a(T) = \frac{1}{T},$$
$$\frac{db(t)}{dt} - a(t) - qb(t) = 0, \qquad b(T) = 0.$$

When $r \neq q$, $a(t)$ and $b(t)$ are found to be

$$a(t) = \frac{e^{-r(T-t)}}{T} \quad \text{and} \quad b(t) = \frac{e^{-q(T-t)} - e^{-r(T-t)}}{T(r-q)}.$$

The Asian option price function for $I \geq XT$ is given by

$$c_{fix}(S, I, t) = \left(\frac{I}{T} - X\right)e^{-r(T-t)} + \frac{e^{-q(T-t)} - e^{-r(T-t)}}{T(r-q)}S. \tag{4.3.28}$$

Though the volatility σ does not appear explicitly in the above price formula, it appears implicitly in S and A. The gamma is easily seen to be zero while the delta is a function of t and $T - t$ but not S or A.

For $I < XT$, there is no closed form analytic solution available. Curran (1994) and Rogers and Shi (1995) proposed the conditioning method to find a *lower bound* on the Asian option price. They both used the approach of projecting the averaging state variable A_T on a $\mathcal{F}_T$-measurable Gaussian random variable Y. By virtue of the Jensen inequality (see Problem 4.36), we obtain

$$\begin{aligned} E_Q[\max(A_T - X, 0)] &= E_Q\big[E_Q\big[\max(A_T - X, 0)|Y\big]\big] \\ &\geq E_Q\big[\max\big(E_Q[A_T - X|Y], 0\big)\big]. \end{aligned} \tag{4.3.29}$$

The resulting expectation involving Y may be solvable in closed form. A natural choice of Y would be the logarithm of the geometric average. The approximation error would be small since the correlation coefficient between the geometric average and arithmetic average is close to one. An application of the conditioning mean method is illustrated in Problem 4.37.

The analytic approximation approach has also been applied to continuously monitored floating strike Asian options. Bouaziz, Briys and Crouhy (1994) used the linear approximation technique to approximate the law of $\{A_T, S_T\}$ by a joint lognormal distribution [see also the extension to quadratic approximation by Chung, Shackleton and Wojakowski (2003)]. Several other analytic approximation methods for pricing Asian options can be found in Milevsky and Posner (1998), Nielsen and Sandmann (2003) and Tsao, Chang and Lin (2003). Some of these results are illustrated in Problem 4.38.

4.3.4 Put-Call Parity and Fixed-Floating Symmetry Relations

It is well known that the difference of the prices of European vanilla call and put options is equal to a European forward contract. Do we have similar put-call parity relations for European Asian options? Also, can we establish symmetry relations between the prices of fixed strike and floating strike Asian options, like those of the lookback options? In this section, we derive these parity and symmetry relations for continuously monitored Asian options under the Black–Scholes framework. Some of these relations can be extended to more general stochastic price dynamics (Hoogland and Neumann, 2000).

Put-Call Parity Relation

Let $c_{fix}(S, I, t)$ and $p_{fix}(S, I, t)$ denote the price function of the fixed strike arithmetic averaging Asian call option and put option, respectively. Their terminal payoff functions are given by

$$c_{fix}(S, I, T) = \max\left(\frac{I}{T} - X, 0\right) \tag{4.3.30a}$$

$$p_{fix}(S, I, T) = \max\left(X - \frac{I}{T}, 0\right), \tag{4.3.30b}$$

where $I = \int_0^T S_u\,du$. Let $D(S, I, t)$ denote the difference of c_{fix} and p_{fix}. Since both c_{fix} and p_{fix} are governed by the same differential equation [see (4.3.25)], so does $D(S, I, t)$. The terminal condition of their difference $D(S, I, t)$ is given by

$$D(S, I, T) = \max\left(\frac{I}{T} - X, 0\right) - \max\left(X - \frac{I}{T}, 0\right) = \frac{I}{T} - X.$$

The above terminal condition is the same as that of the continuously monitored arithmetic averaging option with $I \geq XT$. Hence, when $r \neq q$, the put-call parity relation between the prices of fixed strike Asian options under continuously monitored arithmetic averaging is given by [see also (4.3.28)]

$$\begin{aligned} &c_{fix}(S, I, t) - p_{fix}(S, I, t) \\ &= \left(\frac{I}{T} - X\right)e^{-r(T-t)} + \frac{e^{-q(T-t)} - e^{-r(T-t)}}{T(r-q)}S. \end{aligned} \tag{4.3.31}$$

Similar techniques can be used to derive the put-call parity relations between other types of Asian options (floating/fixed strike and geometric/arithmetic averaging) (see Problems 4.28 and 4.29).

Fixed-Floating Symmetry Relations

By applying a change of measure and identifying a time-reversal of a Brownian process (Henderson and Wojakowski, 2002), it is possible to establish the symmetry relations between the prices of floating strike and fixed strike arithmetic averaging Asian options at the start of the averaging period.

Suppose we write the price functions of various continuously monitored arithmetic averaging option at the start of the averaging period (taken to be time zero) as

$$\begin{aligned} c_{f\ell}(S_0, \lambda, r, q, T) &= e^{-rT}E_Q[\max(\lambda S_T - A_T, 0)] \\ p_{f\ell}(S_0, \lambda, r, q, T) &= e^{-rT}E_Q[\max(A_T - \lambda S_T, 0)] \\ c_{fix}(X, S_0, r, q, T) &= e^{-rT}E_Q[\max(A_T - X, 0)] \\ p_{fix}(X, S_0, r, q, T) &= e^{-rT}E_Q[\max(X - A_T, 0)]. \end{aligned}$$

Under the risk neutral measure Q, the asset price S_t follows the Geometric Brownian process

$$\frac{dS_t}{S_t} = (r - q)\,dt + \sigma\,dZ_t. \tag{4.3.32}$$

Here, Z_t is a Q-Brownian process. Suppose the asset price is used as the numeraire, then

$$\begin{aligned} c_{f\ell}^* = \frac{c_{f\ell}}{S_0} &= \frac{e^{-rT}}{S_0} E_Q\big[\max(\lambda S_T - A_T, 0)\big] \\ &= E_Q\left[\frac{S_T e^{-rT}}{S_0} \frac{\max(\lambda S_T - A_T, 0)}{S_T}\right]. \end{aligned} \tag{4.3.33}$$

To effect the change of numeraire, we define the measure Q^* by

$$\left.\frac{dQ^*}{dQ}\right|_{\mathcal{F}_T} = e^{-\frac{\sigma^2}{2}T + \sigma Z_T} = \frac{S_T e^{-rT}}{S_0 e^{-qT}}. \tag{4.3.34}$$

By virtue of the Girsanov Theorem, $Z_T^* = Z_T - \sigma T$ is Q^*-Brownian [see Problem 3.10]. If we write $A_T^* = A_T / S_T$, then

$$c_{f\ell}^* = e^{-qT} E_{Q^*}\big[\max\big(\lambda - A_T^*, 0\big)\big], \tag{4.3.35}$$

where E_{Q^*} denotes the expectation under Q^*. Now, we consider

$$A_T^* = \frac{1}{T}\int_0^T \frac{S_u}{S_T}\,du = \frac{1}{T}\int_0^T S_u^*(T)\,du, \tag{4.3.36}$$

where

$$S_u^*(T) = \exp\left(-\left(r - q - \frac{\sigma^2}{2}\right)(T - u) - \sigma(Z_T - Z_u)\right).$$

In terms of the Q^*-Brownian process Z_t^*, where $Z_T - Z_u = \sigma(T - u) + Z_T^* - Z_u^*$, we can write

$$S_u^*(T) = \exp\left(\left(r - q + \frac{\sigma^2}{2}\right)(u - T) + \sigma(Z_u^* - Z_T^*)\right). \tag{4.3.37}$$

Furthermore, we define a reflected Q^*-Brownian process starting at zero by $\widehat{Z}_t$, where $\widehat{Z}_t = -Z_t^*$, then $\widehat{Z}_{T-u}$ equals in law to $Z_u^* - Z_T^*$ due to the stationary increment property of a Brownian process. Hence, we establish

$$A_T^* \overset{\text{law}}{=} \widehat{A}_T = \frac{1}{T}\int_0^T e^{\sigma \widehat{Z}_{T-u} + (r-q+\frac{\sigma^2}{2})(u-T)}\,du, \tag{4.3.38}$$

and via time-reversal of $\widehat{Z}_{T-u}$, we obtain

$$\widehat{A}_T = \frac{1}{T}\int_0^T e^{\sigma \widehat{Z}_\xi + (q-r-\frac{\sigma^2}{2})\xi}\,d\xi. \tag{4.3.39}$$

Note that $\widehat{A}_T S_0$ is the arithmetic average of the price process with drift rate $q - r$. Summing the results together, we have

$$c_{f\ell} = S_0 c^*_{f\ell} = e^{-qT} E_{Q^*}\big[\max(\lambda S_0 - \widehat{A}_T S_0, 0)\big], \tag{4.3.40}$$

and from which we deduce the following fixed-floating symmetry relation

$$c_{f\ell}(S_0, \lambda, r, q, T) = p_{fix}(\lambda S_0, S_0, q, r, T). \tag{4.3.41}$$

By combining the put-call parity relations for floating and fixed Asian options and the above symmetry relation, we can derive the following fixed-floating symmetry relation between c_{fix} and $p_{f\ell}$ (see Problem 4.30)

$$c_{fix}(X, S_0, r, q, T) = p_{f\ell}\left(S_0, \frac{X}{S_0}, q, r, T\right). \tag{4.3.42}$$

4.3.5 Fixed Strike Options with Discrete Geometric Averaging

Consider the discrete geometric averaging of the asset prices at evenly distributed discrete times $t_i = i\Delta t, i = 1, 2, \cdots, n$, where Δt is the uniform time interval between fixings and $t_n = T$ is the time of expiration. Define the running geometric averaging by

$$G_k = \left[\prod_{i=1}^{k} S_{t_i}\right]^{1/k}, \quad k = 1, 2, \cdots, n. \tag{4.3.43}$$

The terminal payoff of a European average value call option with discrete geometric averaging is given by $\max(G_n - X, 0)$, where X is the strike price. Suppose the asset price follows the Geometric Brownian process, then the asset price ratio $R_i = \frac{S_{t_i}}{S_{t_{i-1}}}, i = 1, 2, \cdots, n$ is lognormally distributed. Assume that under the risk neutral measure Q

$$\ln R_i \sim N\left(\left(r - \frac{\sigma^2}{2}\right)\Delta t, \sigma^2 \Delta t\right), \quad i = 1, 2, \cdots, n, \tag{4.3.44}$$

where r is the riskless interest rate and $N(\mu, \sigma^2)$ represents a normal distribution with mean μ and variance σ^2.

European Fixed Strike Call Option

The price formula of the European fixed strike call option depends on whether the current time t is prior to or after time t_0. First, we consider $t < t_0$ and write

$$\frac{G_n}{S_t} = \frac{S_{t_0}}{S_t}\left\{\frac{S_{t_n}}{S_{t_{n-1}}}\left[\frac{S_{t_{n-1}}}{S_{t_{n-2}}}\right]^2 \cdots \left[\frac{S_{t_1}}{S_{t_0}}\right]^n\right\}^{1/n},$$

so that

$$\ln \frac{G_n}{S_t} = \ln \frac{S_{t_0}}{S_t} + \frac{1}{n}\big[\ln R_n + 2\ln R_{n-1} + \cdots + n \ln R_1\big], \quad t < t_0. \tag{4.3.45}$$

Since $\ln R_i, i = 1, 2, \cdots n$ and $\ln \frac{S_{t_0}}{S_t}$ represent independent Brownian increments over nonoverlapping time intervals, they are normally distributed and independent. Observe that $\ln \frac{G_n}{S_t}$ is a linear combination of these independent Brownian increments, so it remains to be normally distributed with mean

$$\begin{aligned}&\left(r - \frac{\sigma^2}{2}\right)(t_0 - t) + \frac{1}{n}\left(r - \frac{\sigma^2}{2}\right)\Delta t \sum_{i=1}^{n} i \\ &= \left(r - \frac{\sigma^2}{2}\right)\left[(t_0 - t) + \frac{n+1}{2n}(T - t_0)\right],\end{aligned}$$

and variance

$$\sigma^2(t_0 - t) + \frac{1}{n^2}\sigma^2 \Delta t \sum_{i=1}^{n} i^2 = \sigma^2\left[(t_0 - t) + \frac{(n+1)(2n+1)}{6n^2}(T - t_0)\right].$$

Let $\tau = T - t$, where τ is the time to expiry. Suppose we write

$$\begin{aligned}\sigma_G^2 \tau &= \sigma^2 \left\{\tau - \left[1 - \frac{(n+1)(2n+1)}{6n^2}\right](T - t_0)\right\} \\ \left(\mu_G - \frac{\sigma_G^2}{2}\right)\tau &= \left(r - \frac{\sigma^2}{2}\right)\left[\tau - \frac{n-1}{2n}(T - t_0)\right],\end{aligned}$$

then the transition density function of G_n at time T, given the asset price S_t at an earlier time $t < t_0$, can be expressed as

$$\psi(G_n; S_t) = \frac{1}{G_n\sqrt{2\pi\sigma_G^2\tau}} \exp\left(-\frac{\left\{\ln G_n - \left[\ln S_t + \left(\mu_G - \frac{\sigma_G^2}{2}\right)\tau\right]\right\}^2}{2\sigma_G^2\tau}\right). \quad (4.3.46)$$

By the risk neutral valuation approach, the price of the European fixed strike call with discrete geometric averaging is given by

$$\begin{aligned}c_G(S_t, t) &= e^{-r\tau} E_Q[\max(G_n - X, 0)] \\ &= e^{-r\tau}\left[S_t e^{\mu_G \tau} N(d_1) - XN(d_2)\right], \quad t < t_0 \qquad (4.3.47)\end{aligned}$$

where

$$d_1 = \frac{\ln \frac{S_t}{X} + \left(\mu_G + \frac{\sigma_G^2}{2}\right)\tau}{\sigma_G\sqrt{\tau}}, \quad d_2 = d_1 - \sigma_G\sqrt{\tau}.$$

We consider the two extreme cases where $n = 1$ and $n \to \infty$. When $n = 1$, $\sigma_G^2\tau$ and $(\mu_G - \frac{\sigma_G^2}{2})\tau$ reduce to $\sigma^2\tau$ and $(r - \frac{\sigma^2}{2})\tau$, respectively, so that the call price reduces to that of a European vanilla call option. We observe that $\sigma_G^2\tau$ is a decreasing function of n, which is consistent with the intuition that the more frequent we take the averaging, the lower volatility is resulted. When $n \to \infty$, $\sigma_G^2\tau$ and $(\mu_G - \frac{\sigma_G^2}{2})\tau$

tend to $\sigma^2[\tau - \frac{2}{3}(T - t_0)]$ and $(r - \frac{\sigma^2}{2})(\tau - \frac{T-t_0}{2})$, respectively. Correspondingly, discrete geometric averaging becomes its continuous analog. In particular, the price of a European fixed strike call with continuous geometric averaging at $t = t_0$ is found to be [see also (4.3.13)]

$$c_G(S_{t_0}, t_0) = S_{t_0} e^{-\frac{1}{2}(r+\frac{\sigma^2}{6})(T-t_0)} N(\widehat{d_1}) - Xe^{-r(T-T_0)} N(\widehat{d_2}), \tag{4.3.48}$$

where

$$\widehat{d_1} = \frac{\ln \frac{S_{t_0}}{X} + \frac{1}{2}\left(r + \frac{\sigma^2}{6}\right)(T - t_0)}{\sigma\sqrt{\frac{T-t_0}{3}}}, \quad \widehat{d_2} = \widehat{d_1} - \sigma\sqrt{\frac{T - t_0}{3}}.$$

Next, we consider the in-progress option where the current time t is within the averaging period, that is, $t \geq t_0$. Here, $t = t_k + \xi \Delta t$ for some integer k, $0 \leq k \leq n-1$ and $0 \leq \xi < 1$. Now, $S_{t_1}, S_{t_2} \cdots S_{t_k}, S_t$ are known quantities while the price ratios $\frac{S_{t_{k+1}}}{S_t}, \frac{S_{t_{k+2}}}{S_{t_{k+1}}}, \cdots, \frac{S_{t_n}}{S_{t_{n-1}}}$ are independent lognormal random variables. We may write

$$\begin{aligned} G_n = &\left[S_{t_1} \cdots S_{t_k}\right]^{1/n} S_t^{(n-k)/n} \\ &\left\{\frac{S_{t_n}}{S_{t_{n-1}}}\left[\frac{S_{t_{n-1}}}{S_{t_{n-2}}}\right]^2 \cdots \left[\frac{S_{t_{k+1}}}{S_t}\right]^{n-k}\right\}^{1/n} \end{aligned}$$

so that

$$\ln \frac{G_n}{\widetilde{S}_t} = \frac{1}{n}[\ln R_n + 2\ln R_{n-1} + \cdots + (n-k-1)\ln R_{k+2} + (n-k)\ln R_t], \tag{4.3.49}$$

where

$$\widetilde{S}_t = [S_{t_1} \cdots S_{t_k}]^{1/n} S_t^{(n-k)/n} = G_k^{k/n} S_t^{(n-k)/n} \text{ and } R_t = S_{t_{k+1}}/S_t.$$

Let the variance and mean of $\ln \frac{G_n}{\widetilde{S}_t}$ be denoted by $\widetilde{\sigma}_G^2\tau$ and $(\widetilde{\mu}_G - \frac{\widetilde{\sigma}_G^2}{2})\tau$, respectively. They are found to be

$$\widetilde{\sigma}_G^2\tau = \sigma^2\Delta t\left[\frac{(n-k)^2}{n^2}(1-\xi) + \frac{(n-k-1)(n-k)(2n-2k-1)}{6n^2}\right],$$

and

$$\left(\widetilde{\mu}_G - \frac{\widetilde{\sigma}_G^2}{2}\right)\tau = \left(r - \frac{\sigma^2}{2}\right)\Delta t\left[\frac{n-k}{n}(1-\xi) + \frac{(n-k-1)(n-k)}{2n}\right].$$

Similar to formula (4.3.47), the price formula of the in-progress European fixed strike call option takes the form

$$c_G(S_t, \tau) = e^{-r\tau}\left[\widetilde{S}_t e^{\widetilde{\mu}_G\tau} N(\widetilde{d_1}) - XN(\widetilde{d_2})\right], \quad t \geq t_0, \tag{4.3.50}$$

where

$$\widetilde{d}_1 = \frac{\ln \frac{\widetilde{S}_t}{X} + (\widetilde{\mu}_G + \frac{\widetilde{\sigma}_G^2}{2})\tau}{\widetilde{\sigma}_G \sqrt{\tau}}, \quad \widetilde{d}_2 = \widetilde{d}_1 - \widetilde{\sigma}_G \sqrt{\tau}.$$

Again, by taking the limit $n \to \infty$, the limiting values of $\tilde{\sigma}_G^2$, $\tilde{\mu}_G - \frac{\tilde{\sigma}_G^2}{2}$ and $\tilde{S}(t)$ become

$$\lim_{n\to\infty} \widetilde{\sigma}_G^2 = \left(\frac{T-t}{T-t_0}\right)^2 \frac{\sigma^2}{3}, \quad \lim_{n\to\infty} \widetilde{\mu}_G - \frac{\widetilde{\sigma}_G^2}{2} = \left(r - \frac{\sigma^2}{2}\right) \frac{T-t}{2(T-t_0)}, \tag{4.3.51a}$$

and

$$\lim_{n\to\infty} \widetilde{S}_t = S_t^{\frac{T-t}{T-t_0}} \widetilde{G}_t \text{ where } \widetilde{G}_t = \exp\left(\frac{1}{T-t_0} \int_{t_0}^t \ln S_u \, du\right). \tag{4.3.51b}$$

The price of the corresponding continuous geometric averaging call option can be obtained by substituting these limiting values into the price formula (4.3.50).

European Fixed Strike Put Option

Using a similar derivation procedure, the price of the corresponding European fixed strike put option with discrete geometric averaging can be found to be

$$p_G(S,\tau) = \begin{cases} e^{-r\tau}\left[XN(-d_2) - Se^{\mu_G \tau} N(-d_1)\right], & t < t_0 \\ e^{-r\tau}\left[XN(-\widetilde{d}_2) - \widetilde{S}e^{\widetilde{\mu}_G \tau} N(-\widetilde{d}_1)\right], & t \geq t_0, \end{cases} \tag{4.3.52}$$

where d_1 and d_2 are given by (4.3.47), and $\tilde{d}_1$ and $\tilde{d}_2$ are given by (4.3.50). The put-call parity relation for the European fixed strike Asian options with discrete geometric averaging can be deduced to be

$$c_G(S,\tau) - p_G(S,\tau) = \begin{cases} e^{-r\tau} Se^{\mu_G \tau} - Xe^{-r\tau}, & t < t_0 \\ e^{-r\tau} \widetilde{S}e^{\widetilde{\mu}_G \tau} - Xe^{-r\tau}, & t \geq t_0. \end{cases} \tag{4.3.53}$$

Additional analytic price formulas of the European Asian options with geometric averaging can be found in Boyle (1993).

4.3.6 Fixed Strike Options with Discrete Arithmetic Averaging

The most common type of Asian options are the fixed strike options whose terminal payoff is determined by discrete arithmetic average of past prices. The valuation of these options is made difficult by the choice of Geometric Brownian process for the underlying asset price since the sum of lognormal components has no closed form representation.

Suppose the average of asset prices is calculated over the time interval $[t_0, t_n]$ and at discrete points $t_i = t_0 + i\Delta t, t = 0, 1, \cdots, n$, $\Delta t = \frac{t_n - t_0}{n}$. The running average $A(t)$ is defined for the current time t, $t_m \leq t < t_{m+1}$, by

$$A(t) = \frac{1}{m+1} \sum_{i=0}^{m} S_{t_i}, \quad 0 \leq m \leq n, \tag{4.3.54}$$

and $A(t) = 0$ for $t < t_0$. Let t_n be the time of expiration so that the payoff function at expiry is given by $\max(A(t_n) - X, 0)$ for the fixed strike call, where X is the strike price. It may be more convenient to consider the terminal payoff in terms of $\widetilde{A}(t_n; t)$ as defined by

$$\widetilde{A}(t_n; t) = A(t_n) - \frac{m+1}{n+1} A(t) = \frac{1}{n+1} \sum_{i=m+1}^{n} S_{t_i}, \tag{4.3.55}$$

which is the average of the unknown stochastic components beyond the current time. For example, the terminal payoff of the fixed strike call can be rewritten as $\max(\widetilde{A}(t_n; t) - X^*, 0)$, where

$$X^* = X - \frac{m+1}{n+1} A(t) \tag{4.3.56}$$

is the effective strike price of the option. It is easily seen that if X^* becomes negative, then the Asian call option is surely exercised at expiration.

The probability distribution of either $A(t_n)$ or $\widetilde{A}(t_n; t)$ has no available explicit representation. The best approach for deriving approximate analytic price formulas is to approximate the distribution of $A(t_n)$ [or $\widetilde{A}(t_n; t)$] by an approximate lognormal distribution through the method of *generalized Edgeworth series expansion*. The Edgeworth series expansion is quite similar to the Taylor series expansion for analytic functions in complex function theory. A brief discussion of the Edgeworth series expansion is presented below (Jarrow and Rudd, 1982).

Edgeworth Series Expansion

Given a probability distribution $F_t(s)$, called the true distribution, we would like to approximate $F_t(s)$ using an approximating distribution $F_a(s)$. The distributions considered are restricted to the class where both $\frac{dF_t(s)}{ds} = f_t(s)$ and $\frac{dF_a(s)}{ds} = f_a(s)$ exist, that is, those distributions which have continuous density functions. First, we define the following quantities:

(i) jth moment of distribution F

$$\alpha_j(F) = \int_{-\infty}^{\infty} s^j f(s)\, ds.$$

(ii) jth central moment of distribution F

$$\mu_j(F) = \int_{-\infty}^{\infty} [s - \alpha_1(F)]^j f(s)\, ds.$$

(iii) characteristic function of F:

$$\phi(F, t) = \int_{-\infty}^{\infty} e^{its} f(s)\, ds, \quad i = \sqrt{-1}.$$

Here, it is assumed that the moments $\alpha_j(F)$ exist for $j \leq n$. Next, the cumulants $k_j(F)$ are defined by

$$\ln \phi(F, t) = \sum_{j=1}^{n-1} k_j(F) \frac{(it)^j}{j!} + o(t^{n-1}). \tag{4.3.57}$$

It can be shown by theoretical analysis that the first $n-1$ cumulants exist, provided that $\alpha_n(F)$ exists. The first four cumulants are found to be

$$k_1(F) = \alpha_1(F), k_2(F) = \mu_2(F), k_3(F) = \mu_3(F), k_4(F) = \mu_4(F) - 3[\mu_2(F)]^2.$$

Also, we assume the existence of the derivatives $\frac{d^j F_a(s)}{ds^j}$, $j \leq m$. Let $N = \min(n, m)$, the difference of $\ln \phi(F_t, t)$ and $\ln \phi(F_a, t)$ can be represented by

$$\ln \phi(F_t, t) = \sum_{j=1}^{N-1} \left[k_j(F_t) - k_j(F_a)\right] \frac{(it)^j}{j!} + \ln \phi(F_a, t) + o(t^{N-1}).$$

Taking the exponential of the above equation [note that $e^{o(t^{N-1})} = 1 + o(t^{N-1})$], we obtain

$$\phi(F_t, t) = \exp\left(\sum_{j=1}^{N-1} \left[k_j(F_t) - k_j(F_a)\right] \frac{(it)^j}{j!}\right) \phi(F_a, t) + o(t^{n-1}). \tag{4.3.58}$$

Suppose the above exponential term is expanded in a power series in it, we have

$$\exp\left(\sum_{j=1}^{N-1} [k_j(F_t) - k_j(F_a)] \frac{(it)^j}{j!}\right) = \sum_{j=0}^{N-1} E_j \frac{(it)^j}{j!} + o(t^{N-1}),$$

where the first few coefficients are given by

$$E_0 = 1,\ E_1 = k_1(F_t) - k_1(F_a),\ E_2 = [k_2(F_t) - k_2(F_a)] + E_1^2,$$
$$E_3 = [k_3(F_t) - k_3(F_a)] + 3E_1[k_2(F_t) - k_2(F_a)] + E_1^3, \text{ etc.}$$

In terms of the cumulants, (4.3.58) can be rewritten as

$$\phi(F_t, t) = \left[\sum_{j=0}^{N-1} E_j \frac{(it)^j}{j!}\right] \phi(F_a, t) + o(t^{N-1}). \tag{4.3.59}$$

Finally, we take the inverse Fourier transform of the above equation. Using the following relations

$$f_t(s) = \frac{1}{2\pi} \int_{-\infty}^{\infty} e^{-its} \phi(F_t, t)\, dt,$$
$$(-1)^j \frac{d^j f_a(s)}{ds^j} = \frac{1}{2\pi} \int_{-\infty}^{\infty} e^{-its} (it)^j \phi(F_a, t)\, dt,$$
$$j = 0, 1, \cdots, N-1,$$

we obtain the following representation of the Edgeworth series expansion

$$f_t(s) = f_a(s) + \sum_{j=1}^{N-1} E_j \frac{(-1)^j}{j!} \frac{d^j f_a(s)}{ds^j} + \epsilon(s, N), \tag{4.3.60}$$

where

$$\epsilon(s, N) = \frac{1}{2\pi} \int_{-\infty}^{\infty} e^{its} o(t^{N-1})\, dt.$$

In order that $\epsilon(s, N)$ exists for all s, it is necessary to observe

$$\lim_{N\to\infty} |\epsilon(s, N)| = 0 \quad \text{for all } s.$$

Suppose all moments of the true and approximating distributions can be calculated, one may claim theoretically that a given distribution can be approximated by another distribution to any desired level of accuracy.

It is most convenient to use the lognormal distribution as the approximating distribution in valuation problems for arithmetic averaging Asian options since the resulting approximate price formula resembles the Black–Scholes type formula. Suppose we choose the parameters of the approximating lognormal distribution such that its first two moments match with the first two moments of the true distribution, that is, $\alpha_1(F_t) = \alpha_1(F_a)$ and $\mu_2(F_t) = \mu_2(F_a)$, then the corresponding two-term Edgeworth series expansion becomes

$$f_t(s) = f_a(s) + \epsilon(s, 3), \tag{4.3.61}$$

since $E_1 = \alpha_1(F_t) - \alpha_1(F_a) = 0$ and $E_2 = \mu_2(F_t) - \mu_2(F_a) + E_1^2 = 0$.

Fixed Strike Call Option

Consider the fixed strike call option with discrete arithmetic averaging, the terminal payoff is defined to be $\max(A(t_n) - X, 0)$. By the risk neutral valuation approach, the price of the fixed strike Asian call is given by

$$\begin{aligned} c(S, A, t) &= e^{-r\tau} E_Q[\max(A(t_n) - X, 0)] \\ &= e^{-r\tau} E_Q[\max(\widetilde{A}(t_n; t) - X^*, 0)], \end{aligned} \tag{4.3.62}$$

where the expectation is taken under the risk neutral measure Q conditional on $S_t = S$ and $A(t) = A$, $\tau = t_n - t$, $\widetilde{A}(t_n; t)$ and X^* are defined by (4.3.55)–(4.3.56).

We would like to approximate the distribution of $\widetilde{A}(t_n; t)$ by a lognormal distribution. More specifically, we approximate the distribution of $\ln \widetilde{A}(t_n; t)$ by a normal distribution whose mean and variance are $\mu(t)$ and $\sigma(t)^2$, respectively. The first two moments of the approximating lognormal distribution are then given by (see Problem 2.28)

$$\begin{aligned} \alpha_1(F_a) &= \mu(t) + \frac{\sigma(t)^2}{2} \\ \alpha_2(F_a) &= 2\mu(t) + 2\sigma(t)^2, \end{aligned}$$

respectively. Suppose we adopt the two-term Edgeworth approximation for $\widetilde{A}(t_n; t)$, which can be achieved by equating the first two moments of the approximating log-normal distribution and the distribution of $\widetilde{A}(t_n; t)$, that is,

$$\mu(t) + \frac{\sigma(t)^2}{2} = \ln E_Q[\widetilde{A}(t_n; t)]$$
$$2\mu(t) + 2\sigma(t)^2 = \ln E_Q[\widetilde{A}(t_n; t)^2].$$

Solving for $\mu(t)$ and $\sigma(t)^2$, we obtain

$$\mu(t) = 2\ln E_Q[\widetilde{A}(t_n; t] - \frac{1}{2}\ln E_Q[\widetilde{A}(t_n; t)^2]$$
$$\sigma(t)^2 = \ln E_Q[\widetilde{A}(t_n; t)^2] - 2\ln E_Q(\widetilde{A}[t_n; t)].$$

Assuming $\ln \widetilde{A}(t_n; t)$ to be normally distributed with mean $\mu(t)$ and variance $\sigma(t)^2$, then the price of the fixed strike Asian call would become

$$c(S, A, t) = e^{-r\tau}\{E_Q[\tilde{A}(t_n; t)]N(d_1) - X^* N(d_2)\}, \tag{4.3.63}$$

where $\tau = t_n - t$ and

$$d_1 = \frac{\mu(t) + \sigma(t)^2 - \ln X^*}{\sigma(t)}, \quad d_2 = d_1 - \sigma(t).$$

The remaining procedure amounts to the determination of $E[\widetilde{A}(t_n; t)]$ and $E[\widetilde{A}(t_n; t)^2]$.

Let S_t denote the time-t asset price, where $t = t_m + \xi\Delta t, 0 \leq \xi < 1$. For $t \geq t_0$, we have

$$E_Q[\tilde{A}(t_n; t)] = \frac{1}{n+1}\sum_{i=m+1}^{n} E_Q[S_{t_i}]. \tag{4.3.64}$$

The asset price dynamics under the risk neutral measure Q is assumed to be

$$\frac{dS_t}{S_t} = r\,dt + \sigma\,dZ_t, \tag{4.3.65}$$

where Z_t is Q-Brownian. We then have

$$E_Q[S_{t_i}] = S_t e^{r(i-m-\xi)\Delta t},$$

so that

$$E_Q[\tilde{A}(t_n; t)] = \frac{S_t}{n+1} e^{r(1-\xi)\Delta t}\left[\frac{1 - e^{r(n-m)\Delta t}}{1 - e^{r\Delta t}}\right], \quad t \geq t_0. \tag{4.3.66}$$

For $t < t_0$, it can be shown similarly that

$$E_Q[\tilde{A}(t_n; t)] = \frac{S_t}{n+1} e^{r(t_0-t)}\left[\frac{1 - e^{r(n+1)\Delta t}}{1 - e^{r\Delta t}}\right], \quad t < t_0. \tag{4.3.67}$$

Next, we consider $E[\widetilde{A}(t_n;t)^2]$. For $t \geq t_0$, we have

$$E_Q[\widetilde{A}(t_n;t)^2] = \frac{1}{(n+1)^2} \sum_{i=m+1}^{n} \sum_{j=m+1}^{n} E[S_{t_i} S_{t_j}]$$
$$= \frac{1}{(n+1)^2} \sum_{i=m+1}^{n} \sum_{j=m+1}^{n} e^{r(i+j-2\xi)\Delta t+\sigma^2(\min(i,j)-\xi)\Delta t} S_t^2.$$

After some tedious manipulation, we obtain (Levy, 1992)

$$E_Q[\widetilde{A}(t_n;t)^2] = \frac{S_t^2}{(n+1)^2} e^{-2\xi\left(r+\frac{\sigma^2}{2}\right)\Delta t}(A_1 - A_2 + A_3 - A_4), \quad t \geq t_0, \quad (4.3.68)$$

where

$$A_1 = \frac{e^{(2r+\sigma^2)\Delta t} - e^{(2r+\sigma^2)(N-m+1)\Delta t}}{(1-e^{r\Delta t})[1-e^{(2r+\sigma^2)\Delta t}]},$$
$$A_2 = \frac{e^{[r(N-m+2)+\sigma^2]\Delta t} - e^{(2r+\sigma^2)(N-m+1)\Delta t}}{(1-e^{r\Delta t})[1-e^{(r+\sigma^2)\Delta t}]},$$
$$A_3 = \frac{e^{(3r+\sigma^2)\Delta t} - e^{[r(N-m+2)+\sigma^2]\Delta t}}{(1-e^{r\Delta t})[1-e^{(r+\sigma^2)\Delta t}]},$$
$$A_4 = \frac{e^{2(2r+\sigma^2)\Delta t} - e^{(2r+\sigma^2)(N-m+1)\Delta t}}{[1-e^{(r+\sigma^2)\Delta t}][1-e^{(2r+\sigma^2)\Delta t}]}.$$

The calculation of $E_Q[\widetilde{A}(t_n;t)^2]$ for $t < t_0$ can be performed similarly [see Problem 4.33].

4.4 Problems

4.1 Consider the function

$$f(S,\tau) = \left(\frac{S}{B}\right)^{\lambda} c_E\left(\frac{B^2}{S}, \tau\right),$$

where $c_E(S,\tau)$ is the price of a vanilla European call option and λ is a constant parameter. Show that $f(S,\tau)$ satisfies the Black–Scholes equation

$$\frac{\partial f}{\partial \tau} = \frac{\sigma^2}{2} S^2 \frac{\partial^2 f}{\partial S^2} + rS\frac{\partial f}{\partial S} - rf$$

when λ is chosen to be $-\dfrac{2r}{\sigma^2} + 1$.

Hint: Substitution of $f(S,\tau)$ into the Black–Scholes equation gives

$$\frac{\partial f}{\partial \tau} - \left[\frac{\sigma^2}{2}S^2\frac{\partial^2 f}{\partial S^2} + rS\frac{\partial f}{\partial S} - rf\right]$$
$$= \left(\frac{S}{B}\right)^{\lambda}\left[\frac{\partial c_E}{\partial \tau} - \frac{\sigma^2}{2}\xi^2\frac{\partial^2 c_E}{\partial \xi^2}\right.$$
$$\left. + (\lambda-1)\sigma^2\xi\frac{\partial c_E}{\partial \xi} - \lambda(\lambda-1)\frac{\sigma^2}{2}c_E - r\lambda c_E + r\xi\frac{\partial c_E}{\partial \xi} + rc_E\right],$$

where $c_E = c_E(\xi,\tau)$, $\xi = \frac{B^2}{S}$.

4.2 Consider the European zero-rebate up-and-out put option with an exponential barrier: $B(\tau) = Be^{-\gamma\tau}$, where $B(\tau) > X$ for all τ. Show that the price of this barrier put option is given by

$$p(S,\tau) = p_E(S,\tau) - \left[\frac{B(\tau)}{S}\right]^{\delta-1} p_E\left(\frac{B(\tau)^2}{S},\tau\right), \quad \delta = \frac{2(r-\gamma)}{\sigma^2},$$

where $p_E(S,\tau)$ is the price of the corresponding European vanilla put option. Deduce the price of the corresponding European up-and-in put option with the same barrier.
Hint: Let $y = \ln\frac{S}{B(\tau)}$, show that $p(y,\tau)$ satisfies

$$\frac{\partial p}{\partial \tau} = \frac{\sigma^2}{2}\frac{\partial^2 p}{\partial y^2} + \left(r - \frac{\sigma^2}{2} - \gamma\right)\frac{\partial p}{\partial y} - rp.$$

4.3 By applying the following transformation on the dependent variable in the Black–Scholes equation

$$c = e^{\alpha y + \beta\tau}w,$$

where $\alpha = \frac{1}{2} - \frac{r}{\sigma^2}$, $\beta = -\frac{\alpha^2\sigma^2}{2} - r$, show that (4.1.3a) is reduced to the prototype diffusion equation

$$\frac{\partial w}{\partial \tau} = \frac{\sigma^2}{2}\frac{\partial^2 w}{\partial y^2},$$

while the auxiliary conditions are transformed to become

$$w(0,\tau) = e^{-\beta\tau}R(\tau) \text{ and } w(y,0) = \max(e^{\alpha y}(e^y - X), 0).$$

Consider the following diffusion equation defined in a semi-infinite domain

$$\frac{\partial v}{\partial t} = a^2\frac{\partial^2 v}{\partial x^2}, \quad x > 0 \text{ and } t > 0, \quad a \text{ is a positive constant},$$

with initial condition: $v(x,0) = f(x)$ and boundary condition: $v(0,t) = g(t)$, the solution to the diffusion equation is given by (Kevorkian, 1990)

$$v(x,t) = \frac{1}{2a\sqrt{\pi t}} \int_0^\infty f(\xi)[e^{-x-\xi)^2/4a^2t} - e^{-(x+\xi)^2/4a^2t}]\,d\xi$$
$$+ \frac{x}{2a\sqrt{\pi}} \int_0^t \frac{e^{-x^2/4a^2\omega}}{\omega^{3/2}} g(t-\omega)\,d\omega.$$

Using the above form of solution, show that the price of the European down-and-out call option is given by

$$c(y,\tau) = e^{\alpha y+\beta\tau}\left\{\frac{1}{\sqrt{2\pi\tau}\sigma} \int_0^\infty \max(e^{-\alpha\xi}(e^\xi - X), 0)\right.$$
$$\left[e^{-(y-\xi)^2/2\sigma^2\tau} - e^{-(y+\xi)^2/2\sigma^2\tau}\right]d\xi$$
$$\left. + \frac{y}{\sqrt{2\pi}\sigma} \int_0^\tau \frac{e^{-\beta(\tau-\omega)}e^{-y^2/2\sigma^2\omega}}{\omega^{3/2}} R(\tau-\omega)\,d\omega\right\}.$$

Assuming $B < X$, show that the price of the European down-and-out call option is given by [see (4.1.11)]

$$c(S,\tau) = c_E(S,\tau) - \left(\frac{B}{S}\right)^{\delta-1} c_E\left(\frac{B^2}{S},\tau\right)$$
$$+ \int_0^\tau e^{-r\omega} \frac{\ln\frac{S}{B}}{\sqrt{2\pi}\sigma} \frac{\exp\left(\frac{-[\ln\frac{S}{B}+(r-\frac{\sigma^2}{2})\omega]^2}{2\sigma^2 w}\right)}{\omega^{3/2}} R(\tau-\omega)\,d\omega.$$

The last term represents the additional option premium due to the rebate payment.

4.4 Suppose the asset price follows the Geometric Brownian process with drift rate r and volatility σ under the risk neutral measure Q. Find the density function of the asset price S_T at expiration time T, with time-0 asset price S starting below the barrier B then breaching the barrier but ending below the barrier at expiration.

4.5 We define

$$d_1 = \frac{\ln\frac{S}{K} + \left(r + \frac{\sigma^2}{2}\right)\tau}{\sigma\sqrt{\tau}}, \quad d_2 = d_1 - \sigma\sqrt{\tau},$$
$$d_3 = \frac{2\ln\frac{B}{S}}{\sigma\sqrt{\tau}} + d_1, \quad d_4 = d_3 - \sigma\sqrt{\tau}, \quad \delta = \frac{2r}{\sigma^2},$$

and

$$V_b(S,\tau;K) = [Xe^{-r\tau}N(-d_2) - SN(-d_1)]$$
$$- \left(\frac{B}{S}\right)^{\delta-1}\left[Xe^{-r\tau}N(-d_4) - \frac{B^2}{S}N(-d_3)\right].$$

Show that the price functions of various European barrier put options are given by

$$
\begin{aligned}
p_{uo}(S,\tau;X,B) &= V_b(S,\tau;\min(X,B))\\
p_{do}(S,\tau;X,B) &= \max(V_b(S,\tau;X) - V_b(S,\tau;B), 0)\\
p_{ui}(S,\tau;X,B) &= p_E(S,\tau;X) - p_{uo}(S,\tau;X,B)\\
p_{di}(S,\tau;X,B) &= p_E(S,\tau;X) - p_{do}(S,\tau;X,B),
\end{aligned}
$$

where $p_E(S,\tau;X)$ is the price function of the European put option, B is the barrier and X is the strike price.

4.6 Consider a European down-and-out call option where the terminal payoff depends on the payoff state variable S_1 and knock-out occurs when the barrier state variable S_2 breaches the downstream barrier B_2. Assume that under the risk neutral measure Q, the dynamics of $S_{1,t}$ and $S_{2,t}$ are given by

$$\frac{dS_{i,t}}{S_{i,t}} = r\,dt + \sigma_i\,dZ_{i,t}, \quad i = 1, 2 \quad \text{and} \quad dZ_1\,dZ_2 = \rho\,dt.$$

Let X_1 denote the option's strike price. Show that the price of this down-and-out call with an *external barrier* is given by (Kwok, Wu and Yu, 1998)

$$
\begin{aligned}
&\text{call price}\\
&= e^{-rT} E_Q[(S_{1,T} - X_1)\mathbf{1}_{\{S_{1,T}>X_1\}}\mathbf{1}_{\{m_{2,0}^T > B_2\}}]\\
&= S_{1,0}\left[N_2(d_1, e_1;\rho) - \left(\frac{B_2}{S_{2,0}}\right)^{\delta_2 - 1 + 2\gamma_{12}} N(d_1', e_1';\rho)\right]\\
&\quad - e^{-rT} X_1\left[N(d_2, e_2;\rho) - \left(\frac{B_2}{S_{2,0}}\right)^{\delta_2 - 1} N(d_2', e_2';\rho)\right],
\end{aligned}
$$

where

$$
\begin{aligned}
d_1 &= \frac{\ln\frac{S_{1,0}}{X_1} + \left(r + \frac{\sigma_1^2}{2}\right)T}{\sigma_1\sqrt{T}}, & d_2 &= d_1 - \sigma_1\sqrt{T},\\
d_1' &= d_1 + \frac{2\gamma_{12}\ln\frac{B_2}{S_{2,0}}}{\sigma_1\sqrt{T}}, & d_2' &= d_1' - \sigma_1\sqrt{T},\\
e_1 &= \frac{\ln\frac{S_{2,0}}{B_2} + \left(r - \frac{\sigma_1^2}{2} + \rho\sigma_1\sigma_2\right)T}{\sigma_2\sqrt{T}}, & e_2 &= e_1 - \rho\sigma_1\sqrt{T},\\
e_1' &= e_1 + \frac{2\ln\frac{B_2}{S_{2,0}}}{\sigma_2\sqrt{T}}, & e_2' &= e_1' - \rho\sigma_1\sqrt{T},\\
\delta_2 &= \frac{2r}{\sigma_2^2}, \quad \gamma_{12} = \rho\frac{\sigma_1}{\sigma_2}.
\end{aligned}
$$

4.7 Consider a European down-and-out *partial barrier* call option where the barrier provision is activated only between the option's starting date (time 0) and t_1. Here, t_1 is some time earlier than the expiration date T, where $0 < t_1 < T$. Let B and X denote the down-barrier and strike, respectively, where $B < X$. Let the dynamics of S_t be governed by

$$\frac{dS_t}{S_t} = r\,dt + \sigma Z_t$$

under the risk neutral measure Q. Assuming $S_0 > B$, show that the down-and-out call price is given by (Heynen and Kat, 1994a)

$$\begin{aligned}
&\text{call price} \\
&= e^{-rT} E_Q\big[(S_T - X)\mathbf{1}_{\{S_T > X\}}\mathbf{1}_{\{m_0^{t_1} > B\}}\big] \\
&= S_0\left[N\left(d_1, e_1; \sqrt{\frac{t_1}{T}}\right) - \left(\frac{B}{S}\right)^{\delta+1} N\left(d_1', e_1'; \sqrt{\frac{t_1}{T}}\right)\right] \\
&\quad - e^{-rT} X\left[N\left(d_2, e_2; \sqrt{\frac{t_1}{T}}\right) - \left(\frac{B}{S}\right)^{\delta-1} N\left(d_2', e_2'; \sqrt{\frac{t_1}{T}}\right)\right],
\end{aligned}$$

where

$$\begin{aligned}
d_1 &= \frac{\ln\frac{S_0}{X} + \left(r + \frac{\sigma^2}{2}\right)T}{\sigma\sqrt{T}}, & d_2 &= d_1 - \sigma\sqrt{T}, \\
d_1' &= d_1 + \frac{2\ln\frac{B}{S_0}}{\sigma\sqrt{T}}, & d_2' &= d_1' - \sigma\sqrt{T}, \\
e_1 &= \frac{\ln\frac{S_0}{B} + \left(r + \frac{\sigma^2}{2}\right)t_1}{\sigma\sqrt{t_1}}, & e_2 &= e_1 - \sigma\sqrt{t_1}, \\
e_1' &= e_1 + \frac{2\ln\frac{B}{S_0}}{\sigma\sqrt{t_1}}, & e_2' &= e_1' - \sigma\sqrt{t_1}, \\
\delta &= \frac{2r}{\sigma^2}.
\end{aligned}$$

Show that the above price formula reduces to the price function defined in (4.1.12a,b) when t_1 is set equal to T.

Hint: Modify the price formula in Problem 4.6 by setting $\rho = \sqrt{\frac{t_1}{T}}$, $\sigma_1 = \sigma$ and $\sigma_2 = \sqrt{\frac{t_1}{T}}\sigma$ so that $\gamma_{12} = 1$.

4.8 Consider a typical term in $g(x,t)$

$$\frac{\exp\left(\frac{\mu x}{\sigma^2} - \frac{\mu^2 t}{2\sigma^2}\right)}{\sqrt{2\pi\sigma^2 t}} \exp\left(-\frac{(x-\xi)^2}{2\sigma^2 t}\right),$$

where ξ can be either $2n(u-\ell)$ or $2\ell + 2n(u-\ell)$. Show that the above term can be rewritten as

$$\frac{e^{\frac{\mu\xi}{\sigma^2}}}{\sqrt{2\pi\sigma^2 t}} \exp\left(\frac{-(x-\xi-\mu t)^2}{2\sigma^2 t}\right).$$

Hence, show that

$$\begin{aligned}
&\int_\ell^u g(x,t)\,dx \\
&= \sum_{n=-\infty}^{\infty} \exp\left(\frac{2\mu n(u-\ell)}{\sigma^2}\right) \\
&\quad \left[N\left(\frac{u-\mu t-2u(u-\ell)}{\sigma\sqrt{t}}\right) - N\left(\frac{\ell-\mu t-2n(u-\ell)}{\sigma\sqrt{t}}\right)\right] \\
&\quad - \exp\left(\frac{2\mu[\ell+n(u-\ell)]}{\sigma^2}\right) \\
&\quad \left[N\left(\frac{u-\mu t-2\ell-2n(u-\ell)}{\sigma\sqrt{t}}\right) - N\left(\frac{\ell-\mu t-2\ell-2n(u-\ell)}{\sigma\sqrt{t}}\right)\right].
\end{aligned}$$

Use the above result to derive the price formula of the European double knock-out call option c_{LU}^o [see Sect. 4.1.3].

We generalize the barriers to become exponential functions in time. Suppose the upper and lower barriers are set to be $Ue^{\delta_1 t}$ and $Le^{\delta_2 t}$, $t \in [0,T]$. Here, δ_1 and δ_2 are constant parameters and the barriers do not intersect over $[0,T]$. Show that the price formula of the European double knock-out call option can be expressed as (Kunitomo and Ikeda, 1992)

$$\begin{aligned}
c_{LU}^o = S &\sum_{n=-\infty}^{\infty} \left\{ \left(\frac{U^n}{L^n}\right)^{\mu_1} \left(\frac{L}{S}\right)^{\mu_2} [N(d_1) - N(d_2)] \right. \\
&\left. - \left(\frac{L^{n+1}}{U^n S}\right)^{\mu_3} [N(d_3) - N(d_4)] \right. \\
&- Xe^{-rT} \sum_{n=-\infty}^{\infty} \left\{ \left(\frac{U^n}{L^n}\right)^{\mu_1 - 2} \left(\frac{L}{S}\right)^{\mu_2} N(d_1 - \sigma\sqrt{T}) - N(d_2 - \sigma\sqrt{T}) \right\} \\
&\left. - \left(\frac{L^{n+1}}{U^n S}\right)^{\mu_3 - 2} [N(d_3 - \sigma\sqrt{T}) - N(d_4 - \sigma\sqrt{T})] \right\},
\end{aligned}$$

where

$$d_1 = \frac{\ln \frac{SU^{2n}}{XL^{2n}} + \left(r + \frac{\sigma^2}{2}\right)T}{\sigma\sqrt{T}}, \qquad d_2 = \frac{\ln \frac{SU^{2n}}{FL^{2n}} + \left(r + \frac{\sigma^2}{2}\right)T}{\sigma\sqrt{T}},$$

$$d_3 = \frac{\ln \frac{L^{2n+2}}{XSU^{2n}} + \left(r + \frac{\sigma^2}{2}\right)T}{\sigma\sqrt{T}}, \qquad d_4 = \frac{\ln \frac{L^{2n+2}}{FSU^{2n}} + \left(r + \frac{\sigma^2}{2}\right)T}{\sigma\sqrt{T}},$$

$$\mu_1 = \frac{2[r - \delta_2 - n(\delta_1 - \delta_2)]}{\sigma^2} + 1, \quad \mu_2 = 2n\frac{\delta_1 - \delta_2}{\gamma^2},$$

$$\mu_3 = \frac{2[r - \delta_2 + n(\delta_1 - \delta_2)]}{\sigma^2} + 1, \quad F = Ue^{\delta_1 T}.$$

4.9 Let $P(x, t; x_0, t_0)$ denote the transition density function of the restricted Brownian process $W_t^\mu = \mu t + \sigma Z_t$ with two absorbing barriers at $x = 0$ and $x = \ell$. Using the method of separation of variables (Kevorkian, 1990), show that the solution to $P(x, t; x_0, t_0)$ admits the following eigen-function expansion [which differs drastically in analytic form from that in (4.1.48)]

$$P(x, t; x_0, t_0) = e^{\frac{\mu}{\sigma^2}(x - x_0)} \frac{2}{\ell} \sum_{k=1}^{\infty} e^{-\lambda_k (t - t_0)} \sin \frac{k\pi x}{\ell} \sin \frac{k\pi x_0}{\ell},$$

where the eigenvalues are given by

$$\lambda_k = \frac{1}{2}\left(\frac{\mu^2}{\sigma^2} + \frac{k^2\pi^2\sigma^2}{\ell^2}\right).$$

Hint: $P(x, t; x_0, t_0)$ satisfies the forward Fokker–Planck equation with auxiliary conditions: $P(0, t) = P(\ell, t) = 0$ and $P(x, t_0^+; x_0, t_0) = \delta(x - x_0)$. Pelsser (2000) derived the above solution by performing the Laplace inversion using Bromwich contour integration.

4.10 Let the exit time density $q^+(t; x_0, t_0)$ have dependence on the initial state $X(t_0) = x_0$. We write $\tau = t - t_0$ so that $q^+(t; x_0, t_0) = q^+(x_0, \tau)$. Show that the partial differential equation formulation is given by

$$\frac{\partial q^+}{\partial \tau} = \mu \frac{\partial q^+}{\partial x_0} + \frac{\sigma^2}{2} \frac{\partial^2 q^+}{\partial x_0^2}, \quad \ell < x_0 < u, \quad \tau > 0,$$

with auxiliary conditions:

$$q^+(u, \tau) = 0, \quad q^+(\ell, \tau) = 0 \quad \text{and} \quad q^+(x_0, 0) = \delta(x_0).$$

Solve for $q^+(x_0, \tau)$ using the partial differential equation approach and compare the solution with that given in (4.1.53b). Also, show that

$$\int_{t_0}^{t} q^+(s; x_0, t_0)\, ds + \int_{t_0}^{t} q^-(s; x_0, t_0)\, ds + \int_{\ell}^{u} P(x, t; x_0, t_0)\, dx = 1,$$

where $P(x, t; x_0, t_0)$ is the transition density function defined in Problem 4.9.

4.11 Using the method of path counting, Sidenius (1998) showed that $g^+(x, T)$ defined in (4.1.44b) has the following analytic solution

$$g^+(x,T) = \exp\left(\frac{\mu x}{\sigma^2} - \frac{\mu^2}{\sigma^2}T\right)$$
$$\left[\sum_{n=1}^{\infty} \phi(x;\alpha_n^+,\sigma\sqrt{T}) - \phi(x;\beta_n^+,\sigma\sqrt{T})\right],$$

where

$$\phi(x;\lambda,\nu) = \frac{1}{\sqrt{2\pi\nu^2}}\exp\left(-\frac{(x-\lambda)^2}{2\nu^2}\right)$$
$$\alpha_n^+ = 2n(u-\ell) + 2\ell \quad \text{and} \quad \beta_n^+ = 2n(u-\ell).$$

Find the closed form price formula of the European upper-barrier knock-in call option c_U^i [see Sect. 4.1.3 and Luo (2001)].

4.12 A sequential barrier option is a barrier option with two-sided barriers where nullification occurs only when the barriers are breached at a prespecified order (say, up then down). Show that the price formula of the sequential up-then-down out call option can be inferred from that of the double knock-out call option (see Problem 4.8) except that the infinite summation over n is replaced by summation over two terms only (Li, 1999).

Hint: The sequential up-then-down out call option becomes the corresponding down-and-out call when the asset price hits the upper barrier.

4.13 Consider a discretely monitored down-and-out call option with strike price X and barrier level B_i at discrete time t_i, $i = 1, 2, \cdots, n$. Show that the price of this European barrier call option is given by (Heynen and Kat, 1996)

$$c_{do}(S_0, T; X, B_1, B_2, \cdots, B_n)$$
$$= S_0 N_{n+1}(d_1^1, d_1^2, \cdots, d_1^{n+1}; \Gamma) - e^{-rT} N_{n+1}(d_2^1, d_2^2, \cdots, d_2^{n+1}; \Gamma),$$

where

$$d_1^i = \frac{\ln\frac{S_0}{B_{t_i}} + \left(r + \frac{\sigma^2}{2}\right)t_i}{\sigma\sqrt{t_i}}, \quad d_2^i = d_1^i - \sigma\sqrt{t_i}, \quad i = 1, 2, \cdots, n,$$
$$d_1^{n+1} = \frac{\ln\frac{S_0}{X} + \left(r + \frac{\sigma^2}{2}\right)T}{\sigma\sqrt{T}}, \quad d_2^{n+1} = d_1^{n+1} - \sigma\sqrt{T}.$$

Also, Γ is the $(n+1)\times(n+1)$ correlation matrix whose entries are given by

$$\rho_{jk} = \frac{\min(t_j, t_k)}{\sqrt{t_j}\sqrt{t_k}}, \quad 1 \le j, k \le n; \quad \rho_{j,n+1} = \sqrt{\frac{t_j}{T}}, j = 1, 2, \cdots, n.$$

4.14 Let $f_{min}(y)$ and $f_{max}(y)$ denote the density function of y_T and Y_T, respectively. Show that

$$f_{min}(y) = \frac{1}{\sigma\sqrt{\tau}} n\left(\frac{-y+\mu\tau}{\sigma\sqrt{\tau}}\right) + \frac{2\mu}{\sigma^2} e^{\frac{2\mu y}{\sigma^2}} N\left(\frac{y+\mu\tau}{\sigma\sqrt{\tau}}\right)$$
$$+ e^{\frac{2\mu y}{\sigma^2}} \frac{1}{\sigma\sqrt{\tau}} n\left(\frac{y+\mu\tau}{\sigma\sqrt{\tau}}\right),$$
$$f_{max}(y) = \frac{1}{\sigma\sqrt{\tau}} n\left(\frac{y-\mu\tau}{\sigma\sqrt{\tau}}\right) - \frac{2\mu}{\sigma^2} e^{\frac{2\mu y}{\sigma^2}} N\left(\frac{-y-\mu\tau}{\sigma\sqrt{\tau}}\right)$$
$$+ e^{\frac{2\mu y}{\sigma^2}} \frac{1}{\sigma\sqrt{\tau}} n\left(\frac{-y-\mu\tau}{\sigma\sqrt{\tau}}\right).$$

Compare the results with $f_{down}(x, m, T)$ and $f_{up}(y, M, T)$ in (4.1.26)–(4.1.27).

4.15 As an alternative approach to derive the value of a European floating strike lookback call, we consider

$$c_{f\ell}(S, m, \tau) = e^{-r\tau} E_Q[S_T - \min(m, m_t^T)]$$
$$= S - e^{-r\tau} E_Q[\min(m, m_t^T)],$$

where $S_t = S, m_{T_0}^t = m$ and $\tau = T - t$. We may decompose the above expectation calculation into two terms:

$$E_Q[\min(m, m_t^T)] = mP(m \le m_t^T) + E_Q[m_t^T \mathbf{1}_{\{m > m_t^T\}}].$$

Show that the first term is given by

$$mP\left(\ln \frac{m_t^T}{S} \ge \ln \frac{m}{S}\right)$$
$$= m\left[N\left(\frac{-\ln \frac{m}{S} + \mu\tau}{\sigma\sqrt{\tau}}\right) - \left(\frac{S}{m}\right)^{1-\frac{2r}{\sigma^2}} N\left(\frac{\ln \frac{m}{S} + \mu\tau}{\sigma\sqrt{\tau}}\right)\right].$$

Now, the second term can be expressed as

$$E_Q[m_t^T \mathbf{1}_{\{m_t^T > m\}}] = \int_{-\infty}^{\ln \frac{m}{S}} Se^y f_{min}(y)\, dy.$$

By performing the tedious integration procedure, show that the same price function for $c_{f\ell}(S, m, \tau)$ [see (4.2.9)] is obtained.

4.16 Using the following form of the distribution function of m_t^T [see (4.2.4a)]

$$P(m \le m_t^T) = N\left(\frac{-\ln \frac{m}{S} + \mu\tau}{\sigma\sqrt{\tau}}\right) - \left(\frac{S}{m}\right)^{1-\frac{2r}{\sigma^2}} N\left(\frac{\ln \frac{m}{S} + \mu\tau}{\sigma\sqrt{\tau}}\right),$$

show that $P(m \le m_t^T)$ becomes zero when $S = m$.

4.17 Suppose we use a straddle (combination of a call and a put with the same strike m) in the rollover strategy for hedging the floating strike lookback call and write

$$c_{f\ell}(S, m, \tau) = c_E(S, \tau; m) + p_E(S, \tau; m) + \text{ strike bonus premium.}$$

Find an integral representation of the strike bonus premium in terms of the distribution functions of S_T and m_t^T. How would you compare the strike bonus premium given in (4.2.14) when the European call option is used in the rollover strategy?

4.18 Prove the following put-call parity relation between the prices of the fixed strike lookback call and floating strike lookback put:

$$c_{fix}(S, M, \tau; X) = p_{f\ell}(S, \max(M, X), \tau) + S - Xe^{-r\tau}.$$

Deduce that

$$\frac{\partial c_{fix}}{\partial M} = 0 \quad \text{for} \quad M < X.$$

Give a financial interpretation why c_{fix} is insensitive to M when $M < X$ [see also (4.2.6)].

4.19 Suppose the terminal payoff function of the partial lookback call and put options are $\max(S_T - \lambda m_{T_0}^T, 0)$, $\lambda > 1$ and $\max\left(\lambda M_{T_0}^T - S_T, 0\right)$, $0 < \lambda < 1$, respectively. Show that the price formulas of these lookback options are, respectively (Conze and Viswanathan, 1991),

$$\begin{aligned} c(S, m, \tau) = {} & SN\left(d_m - \frac{\ln \lambda}{\sigma\sqrt{\tau}}\right) - \lambda m e^{-r\tau} N\left(d_m - \frac{\ln \lambda}{\sigma\sqrt{\tau}} - \sigma\sqrt{\tau}\right) \\ & + e^{-r\tau}\frac{\sigma^2}{2r}\lambda S\left[\left(\frac{S}{m}\right)^{-2r/\sigma^2} N\left(-d_m + \frac{2r}{\sigma}\sigma - \frac{\ln \lambda}{\sigma\sqrt{\tau}}\right)\right. \\ & \left. - e^{r\tau}\lambda^{2r/\sigma^2} N\left(-d_m - \frac{\ln \lambda}{\sigma\sqrt{\tau}}\right)\right] \end{aligned}$$

$$\begin{aligned} p(S, M, \tau) = {} & -SN\left(-d_M + \frac{\ln \lambda}{\sigma\sqrt{\tau}}\right) \\ & + \lambda M e^{-r\tau} N\left(-d_M + \frac{\ln \lambda}{\sigma\sqrt{\tau}} + \sigma\sqrt{\tau}\right) \\ & - e^{-r\tau}\frac{\sigma^2}{2r}\lambda S\left[\left(\frac{S}{M}\right)^{-2r/\sigma^2} N\left(d_M - \frac{2r}{\sigma}\sqrt{\tau} + \frac{\ln \lambda}{\sigma\sqrt{\tau}}\right)\right. \\ & \left. - e^{r\tau}\lambda^{2r/\sigma^2} N\left(d_M + \frac{\ln \lambda}{\sigma\sqrt{\tau}}\right)\right], \end{aligned}$$

where d_m and d_M are given by (4.2.9) and (4.2.10), respectively.

4.20 The terminal payoff of the lookback spread option is given by

$$c_{sp}(S, m, M, 0) = \max(M_{T_0}^T - m_{T_0}^T - X, 0).$$

Show that the price of the European lookback spread option can be expressed as (Wong and Kwok, 2003)

(i) currently at- or in-the-money, that is, $M - m - X \geq 0$

$$c_{sp}(S, m, M, \tau) = c_{f\ell}(S, m, \tau) + p_{f\ell}(S, M, \tau) - Xe^{-r\tau};$$

(ii) currently out-of-the-money, that is, $M - m - X < 0$

$$c_{sp}(S, m, M, \tau) = c_{f\ell}(S, m, \tau) + p_{f\ell}(S, M, \tau) - Xe^{-r\tau} + e^{-r\tau} \int_M^{m+X} P(M_t^T < \xi \leq m_t^T + X)\, d\xi.$$

4.21 The holder of a European in-the-money call option may suffer loss in profits if the asset price drops substantially just before expiration. The "limited period" fixed strike lookback feature may help remedy that holder's market exit problem. To achieve an optimal timing for market exit, it is sensible to choose the lookback period starting some time after the option's starting date and ending at expiration. Let $[T_1, T]$ denote the lookback period and consider the pricing of this fixed strike lookback call at time t before the lookback period, with terminal payoff: $\max(M_{T_1}^T - X, 0)$. Note that when $S_{T_1} > X$, the call is guaranteed to be in-the-money at expiration since $M_{T_1}^T \geq S_{T_1}$. Show that the price of the European "limited-period" lookback call option is given by

$$\begin{aligned} c(S, \tau) &= e^{-r\tau} E_Q\big[\max\big(M_{T_1}^T - X, 0\big)\big] \\ &= e^{-r\tau} E_Q\left[\left\{S_{T_1}\left(\frac{M_{T_1}^T}{S_{T_1}}\right) - X\right\}\mathbf{1}_{\{m_{T_1}^T > X, S_{T_1} < X\}}\right] \\ &\quad + e^{-r\tau} E_Q\left[\left\{S_{T_1}\left(\frac{M_{T_1}^T}{S_{T_1}}\right) - X\right\}\mathbf{1}_{\{S_{T_1} > X\}}\right], \quad t < T_1, \end{aligned}$$

where expectation is taken under the risk neutral measure Q and X is the strike price. Assuming the usual Geometric Brownian process for the asset price under Q, show that

$$\begin{aligned} c(S, \tau) &= SN(d_1) - e^{-r\tau} XN(d_2) - SN_2\left(-e_1, d_1; -\sqrt{\frac{T - T_1}{T - t}}\right) \\ &\quad - e^{-r\tau} XN_2\left(f_2, -d_2; -\sqrt{\frac{T_1 - t}{T - t}}\right) + e^{-r\tau}\frac{\sigma^2}{2r} S\left[-\left(\frac{S}{X}\right)^{-\frac{2r}{\sigma^2}}\right. \\ &\quad N_2\left(d_1 - \frac{2r}{\sigma}\sqrt{\tau}, -f_1 + \frac{2r}{\sigma}\sqrt{T_1 - t}; -\sqrt{\frac{T_1 - t}{T - t}}\right) \end{aligned}$$

$$+ e^{r\tau} N_2\left(e_1, d_1; \sqrt{\frac{T - T_1}{T - t}}\right)\Bigg]$$

$$+ e^{-r(T-T_1)}\left(1 - \frac{\sigma^2}{2r}\right)SN(f_1)N(-e_2), \quad t < T_1,$$

where $d_1, e_1, f_1, \cdots$, etc. are the same as those defined in (4.2.16) except that m is replaced by X accordingly (Heynen and Kat, 1994b). Deduce the price formula of the corresponding "limited period" fixed strike lookback put option whose terminal payoff function is $\max(X - m_{T_1}^{T}, 0)$.

4.22 Use (4.2.21) to derive the following partial differential equation for the floating strike lookback put option

$$\frac{\partial V}{\partial \tau} = \frac{\sigma^2}{2}\frac{\partial^2 V}{\partial \xi^2} - \left(r + \frac{\sigma^2}{2}\right)\frac{\partial V}{\partial \xi}, \quad 0 < \xi < \infty, \tau > 0,$$

where $V(\xi, \tau) = p_{f\ell}(S, M, t)/S$ and $\tau = T - t$, $\xi = \ln \frac{M}{S}$. The auxiliary conditions are

$$V(\xi, 0) = e^{\xi} - 1 \quad \text{and} \quad \frac{\partial V}{\partial \xi}(0, \tau) = 0.$$

Solve the above Neumann boundary value problem and check the result with the put price formula given in (4.2.10).
Hint: Define $W = \frac{\partial V}{\partial \xi}$ so that W satisfies the same governing differential equation but the boundary condition becomes $W(0, \tau) = 0$. Solve for $W(\xi, \tau)$, then integrate W with respect to ξ to obtain V. Be aware that an arbitrary function $\phi(t)$ is generated upon integration with respect to ξ. Obtain an ordinary differential equation for $\phi(t)$ by substituting the solution for V into the original differential equation.

4.23 Let $p(S, t; \delta t)$ denote the value of a floating strike lookback put option with discrete monitoring of the realized maximum value of the asset price, where δt is the regular interval between monitoring instants. Suppose we assume the following two-term Taylor expansion of $p(S, t; \delta t)$ in powers of $\sqrt{\delta t}$ (Levy and Mantion, 1997)

$$p(S, t; \delta t) \approx p(S, t; 0) + \alpha\sqrt{\delta t} + \beta\delta t.$$

With $\delta t = 0$, $p(S, t; 0)$ represents the floating strike lookback put value corresponding to continuous monitoring. Let τ denote the time to expiry. By setting $\delta t = \tau$ and $\delta t = \tau/2$, we deduce the following pair of linear equations for α and β:

$$\alpha\sqrt{\frac{\tau}{2}} + \frac{\beta\tau}{2} = p\left(S, t; \frac{\tau}{2}\right) - p(S, t; 0)$$

$$\alpha\sqrt{\tau} + \beta\tau = p(S, t; \tau) - p(S, t; 0).$$

Hence, $p(S, t; \tau)$ is simply the vanilla put value with strike price equal to the current realized maximum asset price M. With only one monitoring instant at the midpoint of the remaining option's life, show that $p(S, t; \frac{\tau}{2})$ is given by

$$\begin{aligned} p\left(S, t; \frac{\tau}{2}\right) &= -e^{-q\tau} S + e^{-r\tau} \left[M N_2\left(-d_M\left(\frac{\tau}{2}\right) + \sigma\sqrt{\frac{\tau}{2}}, -d_M(\tau) + \sigma\sqrt{\tau}; \frac{1}{\sqrt{2}} \right) \right. \\ &\quad + e^{(r-q)\tau} S N_2\left(d_M(\tau), d\left(\frac{\tau}{2}\right); \frac{1}{\sqrt{2}} \right) \\ &\quad \left. + e^{(r-q)\frac{\tau}{2}} S N_2\left(d_M\left(\frac{\tau}{2}\right), -d\left(\frac{\tau}{2}\right) + \sigma\sqrt{\frac{\tau}{2}}; 0 \right) \right], \end{aligned}$$

where

$$d_M(\tau) = \frac{\ln \frac{S}{M} + \left(r - q + \frac{\sigma^2}{2}\right)\tau}{\sigma\sqrt{\tau}} \text{ and } d(\tau) = \frac{\left(r - q + \frac{\sigma^2}{2}\right)\sqrt{\tau}}{\sigma}.$$

Once α and β are determined, we then obtain an approximate price formula of the discretely monitored floating strike lookback put.

4.24 The dynamic fund protection feature in an equity-linked fund product guarantees a predetermined protection level K to an investor who owns the underlying fund. Let S_t denote the value of the underlying fund. The dynamic protection replaces the original value of the underlying fund by an upgraded value F_t so that F_t is guaranteed not to fall below K. That is, whenever F_t drops to K, just enough capital will be added by the sponsor so that the upgraded fund value does not fall below K.

(a) Show that the value of the upgraded fund at maturity time T is given by

$$F_T = S_T \max\left(1, \max_{0 \le u \le T} \frac{K}{S_u}\right).$$

(b) Let X_T denote the terminal value of the derivative that provides the dynamic fund protection. Define the lookback state variable

$$M_t = \max\left(1, \frac{K}{\min\limits_{0 \le u \le t} S_u}\right), \quad 0 \le t \le T,$$

which is known at the current time t. Show that

$$\begin{aligned} X_T &= F_T - S_T \\ &= S_T(M_t - 1) + S_T \max\left(\frac{K}{\min\limits_{t \le u \le T} S_u} - M_t, 0\right). \end{aligned}$$

(c) Under the risk neutral measure Q, let the dynamics of S_t be governed by

$$\frac{dS_t}{S_t} = r\,dt + \sigma\,dZ_t.$$

Show that the fair value of the dynamic fund protection is given by (Imai and Boyle, 2001)

$$\begin{aligned} V(S, M, t) &= E_Q[X_T] \\ &= S[MN(d_1) - 1] + \frac{K}{\alpha}\left(\frac{\widehat{K}}{S}\right)^{\alpha} N(d_2) \\ &\quad + \left(1 - \frac{1}{\alpha}\right) K e^{-r\tau} N(d_3), \quad \tau = T - t, \end{aligned}$$

where E_Q denotes the expectation under Q conditional on $S_t = S$ and $M_t = M$. The other parameter values are defined by

$$\alpha = 2r/\sigma^2, \quad \widehat{K} = K/M, \quad k = \ln\frac{S}{\widehat{K}},$$

$$d_1 = \frac{k + \left(r + \frac{\sigma^2}{2}\right)\tau}{\sigma\sqrt{\tau}}, \quad d_2 = \frac{-k + \left(r + \frac{\sigma^2}{2}\right)\tau}{\sigma\sqrt{\tau}},$$

$$d_3 = \frac{-k - \left(r - \frac{\sigma^2}{2}\right)\tau}{\sigma\sqrt{\tau}}.$$

4.25 Explain why

$$\text{call}_{\text{European}} \geq \text{Asian call}_{\text{arithmetic}} \geq \text{Asian call}_{\text{geometric}}.$$

Hint: An average price is less volatile than the series of prices from which it is computed.

4.26 Apply the exchange option price formula (see Problem 3.34) to price the floating strike Asian call option based on the knowledge of the price formula of the fixed strike Asian call option.
Hint: The covariance between S_T and G_T is equal to $\frac{\sigma^2(T-t)^2}{2}$.

4.27 We define the geometric average of the price path of asset price S_i, $i = 1, 2$, during the time interval $[t, t+T]$ by

$$G_i(t+T) = \exp\left(\frac{1}{T}\int_0^T \ln S_i(t+u)\,du\right).$$

Consider an Asian option involving two assets whose terminal payoff is given by $\max(G_1(t+T) - G_2(t+T), 0)$. Show that the price formula of this European Asian option at time t is given by (Boyle, 1993)

$$V(S_1, S_2, t; T) = \widetilde{S}_1 N(d_1) - \widetilde{S}_2 N(d_2),$$

where

$$\widetilde{S}_i = S_i \exp\left(-\left(\frac{r^2}{2} + \frac{\sigma^2}{12}\right)T\right), i = 1, 2, \quad \sigma^2 = \frac{\sigma_1^2}{3} + \frac{\sigma_2^2}{3} - \frac{2}{3}\rho\sigma_1\sigma_2,$$

$$d_1 = \frac{\ln \frac{\widetilde{S}_1}{\widetilde{S}_2} + \frac{\sigma^2}{2}T}{\sigma\sqrt{T}}, \quad d_2 = d_1 - \sigma\sqrt{T}.$$

Hint: Apply the price formula of the exchange option in Problem 3.34.

4.28 Deduce the following put-call parity relation between the prices of European fixed strike Asian call and put options under continuously monitored geometric averaging

$$\begin{aligned} c(S, G, t) &- p(S, G, t) \\ &= e^{-r(T-t)}\left\{G^{t/T} S^{(T-t)/T} \exp\left((T-t)\left[\frac{\sigma^2}{6}\left(\frac{T-t}{T}\right)^2 \right.\right.\right. \\ &\qquad \left.\left.\left. + \frac{r - q - \frac{\sigma^2}{2}}{2}\frac{T-t}{T}\right] - X\right)\right\}. \end{aligned}$$

4.29 Suppose continuous arithmetic averaging of the asset price is taken from $t = 0$ to T, T is the expiration time. The terminal payoff function of the floating strike call and put options are, respectively,

$$\max\left(S_T - \frac{1}{T}\int_0^T S_u\, du, 0\right) \text{ and } \max\left(\frac{1}{T}\int_0^T S_u\, du - S_T, 0\right).$$

Show that the put-call parity relation for the above pair of European floating strike options is given by

$$c - p = Se^{-q(T-t)} + \frac{S}{(r-q)T}[e^{-r(T-t)} - e^{-q(T-t)}] - e^{-r(T-t)} A_t,$$

where

$$A_t = \frac{1}{T}\int_0^t S_u\, du.$$

Suppose continuous geometric averaging of the asset price is taken, show that the corresponding put-call parity relation is given by

$$\begin{aligned} c - p = Se^{-q(T-t)} &- G^{t/T} S^{(T-t)/T} \\ &\exp\left(\frac{\sigma^2(T-t)^3}{6T^2} + \frac{(r - q - \frac{\sigma^2}{2})(T-t)^2}{2T} - r(T-t)\right), \end{aligned}$$

where

$$G_t = \exp\left(\frac{1}{t}\int_0^t \ln S_u \, du\right).$$

4.30 Show that the put-call parity relations between the prices of floating strike and fixed strike Asian options at the start of the averaging period are given by

$$p_{f\ell}(S_0, \lambda, r, q, T) - c_{f\ell}(S_0, \lambda, r, q, T) = \frac{S(e^{-qT} - e^{-rT})}{(r-q)T} - \lambda S_0$$

$$c_{fix}(X, S_0, r, q, T) - p_{fix}(X, S_0, r, q, T) = \frac{S(e^{-qT} - e^{-rT})}{(r-q)T} - e^{-rT}X.$$

By combining the above put-call parity relations with the fixed-floating symmetry relation between $c_{f\ell}$ and p_{fix}, deduce the following symmetry relation between c_{fix} and $p_{f\ell}$:

$$c_{fix}(X, S_0, r, q, T) = p_{f\ell}\left(S_0, \frac{X}{S_0}, q, r, T\right).$$

4.31 Consider a self-financing trading strategy of a portfolio with a dividend paying asset and a money market account over the time horizon $[0, T]$. Under the risk neutral measure Q, let the dynamics of the asset price S_t be governed by

$$\frac{dS_t}{S_t} = (r-q)dt + \sigma \, dZ_t,$$

where q is the dividend yield, $q \neq r$. We adopt the trading strategy of holding n_t units of the asset at time t, where

$$n_t = \frac{1}{(r-q)T}\left[e^{-q(T-t)} - e^{-r(T-t)}\right].$$

Let X_t denote the portfolio value at time t, whose dynamics is then given by

$$dX_t = n_t \, dS_t + r(X_t - n_t S_t)\, dt + q n_t S_t \, dt.$$

The initial portfolio value X_0 is chosen to be

$$X_0 = n_0 S_0 - e^{-rT} X.$$

Show that

$$X_T = \frac{1}{T}\int_0^T S_t \, dt - X.$$

Defining $Y_t = \frac{X_t}{e^{qt} S_t}$, show that

$$dY_t = -(Y_t - e^{-qt} n_t)\sigma \, dZ_t^*,$$

where $Z_t^* = Z_t - \sigma t$ is a Brownian process under Q^*-measure with $e^{qt} S_t$ as the numeraire. Note that the price function of the fixed strike Asian call option with strike X is given by

$$c_{fix}(S_0, 0; X) = e^{-rT} E_Q[\max(X_T, 0)] = S_0 e^{-rT} E_{Q^*}[\max(Y_T, 0)],$$

with

$$Y_0 = \frac{X_0}{S_0} = \frac{e^{-qT} - e^{-rT}}{(r-q)T} - e^{-rT}\frac{X}{S_0}.$$

Show that

$$c_{fix}(S_0, 0; X) = S_0 u(Y_0, 0),$$

where $u(y, t)$ satisfies the following one-dimensional partial differential equation:

$$\frac{\partial u}{\partial t} + \frac{1}{2}(y - e^{-qt} n_t)^2 \sigma^2 \frac{\partial^2 u}{\partial y^2} = 0$$

with $u(y, T) = \max(y, 0)$.

4.32 Consider the European continuously monitored arithmetic average Asian option with terminal payoff: $\max(A_T - X_1 S_T - X_2, 0)$, where

$$A_T = \frac{1}{T}\int_0^T S_u\, du.$$

At the current time $t > 0$, the average value A_t over the time period $[0, t]$ has been realized. Let $V(S, \tau; X_1, X_2)$ denote the price function of the Asian option at the start of the averaging period. Show that the value of the in-progress Asian option is given by

$$\frac{T-t}{T} V\left(S_t, T-t; \frac{X_1 T}{T-t}, \frac{X_2 T}{T-t} - \frac{A_t - t}{T-t}\right).$$

4.33 Consider $E_Q[\widetilde{A}(t_n; t)^2]$ defined in (4.3.68). Show that when $t < t_0$, we have (Levy, 1992)

$$E_Q[\widetilde{A}(t_n; t)^2] = \frac{S_t^2}{(n+1)^2} e^{(2r+\sigma)(t_0-t)}(B_1 - B_2 + B_3 - B_4),$$

where

$$B_1 = \frac{1 - e^{(2r+\sigma^2)(n+1)\Delta t}}{(1 - e^{r\Delta t})[1 - e^{(2r+\sigma^2)\Delta t}]}, \qquad B_2 = \frac{e^{r(n+1)\Delta t} - e^{(2r+\sigma^2)(n+1)\Delta t}}{(1 - e^{r\Delta t})[1 - e^{(r+\sigma^2)\Delta t}]},$$

$$B_3 = \frac{e^{r\Delta t} - e^{r(n+1)\Delta t}}{(1 - e^{r\Delta t})[1 - e^{(r+\sigma^2)\Delta t}]}, \qquad B_4 = \frac{e^{(2r+\sigma^2)\Delta t} - e^{(2r+\sigma^2)(n+1)\Delta t}}{[1 - e^{(r+\sigma^2)\Delta t}][1 - e^{(2r+\sigma^2)\Delta t}]}.$$

4.34 Under the risk neutral measure Q, let S_t be governed by

$$\frac{dS_t}{S_t} = (r-q)\,dt + \sigma\,dZ_t.$$

Defining

$$A(t,T) = \frac{1}{T-t}\int_t^T S_u\,du,$$

show that (Milevsky and Posner, 1998)

$$E_Q[A(t,T)] = \begin{cases} S_t\frac{\exp((r-q)(T-t))-1}{(r-q)(T-t)}, & \text{if } r \neq q \\ S_t & \text{if } r = q \end{cases};$$

$$E_Q[A(t,T)^2] = \begin{cases} \frac{2S_t^2}{(T-t)^2}\left\{\frac{\exp([2(r-q)+\sigma^2](T-t))}{(r-q+\sigma^2)(2r-2q+\sigma^2)}\right. \\ \left.+\frac{1}{r-q}\left[\frac{1}{2(r-q)+\sigma^2} - \frac{\exp((r-q)(T-t))}{r-q+\sigma^2}\right]\right\}, & \text{if } r \neq q \\ \frac{2S_t^2}{(T-t)^2}\frac{e^{\sigma^2(T-t)}-1-\sigma^2(T-t)}{\sigma^4}, & \text{if } r = q \end{cases}.$$

4.35 Suppose we define the flexible geometric average $G_F(n)$ of asset prices at n evenly spaced time instants by

$$G_F(n) = \prod_{i=1}^n S_i^{\omega_i}, \quad \omega_i = \frac{i^\alpha}{\sum_{i=1}^n i^\alpha}$$

and S_i is the asset price at time t_i. Here, ω_i is the weighting factor associated with S_i. Note that the larger the value of α, the heavier are the weights allocated to the more recent asset price. Under the risk neutral measure, the asset price is assumed to follow the Geometric Brownian process

$$\frac{dS_t}{S_t} = r\,dt + \sigma\,dZ_t.$$

We consider the fixed strike Asian option with terminal payoff

$$V(S, G_F, T) = \max(\phi(G_F(n) - X), 0),$$

where X is the strike price, and ϕ is the binary variable which is set to 1 for a call or -1 for a put. Show that the Asian option value is given by (Zhang, 1994)

$$V(S, G_F, t) = \phi\left[SA_j^f N\left(\phi\left(d_{n-j}^f + \sigma\sqrt{T_{n-j}^f}\right)\right) - Xe^{-r\tau}N\left(\phi d_{n-j}^f\right)\right],$$

where

$$A_j^f = \exp\Big(-r\big(\tau - T^f_{\mu,n-j}\big) - \frac{\sigma^2}{2}\big(T^f_{\mu,n-j} - T^f_{n-j}\big)\Big) B_j^f,$$

$$B_0^f = 1, \quad B_j^f = \prod_{i=1}^{j}\left(\frac{S_{n-i}}{S}\right)^{\omega_i}, \qquad 1 \le j \le n,$$

$$d^f_{n-j} = \frac{\ln\frac{S}{X} + \big(r - \frac{\sigma^2}{2}\big)T^f_{\mu,n-j} + \ln B_j^f}{\sigma\sqrt{T^f_{n-j}}},$$

$$T^f_{\mu,n-j} = \sum_{i=j+1}^{n} \omega_i[\tau - (n-i)\Delta t],$$

$$T^f_{n-j} = \sum_{i=j+1}^{n} \omega_i^2[\tau - (n-i)\Delta t] + 2\sum_{i=2}^{n-j}\sum_{k=1}^{i-1} \omega_i\omega_k[\tau - (n-k)\Delta t],$$

n is the number of asset prices taken for averaging, Δt is the time interval between successive observational instants, j is the number of observations already passed, B_j^f can be considered as the weighted average of the returns of those observations that have already passed.

4.36 Show that for any random variable X, we have

$$0 \le E[\max(X,0)] - \max(E[X],0) \le \frac{1}{2}\sqrt{\operatorname{var}(X)},$$

and apply the result to show the result in (4.3.29).

4.37 Let Z_t denote the standard Brownian process. Show that the covariance matrix of the bivariate Gaussian random variable $(Z_t, \int_0^1 Z_u\,du)$ is given by

$$E\left[\left(Z_t, \int_0^1 Z_u\,du\right)^T \left(Z_t, \int_0^1 Z_u\,du\right)\right] = \begin{pmatrix} t & t(1-\frac{t}{2}) \\ t(1-\frac{t}{2}) & \frac{1}{3} \end{pmatrix}.$$

Also, show that the conditional distribution of Z_t given $\int_0^1 Z_u\,du = z$ is normal with mean $3t(1-\frac{t}{2})z$ and variance $t - 3t^2(1-\frac{t}{2})^2$. Using (4.3.29) with the choice of Y to be $\int_0^1 Z_u\,du$ and $T = 1$, show that (Thompson, 1999)

$$c_{fix}(S, I, 0) \ge e^{-r}\int_{-\infty}^{\infty} \sqrt{3}n(\sqrt{3}z)$$
$$\int_0^1 \max\big(Se^{3\sigma t(1-t/2)z + (r-q)t + \frac{\sigma^2}{2}[t - 3t^2(1-\frac{t}{2})^2]} - X, 0\big)\,dt\,dz,$$

where $n(z) = \frac{1}{\sqrt{2\pi}}\,e^{-z^2/2}$.

4.38 Let $S(t_i)$ denote the asset price at time $t_i, i = 1, 2, \cdots, N$, where $0 = t_0 < t_1 < \cdots < t_N = T$. Define the discretely monitored arithmetic average and geometric average by

$$A(T) = \frac{1}{N}\sum_{i=1}^{N} S(t_i) \quad \text{and} \quad G(T) = \left[\prod_{i=1}^{N} S(t_i)\right]^{1/N}.$$

Let $c_A(0; X)$ and $c_G(0; X)$ denote the time-0 value of the European Asian fixed strike call option with strike price X and whose underlying are $A(T)$ and $G(T)$, respectively. Under the usual Geometric Brownian process assumption of the asset price, show that (Nielsen and Sandmann, 2003)

$$c_G(0; X) \leq c_A(0; X) \leq c_G(0; X) + e^{-rT} E_Q[A(T) - G(T)],$$

where

$$E_Q[A(T)] = \frac{S(0)}{N} e^{-r\Delta} \frac{1 - e^{rN\Delta}}{1 - e^{r\Delta}}, \qquad \Delta = t_{i+1} - t_i, \quad i = 1, 2, \cdots, N-1,$$

$$E_Q[G(T)] = \exp\left(m_G + \frac{\sigma_G^2}{2}\right),$$

$$m_G = \ln S(0) + \left(r - \frac{\sigma^2}{2}\right)\frac{N+1}{2}\Delta, \sigma_G^2 = \sigma^2 \Delta\left[1 + \frac{(N-1)(2N-1)}{6N}\right].$$

5

American Options

The distinctive feature of an American option is its early exercise privilege, that is, the holder can exercise the option prior to the date of expiration. Since the additional right should not be worthless, we expect an American option to be worth more than its European counterpart. The extra premium is called the *early exercise premium*.

First, we recall some of the pricing properties of American options discussed in Sect. 1.2. The early exercise of either an American call or an American put leads to the loss of insurance value associated with holding of the option. For an American call, the holder gains on the dividend yield from the asset but loses on the time value of the strike price. There is no advantage to exercise an American call prematurely when the asset received upon early exercise does not pay dividends. The early exercise right is rendered worthless when the underlying asset does not pay dividends, so in this case the American call has the same value as that of its European counterpart. Furthermore, we showed using the dominance argument that an American option must be worth at least its corresponding intrinsic value, namely, $\max(S - X, 0)$ for a call and $\max(X - S, 0)$ for a put, where S and X are the asset price and strike price, respectively. Although a put-call parity relation exists for European options, we can only obtain lower and upper bounds on the difference of American call and put option values.

When the underlying asset is dividend paying, it may become optimal for the holder to exercise prematurely an American call option when the asset price S rises to some critical asset value, called the *optimal exercise price*. Since the loss of insurance value and time value of the strike price is time dependent, the optimal exercise price depends on time to expiry. For a longer-lived American call option, the insurance value associated with long holding of the American call and the time value of the strike are higher. Hence, the optimal exercise price should assume a higher value so that the chance of regret of early exercise is lower and the dividend amount received from holding the asset is larger. When the underlying asset pays a continuous dividend yield, the collection of these optimal exercise prices for all times constitutes a continuous curve, which is commonly called the *optimal exercise boundary*. For an American put option, the early exercise leads to some gain on time value of strike. Therefore, when the riskless interest rate is positive, there always exists an optimal exercise price below which it becomes optimal to exercise the American put prematurely.

The optimal exercise boundary of an American option is not known in advance but has to be determined as part of the solution process of the pricing model. Since the boundary of the domain of an American option model is a free boundary, the valuation problem constitutes a *free boundary value problem*. In Sect. 5.1, we present the characterization of the optimal exercise boundary at infinite time to expiry and at the moment immediately prior to expiry. We derive the optimality condition in the form of smooth pasting of the option value curve with the intrinsic value line. When the underlying asset pays discrete dividends, the early exercise of the American call may become optimal only at a time right before a dividend date. Since the early exercise policy becomes relatively simple, we manage to derive closed form price formulas for American calls on an asset that pays discrete dividends. We also discuss the optimal exercise policies of American put options on a discrete dividend paying asset.

In Sect. 5.2, we present two pricing formulations of American options, namely, the linear complementarity formulation and the optimal stopping formulation. We show how the early exercise premium can be expressed in terms of the exercise boundary in an integral representation and examine how the determination of the optimal exercise boundary is relegated to the solution of an integral equation. The early exercise premium can be interpreted as the compensation paid to the holder for delaying his early exercise right, otherwise it should have been optimal for him to exercise the option prematurely. The early exercise feature can be combined with other path dependent features in an option contract. We examine the impact of the barrier feature on the early exercise policies of the American barrier options. Also, we obtain the analytic price formula for the Russian option, which is essentially a perpetual American lookback option.

In general, analytic price formulas are not available for American options, except for a few special types. In Sect. 5.3, we present several analytic approximation methods for pricing American options. One approximation approach is to limit the exercise privilege such that the American option is exercisable only at a finite number of time instants. The other method is the solution of the integral equation of the exercise boundary by a recursive integration method. The third method, called the quadratic approximation approach, is based on the reduction of the Black–Scholes equation to an ordinary differential equation so that analytic tractability is enhanced.

The modeling of a financial derivative with voluntary right to reset certain terms in the contract, like resetting the strike price to the prevailing asset price, also constitutes a free boundary value problem. In Sect. 5.4, we construct the pricing model for the reset-strike put option and examine the optimal reset strategy adopted by the option holder. While an American option can be exercised only once, multiple reset may be allowed. We also examine the pricing behavior of multireset put options. Interestingly, when the right to reset is allowed to be infinitely often, the multireset put option becomes a European lookback option.

5.1 Characterization of the Optimal Exercise Boundaries

The characteristics of the optimal early exercise policies of American options depend critically on whether the underlying asset is nondividend paying or dividend paying (discrete or continuous). Throughout our discussion, we assume that the dividends are known in advance, both in amount and time of payment. In this section, we give some detailed quantitative analysis of the properties of the early exercise boundary. We show that the optimal exercise boundary of an American put, with continuous dividend yield or zero dividend, is a *continuous decreasing* function of time of expiry τ. However, the optimal exercise boundary for an American put on an asset that pays discrete dividends may or may not have *jumps of discontinuity*, depending on the size of the discrete dividend payments. For an American call on an asset which pays a continuous dividend yield, we explain why it becomes optimal to exercise the call at sufficiently high value of S. The corresponding optimal exercise boundary is a *continuous increasing* function of τ. When the underlying asset of an American call pays discrete dividends, optimal early exercise of the American call may occur only at those time instants immediately before the asset goes ex-dividend. There are several additional conditions required for optimal early exercise, which include (i) the discrete dividend is sufficiently large relative to the strike price, (ii) the ex-dividend date is fairly close to expiry and (iii) the asset price level prior to the dividend date is higher than some threshold value. Since the possibilities of early exercise are limited to a few discrete dividend dates, the price formula for an American call on an asset paying known discrete dividends can be obtained by relating the American call option to a European compound option.

The auxiliary conditions in the pricing model of an American option include the value matching condition and smooth pasting condition of the American option value across the optimal exercise boundary. The *smooth pasting condition* is a result derived from maximizing the American option value among all possible early exercise policies (see Sect. 5.1.2).

5.1.1 American Options on an Asset Paying Dividend Yield

First, we consider the effects of continuous dividend yield (at the constant yield $q > 0$) on the early exercise policy of an American call. When the asset value S is exceedingly high, it is almost certain that the European call option on a continuous dividend paying asset will be in-the-money at expiry. The American call then behaves almost like the asset but without its dividend income minus the present value of the strike price X. When the call is sufficiently deep in-the-money, by observing

$$N(\widehat{d_1}) \sim 1 \quad \text{and} \quad N(\widehat{d_2}) \sim 1$$

in the European call price formula (3.4.4), we obtain

$$c(S, \tau) \sim e^{-q\tau} S - e^{-r\tau} X \qquad \text{when} \quad S \gg X. \tag{5.1.1}$$

The price of this European call may be below the intrinsic value $S-X$ at a sufficiently high asset value, due to the presence of the factor $e^{-q\tau}$ in front of S. Although it is

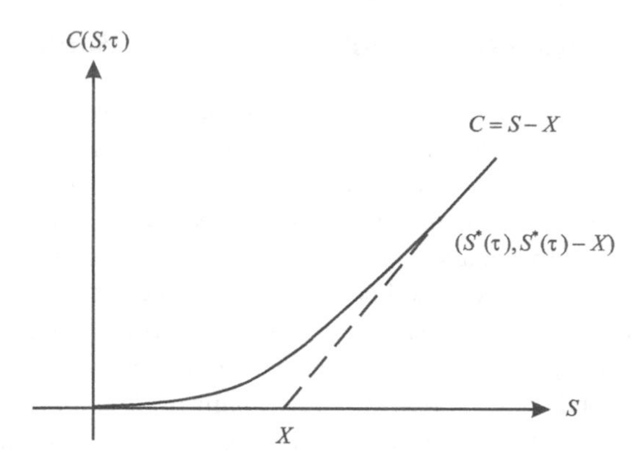

Fig. 5.1. The solid curve shows the price function $C(S, \tau)$ of an American call on an asset paying continuous dividend yield. The price curve touches the dotted intrinsic value line tangentially at the point $(S^*(\tau), S^*(\tau) - X)$, where $S^*(\tau)$ is the optimal exercise price. When $S \geq S^*(\tau)$, the American call value becomes $S - X$.

possible that the value of a European option stays below its intrinsic value, the holder of an American option with embedded early exercise right would not allow the value of his option to fall below the intrinsic value. Hence, at a sufficiently high asset value, it becomes optimal for the American option on a continuous dividend paying asset to be exercised prior to expiry; thus, its value will not fall below the intrinsic value if unexercised.

In Fig. 5.1, the American call option price curve $C(S, \tau)$ touches *tangentially* the dotted line representing the intrinsic value of the call at some optimal exercise price $S^*(\tau)$. Note that the optimal exercise price has dependence on τ, the time to expiry. The tangency behavior of the American price curve at $S^*(\tau)$ (continuity of delta value) will be explained in the next subsection. When $S \geq S^*(\tau)$, the American call value is equal to its intrinsic value $S - X$. The collection of all these points $(S^*(\tau), \tau)$, for all $\tau \in (0, T]$, in the (S, τ)-plane constitutes the optimal exercise boundary. The American call option remains alive only within the *continuation region* $\mathcal{C} = \{(S, \tau) : 0 \leq S < S^*(\tau), 0 < \tau \leq T\}$. The complement of $\mathcal{C}$ is called the *stopping region* $\mathcal{S}$, inside which the American call should be optimally exercised (see Fig. 5.2).

Under the assumption of continuity of the asset price path and dividend yield, we expect that the optimal exercise boundary should also be a continuous function of τ, for $\tau > 0$. While a rigorous proof of the continuity of $S^*(\tau)$ is rather technical, a heuristic argument is provided below. Assume the contrary, suppose $S^*(\tau)$ has a downward jump as τ decreases across the time instant $\widehat{\tau}$. Assume that the asset price S at $\widehat{\tau}$ satisfies $S^*(\widehat{\tau}^-) < S < S^*(\widehat{\tau}^+)$, the American call option value is strictly above the intrinsic value $S - X$ at $\widehat{\tau}^+$ since $S < S^*(\widehat{\tau}^+)$ and becomes equal to the intrinsic value $S - X$ at $\widehat{\tau}^-$ since $S > S^*(\widehat{\tau}^-)$. This indicates a discrete downward jump in option value across $\widehat{\tau}$. As there is no cash flow associated with long holding

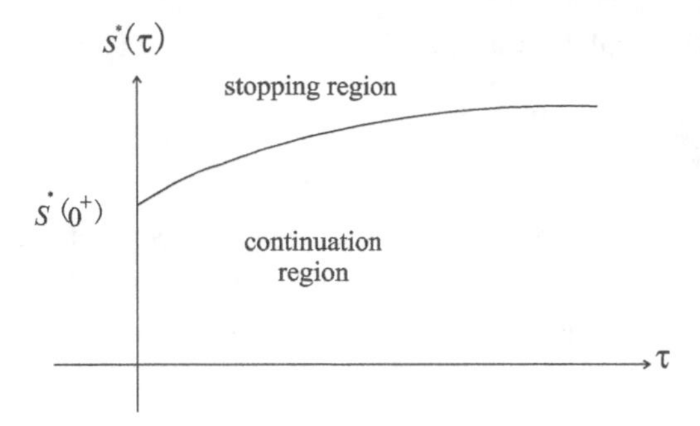

Fig. 5.2. An American call on an asset paying continuous dividend yield remains alive inside the continuation region $\mathcal{C} = \{(S, \tau) : S \in [0, S^*(\tau)), \tau \in (0, T]\}$. The optimal exercise boundary $S^*(\tau)$ is a continuous increasing function of τ.

of the American call across $\widehat{\tau}$, this discrete jump in value would lead to an arbitrage opportunity.

5.1.2 Smooth Pasting Condition

We would like to examine the smooth pasting condition (tangency condition) along the optimal exercise boundary for an American call on a continuous dividend paying asset. At $S = S^*(\tau)$, the value of the exercised American call is $S^*(\tau) - X$ so that

$$C(S^*(\tau), \tau) = S^*(\tau) - X. \tag{5.1.2}$$

For obvious reasons, this is termed the value matching condition.

Supposing $S^*(\tau)$ is a known continuous function, the pricing model becomes a boundary value problem with a time dependent boundary. However, in the American call option model, $S^*(\tau)$ is not known in advance. Rather, it must be determined as part of the solution. An additional auxiliary condition has to be prescribed along $S^*(\tau)$ so as to reflect the nature of optimality of the exercise right embedded in the American option.

We follow Merton's (1973, Chap. 1) argument to show the continuity of the delta of option value of an American call at the optimal exercise price $S^*(\tau)$. Let $f(S, \tau; b(\tau))$ denote the solution to the Black–Scholes equation in the domain $\{(S, \tau)\colon S \in (0, b(\tau)), \tau \in (0, T]\}$, where $b(\tau)$ is a known boundary. The holder of the American call chooses an early exercise policy which maximizes the value of the call. Using such an argument, the American call value is given by

$$C(S, \tau) = \max_{\{b(\tau)\}} f(S, \tau; b(\tau)) \tag{5.1.3}$$

for all possible continuous functions $b(\tau)$. For fixed τ, for convenience, we write $f(S, \tau; b(\tau))$ as $F(S, b)$, where $0 \leq S \leq b$. It is observed that $F(S, b)$ is a differentiable function, concave in its second argument. Further, we write $h(b) = F(b, b)$ which is assumed to be a differentiable function of b. For the American call option,

$h(b) = b - X$. The total derivative of F with respect to b along the boundary $S = b$ is given by

$$\frac{dF}{db} = \frac{dh}{db} = \frac{\partial F}{\partial S}(S, b)\bigg|_{S=b} + \frac{\partial F}{\partial b}(S, b)\bigg|_{S=b},$$

where the property $\frac{\partial S}{\partial b} = 1$ along $S = b$ has been incorporated. Let b^* be the critical value of b that maximizes F. When $b = b^*$, we have $\frac{\partial F}{\partial b}(S, b^*) = 0$ as the first derivative condition at a maximum point. On the other hand, from the exercise payoff function of the American call option, we have

$$\frac{dh}{db}\bigg|_{b=b^*} = \frac{d}{db}(b - X)\bigg|_{b=b^*} = 1.$$

Putting the results together, we obtain

$$\frac{\partial F}{\partial S}(S, b^*)\bigg|_{S=b^*} = 1. \tag{5.1.4}$$

Note that the optimal choice $b^*(\tau)$ is just the optimal exercise price $S^*(\tau)$. The above condition can be expressed in the following form:

$$\frac{\partial C}{\partial S}(S^*(\tau), \tau) = 1. \tag{5.1.5}$$

Condition (5.1.5) is commonly called the *smooth pasting* or *tangency condition*. The two auxiliary conditions (5.1.2) and (5.1.5), respectively, reveal that $C(S, \tau)$ and $\frac{\partial C}{\partial S}(S, \tau)$ are continuous across the optimal exercise boundary (see Fig. 5.1).

The smooth pasting condition is applicable to all types of American options. For an American put option, the slope of the intrinsic value line is -1. The continuity of the delta of the American put value at $S = S^*(\tau)$ gives

$$\frac{\partial P}{\partial S}(S^*(\tau), \tau) = -1. \tag{5.1.6}$$

An alternative proof of the above smooth pasting condition for the American put option is outlined in Problem 5.5.

5.1.3 Optimal Exercise Boundary for an American Call

Consider an American call on a continuous dividend paying asset, where the optimal exercise boundary $S^*(\tau)$ is a continuous increasing function of τ. The increasing property stems from the fact that the losses on the insurance value associated with long holding of the American call and time value of strike are more significant for a longer-lived American call so that the call must be deeper-in-the-money in order to induce early exercise decision. In addition, the compensation from the dividend received from the asset is higher. Hence, the American call should be exercised at a higher optimal exercise price $S^*(\tau)$ when compared to its shorter-lived counterpart.

The increasing property of $S^*(\tau)$ can also be explained by relating to the increasing property of the price curve $C(S, \tau)$ as a function of τ [see (1.2.5a)]. The option price curve of a longer-lived American call plotted against S always stays above that of its shorter-lived counterpart. The upper price curve corresponding to the longer-lived option cuts the intrinsic value line tangentially at a higher critical asset value $S^*(\tau)$.

Moreover, it is obvious from Fig. 5.1 that the price curve of an American call always cuts the intrinsic value line at a critical asset value greater than X. Hence, we have $S^*(\tau) \geq X$ for $\tau \geq 0$. Alternatively, assume the contrary, suppose $S^*(\tau) < X$, then the early exercise proceed $S^*(\tau) - X$ becomes negative. Since the early exercise privilege cannot be a liability, the possibility $S^*(\tau) < X$ is ruled out and so $S^*(\tau) \geq X$.

Next, we present the analysis of the asymptotic behavior of $S^*(\tau)$ at $\tau \to 0^+$ and $\tau \to \infty$.

Asymptotic Behavior of $S^*(\tau)$ Close to Expiry

When $\tau \to 0^+$ and $S > X$, by the continuity of the call price function, the call value tends to the exercise payoff so that $C(S, 0^+) = S - X$. If the American call is alive, then the call value satisfies the Black–Scholes equation. By substituting the above call value into the Black–Scholes equation, given that (S, τ) lies in the continuation region, we have

$$\begin{aligned} \left.\frac{\partial C}{\partial \tau}\right|_{\tau=0^+} &= \frac{\sigma^2}{2} S^2 \left.\frac{\partial^2 C}{\partial S^2}\right|_{\tau=0^+} + (r-q)S\left.\frac{\partial C}{\partial S}\right|_{\tau=0^+} - rC\bigg|_{\tau=0^+} \\ &= (r-q)S - r(S-X) = rX - qS. \end{aligned} \tag{5.1.7}$$

Suppose $\frac{\partial C}{\partial \tau}(S, 0^+) < 0$, $C(S, \tau)$ becomes less than $C(S, 0) = S - X$ (intrinsic value of the American call) immediately prior to expiry. This leads to a contradiction since the American call value is always above the intrinsic value. Therefore, we must have $\frac{\partial C}{\partial \tau}(S, 0^+) \geq 0$ in order that the American call is kept alive until the time close to expiry. The value of S at which $\frac{\partial C}{\partial \tau}(S, 0^+)$ changes sign is $S = \frac{r}{q}X$. Also, $\frac{r}{q}X$ lies in the interval $S > X$ only when $q < r$. We consider the two separate cases, $q < r$ and $q \geq r$.

1. $q < r$
   At time immediately prior to expiry, we argue that the American call should be kept alive when $S < \frac{r}{q}X$. This is because within a short time interval δt prior to expiry, the dividend $qS\delta t$ earned from holding the asset is less than the interest $rX\delta t$ earned from depositing the amount X in a money market account at the riskless interest rate r. The above observation is consistent with nonnegativity of $\frac{\partial C}{\partial \tau}(S, 0^+)$ when $S \leq \frac{r}{q}X$. When $S > \frac{r}{q}X$, the American call should be exercised since the negativity of $\frac{\partial C}{\partial \tau}(S, 0^+)$ would violate the condition that the American call value must be above the intrinsic value $S - X$. Hence, for $q < r$, the optimal exercise price $S^*(0^+)$ is given by the asset value at which $\frac{\partial C}{\partial \tau}(S, 0^+)$ changes sign. We then obtain

$$S^*(0^+) = \frac{r}{q}X.$$

In particular, when $q = 0$, $S^*(0^+)$ becomes infinite. Furthermore, since $S^*(\tau)$ is a monotonically increasing function of τ, we then deduce that $S^*(\tau) \to \infty$ for all values of τ. This result is consistent with the well-known fact that it is always nonoptimal to exercise an American call on a nondividend paying asset prior to expiry.

2. $q \geq r$
When $q \geq r$, $\frac{r}{q}X$ becomes less than X and so the above argument has to be modified. First, we show that $S^*(0^+)$ cannot be greater than X. Assume the contrary, suppose $S^*(0^+) > X$ so that the American call is still alive when $X < S < S^*(0^+)$ at time close to expiry. Given the combined conditions $q \geq r$ and $S > X$, it is observed that the loss in dividend amount $qS\delta t$ not earned is more than the interest amount $rX\delta t$ earned if the American call is not exercised within a short time interval δt prior to expiry. This represents a nonoptimal early exercise policy. Hence, we must have $S^*(0^+) \leq X$. Together with the properties that $S^*(\tau) \geq X$ for $\tau > 0$ and $S^*(\tau)$ is a continuous increasing function of τ, for $q \geq r$, we then have
$$S^*(0^+) = X.$$

In summary, the optimal exercise price $S^*(\tau)$ of an American call on a continuous dividend paying asset at time close to expiry is given by

$$\lim_{\tau \to 0^+} S^*(\tau) = \begin{cases} \frac{r}{q}X & q < r \\ X & q \geq r \end{cases} = X \max\left(1, \frac{r}{q}\right). \tag{5.1.8}$$

At expiry $\tau = 0$, the American call option will be exercised whenever $S \geq X$ and so $S^*(0) = X$. Hence, for $q < r$, there is a jump of discontinuity of $S^*(\tau)$ at $\tau = 0$.

Asymptotic Behavior of $S^*(\tau)$ at Infinite Time to Expiry

Since $S^*(\tau)$ is a monotonically increasing function of τ, the lower bound for the optimal exercise boundary $S^*(\tau)$ for $\tau > 0$ is given by $\lim_{\tau \to 0^+} S^*(\tau)$. It would be interesting to explore whether $\lim_{\tau \to \infty} S^*(\tau)$ has a finite bound or otherwise. An option with infinite time to expiration is called a *perpetual option*. The determination of $\lim_{\tau \to \infty} S^*(\tau)$ is related to the analysis of the price function of corresponding perpetual American option.

Let $C_\infty(S; X, q)$ denote the price of an American perpetual call option that strike price X and on an asset that pays a continuous dividend yield q. Note that there is no time dependence in the price function of the perpetual American call. Since the value of a perpetual option is insensitive to temporal rate of change, so the Black–Scholes equation is reduced to the following ordinary differential equation

$$\frac{\sigma^2}{2} S^2 \frac{d^2 C_\infty}{dS^2} + (r - q)S \frac{dC_\infty}{dS} - rC_\infty = 0, \quad 0 < S < S^*_\infty, \tag{5.1.9a}$$

where S^*_∞ is the optimal exercise price at which the perpetual American call option should be exercised. Note that S^*_∞ is independent of τ since it is simply the asymptotic value $\lim_{\tau \to \infty} S^*(\tau)$. The boundary conditions for the pricing model of the perpetual American call are

$$C_\infty(0) = 0 \quad \text{and} \quad C_\infty(S_\infty^*) = S_\infty^* - X. \tag{5.1.9b}$$

We let $f(S; S_\infty^*)$ denote the solution to (5.1.9a,b) for a given value of S_∞^*. Since (5.1.9a) is a linear equidimensional ordinary differential equation, its general solution is of the form

$$f(S; S_\infty^*) = c_1 S^{\mu_+} + c_2 S^{\mu_-},$$

where c_1 and c_2 are arbitrary constants, μ_+ and μ_- are the respective positive and negative roots of the auxiliary equation

$$\frac{\sigma^2}{2}\mu^2 + \left(r - q - \frac{\sigma^2}{2}\right)\mu - r = 0.$$

Since $f(0; S_\infty^*) = 0$, we must have $c_2 = 0$. Applying the boundary condition at S_∞^*, we have

$$f(S_\infty^*; S_\infty^*) = c_1 S_\infty^{*\mu_+} = S_\infty^* - X,$$

thus giving

$$c_1 = \frac{S_\infty^* - X}{S_\infty^{*\mu_+}}.$$

The solution $f(S; S_\infty^*)$ is now reduced to the form

$$f(S; S_\infty^*) = (S_\infty^* - X)\left(\frac{S}{S_\infty^*}\right)^{\mu_+}, \tag{5.1.10}$$

where

$$\mu_+ = \frac{-(r - q - \frac{\sigma^2}{2}) + \sqrt{(r - q - \frac{\sigma^2}{2})^2 + 2\sigma^2 r}}{\sigma^2} > 0.$$

To complete the solution, S_∞^* has yet to be determined. We find S_∞^* by maximizing the value of the perpetual American call option among all possible optimal exercise prices, that is,

$$C_\infty(S; X, q) = \max_{\{S_\infty^*\}} \left\{(S_\infty^* - X)\left(\frac{S}{S_\infty^*}\right)^{\mu_+}\right\}. \tag{5.1.11}$$

The use of calculus shows that $f(S; S_\infty^*)$ is maximized when

$$S_\infty^* = \frac{\mu_+}{\mu_+ - 1}X. \tag{5.1.12}$$

Suppose we write $S_{\infty,C}^* = \frac{\mu_+}{\mu_+ - 1}X$, then the value of the perpetual American call takes the form

$$C_\infty(S; X, q) = \left(\frac{S_{\infty,C}^*}{\mu_+}\right)\left(\frac{S}{S_{\infty,C}^*}\right)^{\mu_+}. \tag{5.1.13}$$

It can be easily verified that the above solution also satisfies the smooth pasting condition:

$$\left.\frac{dC_\infty}{dS}\right|_{S=S^*_{\infty,C}} = 1. \tag{5.1.14}$$

One may solve for $S^*_{\infty,C}$ by applying the smooth pasting condition directly without going through the above maximization procedure. Indeed, the application of the smooth pasting condition implicity incorporates the procedure of taking the maximum of the option values among all possible choices of $S^*_{\infty,C}$.

5.1.4 Put-Call Symmetry Relations

The behavior of the optimal exercise boundary for an American put option on a continuous dividend paying asset can be inferred from that of its call counterpart once the put-call symmetry relations between their price functions and optimal exercise prices are established. The plot of the price function $P(S,\tau)$ of an American put against S is shown in Fig. 5.3.

We may consider an American call option as providing the right at any time during the option's life to exchange X dollars of cash (in the form of money market account) for one unit of the underlying asset which is worth S dollars. If we take asset one to be the underlying asset, asset two to be the cash, then asset one and asset two have their dividend yield q and r, respectively. The above call option can be considered an exchange option which exchanges asset two for asset one. Similarly, we may consider an American put option as providing the right to exchange one unit of the underlying asset which is worth S dollars for X dollars of cash at any time. What would happen if we interchange the role of the underlying asset and

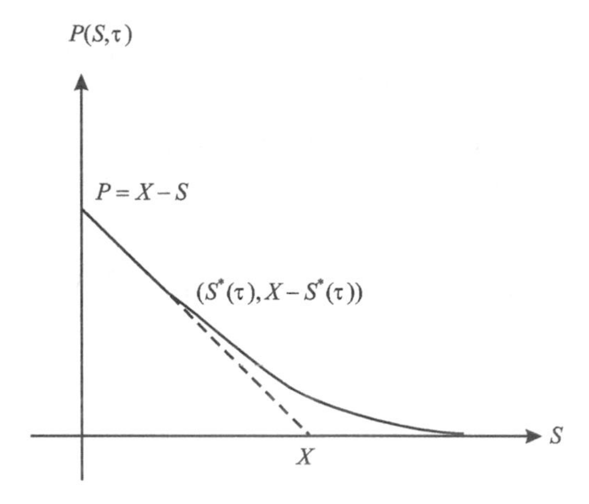

Fig. 5.3. The solid curve shows the price function of an American put at a given time to expiry τ. The price curve touches the dotted intrinsic value line tangentially at the point $(S^*(\tau), X - S^*(\tau))$, where $S^*(\tau)$ is the optimal exercise price. When $S \le S^*(\tau)$, the American put value becomes $X - S$.

cash in the American put option? Now, this new American put can be considered to be equivalent to the usual American call since both options confer to their holders the same right of exchanging cash for underlying asset. If we use $P(S, \tau; X, r, q)$ to denote the price function of the American put, then the price function of the modified American put (after interchanging the role of the underlying asset and cash) is given by $P(X, \tau; S, q, r)$, where S and X are interchanged and also for r and q. Since the modified American put is equivalent to the American call, we then have

$$C(S, \tau; X, r, q) = P(X, \tau; S, q, r). \tag{5.1.15}$$

This symmetry between the price functions of American call and put is called the *put-call symmetry relation.*

Next, we establish the put-call symmetry relation for the optimal exercise prices for American put and call options. Let $S_P^*(\tau; r, q)$ and $S_C^*(\tau; r, q)$ denote the optimal exercise boundary for the American put and call options on a continuous dividend paying asset, respectively. When $S = S_C^*(\tau; r, q)$, the call owner is willing to exchange X dollars of cash for one unit of the underlying asset which is worth S_C^* dollars or *one dollar* of cash for $\frac{1}{X}$ units of the asset which is worth $\frac{S_C^*}{X}$ dollars. Similarly, when $S = S_P^*(\tau; r, q)$, the put owner is willing to exchange $\frac{1}{S_P^*}$ units of the asset which is worth *one dollar* for $\frac{X}{S_P^*}$ dollars of cash. If both of these American call and put options can be considered as exchange options and the roles of cash and underlying asset are interchangeable, then the corresponding put-call symmetry relation for the optimal exercise prices is deduced to be

$$S_C^*(\tau; r, q) = \frac{X^2}{S_P^*(\tau; q, r)}. \tag{5.1.16}$$

A mathematical proof of the symmetry relation between American option prices can be established quite easily (see Problem 5.7). Indeed, more exotic forms of symmetry relations between the price functions of American call and put options can be derived (see Problems 5.8–5.9).

Behavior of $S_P^*(\tau)$ Near Expiry

From (5.1.16) and the monotonically increasing property of $S_C^*(\tau)$, we can deduce that $S_P^*(\tau)$ is a monotonically decreasing function of τ. Since (5.1.16) remains valid as $\tau \to 0^+$, the upper bound for $S_P^*(\tau)$ is given by

$$\lim_{\tau \to 0^+} S_P^*(\tau; r, q) = \frac{X^2}{\lim_{\tau \to 0^+} S_C^*(\tau; q, r)} = \frac{X^2}{X \max\left(1, \frac{q}{r}\right)} = X \min\left(1, \frac{r}{q}\right). \tag{5.1.17}$$

From (5.1.17), we observe that when $q \leq r$, we have $\lim_{\tau \to 0^+} S_P^*(\tau) = X$. Now, even when $q = 0$, $S_P^*(\tau)$ is nonzero since $S_P^*(\tau)$ is a continuous decreasing function of τ for $\tau > 0$ and its upper bound equals X. Hence, it is always optimal to exercise an American put even when the underlying asset pays no dividend. On the other hand, at zero interest rate, $\lim_{\tau \to 0^+} S_P^*(\tau)$ becomes zero. It then follows that $S_P^*(\tau) = 0$ for $\tau > 0$ since $S_P^*(\tau)$ is a decreasing function of τ. Therefore, it is never optimal to exercise an American put prematurely when the interest rate is zero. From financial intuition, such a conclusion is obvious since there is no time value gained on the strike price from the early exercise of the American put when there is null interest.

The understanding of more refined asymptotic behavior of $S_P^*(\tau)$ when $\tau \to 0^+$ poses great mathematical challenges. Evans, Kuske and Keller (2002) showed that at time close to expiry the optimal exercise boundary is parabolic when $q > r$ but it becomes parabolic-logarithmic when $q \leq r$. The asymptotic expansion of $S_P^*(\tau)$ as $\tau \to 0^+$ has the following analytic representation:

(i) $0 \leq q < r$

$$S_P^*(\tau) \sim X - X\sigma\sqrt{\tau \ln\left(\frac{\sigma^2}{8\pi\tau(r-q)^2}\right)}; \tag{5.1.18a}$$

(ii) $q = r$

$$S_P^*(\tau) \sim X - X\sigma\sqrt{2\tau \ln\left(\frac{1}{4\sqrt{\pi q \tau}}\right)}; \text{ and} \tag{5.1.18b}$$

(iii) $q > r$

$$S_P^*(\tau) \sim \frac{r}{q} X(1 - \sigma\alpha\sqrt{2\tau}). \tag{5.1.18c}$$

Here, α is a numerical constant that satisfies the following transcendental equation

$$-\alpha^3 e^{\alpha^2} \int_\alpha^\infty e^{-u^2}\, du = \frac{1 - 2\alpha^2}{4}.$$

Behavior of $S_P^*(\tau)$ at Infinite Time to Expiry

Following a similar derivation procedure as that for the perpetual American call option, the price of the perpetual American put option can be deduced to be

$$P_\infty(S; X, q) = -\frac{S_{\infty,P}^*}{\mu_-}\left(\frac{S}{S_{\infty,P}^*}\right)^{\mu_-}. \tag{5.1.19}$$

Here, $S_{\infty,P}^*$ denotes the optimal exercise price at infinite time to expiry and its value is given by

$$S_{\infty,P}^* = \frac{\mu_-}{\mu_- - 1} X, \tag{5.1.20}$$

where

$$\mu_- = \frac{-(r - q - \frac{\sigma^2}{2}) - \sqrt{(r - q - \frac{\sigma^2}{2})^2 + 2\sigma^2 r}}{\sigma^2} < 0.$$

One can easily verify that

$$S_{\infty,P}^*(r, q) = \frac{X^2}{S_{\infty,C}^*(q, r)}, \tag{5.1.21}$$

a result that is consistent with the relation given in (5.1.16).

5.1.5 American Call Options on an Asset Paying Single Dividend

It was explained in Sect. 1.2 that when an asset pays discrete dividend payments, the asset price declines by the same amount as the dividend right after the dividend date if there are no other factors affecting the income proceeds. Empirical studies show that the relative decline of the stock price as a proportion of the amount of the dividend is shown to be not meaningfully different from one. In our subsequent discussion, for simplicity, we assume that the asset price falls by the same amount as the discrete dividend right after an ex-dividend date. An option is said to be *dividend protected* if the value of the option is invariant to the choice of the dividend policy. This is done by adjusting the strike price in relation to the dividend amount. Here, we consider the effects of discrete dividends on the early exercise policy of American options which are not protected against the dividend, that is, the strike price is not marked down (for calls) or marked up (for puts) by the same amount as the dividend.

Early Exercise Policies

Since the holder of an American call on an asset paying discrete dividends will not receive any dividends between successive ex-dividend dates, it is never optimal to exercise the American call on any nondividend paying date. For those times between dividend dates, the early exercise right is noneffective. If the American call is exercised at all, the possible choices of exercise times are those instants immediately before the asset goes ex-dividend. As a result, the holder owns the asset right before the asset goes ex-dividend and receives the dividend in the next instant. We explore the conditions under which the holder of such American call would optimally choose to exercise his or her option.

In the following discussion, it is more convenient to characterize the time dependence of the optimal exercise boundary using the calendar time t. We consider an American call on an asset which pays only one discrete dividend of deterministic amount D at the known dividend date t_d. The generalization to multidividend models can be found in Problems 5.15–5.17. Let S_d^- (S_d^+) denote the asset price at time t_d^- (t_d^+) which is immediately before (after) the dividend date t_d. If the American call is exercised at t_d^-, the call value becomes $S_d^- - X$. Otherwise, the asset value drops to $S_d^+ = S_d^- - D$ right after the asset goes ex-dividend. Since there is no further discrete dividend after time t_d, the American price function behaves like that of its European counterpart for $t > t_d^+$. To preclude arbitrage opportunities, the call price function must be continuous across the ex-dividend instant since the holder of the call option does not receive any dividend payment on the dividend date (unlike holding the asset).

From (1.2.11), the lower bound of the American call value at t_d^+ is $S_d^+ - Xe^{-r(T-t_d^+)}$, where $T - t_d^+$ is the time to expiry. As far as time to expiry is concerned, the quantities $T - t_d$, $T - t_d^+$ and $T - t_d^-$ are considered equal. By virtue of the continuity of the call value across the dividend date, the lower bound for the call value at time t_d^- should also be equal to $S_d^+ - Xe^{-r(T-t_d)} = (S_d^- - D) - Xe^{-r(T-t_d)}$. Note that the lower bound for the call value at t_d^- is driven down by D in anticipation of the known discrete dividend amount D in the next instant. Now, it may occur that the lower bound value at t_d^- becomes less than the exercise payoff of $S_d^- - X$ when D is sufficiently large. We compare the following two quantities: exercise payoff $E = S_d^- - X$ and lower bound of the call value $B = (S_d^- - D) - Xe^{-r(T-t_d)}$. Suppose $E \leq B$, that is

$$S_d^- - X \leq (S_d^- - D) - Xe^{-r(T-t_d)} \quad \text{or} \quad D \leq X\,[1 - e^{-r(T-t_d)}], \tag{5.1.22}$$

then it is never optimal to exercise the American call at t_d^-. This is because at any value of asset price S_d^- the American call is worth more when it is held than when it is exercised. However, when the discrete dividend D is deep enough, in particular when $D > X[1 - e^{-r(T-t_d)}]$, then it may become optimal to exercise at t_d^- when the asset price S_d^- is above some threshold value. This requirement on D gives one of the necessary conditions for the commencement of early exercise. The dividend amount D must be sufficiently deep to offset the loss amount in the time value of the strike price, where the loss amount is given by $X[1 - e^{-r(T-t_d)}]$.

Let $C_d(S, t)$ denote the price function of the one-dividend American call option with the calendar time t as the time variable. By virtue of the continuity property of the call value across the dividend date, we have

$$C_d(S_d^-, t_d^-) = c(S_d^- - D, t_d^+), \tag{5.1.23}$$

where $c(S_d^- - D, t_d^+)$ is the European call price given by the Black–Scholes formula with asset price $S_d^- - D$ and calendar time t_d^+. To better understand the decision of early exercise at t_d^-, we plot the call price function, the exercise payoff E (corresponds to line ℓ_1: $E = S_d^- - X$) and the lower bound value B (corresponds to line ℓ_2: $B = S_d^- - D - Xe^{-r(T-t_d)}$) versus the asset price S_d^- (see Fig. 5.4). The exercise payoff line l_1 lies on the left side of the lower bound value line l_2 when $D > X[1 - e^{-r(T-t_d)}]$. Now, the call price curve may intersect (not tangentially) the

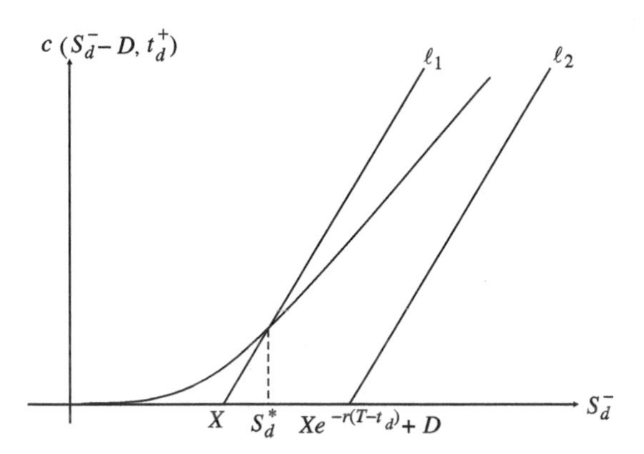

Fig. 5.4. For $S_d^- > S_d^*$, the European call price curve $V = c(S_d^- - D, t_d^+)$ stays below the exercise payoff line $\ell_1 : E = S_d^- - X$ when ℓ_1 lies on the left side of the lower bound value line $\ell_2 : B = S_d^- - D - Xe^{-r(T-t_d)}$. Here, S_d^* is the value of S_d^- at which the European call price curve cuts the exercise payoff line ℓ_1.

exercise payoff line l_1 at some critical asset price S_d^*, which is given by the solution to the following algebraic equation

$$c(S_d^- - D, t_d) = S_d^- - X. \tag{5.1.24}$$

It can be shown mathematically that when $D \leq X[1-e^{-r(T-t_d)}]$, there is no solution to (5.1.24), a result that is consistent with the necessary condition on D discussed earlier (see Problem 5.13). When the discrete dividend is sufficiently deep such that $D > X[1 - e^{-r(T-t_d)}]$, the American call remains alive beyond the dividend date only if $S_d^- < S_d^*$. When S_d^- is at or above S_d^*, the call should be optimally exercised at t_d^-. Hence, the American call price at time t_d^- is given by

$$C_d(S_d^-, t_d^-) = \begin{cases} c(S_d^- - D, t_d^+) & \text{when} \quad S_d^- < S_d^* \\ S_d^- - X & \text{when} \quad S_d^- \geq S_d^*. \end{cases} \tag{5.1.25}$$

If the American call is not optimally exercised at t_d^-, then its value remains unchanged as time lapses across the dividend date. Note that S_d^* depends on D, which decreases in value when D increases (see Problem 5.13). This agrees with the financial intuition that the propensity of optimal early exercise becomes higher (corresponding to a lower value of S_d^*) with a deeper discrete dividend payment.

In summary, the holder of an American call option on an asset paying single discrete dividend exercises the call optimally only at the instant immediately prior to the dividend date, provided that $S_d^- \geq S_d^*$, where S_d^* satisfies (5.1.24). Also, S_d^* exists only when $D > X[1 - e^{-r(T-t_d)}]$, implying that the dividend is sufficiently deep to offset the loss on time value of strike.

Analytic Price Formula for an One-Dividend American Call

Since the American call on an asset paying known discrete dividends will be exercised only at instants immediately prior to ex-dividend dates, the American call can be replicated by a European compound option with the expiration dates of the compound option coinciding with the ex-dividend dates. The presence of this replication strategy makes possible the derivation of an analytic price formula for an American call on an asset paying discrete dividends.

If the whole asset price S follows the Geometric Brownian process, then there exists the possibility that the dividends cannot be paid since the asset value may fall below the dividend payment on a dividend date. The difficulty can be resolved if we modify the assumption on the diffusion process. Now, we assume the asset price net of the present value of the escrowed dividends, denoted by $\widetilde{S}$, to follow the Geometric Brownian process. We call $\widetilde{S}$ to be the risky component of the asset price.

Suppose the asset pays a single discrete dividend of amount D at time t_d, then the risky component of S is defined by

$$\widetilde{S} = \begin{cases} S & \text{for} \quad t_d^+ \leq t \leq T \\ S - De^{-r(t_d - t)} & \text{for} \quad t \leq t_d^-. \end{cases} \tag{5.1.26}$$

Note that $\widetilde{S}$ is continuous across the dividend date. The Black–Scholes assumption on the asset price movement is modified such that under the risk neutral measure the risky component $\widetilde{S}$ follows the Geometric Brownian process:

$$\frac{d\widetilde{S}}{\widetilde{S}} = r\,dt + \sigma\,dZ, \tag{5.1.27}$$

where σ is the volatility of $\widetilde{S}$.

Now, we would like to derive the price formula of an American call option on an asset paying single discrete dividend D at time t_d, where $D > X[1 - e^{-r(T-t_d)}]$. Let $C_d(\widetilde{S}, t)$ denote the price of this one-dividend American call and $c(\widetilde{S}, t)$ denote the Black–Scholes call price function, where t is the calendar time. Let $\widetilde{S}_d$ denote the risky component of the asset value on the ex-dividend date t_d. Let $\widetilde{S}_d^*$ denote the critical value of the risky component at $t = t_d$, above which it is optimal to exercise. This critical value $\widetilde{S}_d^*$ is the solution to the following equation [see (5.1.24)]

$$\widetilde{S}_d + D - X = c(\widetilde{S}_d, t_d). \tag{5.1.28}$$

The one-dividend American call option can be replicated by a European compound option whose first expiration date coincides with the ex-dividend date t_d. The compound option pays at t_d either $\widetilde{S}_d + D - X$ if $\widetilde{S}_d \geq \widetilde{S}_d^*$ or a European call option with strike price X and time to expiry $T - t_d$ if $\widetilde{S}_d < \widetilde{S}_d^*$. Let $\psi(\widetilde{S}_d, \widetilde{S}; t_d, t)$ denote the transition density function of $\widetilde{S}_d$ at time t_d, given the asset price $\widetilde{S}$ at an earlier time $t < t_d$. For $t < t_d$, the time t price of the one-dividend American call option is given by (Whaley, 1981)

$$\begin{aligned} C_d(\widetilde{S}, t) = e^{-r(t_d-t)} \bigg[& \int_{\widetilde{S}_d^*}^{\infty} [\widetilde{S}_d - (X - D)]\, \psi(\widetilde{S}_d, \widetilde{S}; t_d, t)\, d\widetilde{S}_d \\ & + \int_0^{\widetilde{S}_d^*} c(\widetilde{S}_d, t_d)\, \psi(\widetilde{S}_d, \widetilde{S}; t_d, t)\, d\widetilde{S}_d \bigg], \quad t < t_d. \end{aligned} \tag{5.1.29}$$

The first term may be interpreted as the price of a European call with two different strike prices. The strike price $\widetilde{S}_d^*$ determines the moneyness of the call option at expiry and the other strike price $X - D$ is the amount paid in exchange of the asset at expiry. The second term represents the price of a European put-on-call with strike price $\widetilde{S}_d^*$ at t_d and strike price X at T. The price formula of the one-dividend American call option is given by

$$\begin{aligned} & C_d(\widetilde{S}, t) \\ &= \widetilde{S}N(a_1) - (X - D)e^{-r(t_d-t)}N(a_2) - Xe^{-r(T-t)}N_2\left(-a_2, b_2; -\sqrt{\frac{t_d - t}{T - t}}\right) \\ &\quad + \widetilde{S}N_2\left(-a_1, b_1; -\sqrt{\frac{t_d - t}{T - t}}\right) \\ &= \widetilde{S}\left[1 - N_2\left(-a_1, -b_1; \sqrt{\frac{t_d - t}{T - t}}\right)\right] + De^{-r(t_d-t)}N(a_2) \\ &\quad - X\left[e^{-r(t_d-t)}N(a_2) + e^{-r(T-t)}N_2\left(-a_2, b_2; -\sqrt{\frac{t_d - t}{T - t}}\right)\right], \end{aligned} \tag{5.1.30}$$

where

$$a_1 = \frac{\ln \frac{\widetilde{S}}{S_d^*} + (r + \frac{\sigma^2}{2})(t_d - t)}{\sigma\sqrt{t_d - t}}, \qquad a_2 = a_1 - \sigma\sqrt{t_d - t},$$

$$b_1 = \frac{\ln \frac{\widetilde{S}}{X} + (r + \frac{\sigma^2}{2})(T - t)}{\sigma\sqrt{T - t}}, \qquad b_2 = b_1 - \sigma\sqrt{T - t}.$$

The generalization of the pricing procedure to the two-dividend American call option model is considered in Problem 5.17.

Black's Approximation Formula
Let $c(S, \tau)$ denote the price function of a European call, where the temporal variable τ is the time to expiry. Black (1975) proposed the approximate value of the one-dividend American call to be given by

$$\max\{c(\widetilde{S}, T - t; X), c(S, t_d - t; X)\}.$$

The first term gives the one-dividend American call value when the probability of early exercise is zero while the second term assumes the probability of early exercise to be one. Since both cases represent suboptimal early exercise policies, it is obvious that

$$C_d(\widetilde{S}, T - t; X) \geq \max\{c(\widetilde{S}, T - t; X), c(S, t_d - t; X)\}, \quad t < t_d. \qquad (5.1.31)$$

5.1.6 One-Dividend and Multidividend American Put Options

Consider an American put on an asset which pays out discrete dividends with certainty during the life of the option. The corresponding optimal exercise policy exhibits more complicated behavior compared to its call counterpart. Within some short time period prior to a dividend payment date, the put holder may choose not to exercise at any asset price level due to the anticipation of the dividend payment. That is, the holder prefers to defer early exercise until immediately after an ex-dividend date in order to receive the dividend. During the time interval from the last dividend date to expiration, the optimal exercise boundary behaves like that of an American put on a nondividend payment asset, so the optimal exercise price $S^*(t)$ increases monotonically with increasing calendar time t. For times in between the dividend dates and before the first dividend date, $S^*(t)$ may rise or fall with increasing t or even becomes zero (see Figs. 5.5 and 5.6). Due to the complicated nature of the optimal exercise policy, no analytic price formula exists for an American put on an asset paying discrete dividends.

One-Dividend American Put
First, we consider the early exercise policy for the one-dividend American put model. Let the ex-dividend date be t_d, the expiration date be T and the dividend amount be D. Since the exercise policy at $t > t_d$ is identical to that of the American put on

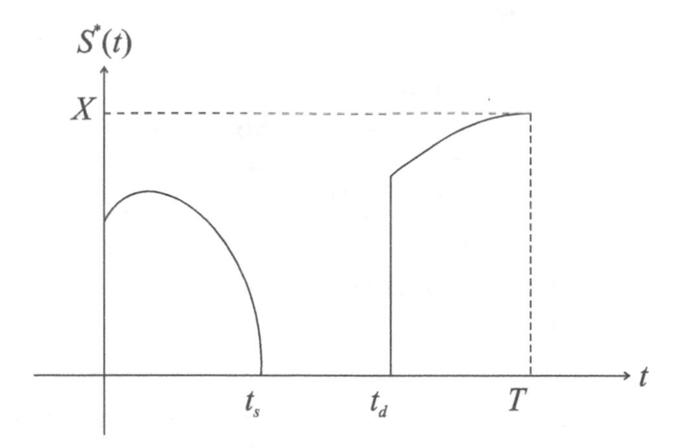

Fig. 5.5. The plot of the optimal exercise boundary $S^*(t)$ as a function of t for the one-dividend American put option.

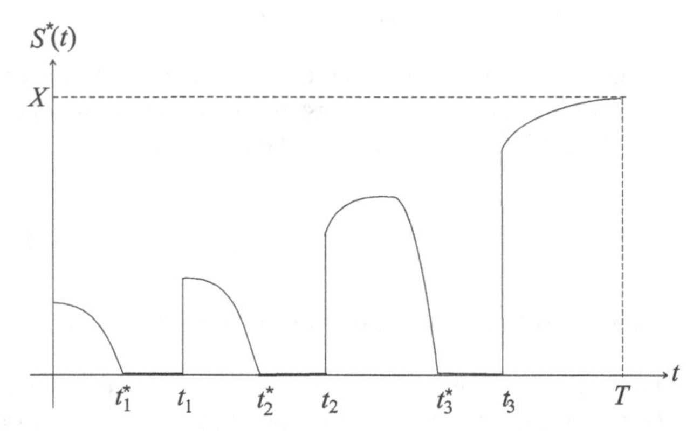

Fig. 5.6. The characterization of the optimal exercise boundary $S^*(t)$ as a function of the calendar time t for a three-dividend American put option model. Observe that $S^*(t)$ is monotonically increasing in (t_3, T) and $S^*(T) = X$. It stays at the zero value in $[t_3^*, t_3]$. Furthermore, $S^*(t)$ can be increasing to some peak value then decreasing as in (t_2, t_3^*), or simply decreasing monotonically as in (t_1, t_2^*).

the same asset with zero dividend, it suffices to consider the exercise policy at time t before the ex-dividend date. Suppose the American put is exercised at time t, then the interest received from t to t_d as the gain on the time value of the strike price X is $X[e^{r(t_d-t)} - 1]$, where r is the riskless interest rate. When the gain on the time value of the strike price is less than the discrete dividend, that is, $X[e^{r(t_d-t)} - 1] < D$, the early exercise of the American put is never optimal. One observes that the interest income $X[e^{r(t_d-t)} - 1]$ depends on $t_d - t$, and its value increases when $t_d - t$ increases. There exists a critical value t_s such that

$$X[e^{r(t_d-t_s)} - 1] = D. \tag{5.1.32}$$

Solving for t_s, we obtain

$$t_s = t_d - \frac{1}{r} \ln\left(1 + \frac{D}{X}\right). \tag{5.1.33}$$

Over the interval $[t_s, t_d]$, it is never optimal to exercise the American put.

When $t < t_s$, we have $X[e^{r(t_d - t)} - 1] > D$. Under such conditions, early exercise may become optimal when the asset price is below a certain critical value. The optimal exercise price $S^*(t)$ is governed by two offsetting effects, the time value of the strike and the discrete dividend. When t is approaching t_s, the dividend effect becomes more dominant so that the American put would be exercised only when it is deeper-in-the-money. Thus, $S^*(t)$ decreases as t is increasing and approaching t_s. When t is sufficiently far from t_s, the dividend effect diminishes so that the optimal exercise policy behaves more like an American put on a zero-dividend asset. In this case, $S^*(t)$ becomes an increasing function in t. Combining these results, the plot of $S^*(t)$ against t resembles a hump-shape curve for the time period prior to t_s (see Fig. 5.5).

From (5.1.33), t_s is seen to increase with increasing r so that the time interval of "no-exercise" $[t_s, t_d]$ shrinks with a higher interest rate. Since the early exercise of an American put results in gain on the time value of strike, a higher interest rate implies a higher opportunity cost of holding an in-the-money American put. Hence, the propensity of early exercise increases.

In summary, the optimal exercise boundary $S^*(t)$ of the one-dividend American put model exhibits the following behavior (see Fig. 5.5).

(i) When $t < t_s$, $S^*(t)$ first increases then decreases smoothly with increasing t until it drops to the zero value at t_s.
(ii) $S^*(t)$ stays at the zero value in the time interval $[t_s, t_d]$.
(iii) When $t \in (t_d, T]$, $S^*(t)$ is a monotonically increasing function of t with $S^*(T) = X$.

Multidividend American Put

Analysis of the optimal exercise policy for the multidividend American put model can be performed in a similar manner. Suppose dividends of amount $D_1, D_2, \cdots, D_n$ are paid on the ex-dividend dates $t_1, t_2, \cdots, t_n$, there is an interval $[t_j^*, t_j]$ before the ex-dividend time t_j, $j = 1, 2, \cdots, n$ such that it is never optimal to exercise the put prematurely. That is, $S^*(t) = 0$ for $t \in [t_j^*, t_j]$, $j = 1, 2, \cdots, n$. The critical time t_j^* is given by

$$t_j^* = t_j - \frac{1}{r} \ln\left(1 + \frac{D_j}{X}\right), \quad j = 1, 2, \cdots, n. \tag{5.1.34}$$

Note that t_j^* decreases when D_j increases. It may occur that t_j^* becomes less than t_{j-1} when D_j is sufficiently deep.

Here, we use the calendar time t in the description of the optimal exercise boundary. When t falls inside the time interval (t_{j-1}, t_j^*), $j = 2, \cdots, n$ or $t \leq t_1^*$, the optimal exercise price $S^*(t)$ may first increase with time to some peak value, then decreases and eventually drops to the zero value when the time reaches t_j^*. This corresponds to the scenario where the dividend amount is small. When the dividend becomes sufficiently deep, $S^*(t)$ may decrease monotonically throughout the interval (t_{j-1}, t_j^*) from some peak value to the zero value. When D_j increases further, it may be possible that t_j^* is less than t_{j-1}. As a consequence, $S^*(t) = 0$ for the whole time interval $[t_{j-1}, t_j]$. For the last time interval $(t_n, T]$, $S^*(t)$ increases monotonically to X as expiration is approached.

The plot of the optimal exercise boundary $S^*(t)$ of a three-dividend American put model as a function of the calendar time t is depicted in Fig. 5.6. Meyer (2001) performed careful numerical studies on the optimal exercise policies of the multidividend American put options. His results are consistent with the characterization of $S^*(t)$ described above.

5.2 Pricing Formulations of American Option Pricing Models

In this section, we consider two pricing formulations of the American option pricing models, namely, the linear complementarity formulation and the formulation as an optimal stopping problem. First, we develop the variational inequalities that are satisfied by the American option price function, and from which we derive the linear complementarity formulation. Alternatively, the American option price can be seen as the supremum of the expectation of the discounted exercise payoff among all possible stopping times. It can be shown that the solution to the optimal stopping formulation satisfies the linear complementarity formulation. From the theory of controlled diffusion process, we are able to derive the integral representation of an American price formula in terms of the optimal exercise boundary. We also show how to obtain the integral representation of the early exercise premium using the financial argument of delay exercise compensation. Using the fact that the optimal exercise price is the asset price at which one is indifferent between exercising or nonexercising, we deduce the integral equation for the optimal exercise price. This section ends with a discussion of two types of American path dependent option models, namely, the pricing of the American barrier option and a special form of the perpetual American lookback option called the "Russian option".

5.2.1 Linear Complementarity Formulation

The valuation of an American option can be formulated as a free boundary value problem, where the free boundary is the optimal exercise boundary which separates the continuation and stopping regions. When the asset price falls within the stopping region, the American call option should be exercised optimally so that its value is given by

$$C(S, \tau) = S - X, \quad S \geq S^*(\tau). \tag{5.2.1}$$

The exercise payoff function, $C = S - X$, does not satisfy the Black–Scholes equation since

$$\left[\frac{\partial}{\partial \tau} - \frac{\sigma^2}{2}S^2\frac{\partial^2}{\partial S^2} - (r-q)S\frac{\partial}{\partial S} + r\right](S-X) = qS - rX. \tag{5.2.2}$$

In the stopping region, we observe $S \geq S^*(\tau) > S^*(0^+) = X\max(1, \frac{r}{q})$ so that $qS - rX > 0$. The call value $C(S, \tau)$ then observes the following inequality

$$\frac{\partial C}{\partial \tau} - \frac{\sigma^2}{2}S^2\frac{\partial^2 C}{\partial S^2} - (r-q)S\frac{\partial C}{\partial S} + rC > 0 \quad \text{for} \quad S \geq S^*(\tau). \tag{5.2.3}$$

The above inequality can also be deduced from the following financial argument. Let Π denote the value of the riskless hedging portfolio defined by

$$\Pi = C - \Delta S \text{ where } \Delta = \frac{\partial C}{\partial S}.$$

In the continuation region where the American option is optimally held alive, by virtue of the no-arbitrage argument, we have $d\Pi = r\Pi\, dt$. However, the optimal early exercise of the American call occurs if and only if the rate of return from the riskless hedging portfolio is less than the riskless interest rate, that is,

$$d\Pi < r\Pi\, dt. \tag{5.2.4}$$

By computing $d\Pi$ using Ito's lemma, the above inequality can be shown to be equivalent to (5.2.3). We then conclude that

$$\frac{\partial C}{\partial \tau} - \frac{\sigma^2}{2}S^2\frac{\partial^2 C}{\partial S^2} - (r-q)S\frac{\partial C}{\partial S} + rC \geq 0, \quad S > 0 \text{ and } \tau > 0, \tag{5.2.5a}$$

where equality holds when (S, τ) lies in the continuation region. On the other hand, the American call value is always above the intrinsic value $S - X$ when $S < S^*(\tau)$ and equal to the intrinsic value when $S \geq S^*(\tau)$, that is,

$$C(S, \tau) \geq S - X, \quad S > 0 \text{ and } \tau > 0. \tag{5.2.5b}$$

In the above inequality, equality holds when (S, τ) lies in the stopping region. Since (S, τ) is either in the continuation region or stopping region, equality holds in one of the above pair of variational inequalities. We then deduce that

$$\left[\frac{\partial C}{\partial \tau} - \frac{\sigma^2}{2}S^2\frac{\partial^2 C}{\partial S^2} - (r-q)S\frac{\partial C}{\partial S} + rC\right][C - (S-X)] = 0, \tag{5.2.6}$$

for all values of $S > 0$ and $\tau > 0$. To complete the formulation of the model, we have to include the terminal payoff condition in the model formulation

$$C(S, 0) = \max(S - X, 0). \tag{5.2.7}$$

Inequalities (5.2.5a,b) and (5.2.6) together with the auxiliary condition (5.2.7) constitute the *linear complementarity formulation* of the American call option pricing model (Dewynne et al., 1993).

From the above linear complementarity formulation, we can deduce the following two properties for the optimal exercise price $S^*(\tau)$ of an American call.

1. It is the lowest asset price for which the American call value is equal to the exercise payoff.
2. It is the asset price at which one is indifferent between exercising and not exercising the American call.

Bunch and Johnson (2000) presented another interesting property of $S^*(\tau)$. It is the lowest asset price at which the American call value does not depend on the time to expiry, that is,

$$\frac{\partial C}{\partial \tau} = 0 \quad \text{at} \quad S = S^*(\tau). \tag{5.2.8}$$

This agrees with the financial intuition that at the moment when it is optimal to exercise immediately, it does not matter how much time is left to maturity. A simple mathematical proof can be constructed as follows. On the optimal exercise boundary $S^*(\tau)$, we have

$$C(S^*(\tau), \tau) = S^*(\tau) - X.$$

Differentiating both sides with respect to τ, we obtain

$$\frac{\partial C}{\partial \tau}(S^*(\tau), \tau) + \frac{\partial C}{\partial S}(S^*(\tau), \tau)\frac{dS^*(\tau)}{d\tau} = \frac{dS^*(\tau)}{\partial \tau}.$$

Using the smooth pasting condition $\frac{\partial C}{\partial S}(S^*(\tau), \tau) = 1$, we then obtain the result in (5.2.8).

5.2.2 Optimal Stopping Problem

The pricing of an American option can also be formulated as an *optimal stopping problem*. A stopping time t^* can be considered as a function assuming value over an interval $[0, T]$ such that the decision to "stop at time t^*" is determined by the information on the asset price path $S_u, 0 \leq u \leq t^*$ (see Sect. 2.2.3). Consider an American put option and suppose that it is exercised at time t^*, $t^* < T$, the payoff is $\max(X - S_{t^*}, 0)$. The fair value of the put option with payoff at t^* defined above is given by

$$E_Q^t[e^{-r(t^*-t)} \max(X - S_{t^*}, 0)],$$

where E_Q^t denotes the expectation under the risk neutral measure Q conditional on $S_t = S$. This is valid provided that t^* is a stopping time, independent of whether it is deterministic or random.

Since the holder can exercise at any time during the life of the option and he or she chooses the exercise time optimally such that the above expectation of discounted payoff is maximized, we deduce that the American put value is given by (Karatzas, 1988; Jacka, 1991; Myneni, 1992)

$$P(S, t) = \sup_{t \leq t^* \leq T} E_Q^t[e^{-r(t^*-t)} \max(X - S_{t^*}, 0)], \tag{5.2.9}$$

where t is the calendar time and the supremum is taken over all possible stopping times. Recall that $P(S, t)$ always stays at or above the payoff and $P(S, t)$ equals the payoff at the stopping time t^*. The above supremum is reached at the optimal stopping time (Krylov, 1980) so that

$$t^*_{opt} = \inf_u\{t \le u \le T : P(S_u, u) = \max(X - S_u, 0)\}, \tag{5.2.10}$$

which is the first time that the American put value drops to its payoff value.

We would like to verify that the solution to the linear complementarity formulation gives the American put value as stated in (5.2.9), where the optimal stopping time is determined by (5.2.10). We recall the renowned *Optional Sampling Theorem*. In one of its forms, it states that if $(M_t)_{t\ge 0}$ is a continuous martingale with respect to the filtration $(\mathcal{F}_t)_{t\ge 0}$ and if t^* is a bounded stopping time with $t^* > t$, then (Lamberton and Lapeyre, 1996, Chap. 2)

$$E[M_{t^*}|\mathcal{F}_t] = M_t.$$

For any stopping time t^*, $t < t^* < T$, we apply Ito's lemma to the solution $P(S, t)$ of the linear complementarity formulation and obtain

$$\begin{aligned} &e^{-rt^*} P(S_{t^*}, t^*) \\ &= e^{-rt} P(S, t) \\ &\quad + \int_t^{t^*} e^{-ru}\left[\frac{\partial}{\partial u} + \frac{\sigma^2}{2} S^2 \frac{\partial^2}{\partial S^2} + (r - q)S\frac{\partial}{\partial S} - r\right] P(S_u, u)\, du \\ &\quad + \int_t^{t^*} e^{-ru}\sigma S \frac{\partial P}{\partial S}(S_u, u)\, dZ_u. \end{aligned}$$

Now, the integrand of the first integral is nonpositive as deduced from one of the variational inequalities [see (5.2.4)]. When we take the expectation of the martingale as represented by the second integral, we obtain

$$E^t_Q\left[\int_t^{t^*} e^{-ru}\sigma S \frac{\partial P}{\partial S}(S_u, u)\, dZ_u\right] = 0,$$

by virtue of the Optional Sampling Theorem. These results lead to

$$P(S, t) \ge E^t_Q[e^{-r(t^*-t)} P(S_{t^*}, t^*)].$$

Furthermore, since the above result is valid for any stopping time and $P(S_{t^*}, t^*) \ge \max(X - S_{t^*}, 0)$, we can deduce

$$\begin{aligned} P(S, t) &\ge \sup_{t\le t^*\le T} E^t_Q\left[e^{-r(t^*-t)} P(S_{t^*}, t^*)\right] \\ &\ge \sup_{t\le t^*\le T} E^t_Q[e^{-r(t^*-t)} \max(X - S_{t^*}, 0)]. \end{aligned} \tag{5.2.11a}$$

On the other hand, suppose the stopping time is chosen to be t^*_{opt} as defined in (5.2.10), then for u between t and t^*_{opt}, we observe that $P(S_u, u)$ lies in the continuation region since t^*_{opt} is the first time that $P(S_{t^*_{opt}}, t^*_{opt}) = \max(X - S_{t^*_{opt}})$. Hence, we have

$$\left[\frac{\partial}{\partial u}+\frac{\sigma^2}{2}S^2\frac{\partial^2}{\partial S^2}+(r-q)S\frac{\partial}{\partial S}-r\right]P(S_u,u)=0.$$

Applying the Optimal Sampling Theorem again, we obtain

$$E_Q^t\left[\int_t^{t^*_{opt}} e^{-ru}\sigma S\frac{\partial P}{\partial S}(S_u,u)\,dZ_u\right]=0.$$

Putting these results together, we observe that the lower bound on $P(S,t)$ as depicted in (5.2.11a) is achieved at $t^*=t^*_{opt}$, where

$$P(S,t)=E_Q^t\big[e^{-r(t^*-t^*_{opt})}\max(X-S_{t^*_{opt}},0)\big]. \tag{5.2.11b}$$

Combining (5.2.11a,b), we deduce that t^*_{opt} is an optimal stopping time and the results in (5.2.9)–(5.2.10) are then obtained.

5.2.3 Integral Representation of the Early Exercise Premium

From the theory of controlled diffusion process, the American put price is given by [a rigorous proof is presented in Krylov (1980)]

$$\begin{aligned}P(S,t)&=E_Q^t[e^{-r(T-t)}\max(X-S_T,0)]\\&\quad+\int_t^T e^{-r(u-t)}E_Q^u\big[(rX-qS_u)\mathbf{1}_{\{S_u<S^*(u)\}}\big]\,du.\end{aligned} \tag{5.2.12a}$$

The first term represents the usual European put price while the second term represents the early exercise premium. Let $\psi(S_u;S)$ denote the transition density function of S_u conditional on $S_t=S$ under the risk neutral measure Q. We may rewrite the above put price formula as

$$\begin{aligned}P(S,t)&=e^{-r(T-t)}\int_0^X(X-S_T)\psi(S_T;S)\,dS_T\\&\quad+\int_t^T e^{-r(u-t)}\int_0^{S^*(u)}(rX-qS_u)\psi(S_u;S)\,dS_u\,du.\end{aligned} \tag{5.2.12b}$$

The early exercise premium is seen to be positive since

$$rX-qS_u>0\quad\text{as}\quad S_u<S^*(u)<\frac{rX}{q}.$$

Delay Exercise Compensation

We would like to provide an intuitive argument that the early exercise premium can be interpreted as the cost of delay exercise compensation (Jamshidian, 1992). So that the American put option is kept alive for all values of asset price until expiration, the holder needs to be compensated by a continuous cash flow when the put should have been exercised optimally. Consider the time interval between u and $u+du$ and suppose S_u falls within the stopping region, the amount of compensation paid to the holder of the American put should be $(rX-qS_u)\,du$ in order that the holder agrees not to exercise even when it is optimal to do so. This is because the holder would have earned interest $rX\,du$ from the strike price received and lost dividend $qS_u\,du$ from the short position of the asset if he were to choose to exercise his put. The discounted expectation for the above continuous cash flow compensation is given by

$$e^{-r(u-t)}\int_0^{S^*(u)}(rX - qS_u)\psi(S_u; S)\,dS_u.$$

The integration of the above discounted cash flow from $u = t$ to $u = T$ gives the early exercise premium of the American put option, which is precisely the last term in (5.2.12b).

Value Matching and Smooth Pasting Conditions

Carr, Jarrow and Myneni (1992) used the dynamic trading strategy to present the financial interpretation of the necessity of the continuity of P and $\frac{\partial P}{\partial S}$ across the optimal exercise boundary $S^*(u)$. After purchasing the American put at the current time t, the investor would instantaneously exercise the put whenever the asset price falls from above to the optimal exercise price $S^*(u)$ at time u and purchase back the put whenever the asset price rises from below to $S^*(u)$. Since the transactions of converting the put option into holding of cash plus short position in asset and vice versa all occur on the early exercise boundary, we require the "value matching" and "smooth pasting" conditions in order to ensure that these transactions are self-financing, that is, each portfolio revision undertaken is exactly financed by the proceeds from the sale of the previous position.

Analytic Representation of the American Put Price Function

We now use the time to expiry τ as the temporal variable in optimal exercise boundary $S^*(\tau)$. The integrals in (5.2.12) can be evaluated to give the following representation of the American put price formula:

$$\begin{aligned} P(S,\tau) &= Xe^{-r\tau}N(-d_2) - Se^{-q\tau}N(-d_1) \\ &\quad + \int_0^\tau [rXe^{-r\xi}N(-d_{\xi,2}) - qSe^{-q\xi}N(-d_{\xi,1})]\,d\xi, \end{aligned} \tag{5.2.13}$$

where $\tau = T - t$ and

$$d_1 = \frac{\ln\frac{S}{X} + \left(r - q + \frac{\sigma^2}{2}\right)\tau}{\sigma\sqrt{\tau}}, \qquad d_2 = d_1 - \sigma\sqrt{\tau},$$

$$d_{\xi,1} = \frac{\ln\frac{S}{S^*(\tau-\xi)} + \left(r - q + \frac{\sigma^2}{2}\right)\xi}{\sigma\sqrt{\xi}}, \qquad d_{\xi,2} = d_{\xi,1} - \sigma\sqrt{\xi}.$$

The dummy time variable ξ can be considered as the time period lapsed from the current time. Accordingly, $\xi = 0$ and $\xi = \tau$ correspond to the current time and expiration date, respectively.

Taking the interest rate r to be zero, the early exercise premium becomes

$$-\int_0^\tau qSe^{-q\xi}N(-d_{\xi,1})\,d\xi,$$

which is seen to be a nonpositive quantity. However, the early exercise premium must be nonnegative. To satisfy both conditions, we must have

$$\int_0^\tau qSe^{-q\xi} N(-d_{\xi,1})\, d\xi = 0,$$

which is satisfied only by setting $S^*(\xi) = 0$ for all values of ξ. The zero value of the optimal exercise price infers that the American put is never exercised. Under zero interest rate, the advantage of gaining the time value of the strike price through early exercise disappears. In this case, the value of the American put is the same as that of its European counterpart since the early exercise right is forfeited.

Integral Equations for the Optimal Exercise Boundary

If we apply the boundary condition: $P(S^*(\tau), \tau) = X - S^*(\tau)$ to the put price formula (5.2.13), we obtain the following integral equation for $S^*(\tau)$

$$\begin{aligned} X - S^*(\tau) &= Xe^{-r\tau} N(-\widehat{d}_2) - S^*(\tau)e^{-q\tau} N(-\widehat{d}_1) \\ &\quad + \int_0^\tau [rXe^{-r\xi} N(-\widehat{d}_{\xi,2}) - qS^*(\tau)e^{-q\xi} N(-\widehat{d}_{\xi,1})]\, d\xi, \end{aligned} \tag{5.2.14}$$

where

$$\widehat{d}_1 = \frac{\ln \frac{S^*(\tau)}{X} + \left(r - q + \frac{\sigma^2}{2}\right)\tau}{\sigma\sqrt{\tau}}, \qquad \widehat{d}_2 = \widehat{d}_1 - \sigma\sqrt{\tau}$$

$$\widehat{d}_{\xi,1} = \frac{\ln \frac{S^*(\tau)}{S^*(\tau-\xi)} + \left(r - q + \frac{\sigma^2}{2}\right)\xi}{\sigma\sqrt{\xi}}, \qquad \widehat{d}_{\xi,2} = \widehat{d}_{\xi,1} - \sigma\sqrt{\xi}.$$

The solution for $S^*(\tau)$ requires the knowledge of $S^*(\tau - \xi),\ 0 < \xi \le \tau$. The solution procedure starts with $S^*(0)$ and integrates backward in calendar time (that is, increasing τ).

The integral equation defined in (5.2.14) may be used to find the optimal exercise price at the limiting case $\tau \to \infty$ [see (5.1.20)]. Let $S_P^*(\infty)$ denote $\lim_{\tau\to\infty} S_P^*(\tau)$, which corresponds to the optimal exercise price for the perpetual American put. Taking the limit $\tau \to \infty$ in (5.2.14), and observing that the value of the perpetual European put is zero, we obtain

$$\begin{aligned} X - S_P^*(\infty) = \int_0^\infty \Bigg[& rXe^{-r\xi} N\left(-\frac{r - q - \frac{\sigma^2}{2}}{\sigma}\sqrt{\xi}\right) \\ & - qS_P^*(\infty)e^{-q\xi} N\left(-\frac{r - q + \frac{\sigma^2}{2}}{\sigma}\sqrt{\xi}\right)\Bigg]\, d\xi. \end{aligned} \tag{5.2.16}$$

The first and second terms in the above integral can be simplified as follows:

$$\begin{aligned} \int_0^\infty e^{-r\xi} N(-\rho\sqrt{\xi})\, d\xi &= -\frac{e^{-r\xi}}{r} N(-\rho\sqrt{\xi})\Big|_0^\infty - \frac{\rho}{2r}\frac{1}{\sqrt{2\pi}} \int_0^\infty \frac{e^{-\rho^2\xi/2} e^{-r\xi}}{\sqrt{\xi}}\, d\xi \\ &= \frac{1}{2r}\left[1 - \frac{\rho}{\sqrt{\rho^2 + 2r}}\right], \quad \rho = \frac{r - q - \frac{\sigma^2}{2}}{\sigma}; \end{aligned}$$

$$\int_0^\infty e^{-q\xi} N(-\rho'\sqrt{\xi})\, d\xi = \frac{1}{2q}\left[1 - \frac{\rho'}{\sqrt{\rho'^2 + 2q}}\right], \quad \rho' = \frac{r - q + \frac{\sigma^2}{2}}{\sigma}.$$

Substituting the above results into (5.2.16), we obtain

$$X - S_P^*(\infty) = \frac{X}{2}\left[1 - \frac{\rho}{\sqrt{\rho^2 + 2r}}\right] - \frac{S_P^*(\infty)}{2}\left[1 - \frac{\rho'}{\sqrt{\rho'^2 + 2q}}\right].$$

Rearranging the terms, we have

$$S_P^*(\infty) = \frac{1 + \frac{\rho}{\sqrt{\rho^2+2r}}}{1 + \frac{\rho'}{\sqrt{\rho'^2+2q}}} X = \frac{\mu_-}{\mu_- - 1} X, \tag{5.2.17}$$

where μ_- is defined by (5.1.20).

Alternatively, we may use the smooth pasting condition: $\frac{\partial P}{\partial S}(S^*(\tau), \tau) = -1$ along $S^*(\tau)$ to derive another integral equation for $S^*(\tau)$. Taking the partial derivative with respect to S of the terms in (5.2.13) and setting $S = S^*(\tau)$, we have

$$\begin{aligned}
0 &= 1 + \frac{\partial P}{\partial S}(S^*(\tau), \tau) \\
&= 1 + \frac{\partial p}{\partial S}(S^*(\tau), \tau) \\
&\quad + \int_0^\tau \left[rXe^{-r\xi} \frac{\partial}{\partial S} N(-d_{\xi,2})\Big|_{S=S^*(\tau)} - qe^{-q\xi} N(-d_{\xi,1})\Big|_{S=S^*(\tau)} \right. \\
&\qquad \left. - qSe^{-q\xi} \frac{\partial}{\partial S} N(-d_{\xi,1})\Big|_{S=S^*(\tau)} \right] d\xi \\
&= N(\widehat{d}_1) - \int_0^\tau \left[\frac{(r-q)e^{-q\xi}}{\sigma\sqrt{2\pi\xi}} e^{-\frac{\widehat{d}_{\xi,1}^2}{2}} + qe^{-q\xi} N(-\widehat{d}_{\xi,1}) \right] d\xi.
\end{aligned} \tag{5.2.15}$$

Various versions of the integral equation for the optimal exercise price can also be derived (Little and Pant, 2000), and some of these alternative forms may provide an easier analysis of the properties of the optimal exercise boundary. The direct analytic solution to any one of these integral equations is definitely intractable and one must resort to numerical methods for their solution. In Sect. 5.3, we discuss the recursive integration method for solving the above integral equations.

Analytic Representation of the American Call Price Function

Similar to the American put price as given in (5.2.13), the analytic representation of the American call price function is given by

$$\begin{aligned}
C(S, \tau) &= Se^{-q\tau} N(d_1) - Xe^{-r\tau} N(d_2) \\
&\quad + \int_0^\tau \left[qSe^{-q\xi} N(d_{\xi,1}) - rXe^{-r\xi} N(d_{\xi,2}) \right] d\xi.
\end{aligned} \tag{5.2.18}$$

The corresponding integral equation for the early exercise boundary $S_C^*(\tau)$ can be deduced similarly by setting $C(S_C^*(\tau), \tau) = S_C^*(\tau) - X$. This gives

$$S_C^*(\tau) - X = S_C^*(\tau)e^{-q\tau}N(\widehat{d}_1) - Xe^{-r\tau}N(\widehat{d}_2) + \int_0^\tau [qS^*(\tau)e^{-q\xi}N(\widehat{d}_{\xi,1}) - rXe^{-r\xi}N(\widehat{d}_{\xi,2})]\, d\xi. \quad (5.2.19)$$

Similarly, by taking the limit $\tau \to \infty$ in (5.2.19), one can also deduce the corresponding asymptotic upper bound of the early exercise boundary of the American call option (see Problem 5.22).

5.2.4 American Barrier Options

An American barrier option is a barrier option embedded with the early exercise right. For example, an American down-and-out call becomes nullified when the down-barrier is breached by the asset price or prematurely terminated due to the optimal early exercise of the holder. Like an usual American option, the option value of an American out-barrier option can be decomposed into the sum of the value of the European barrier option and the early exercise premium. In this section, we derive the price formula of an American down-and-out call and examine some of its pricing behavior. As a remark, the pricing of an American in-barrier option is much more complicated. This is because the in-trigger region associated with the knock-in feature may intersect with the stopping region of the underlying American option. The pricing models of the American in-barrier options have been discussed by Dai and Kwok (2004) (see also Problem 5.26).

For the American down-and-out call option model, we assume that the underlying asset pays a constant dividend yield q and the constant down-barrier B satisfies the condition $B < X$. Assuming zero rebate paid upon nullification of the option, the price function $C_B(S, \tau; X, B)$ of the American down-and-out call option is given by

$$C_B(S, \tau; X, B) = e^{-r\tau}E_Q^t\big[\max(S_T - X, 0)\mathbf{1}_{\{m_t^T > B\}}\big] + \int_t^T e^{-ru}E_Q^u\big[(qS_u - rX)\mathbf{1}_{\{m_t^u > B, (S_u, u)\in\mathcal{S}\}}\big]\, du, \quad (5.2.20)$$

where m_t^u is the realized minimum value of the asset price over the time period $[t, u]$, $\tau = T - t$ and $\mathcal{S}$ denotes the stopping region. The first term gives the value of the European down-and-out barrier option. The second term represents the early exercise premium of the American down-and-out call. The delay exercise compensation is received only when (S_u, u) lies inside the stopping region $\mathcal{S}$ and the barrier option has not been knocked out. To effect the expectation calculations, it is necessary to use the transition density function of the restricted (with down absorbing barrier B) asset price process. After performing the integration procedure, the early exercise premium $e_C(S, \tau; B)$ can be expressed as

$$e_C(S, \tau; B) = \int_0^\tau \left\{K_C(S, \tau; S^*(\tau - \omega), \omega) - \left(\frac{S}{B}\right)^{\delta+1} K_C\left(\frac{B^2}{S}, \tau; S^*(\tau - \omega), \omega\right)\right\} d\omega, \quad (5.2.21)$$

where $\delta = 2(q-r)/\sigma^2$ and $S^*(\tau)$ is the optimal exercise price above which the American down-and-out call option should be exercised. The analytic expression for K_C is given by

$$K_C(S, \tau; S^*(\tau - \omega), \omega) = qSe^{-q\omega}N(d_{\omega,1}) - rXe^{-r\omega}N(d_{\omega,2}), \qquad (5.2.22)$$

where

$$d_{\omega,1} = \frac{\ln \frac{S}{S^*(\tau-\omega)} + \left(r - q + \frac{\sigma^2}{2}\right)\omega}{\sigma\sqrt{\omega}}, \qquad d_{\omega,2} = d_{\omega,1} - \sigma\sqrt{\omega}.$$

It can be shown mathematically that

$$K_C(S, \tau; S^*(\tau - \omega), \omega) > \left(\frac{S}{B}\right)^{\delta+1} K_C\left(\frac{B^2}{S}, \tau; S^*(\tau - \omega), \omega\right) > 0. \qquad (5.2.23)$$

This agrees with the intuition that the early exercise premium is reduced by the presence of the barrier and it always remains positive. Though $e_C(S, \tau)$ apparently becomes negative when $q = 0$, the premium term in fact becomes zero since the early exercise premium must be nonnegative. This is made possible by choosing $S^*(\tau - \omega) \to \infty$ for $0 \leq \omega \leq \tau$. Even with the embedded barrier feature, an American call is never exercised when the underlying asset is nondividend paying.

Next, we explore the effects of the barrier level and rebate on the early exercise policies. Additional pricing properties of the American out-barrier options can be found in Gao, Huang and Subrahmanyam (2000).

Effects of Barrier Level on Early Exercise Policies

From intuition, it is expected that the optimal exercise price $S^*(\tau; B)$ for an American down-and-out call option decreases with an increasing barrier level B. For an in-the-money American down-and-out call option, the holder should consider to exercise the call at a lower optimal exercise price when the barrier level is higher since the adverse chance of asset price dropping to a level below the barrier is higher.

A semi-rigorous explanation of the above intuition can be argued as follows. Since the price curve of the American barrier call option with a lower barrier level is always above that with higher barrier level, it then intersects tangentially the intrinsic value line $C = S - X$ at a higher optimal exercise price (see Fig. 5.7). Therefore, $S^*(\tau; B)$ is a decreasing function of B.

Effects of Rebate on Early Exercise Policies

With the presence of rebate, the holder of an American down-and-out call option will choose to exercise optimally at a higher asset price level since the rebate will lessen the penalty of adverse movement of asset price dropping below the barrier. Mathematically, we argue that the price curve of the American down-and-out call option with rebate should be above that of the counterpart without rebate, so it intersects tangentially the intrinsic value line $C = S - X$ at a higher optimal exercise price. Hence, the optimal exercise price is an increasing function of rebate.

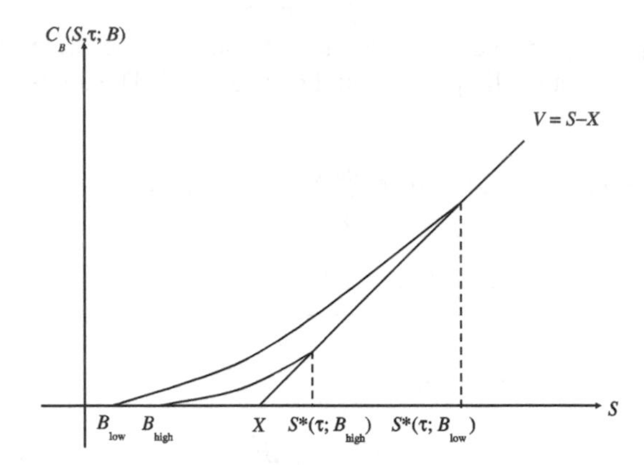

Fig. 5.7. The price curve for an American down-and-out call option with a lower barrier level B_{low} is always above that of the counterpart with a higher barrier level B_{high}.

5.2.5 American Lookback Options

The studies of the optimal exercise policies for various types of finite-lived American lookback options remain challenging problems. Some of the theoretical results on this topic can be found in a series of papers by Dai and Kwok (2004, 2005b, 2005c, 2006). In this section, we consider a special type of a perpetual American option with lookback payoff, called the "Russian option".

The Russian option contract on an asset guarantees that the holder of the option receives the historical maximum value of the asset price path upon exercising the option. Premature exercise of the Russian option can occur at any time chosen by the holder. Let M denote the historical realized maximum of the asset price (the starting date of the lookback period is immaterial for a perpetual option) and S be the asset price, both quantities are taken at the same time. Since it is a perpetual option, the option value is independent of time. Let $V = V(S, M)$ denote the option value and let S^* denote the optimal exercise price at which the Russian option should be exercised. At a sufficiently low asset price, it becomes more attractive to exercise the Russian option and receive the dollar amount M rather than to hold and wait. Therefore, the Russian option is alive when $S^* < S \leq M$ and will be exercised when $S \leq S^*$. The payoff function of the Russian option upon exercising is

$$V(S^*, M) = M. \tag{5.2.24}$$

Like any American option, the Russian option value stays above its exercise payoff when the option is alive.

The asset is assumed to pay a continuous dividend yield q, $q \geq 0$. The special case $q = 0$ will be considered later. By dropping the temporal derivative term in the Black–Scholes equation, the governing equation for the Russian option model is given by

$$\frac{\sigma^2}{2}S^2\frac{\partial^2 V}{\partial S^2} + (r - q)S\frac{\partial V}{\partial S} - rV = 0, \quad S^* < S < M. \tag{5.2.25}$$

The boundary condition at $S = S^*$ was given by (5.2.24). We explained in Sect. 4.2 that the lookback option value is insensitive to M when $S = M$. Therefore, the other boundary condition at $S = M$ is given by

$$\frac{\partial V}{\partial M} = 0 \quad \text{at} \quad S = M. \tag{5.2.26}$$

The optimal exercise price S^* is chosen such that the option value is maximized among all possible values of S^*. The governing equation and boundary conditions can be recast in a more succinct form when the similarity variables

$$W = V/M \quad \text{and} \quad \xi = S/M \tag{5.2.27}$$

are employed. In terms of the new similarity variables, the value of the Russian option is governed by

$$\frac{\sigma^2}{2}\xi^2\frac{d^2W}{d\xi^2} + (r-q)\xi\frac{dW}{d\xi} - rW = 0, \quad \xi^* < \xi < 1, \tag{5.2.28}$$

where $W = W(\xi)$ and $\xi^* = S^*/M$. The boundary conditions become

$$\frac{dW}{d\xi} = W \quad \text{at} \quad \xi = 1, \tag{5.2.29a}$$

$$W = 1 \quad \text{at} \quad \xi = \xi^*. \tag{5.2.29b}$$

First, we solve for the option value in terms of ξ^*, then determine ξ^* such that the option value is maximized. By substituting the assumed form of the solution $A\xi^\lambda$ into (5.2.28), we observe that λ should satisfy the following quadratic equation:

$$\frac{\sigma^2}{2}\lambda(\lambda-1) + (r-q)\lambda - r = 0. \tag{5.2.30}$$

The two roots of the above quadratic equation are

$$\lambda_\pm = \frac{1}{\sigma^2}\left[-(r-q) + \frac{\sigma^2}{2} \pm \sqrt{\left(r-q-\frac{\sigma^2}{2}\right)^2 + 2\sigma^2 r}\,\right],$$

where $\lambda_+ > 0$ and $\lambda_- < 0$. The general solution to (5.2.28) can be expressed as

$$W(\xi) = A_+\xi^{\lambda_+} + A_-\xi^{\lambda_-}, \quad \xi^* < \xi < 1,$$

where A_+ and A_- are arbitrary constants. Applying the boundary conditions (5.2.29a,b), the solution for $W(\xi)$ is found to be

$$W(\xi) = \frac{(1-\lambda_-)\xi^{\lambda_+} - (1-\lambda_+)\xi^{\lambda_-}}{(1-\lambda_-)\xi^{*\lambda_+} - (1-\lambda_+)\xi^{*\lambda_-}}, \quad \xi^* \le \xi \le 1. \tag{5.2.31}$$

The use of calculus reveals that $W(\xi)$ is maximized when ξ^* is chosen to be

$$\xi^* = \left[\frac{\lambda_+(1-\lambda_-)}{\lambda_-(1-\lambda_+)}\right]^{1/(\lambda_- - \lambda_+)}. \tag{5.2.32}$$

Besides the above differential equation approach, one may apply the martingale pricing approach to derive the price formula for the Russian option. Interested readers should see Shepp and Shiryaev (1993) and Gerber and Shiu (1994) for details.

Nondividend Paying Underlying Asset

How does the price function of the Russian option behave when $q = 0$? The two roots then become $\lambda_+ = 1$ and $\lambda_- = -\frac{2r}{\sigma^2}$. The solution for $W(\xi)$ is reduced to

$$W(\xi) = \frac{\xi}{\xi^*}, \quad \xi^* \leq \xi \leq 1, \tag{5.2.33}$$

which is maximized when ξ^* is chosen to be zero. The Russian option value becomes infinite when the underlying asset is nondividend paying. Can you provide a financial argument for the result?

5.3 Analytic Approximation Methods

Except for a few special cases—like the American call on an asset with no dividend or discrete dividends and the perpetual American options—analytic price formulas do not exist for most types of finite-lived American options. In this section, we present three effective analytic approximation methods for finding the American option values and the associated optimal exercise boundaries.

The *compound option approximation method* treats an American option as a compound option by limiting the opportunity set of optimal exercises to only a few discrete times rather than at any time during the life of the option. The compound option approach requires the valuation of multivariate normal integrals in the corresponding approximation formulas, where the dimension of the multivariate integrals is the same as the number of exercise opportunities allowed. We have seen that one may express the early exercise premium in terms of the optimal exercise boundary in an integral representation. This naturally leads to an integral equation for the optimal exercise boundary. The *recursive integration method* considers the direct solution of the integral equation for the early exercise boundary by recursive iterations. The iterative algorithm only involves computation of one-dimensional integrals. Even when we take only a few points on the optimal exercise boundary, the numerical accuracy of both the compound option method and recursive integration method can be improved quite effectively by an extrapolation procedure. The *quadratic approximation method* employs an ingenious transformation of the Black–Scholes equation so that the temporal derivative term can be considered as a quadratic small term and then dropped as an approximation. Once the approximate ordinary differential equation is derived, we only need to determine one optimal exercise point rather than the solution of the whole optimal exercise curve as in the original partial differential equation formulation.

It is commonly observed that American option values are not too sensitive to the location of the optimal exercise boundary. This may explain why the above analytic approximation methods are quite accurate in calculating the American option values even when only a few points on the optimal exercise boundary are estimated. Evaluation of these analytic approximation formulas normally requires the use of a computer, some of them even require further numerical procedures, like numerical approximation of integrals, iteration and extrapolation. However, they do distinguish from direct numerical methods like the binomial method, finite difference method and Monte Carlo simulation (these numerical methods are discussed in full detail in the next chapter). In the process of deriving the analytic approximation methods, the analytic properties of the American option model are fully explored and ingenious approximations are subsequently applied to reduce the complexity of the problems.

5.3.1 Compound Option Approximation Method

An American option contract normally allows for early exercise at any time prior to expiration. However, by limiting the early exercise privilege to commence only at a few predetermined instants between now and expiration, the American option then resembles a compound option. It then becomes plausible to derive the corresponding analytic price formulas. The approximate price formula will converge to the price formula of the American option in the limit when the number of exercisable instants grows to infinity since the continuously exercisable property of the American option is then recovered.

First, we derive the formula for a limited exercisable American put option on a nondividend paying asset where early exercise can only occur at a single instant which is halfway to expiration. Let the current time be zero and T be the expiration time. Let $S_{T/2}$ and S_T denote the asset price at times $T/2$ and T, respectively. Between time $T/2$ to the expiration date, the option behaves like an ordinary European option since there is no early exercise privilege. We determine the critical asset price $S^*_{T/2}$ at $T/2$ such that it is indifferent between exercising the put or otherwise at the asset price $S^*_{T/2}$. Accordingly, $S^*_{T/2}$ is obtained by solving the following nonlinear algebraic equation

$$p(S^*_{T/2}, T/2; X) = X - S^*_{T/2}, \tag{5.3.1}$$

where X is the strike price of the put. Here, $p(S^*_{T/2}, T/2; X)$ is the Black–Scholes price formula for a European put, with $\tau = T/2$.

When $S_{T/2} \leq S^*_{T/2}$, the put option will be exercised with payoff $X - S_{T/2}$. The discounted expectation of $X - S_{T/2}$, conditional on $S_{T/2} \leq S^*_{T/2}$, is found to be

$$\begin{aligned} & e^{-rT/2} \int_0^{S^*_{T/2}} (X - S_{T/2}) \psi(S_{T/2}; S)\, dS_{T/2} \\ &= Xe^{-rT/2} N(-d_2(S, S^*_{T/2}; T/2)) - SN(-d_1(S, S^*_{T/2}; T/2)), \end{aligned} \tag{5.3.2}$$

where

$$d_1(S_1, S_2; T) = \frac{\ln \frac{S_1}{S_2} + \left(r + \frac{\sigma^2}{2}\right) T}{\sigma\sqrt{T}}, \quad d_2(S_1, S_2; T) = d_1(S_1, S_2; T) - \sigma\sqrt{T},$$

and $\psi(S_{T/2}; S)$ is the transition density function. On the other hand, when $S_{T/2} > S^*_{T/2}$, the put option survives until expiry. At expiry, it will be exercised only when $S_T < X$. The discounted expectation of $X - S_T$, conditional on $S_{T/2} > S^*_{T/2}$ and $S_T < X$, is given by

$$\begin{aligned} &e^{-rT} \int_{S^*_{T/2}}^{\infty} \int_0^X (X - S_T)\psi(S_T; S_{T/2})\psi(S_{T/2}; S)\, dS_T dS_{T/2} \\ &= Xe^{-rT} N_2\left(d_2(S, S^*_{T/2}; T/2), -d_2(S, X; T); -1/\sqrt{2}\right) \\ &\quad - SN_2\left(d_1(S, S^*_{T/2}; T/2), -d_1(S, X; T); -1/\sqrt{2}\right). \end{aligned} \tag{5.3.3}$$

Note that the correlation coefficient between overlapping Brownian increments over the time intervals $[0, T/2]$ and $[0, T]$ is found to be $1/\sqrt{2}$. The price of the put option with two exercisable instants $T/2$ and T is given by the sum of the these two expectation integrals. We then have

$$\begin{aligned} P_2(S, X; T) = {} & Xe^{-rT/2}N(-d_2(S, S^*_{T/2}; T/2)) - SN(-d_1(S, S^*_{T/2}; T/2)) \\ & + Xe^{-rT} N_2\big(d_2(S, S^*_{T/2}; T/2), -d_2(S, X; T); -1/\sqrt{2}\big) \\ & - SN_2\big(d_1(S, S^*_{T/2}; T/2), -d_1(S, X; T); -1/\sqrt{2}\big). \end{aligned} \tag{5.3.4}$$

Extension to the general case with N exercisable instants (not necessarily equally spaced) can also be derived in a similar manner (see Problem 5.30).

Let P_n denote the value of the put option with n exercisable instants. We expect that the limit of the sequence $P_1, P_2, \cdots, P_n, \cdots$ tends to the American put value. One may apply the acceleration technique to extrapolate the limit based on the first few members of the sequence. Geske and Johnson (1984) proposed the following Richardson extrapolation scheme when $n = 3$

$$P \approx \frac{9P_3 - 8P_2 + P_1}{2}. \tag{5.3.5}$$

Judging from their numerical experiments, reasonable accuracy is observed for most cases based on extrapolation formula (5.3.5). Improved accuracy can be achieved by relaxing the requirement of equally spaced exercisable instants and seeking for appropriate exercisable instants such that the approximate put value is maximized (Bunch and Johnson, 1992).

5.3.2 Numerical Solution of the Integral Equation

In the integral equation for the optimal exercise boundary for an American put option [see (5.2.19)], the variable τ appears both in the integrand and the upper limit of the integral. A recursive scheme can be derived to solve the integral equation for a given value of τ. In the numerical procedure, all integrals are approximated by the trapezoidal rule. First, we divide τ into n equally spaced subintervals with end points $\tau_i, i = 0, 1, \cdots, n$, where $\tau_0 = 0, \tau_n = \tau$ and $\Delta\tau = \tau/n$. For convenience, we denote the integrand function by

$$f(S^*(\tau), S^*(\tau - \xi); \tau, \xi) = rXe^{-r\xi}N(-\widehat{d}_{\xi,2}) - qS^*(\tau)e^{-q\xi}N(-\widehat{d}_{\xi,1}), \quad (5.3.6)$$

where

$$\widehat{d}_{\xi,1} = \frac{\ln\frac{S^*(\tau)}{S^*(\tau-\xi)} + \left(r - q + \frac{\sigma^2}{2}\right)\xi}{\sigma\sqrt{\xi}}, \quad \widehat{d}_{\xi,2} = \widehat{d}_{\xi,1} - \sigma\sqrt{\xi}.$$

Let S_i^* denote the numerical approximation to $S^*(\tau_i), i = 0, 1, \cdots, n$. Setting $\tau = \tau_1$ in the integral equation and approximating the integral by

$$\begin{aligned}&\int_0^{\tau_1} \left[rXe^{-r\xi}N(-\widehat{d}_{\xi,2}) - qS^*(\tau)e^{-q\xi}N(-\widehat{d}_{\xi,1})\right] d\xi \\ &\approx \frac{\Delta\tau}{2}\left[f(S_1^*, S_1^*; \tau_1, \tau_0) + f(S_1^*, S_0^*; \tau_1, \tau_1)\right], \end{aligned} \quad (5.3.7)$$

we obtain the following nonlinear algebraic equation for S_1^*:

$$X - S_1^* = p(S_1^*, \tau_1) + \frac{\Delta\tau}{2}\left[f(S_1^*, S_1^*; \tau_1, \tau_0) + f(S_1^*, S_0^*; \tau_1, \tau_1)\right]. \quad (5.3.8)$$

Since S_0^* is known to be $\min(X, \frac{r}{q}X)$, one can solve for S_1^* by any root-finding method. Once S_1^* is known, we proceed to set $\tau = \tau_2$ and approximate the integral over the two subintervals: (τ_0, τ_1) and (τ_1, τ_2). The corresponding nonlinear algebraic equation for S_2^* is then given by

$$\begin{aligned}X - S_2^* = p(S_2^*, \tau_2) + \frac{\Delta\tau}{2} \; &[f(S_2^*, S_2^*; \tau_2, \tau_0) + 2f(S_2^*, S_1^*; \tau_2, \tau_1) \\ &+ f(S_2^*, S_0^*; \tau_2, \tau_2)]. \end{aligned} \quad (5.3.9)$$

Recursively, the general algebraic equation for $S_k^*, k = 2, 3, \cdots, n$ can be deduced to be (Huang, Subrahmanyam and Yu, 1996)

$$\begin{aligned}X - S_k^* = p(S_k^*, \tau_k) + \frac{\Delta\tau}{2}\Bigg[&f\ (S_k^*, S_k^*; \tau_k, \tau_0) + f(S_k^*, S_0^*; \tau_k, \tau_k) \\ &+ 2\sum_{i=1}^{k-1} f(S_k^*, S_{k-i}^*; \tau_k, \tau_i)\Bigg], k = 2, 3, \cdots, n, \end{aligned} \quad (5.3.10)$$

where $S_k^*, k = 1, 2, \cdots, n$, are solved sequentially. By choosing n to be sufficiently large, the optimal exercise boundary $S^*(\tau)$ can be approximated to sufficient accuracy as desired.

Once S_k^*, $k = 1, 2, \cdots, n$, are known, the American put value can be approximated by

$$P(S,\tau) \approx P_n = p(S,\tau) + \frac{\Delta\tau}{2}\left[f\ (S, S_n^*; \tau_n, \tau_0) + f(S, S_0^*; \tau_n, \tau_n) + 2\sum_{i=1}^{n-1} f(S, S_{n-i}^*; \tau_n, \tau_i)\right], \tag{5.3.11}$$

where $\tau = \tau_n$. Obviously, the limit of P_n tends to $P(S,\tau)$ as n tends to infinity. Similar to the compound option approximation method, one may apply the following extrapolation scheme

$$P(S,\tau) \approx \frac{9P_3 - 8P_2 + P_1}{2}, \tag{5.3.12}$$

where P_n is defined in (5.3.11). The numerical procedure of the recursive integration method is seen to be much less tedious compared to the compound option approximation method since only one-dimensional integrals are involved. Various versions of numerical schemes for more effective numerical valuation of American option values have been reported in the literature. For example, Ju (1998) proposed pricing an American option by approximating its optimal exercise boundary as a multipiece exponential function. The method is claimed to have the advantage of easy implementation since closed form formulas can be obtained in terms of the bases and exponents of the multipiece exponential function.

One advantage of the recursive integration method is that the Greeks of the American option values can also be found effectively without much additional effort. For example, from the following formula for the delta of the American option price (see Problem 5.20):

$$\Delta = \frac{\partial P}{\partial S} = -N(-d_1) - \int_0^\tau \left[\frac{(r-q)e^{-q\xi}}{\sigma\sqrt{2\pi\xi}} e^{-\frac{d_{\xi,1}^2}{2}} + qe^{-q\xi}N(-d_{\xi,1})\right] d\xi, \tag{5.3.13}$$

one can easily deduce the numerical approximation to the delta Δ by approximating the above integral using the trapezoidal rule as follows:

$$\Delta \approx \Delta_n = -N(-d_1) - \frac{\Delta\tau}{2}\left[g(S, S_n^*;\ \tau_n, \tau_0) + g(S, S_0^*; \tau_n, \tau_n) + 2\sum_{i=1}^{n-1} g(S, S_{n-i}^*; \tau_n, \tau_i)\right], \tag{5.3.14}$$

where

$$g(S, S^*(\tau-\xi); \tau, \xi) = \frac{(r-q)e^{-q\xi}}{\sigma\sqrt{2\pi\xi}} e^{-\frac{d_{\xi,1}^2}{2}} + qe^{-q\xi}N(-d_{\xi,1})$$

$$d_{\xi,1} = \frac{\ln\frac{S}{S^*(\tau-\xi)} + \left(r - q + \frac{\sigma^2}{2}\right)\xi}{\sigma\sqrt{\xi}}.$$

5.3.3 Quadratic Approximation Method

The quadratic approximation method was first proposed by MacMillan (1986) for nondividend paying stock options and later extended to commodity options by Barone-Adesi and Whaley (1987). This method has been proven to be quite efficient with reasonably good accuracy for valuation of American options, particularly for shorter lived options.

The governing equation for the price of a commodity option with a constant cost of carry b and riskless interest rate r is given by

$$\frac{\partial V}{\partial \tau} = \frac{\sigma^2}{2} S^2 \frac{\partial^2 V}{\partial S^2} + bS \frac{\partial V}{\partial S} - rV, \tag{5.3.15}$$

where σ is the constant volatility of the asset price. We consider an American call option written on a commodity and define the early exercise premium by

$$e(S, \tau) = C(S, \tau) - c(S, \tau).$$

Inside the continuation region, (5.3.15) holds for both $C(S, \tau)$ and $c(S, \tau)$. Since the differential equation is linear, the same equation holds for $e(S, \tau)$. By writing $k_1 = 2r/\sigma^2$ and $k_2 = 2b/\sigma^2$, and defining

$$e(S, \tau) = K(\tau) f(S, K),$$

where $K(\tau)$ will be determined. Now, (5.3.15) can be transformed into the form

$$S^2 \frac{\partial^2 f}{\partial S^2} + k_2 S \frac{\partial f}{\partial S} - k_1 f \left[1 + \frac{\frac{dK}{d\tau}}{rK} \left(1 + \frac{K \frac{\partial f}{\partial K}}{f} \right) \right] = 0. \tag{5.3.16}$$

A judicious choice for $K(\tau)$ is

$$K(\tau) = 1 - e^{-r\tau},$$

so that (5.3.16) becomes

$$S^2 \frac{\partial^2 f}{\partial S^2} + k_2 S \frac{\partial f}{\partial S} - \frac{k_1}{K} \left[f + (1 - K) K \frac{\partial f}{\partial K} \right] = 0. \tag{5.3.17}$$

Note that the last term in the above equation contains the factor $(1 - K)K$, and it becomes zero at $\tau = 0$ and $\tau \to \infty$. Further, it has a maximum value of $1/4$ at $K = 1/2$. Suppose we drop the quadratic term $(1 - K)K \frac{\partial f}{\partial K}$, (5.3.17) is then reduced to an ordinary differential equation with the error being controlled by the magnitude of the quadratic term $(1 - K)K$. This is how the name of this approximation method is derived. The approximate equation for f now becomes

$$S^2 \frac{\partial^2 f}{\partial S^2} + k_2 S \frac{\partial f}{\partial S} - \frac{k_1}{K} f = 0, \tag{5.3.18}$$

where K is assumed to be nonzero. The special case where $K = 0$ can be considered separately (see Problem 5.31).

When K is treated as a parameter, (5.3.18) becomes an equi-dimensional differential equation. The general solution for $f(S)$ is given by

$$f(S) = c_1 S^{q_1} + c_2 S^{q_2}, \tag{5.3.19}$$

where c_1 and c_2 are arbitrary constants, q_1 and q_2 are roots of the auxiliary equation

$$q^2 + (k_2 - 1)q - \frac{k_1}{K} = 0. \tag{5.3.20}$$

Solving the above quadratic equation, we obtain

$$q_1 = -\frac{1}{2}\left[(k_2 - 1) + \sqrt{(k_2 - 1)^2 + 4\frac{k_1}{K}}\right] < 0, \tag{5.3.21a}$$

$$q_2 = \frac{1}{2}\left[-(k_2 - 1) + \sqrt{(k_2 - 1)^2 + 4\frac{k_1}{K}}\right] > 0. \tag{5.3.21b}$$

The term $c_1 S^{q_1}$ in (5.3.19) should be discarded since $f(S)$ tends to zero as S approaches 0. The approximate value $\widetilde{C}(S, \tau)$ of the American call option is then given by

$$C(S, \tau) \approx \widetilde{C}(S, \tau) = c(S, \tau) + c_2 K S^{q_2}. \tag{5.3.22}$$

Finally, the arbitrary constant c_2 is determined by applying the value matching condition at the critical asset value S^*, namely, $\widetilde{C}(S^*, \tau) = S^* - X$. However, S^* itself is not yet known. The additional equation required to determine S^* is provided by the smooth pasting condition $\frac{\partial \widetilde{C}}{\partial S}(S^*, \tau) = 1$ along the optimal exercise boundary. These two conditions together lead to the following pair of equations for c_2 and S^*

$$S^* - X = c(S^*, \tau) + c_2 K S^{*q_2} \tag{5.3.23a}$$

$$1 = e^{(b-r)\tau} N(d_1(S^*)) + c_2 K q_2 S^{*q_2 - 1}, \tag{5.3.23b}$$

where

$$d_1(S^*) = \frac{\ln \frac{S^*}{X} + \left(b + \frac{\sigma^2}{2}\right)\tau}{\sigma\sqrt{\tau}}.$$

By eliminating c_2 in (5.3.23a,b), we obtain the following nonlinear algebraic equation for $S^*(\tau)$:

$$S^* - X = c(S^*, \tau) + \left[1 - e^{(b-r)\tau} N(d_1(S^*))\right]\frac{S^*}{q_2}. \tag{5.3.24}$$

In summary, for $b < r$, the approximate value of the American commodity call option can be expressed as

$$\widetilde{C}(S, \tau) = c(S, \tau) + \frac{S^*}{q_2}\left[1 - e^{(b-r)\tau} N(d_1(S^*))\right]\left(\frac{S}{S^*}\right)^{q_2}, \quad S < S^*, \tag{5.3.25}$$

where S^* is obtained by solving (5.3.24). The last term in (5.3.25) gives an approximate value for the early exercise premium, which can be shown to be positive for $b < r$. When $b \geq r$, the American call will never be exercised prematurely (see Problem 5.2) so that the American call option value is the same as that of its European counterpart.

5.4 Options with Voluntary Reset Rights

The reset right embedded in a financial derivative refers to the privilege given to the derivative holder to reset certain terms in the contract according to some specified rules. The reset may be done on the strike price or the maturity date of the derivative or both. The number of resets allowed within the life of the contract may be more than one. Usually there are some predetermined conditions that have to be met in order to activate a reset. The reset may be automatic upon the fulfilment of certain conditions or activated voluntarily by the holder. In this section, we confine our discussion to options with strike reset right and the holder of which can choose optimally the reset moment. We would like to analyze the *optimal reset policies* adopted by the option holder.

We consider the reset-strike put option, where the strike price can be reset to the prevailing asset price at the reset moment. Let X denote the original strike price set at initiation of the option, S_{t^*} and S_T denote the asset price at the reset date t^* and expiration date T, respectively. Suppose there is only one reset right allowed, the terminal payoff of the reset put option is given by $\max(X - S_T, 0)$ if no reset occurs throughout the option's life, and modified to $\max(S_{t^*} - S_T, 0)$ if the reset occurs at time $t^* < T$. Upon reset, the reset-strike put option effectively becomes an at-the-money put option.

The shout options are closely related to the reset-strike put options. Consider the shout option with the call payoff with only one shout right. Suppose the holder has chosen to shout at time t^*, then the terminal payoff is guaranteed to have the floor value $S_{t^*} - X$. More precisely, the terminal payoff is given by $\max(S_T - X, S_{t^*} - X)$ if the holder has shouted at t^* prior to maturity, but stays at the usual call payoff $\max(S_T - X, 0)$ if no shout occurs throughout the option's life. It will be shown later that the shout call can be replicated by a reset-strike put and a forward so that the reset-strike put option and its shout call counterpart follow the same optimal stopping policy [see (5.4.11)].

Another example of reset right is the *shout floor* feature in an index fund with a protective floor. Essentially, the shout floor feature gives the holder the right to shout at any time during the life of the contract to receive an at-the-money put option. In Sect. 5.4.1, we show how to obtain the closed form price formula of the shout floor feature (Dai, Kwok and Wu, 2004). A similar feature of fund value protection can be found in equity-linked annuities. For example, the dynamic fund protection embedded in an investment fund provides a floor level of protection against a reference stock index, where the investor has the right to reset the fund value to that of the reference stock index. The protected fund may allow a finite number of resets throughout the life of the fund. The reset instants can be chosen optimally by the investor. The fund holder also has the right to withdraw the fund prematurely. Details of the pricing of the reset and withdrawal rights in a dynamic fund protection can be found in Chu and Kwok (2004, Chap. 4).

There are a wide variety of derivative instruments in the financial markets with embedded reset features. For example, the Canadian segregated funds are mutual fund investments embedded with a long-term maturity guarantee. These fund contracts contain multiple reset options that allow the holder to reset the guarantee level and the maturity date during the life of the contract. The optimal reset policies of options with combined reset rights on strike and maturity were analyzed in details by Dai and Kwok (2005a, Chap. 3).

5.4.1 Valuation of the Shout Floor

The shout floor feature in an index fund gives the holder the right to shout at any time during the life of the contract to install a floor on the return of the fund, where the floor value is set at the prevailing index value S_{t^*} at the shouting time t^*. This shout floor feature gives the fund holder the upside potential of the index fund, while it also provides a guarantee on the return of the index at the floor value. In essence, the holder receives an at-the-money put option at the shout moment. By virtue of the guarantee on the return, the holder has the right to sell the index fund for the floor value at maturity of the contract. If no shout occurs throughout the life of the contract, then the fund value becomes zero. In summary, the terminal payoff of the shout floor is

$$\begin{cases} \max(S_{t^*} - S_T, 0) & \text{if shout has occurred} \\ 0 & \text{if no shout has occurred,} \end{cases}$$

where S_{t^*} and S_T are the index value at the shout moment t^* and maturity date T, respectively.

Formulation as a Free Boundary Value Problem

Interestingly, a closed form price formula of the shout floor feature under the usual Black–Scholes pricing framework can be obtained. As usual, the stochastic process for the index value S_t under the risk neutral measure Q is assumed to follow the Geometric Brownian process

$$\frac{dS_t}{S_t} = (r - q)dt + \sigma\, dZ_t, \tag{5.4.1}$$

where r and q are the constant riskless interest rate and dividend yield, respectively, and σ is the constant volatility.

Let $V(S, \tau)$ denote the value of the shout floor feature. At the shout moment, the shout floor right is transformed into the ownership of an at-the-money European put option. The price function of an at-the-money put option is seen to be linearly homogeneous in S, which can be written as $Sp^*(\tau)$. By setting the strike price be the current asset price in the Black–Scholes put option price formula, we obtain

$$p^*(\tau) = e^{-r\tau} N(-d_2^*) - e^{-q\tau} N(-d_1^*), \tag{5.4.2}$$

where

$$d_1^* = \frac{r - q + \frac{\sigma^2}{2}}{\sigma}\sqrt{\tau} \quad \text{and} \quad d_2^* = d_1^* - \sigma\sqrt{\tau}.$$

The linear complementarity formulation of the free boundary value problem for the shout floor feature takes a form similar to that of an American option. Recalling that the exercise payoff is $Sp^*(\tau)$ and the terminal payoff is zero, we obtain the following linear complementarity formulation for $V(S, \tau)$:

$$\begin{aligned}
&\frac{\partial V}{\partial \tau} - \frac{\sigma^2}{2}S^2\frac{\partial^2 V}{\partial S^2} - (r-q)S\frac{\partial V}{\partial S} + rV \geq 0, \qquad V \geq Sp^*(\tau), \\
&\left[\frac{\partial V}{\partial \tau} - \frac{\sigma^2}{2}S^2\frac{\partial^2 V}{\partial S^2} - (r-q)S\frac{\partial V}{\partial S} + rV\right]\left[V - Sp^*(\tau)\right] = 0, \\
&V(S, 0) = 0.
\end{aligned} \tag{5.4.3}$$

Since there is no strike price X appearing in the shout floor payoff, the pricing function $V(S, \tau)$ then becomes linearly homogeneous in S. We may write $V(S, \tau) = Sg(\tau)$, where $g(\tau)$ is to be determined. By substituting this assumed form of $V(S, \tau)$ into (5.4.3), we obtain the following set of variational inequalities for $g(\tau)$:

$$\begin{aligned}
&\frac{d}{d\tau}\left[e^{q\tau}g(\tau)\right] \geq 0, \quad g(\tau) \geq p^*(\tau), \\
&\frac{d}{d\tau}\left[e^{q\tau}g(\tau)\right]\left[g(\tau) - p^*(\tau)\right] = 0, \\
&g(0) = 0.
\end{aligned} \tag{5.4.4}$$

The form of solution for $g(\tau)$ depends on the analytic properties of the function $e^{q\tau}p^*(\tau)$. The derivative of $e^{q\tau}p^*(\tau)$ observes the following properties:

(i) If $r \leq q$, then

$$\frac{d}{d\tau}\left[e^{q\tau}p^*(\tau)\right] > 0 \quad \text{for} \quad \tau \in (0, \infty). \tag{5.4.5}$$

(ii) If $r > q$, then there exists a unique critical value $\tau^* \in (0, \infty)$ such that

$$\left.\frac{d}{d\tau}\left[e^{q\tau}p^*(\tau)\right]\right|_{\tau=\tau^*} = 0, \tag{5.4.6a}$$

and

$$\frac{d}{d\tau}\left[e^{q\tau}p^*(\tau)\right] > 0 \quad \text{for} \quad \tau \in (0, \tau^*), \tag{5.4.6b}$$

$$\frac{d}{d\tau}\left[e^{q\tau}p^*(\tau)\right] < 0 \quad \text{for} \quad \tau \in (\tau^*, \infty). \tag{5.4.6c}$$

The hints for the proof of these properties are given in Problem 5.34 [also see Dai, Kwok and Wu, 2004]. The schematic plots of $e^{q\tau}p^*(\tau)$ are shown in Fig. 5.8 for both cases: $r \leq q$ and $r > q$.

The price function $V(S, \tau)$ of the shout floor takes different analytic forms, depending on $r \leq q$ or $r > q$.

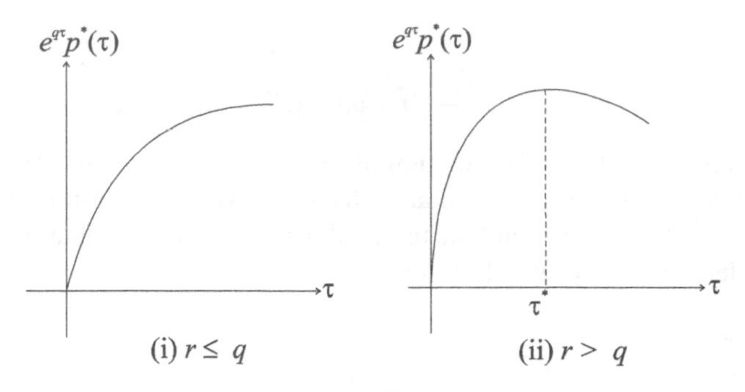

Fig. 5.8. Properties of the function $e^{q\tau} p^*(\tau)$ under (i) $r \leq q$, (ii) $r > q$.

(i) $r \leq q$
By (5.4.5), $\frac{d}{d\tau}[e^{q\tau} p^*(\tau)]$ is strictly positive for all $\tau > 0$ and $p^*(0) = 0$. One then deduces that the solution to $g(\tau)$ is given by

$$g(\tau) = p^*(\tau), \qquad \tau \in (0, \infty). \tag{5.4.7}$$

(ii) $r > q$
By (5.4.6b,c), in a similar manner we obtain

$$g(\tau) = p^*(\tau) \quad \text{for} \quad \tau \in (0, \tau^*]. \tag{5.4.8}$$

However, when $\tau > \tau^*$, we cannot have $g(\tau) = p^*(\tau)$ since this would lead to $\frac{d}{d\tau}[e^{q\tau} g(\tau)] = \frac{d}{d\tau}[e^{q\tau} p^*(\tau)] \geq 0$, contradicting the result in (5.4.6c). By (5.4.4), we must have $\frac{d}{d\tau}[e^{q\tau} g(\tau)] = 0$ for $\tau \in (\tau^*, \infty)$. Together with the auxiliary condition: $g(\tau^*) = p^*(\tau^*)$, the solution is given by

$$g(\tau) = e^{-q(\tau-\tau^*)} p^*(\tau^*) \quad \text{for} \quad \tau \in (\tau^*, \infty). \tag{5.4.9}$$

In summary, the optimal shouting policy adopted by the holder of the shout floor depends on the relative magnitude of r and q. When $r \leq q$, the holder should shout at once—at any time and at any index value level—to install the protective floor. When $r > q$, there exists a critical time earlier than which it is never optimal for the holder to shout. The holder should shout at once at any index value level once τ falls to the critical value τ^*.

5.4.2 Reset-Strike Put Options

The reset feature embedded in the reset-strike put option allows the holder to reset the original strike price to the prevailing asset price at any reset moment chosen by the holder. The reset-strike put is very similar to the shout floor since the holder receives an at-the-money put option upon reset, except that the reset-strike put has an *initial strike price* X at which reset does not occur throughout the life of the contract. Similar to (5.4.3), the linear complementarity formulation for the price function $U(S, \tau)$ of the reset-strike put option is given by

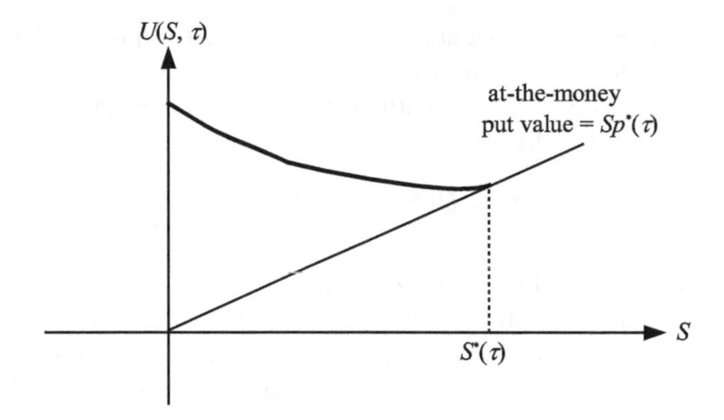

Fig. 5.9. The price curve of the reset-strike put touches tangentially the line representing the at-the-money put value at $S = S^*(\tau)$.

$$\begin{aligned}
&\frac{\partial U}{\partial \tau} - \frac{\sigma^2}{2}S^2\frac{\partial^2 U}{\partial S^2} - (r-q)S\frac{\partial U}{\partial S} + rU \geq 0 \quad U \geq Sp^*(\tau),\\
&\left[\frac{\partial U}{\partial \tau} - \frac{\sigma^2}{2}S^2\frac{\partial^2 U}{\partial S^2} - (r-q)S\frac{\partial U}{\partial S} + rU\right][U - Sp^*(\tau)] = 0, \qquad (5.4.10)\\
&\quad U(S,0) = \max(X - S, 0).
\end{aligned}$$

Unlike the shout floor, the terminal payoff of the reset-strike put option contains the initial strike price X. Now, $U(S, \tau)$ is no longer linear homogeneous in S. The holder should shout to install a new strike only when the asset price reaches some sufficiently high critical level $S^*(\tau)$. Obviously, $S^*(\tau)$ must be greater than the initial strike price X. Similar to the American option models, the optimal reset boundary is not known a priori but has to be solved as part of the above free boundary value problem. Similar to the American option models, the price function $U(S, \tau)$ observes the value matching and smooth pasting conditions, namely,

$$U(S^*(\tau), \tau) = Sp^*(\tau), \qquad (5.4.11a)$$

$$\frac{\partial U}{\partial S}(S^*(\tau), \tau) = p^*(\tau). \qquad (5.4.11b)$$

A schematic plot of $U(S, \tau)$ against S is shown in Fig. 5.9.

Parity Relation between Reset-Strike Put and Shout Call

Consider the portfolio of holding a reset-strike put and a forward contract. Both derivatives have the same maturity date, and the forward price is taken to be the same as the strike price. The terminal payoff of this portfolio is given by

$$\begin{cases} \max(X - S_T, 0) + S_T - X = \max(S_T - X, 0) & \text{if no reset occurs} \\ \max(S_{t^*} - S_T, 0) + S_T - X = \max(S_T - X, S_{t^*} - X) & \text{if reset occurs} \end{cases}.$$

Here, S_{t^*} is the prevailing asset price at the reset moment t^*. The above payoff structure is identical to that of a shout call. Hence, the shout call can be replicated by a combination of a reset-strike put and a forward. As a consequence, the reset-strike put and shout call should share the same optimal reset/shout policy. Let $W(S, \tau)$ denote the price of the shout call. The parity relation between the prices of reset-strike put and shout call is given by

$$W(S, \tau) = U(S, \tau) + Se^{-q\tau} - Xe^{-r\tau}. \tag{5.4.12}$$

Characterization of the Optimal Reset Policy

We examine the characterization of the optimal reset boundary $S^*(\tau)$ of the strike-reset put option, in particular, the asymptotic behavior at $\tau \to 0^+$ and $\tau \to \infty$. Since the new strike price upon reset should not be lower than the original strike price, we should have

$$S^*(\tau) \geq X. \tag{5.4.13}$$

Similar to the American call, $S^*(\tau)$ of the reset-strike put is monotonically increasing with respect to τ. Unlike the American call, $S^*(\tau)$ always starts at X at $\tau \to 0^+$, independent of r and q. To show the claim, we define

$$D(S, \tau) = U(S, \tau) - Sp^*(\tau)$$

and note that $D(S, \tau) \geq 0$ for all S and τ. In the continuation region, $D(S, \tau)$ satisfies

$$\frac{\partial D}{\partial \tau} - \frac{\sigma^2}{2} S^2 \frac{\partial^2 D}{\partial S^2} - (r - q)S\frac{\partial D}{\partial S} + rD = -S[p^{*\prime}(\tau) + qp^*(\tau)],$$
$$0 < S < S^*(\tau), \quad \tau > 0. \tag{5.4.14}$$

Recall that $Sp^*(\tau)$ is the value of the at-the-money put. At $\tau \to 0^+$, $\frac{\partial}{\partial \tau}[Sp^*(\tau)]$ is seen to tend to a large negative value. Hence, we have

$$-S[p^{*\prime}(\tau) + qp^*(\tau)] \to \infty$$

as $\tau \to 0^+$.

Supposing $S^*(0^+) > X$ and considering $S \in (X, S^*(0^+))$. The value of the reset put tends to its exercise value as $\tau \to 0^+$. Hence, we have $D(S, 0^+) = 0$ so that

$$\frac{\partial D}{\partial \tau}(S, 0^+) = -S[p^{*\prime}(0^+) + qp(0^+)] < 0. \tag{5.4.15}$$

This would imply $D(S, 0^+) < 0$, a contradiction to $D(S, \tau) \geq 0$ for all τ. Hence, we must have $S^*(0^+) \leq X$. Together with (5.4.12), we conclude that $S^*(0^+) = X$.

Next, we examine the asymptotic behavior of $S^*(\tau)$ at $\tau \to \infty$. Let $W^\infty(S) = \lim_{\tau \to \infty} e^{r\tau} U(S, \tau)$. The existence of $W^\infty(S)$ requires the existence of $\lim_{\tau \to \infty} e^{r\tau} p^*(\tau)$. It can be shown that when $r \leq q$, we have

$$\lim_{\tau \to \infty} e^{r\tau} p^*(\tau) = 1, \tag{5.4.16}$$

while the limit does not exist when $r > q$. The governing differential equation formulation for $W^\infty(S)$ is given by

$$\frac{\sigma^2}{2}S^2\frac{d^2W^\infty}{dS^2} + (r-q)S\frac{dW^\infty}{dS} = 0, \quad 0 < S < S_\infty^*,$$
$$W^\infty(0) = X, \quad W^\infty(S_\infty^*) = S_\infty^* \quad \text{and} \quad \frac{dW^\infty}{dS}(S_\infty^*) = 1. \qquad (5.4.17)$$

The solution to $W^\infty(S)$ takes the form:

$$W^\infty(S) = A + BS^{1+\alpha}, \quad \alpha = \frac{2(q-r)}{\sigma^2},$$

where A and B are arbitrary constants. The optimal reset boundary S_∞^* is determined by the smooth pasting condition $\frac{dW^\infty}{dS}(S_\infty^*) = 1$. The solution to $W^\infty(S)$ is found to be (see Problem 5.36)

$$W^\infty(S) = X + \frac{\alpha^\alpha}{(1+\alpha)^{1+\alpha}}\frac{S^{1+\alpha}}{X^\alpha}, \quad 0 < S < S_\infty^*, \qquad (5.4.18)$$

where

$$S_\infty^* = \left(1 + \frac{1}{\alpha}\right)X.$$

When $r < q$, $S^*(\tau)$ is defined for all $\tau > 0$ with the asymptotic limit $(1 + \frac{1}{\alpha})X$ at $\tau \to \infty$. In particular, S_∞^* becomes infinite when $r = q$.

Next, we consider the case $r > q$. Recall that it is never optimal to exercise the shout floor when $\tau > \tau^*$ [τ^* can be obtained by solving (5.4.6a)]. Since the reset-strike put is more expensive than the shout floor and their exercise payoffs are the same, it is never optimal to exercise the reset-strike put when $\tau > \tau^*$. We write the optimal reset boundary of the reset-strike put as $S^*(\tau; X)$, with dependence on the strike price X. When $X = 0$, it corresponds to the shout floor and $S^*(\tau; 0)$ is known to be zero. When $X \to \infty$, $S^*(\tau; \infty)$ becomes infinite since it is never optimal to reset at any asset value when the strike price is already at infinite value. One then argues that $S^*(\tau; X)$ is finite when X is finite, $\tau < \tau^*$. When $\tau \to \tau^{*-}$, $S(\tau; X)$ becomes infinite. In Fig. 5.10, we illustrate the behavior of $S^*(\tau)$ under the two separate cases: $r < q$ and $r > q$. More detailed discussion of the pricing behavior of the reset-strike put options can be found in Dai, Kwok and Wu (2004).

Multireset Put Options

We consider the pricing formulation of a put option with multiple rights to reset the strike price throughout the option's life. Let $U_n(S, \tau; X)$ denote the price function of the n-reset put option. Upon the jth reset, the reset put becomes an at-the-money $(j-1)$-reset put, where the strike price equals the prevailing asset price at the reset instant. Let t_j denote the time of the jth reset and S_j^* denote the critical asset value at the reset instant t_j^*. The strike price of the reset put with j reset rights remaining is denoted by S_{j+1}^*. For notational convenience, we write $S_{n+1}^* = X$. It is obvious that $S_{j+1}^* < S_j^*$, $j = 1, 2, \cdots, n$, and $U_{j+1}(S, \tau; X) > U_j(S, \tau; X)$ for all S and τ.

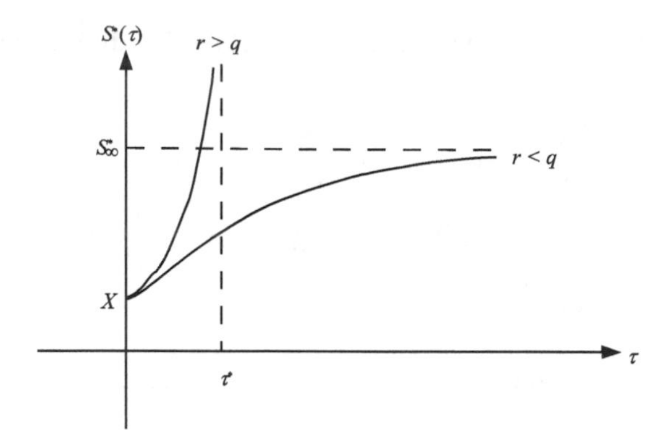

Fig. 5.10. Plot of the optimal reset boundary $S^*(\tau)$ of the reset-strike put against τ. When $r < q$, $S^*(\tau)$ is defined for all τ and a finite asymptotic limit S^*_∞ exists. When $r > q$, $S^*(\tau)$ is defined only for $\tau \in (0, \tau^*)$.

The price function $U_j(S, \tau; X)$ observes linear homogeneity in S and X

$$U_j(S, \tau; X) = XU_j\left(\frac{S}{X}, \tau; 1\right). \tag{5.4.19}$$

When the reset put is at-the-money, $S/X = 1$ and this leads to

$$U_j(S, \tau; S) = SU_j(1, \tau; 1).$$

We write $p_j(\tau) = U_j(1, \tau; 1)$, $j = 0, 1, \cdots, n-1$. The linear complementarity formulation of the pricing model of the n-reset put option is given by

$$\begin{aligned}
&\frac{\partial U_n}{\partial \tau} - \frac{\sigma^2}{2}S^2\frac{\partial^2 U_n}{\partial S^2} - (r-q)S\frac{\partial U_n}{\partial S} + rU_n \geq 0, \quad U_n \geq Sp_{n-1}(\tau),\\
&\left[\frac{\partial U_n}{\partial \tau} - \frac{\sigma^2}{2}S^2\frac{\partial^2 U_n}{\partial S^2} - (r-q)S\frac{\partial U_n}{\partial S} + rU_n\right][U_n - Sp_{n-1}(\tau)] = 0,\\
&U_n(S, 0) = \max(X - S, 0).
\end{aligned} \tag{5.4.20}$$

One has to solve recursively for U_n, starting from $U_1, U_2, \cdots$. For the perpetual n-reset strike put, it is possible to obtain the optimal reset price in closed form when $r < q$. Let $S^*_{n,\infty}$ denote $\lim_{\tau\to\infty} S^*_n(\tau)$. For $r < q$, we have

$$S^*_{n,\infty} = \left(1 + \frac{1}{\alpha}\right)\frac{X}{\beta_n}, \tag{5.4.21}$$

where $\alpha = \frac{2(q-r)}{\sigma^2}$, $\beta_1 = 1$ and

$$\beta_n = 1 + \frac{\alpha^\alpha}{(1+\alpha)^{1+\alpha}}\beta_{n-1}^{1+\alpha}.$$

The hints used to derive $S^*_{n,\infty}$ are outlined in Problem 5.36. Taking the limit $n \to \infty$, we obtain

$$\lim_{n\to\infty} \beta_n = 1 + \frac{1}{\alpha},$$

giving

$$\lim_{n\to\infty} S^*_{n,\infty} = X. \tag{5.4.22}$$

Together with the properties that $S^*_n(\tau)$ is an increasing function of τ and $S^*_n(\tau) \geq X$, we then deduce that

$$\lim_{n\to\infty} S^*_n(\tau) = X \quad \text{for all } \tau. \tag{5.4.23}$$

What is the financial interpretation of the above result? When $r < q$, the holder of an infinite-reset put should exercise the reset right whenever the option becomes in-the-money. More precisely, the holder always resets whenever a new maximum value of the asset value is realized. The terminal payoff of the infinite-reset put then becomes $\max(M_0^T - S_T, X - S_T)$, a payoff involving the lookback variable M_0^T, where $M_0^T = \max_{0\leq t\leq T} S_t$. Since the optimal reset of the infinite-reset put becomes deterministic, the pricing model of this put option is no longer a free boundary value problem. Indeed, it becomes a lookback option model (Dai, Kwok and Wu, 2003).

5.5 Problems

5.1 Find the value of an American vanilla put option when (i) riskless interest rate $r = 0$, (ii) volatility $\sigma = 0$, (iii) strike price $X = 0$, (iv) asset price $S = 0$.

5.2 Find the lower and upper bounds on the difference of the values of the American put and call options on a commodity with cost of carry b.

5.3 Consider an American call option whose underlying asset price follows a Geometric Brownian process. Show that

$$C(\lambda S, \tau) - C(S, \tau) \leq (\lambda - 1)S, \quad \lambda \geq 1.$$

5.4 Explain why an American call (put) futures option is worth more (less) than the corresponding American call (put) option on the same underlying asset when the cost of carry of the underlying asset is positive. Also, why the difference in prices widens when the maturity date of the futures goes beyond the expiration date of the option.

5.5 We would like to show by heuristic arguments that the American price function $P(S, \tau)$ satisfies the smooth pasting condition

$$\left.\frac{\partial P}{\partial S}\right|_{S=S^*(\tau)} = -1$$

at the optimal exercise price $S^*(\tau)$. Consider the behaviors of the American price curve near $S^*(\tau)$ under the following two scenarios:

(i) $\left.\frac{\partial P}{\partial S}\right|_{S=S^*(\tau)} < -1$ and (ii) $\left.\frac{\partial P}{\partial S}\right|_{S=S^*(\tau)} > -1$.

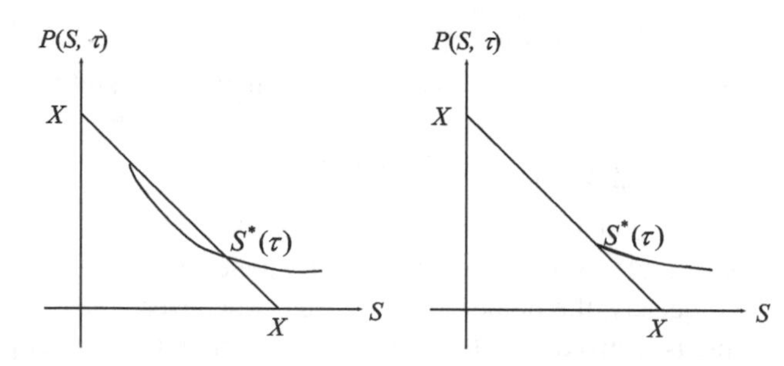

(a) When $\frac{\partial P}{\partial S}|_{S=S^*(\tau)} < -1$, the price curve $P(S,\tau)$ at value of S close to but greater than $S^*(\tau)$ falls below the intrinsic value line (see the top left figure).
(b) When $\frac{\partial P}{\partial S}|_{S=S^*(\tau)} > -1$, argue why the value of the American put option at asset price level close to $S^*(\tau)$ can be increased by choosing a smaller value for $S^*(\tau)$ (see the top right figure).

Explain why both cases do not correspond to the optimal exercise strategy of an American put. Hence, the slope of the American put price curve at $S^*(\tau)$ must satisfy the smooth pasting condition.

5.6 When $q \geq r$, explain why an American call on a continuous dividend paying asset, which is optimally held to expiration, will have zero value at expiration (Kim, 1990).

5.7 Let $P(S, \tau; X, r, q)$ denote the price function of an American put option. Show that $P(X, \tau; S, q, r)$ also satisfies the Black–Scholes equation:

$$\frac{\partial P}{\partial \tau} = \frac{\sigma^2}{2} S^2 \frac{\partial^2 P}{\partial S^2} + (r - q) S \frac{\partial P}{\partial S} - rP$$

together with the auxiliary conditions:

$$P(X, 0; S, q, r) = \max(S - X, 0)$$
$$P(X, \tau; S, q, r) \geq \max(S - X, 0) \quad \text{for } \tau > 0.$$

Note that the auxiliary conditions are identical to those of the price function of the American call option. Hence, we can conclude that

$$C(S, \tau; X, r, q) = P(X, \tau; S, q, r).$$

Hint: Write $P(S', \tau) = P\left(\frac{1}{S}, \tau; \frac{1}{X}, q, r\right) = \frac{1}{SX} P(X, \tau; S, q, r)$, and show that

$$\begin{aligned}
&\frac{\partial}{\partial \tau}[SXP(S', \tau)] - \frac{\sigma^2}{2} S^2 \frac{\partial^2}{\partial S^2}[SXP(S', \tau)] \\
&\quad (r - q)S \frac{\partial}{\partial S}[SXP(S', \tau)] + rSXP(S', \tau) \\
&= SX\left[\frac{\partial P}{\partial \tau}(S', \tau) - \frac{\sigma^2}{2} S'^2 \frac{\partial^2 P}{\partial S'^2}(S', \tau)\right. \\
&\quad \left. - (q - r)S' \frac{\partial P}{\partial S'}(S', \tau) + qP(S', \tau)\right].
\end{aligned}$$

5.8 From the put-call symmetry relation for the prices of American call and put options derived in Problem 5.7, show that

$$\begin{aligned}
\frac{\partial C}{\partial S}(S, \tau; X, r, q) &= \frac{\partial P}{\partial X}(X, \tau; S, q, r) \\
\frac{\partial C}{\partial q}(S, \tau; X, r, q) &= \frac{\partial P}{\partial r}(X, \tau; S, q, r).
\end{aligned}$$

Give financial interpretation of the results.

5.9 Consider the pair of American call and put options with the same time to expiry τ and on the same underlying asset. Assume the volatility of the asset price to be at most time dependent. Let S_C and S_P be the spot asset price corresponding to the call and put, respectively (S_C and S_P need not be the same since the calendar times at which we are comparing values need not be the same). Suppose the two options have the same moneyness, that is,

$$\frac{S_C}{X_C} = \frac{X_P}{S_P},$$

where X_C and X_P are the strike price corresponding to the call and put, respectively. Let $C(S_C, \tau; X_C, r, q)$ and $P(S_P, \tau; X_P, r, q)$ denote the price function of the American call and put, respectively. Derive the generalized put-call symmetry relation (Carr and Chesney, 1996)

$$\frac{C(S_C, \tau; X_C, r, q)}{\sqrt{S_C X_C}} = \frac{P(S_P, \tau; X_P, q, r)}{\sqrt{S_P X_P}}.$$

Furthermore, let $S_C^*(\tau; X_C, r, q)$ and $S_P^*(\tau; X_P, r, q)$ denote the optimal exercise price of the American call and put, respectively. Show that

$$S_C^*(\tau; X_C, r, q) S_P^*(\tau; X_P, q, r) = X_C X_P.$$

This relation is a generalization of the result given in (5.1.16).

5.10 Let H denote the barrier of a perpetual American down-and-out call option. The governing equation for the price of the perpetual American barrier option $C_\infty(S; r, q)$ is given by

$$\frac{\sigma^2}{2}S^2\frac{d^2C_\infty}{dS^2} + (r-q)S\frac{dC_\infty}{dS} - rC_\infty = 0, \quad H < S < S^*_\infty,$$

where S^*_∞ is the optimal exercise price. Determine S^*_∞ and find the option price $C_\infty(S; r, q)$.

Hint: The optimal exercise price is determined by maximizing the solution for the perpetual American call price among all possible exercise prices, that is,

$$C_\infty(S; r, q) = \max_{S^*_\infty}\left\{\frac{S^*_\infty - X}{H^{\lambda_+}{S^*_\infty}^{\lambda_-} - {S^*_\infty}^{\lambda_+}H^{\lambda_-}}(H^{\lambda_+}S^{\lambda_-} - H^{\lambda_-}S^{\lambda_+})\right\},$$

where λ_+ and λ_- are roots of the quadratic equation:

$$\frac{\sigma^2}{2}\lambda(\lambda-1) + (r-q)\lambda - r = 0.$$

5.11 Suppose the continuous dividend paid by an asset is at the constant rate d but not proportional to the asset price S. Show that the American call option on the above asset would not be exercised prematurely if $d < rX$ where r is the riskless interest rate and X is the strike price. Under the above condition, show that the price of the perpetual American call option is given by (Merton, 1973, Chap. 1)

$$C(S, \infty; X) = S - \frac{d}{r}\left[1 - \frac{(\frac{2d}{\sigma^2 S})^{\frac{2r}{\sigma^2}}}{\Gamma(2+\frac{2r}{\sigma^2})}M\left(\frac{2r}{\sigma^2}, 2+\frac{2r}{\sigma^2}, -\frac{2d}{\sigma^2 S}\right)\right],$$

where Γ and M denote the Gamma function and the confluent hypergeometric function, respectively.

5.12 Consider an American call option with a continuously changing strike price $X(\tau)$ where $\frac{dX(\tau)}{d\tau} < 0$. The auxiliary conditions for the American call option model are given by

$$C(S, \tau; X(\tau)) \geq \max(S - X(\tau), 0)$$

and

$$C(S, \tau; X(0)) = \max(S - X(0), 0).$$

Define the following new set of variables:

$$\xi = \frac{S}{X(\tau)} \text{ and } F(\xi, \tau) = \frac{C(S, \tau; X(\tau))}{X(\tau)}.$$

Show that the governing equation for the price of the above American call is given by

$$\frac{\partial F}{\partial \tau} = \frac{\sigma^2}{2}\xi^2 \frac{\partial^2 F}{\partial \xi^2} + \eta(\tau)\xi \frac{\partial F}{\partial \xi} - \eta(\tau)F,$$

where $\eta(\tau) = r + \frac{1}{X}\frac{\partial X}{\partial \tau}$ and r is the riskless interest rate. The auxiliary conditions become

$$F(\xi, 0) = \max(\xi - 1, 0) \quad \text{and} \quad F(\xi, \tau) \geq \max(\xi - 1, 0).$$

Show that if $X(\tau) \geq X(0)e^{-r\tau}$, then it is never optimal to exercise the American call prematurely. In such a case, show that the value of the above American call is the same as that of a European call with a fixed strike price $X(0)$ (Merton, 1973, Chap. 1).

Hint: Show that when the time dependent function $\eta(\tau)$ satisfies the condition $\int_0^\tau \eta(s)\, ds \geq 0$, it is then never optimal to exercise the American call prematurely.

5.13 Consider the one-dividend American call option model. Explain why the exercise price S_d^*, which is obtained by solving (5.1.24), decreases when the dividend amount D increases. Also, show that S_d^* tends to infinity when D falls to the value $X[1 - e^{-r(T-t_d)}]$.

5.14 Give a mathematical proof to the following inequality

$$C_d(\widetilde{S}, T - t; X) \geq \max\{c(\widetilde{S}, T - t; X), c(S, t_d - t; X)\}, \quad t < t_d,$$

which arises from the Black approximation formula for the one-dividend American call (see Sect. 5.1.5). Here, t_d and T are the ex-dividend date and expiration date, respectively; S and $\widetilde{S}$ are the market asset price and the asset price net of the present value of the escrowed dividend, respectively.

5.15 Suppose discrete dividends of amount $D_1, D_2, \cdots, D_n$ are paid at the respective ex-dividend dates $t_1, t_2, \cdots, t_n$ and let t_{n+1} denote the date of expiration T. Show that the risky component is given by

$$\widetilde{S} = S - \sum_{k=j+1}^{n} D_k e^{-r(t_k - t)} \text{ for } t_j^+ \leq t \leq t_{j+1}^-, \quad j = 0, 1, \cdots, n,$$

and $t_0 = 0$.

Hint: Extend the result in (5.1.26).

5.16 Consider an American call option on an asset that pays discrete dividends at anticipated dates $t_1 < t_2 < \cdots < t_n$. Let the size of the dividends be, respectively, $D_1, D_2, \cdots, D_n$, and $T = t_{n+1}$ be the time of expiration. Show that it is never optimal to exercise the American call at any time prior to expiration if all the discrete dividends are not sufficiently deep, as indicated by the following inequality

$$D_i \leq X[1 - e^{-r(t_{i+1}-t_i)}], \quad i = 1, 2, \cdots, n.$$

5.17 In the two-dividend American call option model, we assume discrete dividends of amount D_1 and D_2 are paid out by the underlying asset at times t_1 and t_2, respectively. Let $\widetilde{S}_t$ denote the asset price at time t, net of the present value of escrowed dividends and $\widetilde{S}^*_{t_1}$ ($\widetilde{S}^*_{t_2}$) denote the optimal exercise price at time t_1 (t_2) above which the American call should be exercised prematurely. Let r, σ, X and T denote the riskless interest rate, volatility of $\widetilde{S}$, strike price and expiration time, respectively. Let $C(\widetilde{S}_t, t)$ denote the value of the American call at time t. Show that $\widetilde{S}^*_{t_1}$ and $\widetilde{S}^*_{t_2}$ are given by the solution of the following nonlinear algebraic equations

$$\begin{aligned} C(\widetilde{S}^*_{t_1}, t_1) &= \widetilde{S}^*_{t_1}[1 - N_2(-a_1, -b_1; \rho)] + D_2 e^{-r(t_2-t_1)} N(a_2) \\ &\quad - X[e^{-r(t_2-t_1)} N(a_2) + e^{-r(T-t_1)} N_2(-a_2, b_2; -\rho)] \\ C(\widetilde{S}^*_{t_2}, t_2) &= \widetilde{S}^*_{t_2} N(v_1) - X e^{-r(T-t_2)} N(v_2), \end{aligned}$$

where

$$a_2 = \frac{\ln \frac{\widetilde{S}^*_{t_1}}{\widetilde{S}^*_{t_2}} + \left(r - \frac{\sigma^2}{2}\right)(t_2 - t_1)}{\sigma\sqrt{t_2 - t_1}}, \qquad a_1 = a_2 + \sigma\sqrt{t_2 - t_1},$$

$$b_2 = \frac{\ln \frac{\widetilde{S}^*_{t_1}}{X} + \left(r - \frac{\sigma^2}{2}\right)(T - t_1)}{\sigma\sqrt{T - t_1}}, \qquad b_1 = b_2 + \sigma\sqrt{T - t_1},$$

$$v_2 = \frac{\ln \frac{\widetilde{S}^*_{t_2}}{X} + \left(r - \frac{\sigma^2}{2}\right)(T - t_2)}{\sigma\sqrt{T - t_2}}, \qquad v_1 = v_2 + \sigma\sqrt{T - t_2}.$$

The American call price is given by (Welch and Chen, 1988)

$$\begin{aligned} C(\widetilde{S}_t, t) = \widetilde{S}_t &[1 - N_3(-f_1, -g_1, -h_1; \rho_{12}, \rho_{13}, \rho_{23})] \\ &- X[e^{-r(t_1-t)} N(f_2) + e^{-r(t_2-t)} N_2(-f_2, g_2; -\rho_{12}) \\ &+ e^{-r(T-t)} N_3(-f_2, -g_2, h_2; \rho_{12}, -\rho_{13}, -\rho_{23})] \\ &+ D_1 e^{-r(t_1-t)} N(f_2) \\ &+ D_2 e^{-r(t_2-t)} [N(f_2) + N_2(-f_2, g_2; -\rho_{12})], \end{aligned}$$

where

$$\rho_{12} = \sqrt{\frac{t_1 - t}{t_2 - t}}, \quad \rho_{13} = \sqrt{\frac{t_1 - t}{T - t}}, \quad \rho_{23} = \sqrt{\frac{t_2 - t}{T - t}},$$

$$f_2 = \frac{\ln \frac{\widetilde{S}_t}{S^*_{t_1}} + \left(r - \frac{\sigma^2}{2}\right)(t_1 - t)}{\sigma\sqrt{t_1 - t}}, \qquad f_1 = f_2 + \sigma\sqrt{t_1 - t},$$

$$g_2 = \frac{\ln \frac{\widetilde{S}_t}{S^*_{t_2}} + \left(r - \frac{\sigma^2}{2}\right)(t_2 - t)}{\sigma\sqrt{t_2 - t}}, \qquad g_1 = g_2 + \sigma\sqrt{t_2 - t},$$

$$h_2 = \frac{\ln \frac{\widetilde{S}_t}{X} + \left(r - \frac{\sigma^2}{2}\right)(T - t)}{\sigma\sqrt{T - t}}, \qquad h_1 = h_2 + \sigma\sqrt{T - t}.$$

5.18 Consider the one-dividend American put option model where the discrete dividend at time t_d is paid at the known rate λ, that is, the dividend payment is λS_{t_d}. Show that the slope of the optimal exercise boundary of the American put at time right before t_d is given by (Meyer, 2001)

$$\lim_{t \to t_d^-} \frac{dS^*(t)}{dt} = \frac{r}{\lambda} X,$$

where r is the riskless interest rate.

Hint: Consider the balance of the gain in interest income from the strike price and the loss in dividend over the differential time interval δt right before t_d.

5.19 Bunch and Johnson (2000) gave the following three different definitions of the optimal exercise price of an American put.

1. It is the value of the asset price at which one is indifferent between exercising and not exercising the put.
2. It is the highest value of the asset price for which the value of the put is equal to the exercise price less the stock price.
3. It is the highest value of the asset price at which the put value does not depend on time to maturity.

Give the financial interpretation to the above three definitions.

5.20 Show that the delta of the price of an American put option on an asset which pays a continuous dividend yield at the rate q is given by

$$\frac{\partial P}{\partial S} = -N(-d_1) - \int_0^\tau \left[\frac{(r-q)e^{-q\xi}}{\sigma\sqrt{2\pi\xi}} e^{-\frac{d_{\xi,1}^2}{2}} + qe^{-q\xi} N(-d_{\xi,1}) \right] d\xi,$$

where

$$d_1 = \frac{\ln \frac{S}{X} + \left(r - q + \frac{\sigma^2}{2}\right)\tau}{\sigma\sqrt{\tau}},$$

$$d_{\xi,1} = \frac{\ln \frac{S}{S^*(\tau-\xi)} + \left(r - q + \frac{\sigma^2}{2}\right)\xi}{\sigma\sqrt{\xi}}, \qquad d_{\xi,2} = d_{\xi,1} - \sigma\sqrt{\xi}.$$

Examine the sign of the delta of the early exercise premium when $r \geq q$ and $r < q$. Give financial interpretation of the sign behavior of the above delta. Furthermore, show that

$$\frac{\partial^2 P}{\partial S^2} = \frac{1}{S\sigma\sqrt{2\pi\tau}} e^{-d_1^2/2} + \int_0^\tau \left[\frac{(r-q)e^{-q\xi}}{S\sigma^2\xi\sqrt{2\pi}} d_{\xi,1} e^{-\frac{d_{\xi,1}^2}{2}} + \frac{qe^{-q\xi}}{S\sigma\sqrt{2\pi\xi}} e^{-\frac{d_{\xi,1}^2}{2}} \right] d\xi.$$

Find similar expressions for $\frac{\partial P}{\partial \sigma}$, $\frac{\partial P}{\partial r}$ and $\frac{\partial P}{\partial X}$ (Huang, Subrahmanyam and Yu 1996).

5.21 Consider an American put option on an asset which pays no dividend. Show that the early exercise premium $e(S, \tau; X)$ is bounded by

$$rX \int_0^\tau e^{-r\xi} N(-\widetilde{d_\xi})\, d\xi \leq e(S, \tau; X) \leq rX \int_0^\tau e^{-r\xi} N(-\widehat{d_\xi})\, d\xi,$$

where

$$\widetilde{d_\xi} = \frac{\ln \frac{S^*(\tau)}{S^*(0)} + \left(r - \frac{\sigma^2}{2}\right)\xi}{\sigma\sqrt{\xi}}, \qquad \widehat{d_\xi} = \frac{\ln \frac{S^*(\tau)}{S^*(\infty)} + \left(r - \frac{\sigma^2}{2}\right)\xi}{\sigma\sqrt{\xi}},$$

$$S^*(\infty) = \frac{X}{1 + \frac{\sigma^2}{2r}}, \qquad S^*(0) = X.$$

5.22 Let $S_C^*(\infty)$ denote $\lim_{\tau\to\infty} S_C^*(\tau)$, where $S_C^*(\tau)$ is the solution to the integral equation defined in (5.2.19). By taking the limit $\tau \to \infty$ of the above integral equation, solve for $S_C^*(\infty)$. Compare the result given in (5.1.13).

5.23 By considering the corresponding integral representation of the early exercise premium of an American commodity option with cost of carry b, show that
   (a) when $b \geq r$, r is the riskless interest rate, there is no advantage of early exercise for the American commodity call option;
   (b) advantage of early exercise always exists for the American commodity put option for all values of b.

5.24 Let $C_{do}(S, \tau; X, H, r, q)$ and $P_{uo}(S, \tau; X, H, r, q)$ denote the price function of an American down-and-out barrier call and an American up-and-out barrier put, respectively, both with constant barrier level H. Show that the put-call symmetry relation for the prices of the American barrier call and put options is given by (Gao, Huang and Subrahmanyam, 2000)

$$C_{do}(S, \tau; X, H, r, q) = P_{uo}(X, \tau; SX/H, q, r).$$

Let $S_{do,call}^*(\tau; X, H, r, q)$ and $S_{uo,put}^*(\tau; X, H, r, q)$ denote the optimal exercise price of the American down-and-out call and American up-and-out put, respectively. Show that

$$S^*_{do,call}(\tau; X, H, r, q) = \frac{X^2}{S^*_{uo,put}(\tau; X, X^2/H, q, r)}.$$

5.25 Consider an American up-and-out put option with barrier level $B(\tau) = B_0 e^{-\alpha\tau}$ and strike price X. Assuming that the underlying asset pays a continuous dividend yield q, find the integral representation of the early exercise premium. What would be the effect on the optimal exercise price $S^*(\tau; B(\tau))$ when B_0 decreases?

5.26 Consider a down-and-in American call $C_{di}(S, \tau; X, B)$, where the down-and-in trigger clause entitles the holder to receive an American call option with strike price X when the asset price S falls below the threshold level B. The underlying asset pays dividend yield q and let r denote the riskless interest rate. Let $C(S, \tau; X)$ and $c(S, \tau; X)$ denote the price function of the American call and European call with strike price X, respectively. Show that when $B \leq \max(X, \frac{r}{q}X)$, we have

$$\begin{aligned} &C_{di}(S, \tau; X, B) \\ &= \left(\frac{S}{B}\right)^{1-\frac{2(r-q)}{\sigma^2}} \left[C\left(\frac{B^2}{S}, \tau; X\right) - c\left(\frac{B^2}{S}, \tau; X\right)\right] + c_{di}(S, \tau; X, B), \end{aligned}$$

where $c_{di}(S, \tau; X, B)$ is the price function of the European down-and-in call counterpart. Find the corresponding form of the price function $C_{di}(S, \tau; X, B)$ when (i) $B \geq S^*(\infty)$ and (ii) $S^*(0^+) < B < S^*(\infty)$, where $S^*(\tau)$ is the optimal exercise boundary of the American non-barrier call $C(S, \tau; X)$ (Dai and Kwok, 2004).

5.27 The exercise payoff of an American capped call with the cap L is given by $\max(\min(S, L) - X, 0)$, $L > X$. Let $S^*_{cap}(\tau)$ and $S^*(\tau)$ denote the early exercise boundary of the American capped call and its noncapped counterpart, respectively. Show that (Broadie and Detemple, 1995)

$$S^*_{cap}(\tau) = \min(S^*(\tau), L).$$

5.28 Consider an American call option with the callable feature, where the issuer has the right to recall throughout the whole life of the option. Upon recall by the issuer, the holder of the American option can choose either to exercise his option or receive the constant cash amount K. Let $S^*_{call}(\tau)$ and $S^*(\tau)$ denote the optimal exercise boundary of the callable American call and its noncallable counterpart, respectively. Show that

$$S^*_{call}(\tau) = \min(S^*(\tau), K + X),$$

where X is the strike price. Furthermore, suppose the holder is given a notice period of length τ_n, where his or her decision to exercise the option or receive the cash amount K is made at the end of the notice period. Show that the optimal exercise boundary $S^*_{call}(\tau)$ now becomes

$$S^*_{call}(\tau) = \min(S^*(\tau), \widehat{S}^*(\tau_n)),$$

where $\widehat{S}^*(\tau_n)$ is the solution to the algebraic equation

$$\widehat{S}^*(\tau_n) - X - Ke^{-r\tau_n} = c(S, \tau_n; K + X).$$

Here, $c(S, \tau_n; K + X)$ is the price of the European option with time to expiry τ_n and strike price $K + X$ (Kwok and Wu, 2000; Dai and Kwok, 2005b).

Hint: Note that $S^*_{call}(\tau)$ cannot be greater than $K + X$. If otherwise, at asset price level satisfying $K + X < S < S^*_{call}(\tau)$, the intrinsic value of the American call is above K. This represents a nonoptimal recall policy of the issuer.

5.29 Unlike usual option contracts, the holder of an *installment option* pays the option premium throughout the life of the option. The installment option is terminated if the holder chooses to discontinue the installment payment. In normal cases, the installments are paid at predetermined time instants within the option's life. In this problem, we consider the two separate cases: continuous payment stream and discrete payments.

First, we let s denote the continuous rate of installment payment so that the amount $s\Delta t$ is paid over the interval Δt. Let $V(S,t)$ denote the value of a European installment call option. Show that $V(S,t)$ is governed by

$$\begin{cases} \frac{\partial V}{\partial t} + \frac{\sigma^2}{2}S^2\frac{\partial^2 V}{\partial S^2} + r\frac{\partial V}{\partial S} - rV - s = 0 & \text{if } S > S^*(t) \\ V = 0 & \text{if } S \le S^*(t) \end{cases},$$

where $S^*(t)$ is the critical asset price at which the holder discontinues the installment payment optimally. Solve for the analytic price formula when the installment option has infinite time to expiration (perpetual installment option).

Next, suppose that installments of equal amount d are paid at discrete instants t_j, $j = 1, \cdots, n$. Explain the validity of the following jump condition across the payment date t_j

$$V(S, t_j^-) = \max(V(S, t_j^+) - d, 0).$$

Finally, give a sketch of the variation of the option value $V(S,t)$ as a function of the calendar time t at varying values of asset value S under discrete installment payments.

Hint: There is an increase in the option value of amount d right after the installment payment. Also, it is optimal not to pay the installment at time t_j if $V(S, t_j^+) \le d$.

5.30 Suppose an American put option is only allowed to be exercised at N time instants between now and expiration. Let the current time be zero and denote the exercisable instants by the time vector $\boldsymbol{t} = (t_1 \; t_2 \; \cdots \; t_N)^T$. Let $N_i(\boldsymbol{d}_i; R_i)$ denote the i-dimensional multi-variate normal integral with upper limits of integration given by the i-dimensional vector $\boldsymbol{d}_i$ and correlation matrix R_i. Define the diagonal matrix $D_i = \text{diag}\ (1, \cdots 1, -1)$, and let $\boldsymbol{d}_i^* = D_i \boldsymbol{d}_i$ and $R_i^* = D_i R_i D_i$. Show that the value of the above American put with N exercisable instants is found to be (Bunch and Johnson, 1992)

$$P = X \sum_{i=1}^{N} e^{-rt_i} N_i(\boldsymbol{d}_{i_2}^*; R_i^*) - S \sum_{i=1}^{N} N_i(\boldsymbol{d}_{i_1}^*; R_i^*),$$

where

$$\boldsymbol{d}_{i_1} = (d_{11}, d_{21}, \cdots, d_{i1})^T, \quad \boldsymbol{d}_{i_1}^* = D_i \boldsymbol{d}_{i_1},$$

$$\boldsymbol{d}_{i_2} = \boldsymbol{d}_{i_1} - \sigma(\sqrt{t_1} \; \sqrt{t_2} \; \cdots \; \sqrt{t_i})^T, \quad \boldsymbol{d}_{i_2}^* = D_i \boldsymbol{d}_{i_2},$$

$$d_{j1} = \frac{\ln \frac{S}{S_{t_j}^*} + \left(r + \frac{\sigma^2}{2}\right) t_j}{\sigma \sqrt{t_j}}, \quad j = 1, \cdots, i,$$

and $S_{t_j}^*$ is the optimal exercise price at t_j. Also, find the expression for the correlation matrix R_i.

Hint: When $N = 3$ and the exercisable instants are equally spaced, the correlation matrix R_3 is found to be

$$R_3 = \begin{pmatrix} 1 & 1/\sqrt{2} & 1/\sqrt{3} \\ 1/\sqrt{2} & 1 & \sqrt{2/3} \\ 1/\sqrt{3} & \sqrt{2/3} & 1 \end{pmatrix}.$$

5.31 The approximate equation for f in the quadratic approximation method becomes undefined when $K(\tau) = 1 - e^{-r\tau} = 0$, which corresponds to $r = 0$. Following a similar derivation procedure as in the quadratic approximation method, solve approximately the American option valuation problem for this degenerate case of zero riskless interest rate.

5.32 Show that the approximate value of the American commodity put option based on the quadratic approximation method is given by

$$\widetilde{P}(S, \tau) = p(S, \tau) - \frac{S^*}{q_1}\left[1 - e^{(b-r)}N(-d_1(S^*))\right]\left(\frac{S}{S^*}\right)^{q_1}, \quad S > S^*.$$

Explain why the formula holds for all values of b.

Hint: Show that

$$\widetilde{P}(S, \tau) = p(S, \tau) + c_1 K S^{q_1},$$

and

$$\frac{\partial p}{\partial S}(S^*, \tau) = e^{(b-r)\tau} N(-d_1(S^*)).$$

5.33 Consider the shout call option discussed in Sect. 5.4.2 (Dai, Kwok and Wu, 2004). Explain why the value of the shout call is bounded above by the fixed strike lookback call option with the same strike X.

5.34 Show that

$$e^{q\tau} p^*(\tau; r, q) = p^*(\tau; r - q, 0),$$

where $p^*(\tau)$ is defined in (5.4.2). To prove the results in (5.4.6a,b,c), it suffices to consider the sign behavior of

$$\frac{d}{d\tau} p^*(\tau; r, 0) = e^{-r\tau} f(\tau),$$

where

$$f(\tau) = -rN(-d_2) + \frac{\sigma}{2\sqrt{\tau}} n(-d_2),$$

$$d_2 = \alpha\sqrt{\tau}, \quad d_1 = d_2 + \sigma\sqrt{\tau} \quad \text{and} \quad \alpha = \frac{r - \frac{\sigma^2}{2}}{\sigma}.$$

Consider the following two cases (Dai, Kwok and Wu, 2004).
(a) For $r \leq 0$, show that

$$\frac{d}{d\tau} p^*(\tau; r, 0) > 0.$$

(b) For $r > 0$, show that

$$f'(\tau) = \frac{\sigma n(-d_2)}{4\sqrt{\tau}} \left[\alpha(\alpha + \sigma) - \frac{1}{\tau} \right],$$

hence deduce the results in (5.4.6b,c).

5.35 For the reset-strike put option, assuming $r \leq q$, show that the early reset premium is given by (Dai, Kwok and Wu, 2004)

$$e(S, \tau) = Se^{-q\tau} \int_0^\tau N(d_{1,\tau-u}) \frac{d}{du} [e^{qu} p^*(u)]\, du,$$

where

$$d_{1,\tau-u} = \frac{\ln \frac{S}{S^*(u)} + \left(r - q + \frac{\sigma^2}{2}\right)(\tau - u)}{\sigma\sqrt{\tau - u}}.$$

How do we modify the formula when $r > q$?

5.36 Let $W_n^\infty(S; X) = \lim_{\tau \to \infty} e^{r\tau} U_n(S, \tau; X)$, where $U_n(S, \tau; X)$ is the value of the n-reset put option [see (5.4.20)]. For $r < q$, show that the governing equation for $W_n^\infty(S)$ is given by (Dai, Kwok and Wu, 2003)

$$\frac{\sigma^2}{2} S^2 \frac{d^2 W_n^\infty}{dS^2} + (r - q) S \frac{dW_n^\infty}{dS} = 0, \quad 0 < S < S_{n,\infty}^*.$$

The auxiliary conditions are given by

$$W_n^\infty(S_{n,\infty}^*) = \beta_n S_{n,\infty}^* \quad \text{and} \quad \frac{dW_n^\infty}{dS}(S_{n,\infty}^*) = \beta_n,$$

where $\beta_n = W_{n-1}^\infty(1; 1)$. Show that

$$W_n^\infty(S; X) = X + \frac{\alpha^\alpha}{(1+\alpha)^{1+\alpha}} \frac{\beta_n^{1+\alpha}}{X^\alpha} S^{1+\alpha}$$

and

$$S_{n,\infty}^* = \left(1 + \frac{1}{\alpha}\right)\frac{X}{\beta_n},$$

where $\alpha = 2(q-r)/\sigma^2$. The recurrence relation for β_n is deduced to be

$$\beta_n = 1 + \frac{\alpha^\alpha}{(1+\alpha)^{1+\alpha}} \beta_{n-1}^{1+\alpha}.$$

Show that $\beta_1 = 1$ and $\lim_{n\to\infty} \beta_n = 1 + \frac{1}{\alpha}$. Also, find the first few values of $S_{n,\infty}^*$.

5.37 The reload provision in an employee stock option entitles its holder to receive $\frac{X}{S^*}$ units of newly "reloaded" at-the-money options from the employer upon exercise of the stock option. Here, X is the original strike price and S^* is the prevailing stock price at the exercise moment. The "reloaded" option has the same date of expiration as the original option. The exercise payoff is given by $S^* - X + \frac{X}{S}c(S^*, \tau; S^*, r, q)$. By the linear homogeneity property of the call price function, we can express the exercise payoff as $S - X + S\widehat{c}(\tau; r, q)$, where

$$\widehat{c}(\tau; r, q) = e^{-q\tau}N(\widehat{d}_1) - e^{-r\tau}N(\widehat{d}_2),$$

and

$$\widehat{d}_1 = \frac{r - q + \frac{\sigma^2}{2}}{\sigma}\sqrt{\tau} \quad \text{and} \quad \widehat{d}_2 = \frac{r - q - \frac{\sigma^2}{2}}{\sigma}\sqrt{\tau}.$$

Let $S^*(\tau; r, q)$ denote the optimal exercise boundary that separates the stopping and continuation regions. The stopping region and the optimal exercise boundary $S^*(\tau)$ observe the following properties (Dai and Kwok, 2008).

1. The stopping region is contained inside the region defined by

$$\{(S, \tau) : S \geq X, \quad 0 \leq \tau \leq T\}.$$

2. At a time close to expiry, the optimal stock price is given by

$$S^*(0^+; r, q) = X, \quad q \geq 0, r > 0.$$

3. When the stock pays dividend at constant yield $q > 0$, the optimal stock price at infinite time to expiry is given by

$$S^*(\infty; r, q) = \frac{\mu_+}{\mu_+ - 1} X,$$

where μ_+ is the positive root of the equation:

$$\frac{\sigma^2}{2}\mu^2 + \left(r - q - \frac{\sigma^2}{2}\right)\mu - r = 0.$$

4. If the stock pays no dividend, then
(a) for $r \leq \frac{\sigma^2}{2}$, $S^*(\tau; r, 0)$ is defined for all $\tau > 0$ and $S^*(\infty; r, 0) = \infty$;
(b) for $r > \frac{\sigma^2}{2}$, $S^*(\tau; r, 0)$ is defined only for $0 < \tau < \tau^*$, where τ^* is the unique solution to the algebraic equation

$$\frac{\sigma}{2\sqrt{\tau}} n\left(-\frac{r + \frac{\sigma^2}{2}}{\sigma}\sqrt{\tau}\right) - rN\left(\frac{r + \frac{\sigma^2}{2}}{\sigma}\sqrt{\tau}\right) = 0.$$

5.38 Consider a landowner holding a piece of land who has the right to build a developed structure on the land or abandon the land. Let S be the value of the developed structure and H be the constant rate of holding costs (which may consist of property taxes, property maintenance costs, etc.). Assuming there is no fixed time horizon beyond which the structure cannot be developed, so the value of the land can be modeled as a perpetual American call, whose value is denoted by $C(S)$. Let σ_S denote the volatility of the Brownian process followed by S and r be the riskless interest rate. Suppose the asset value of the developed structure can be hedged by other tradeable asset, use the riskless hedging principle to show that the governing equation for $C(S)$ is given by

$$\frac{\sigma_S^2}{2} S^2 \frac{\partial^2 C}{\partial S^2} + rS\frac{\partial C}{\partial S} - rC - H = 0.$$

Let Z denote the lower critical value of S below which it is optimal to abandon the land. Let W be the higher critical value of S at which it is optimal to build the structure. Let X be the amount of cash investment required to build the structure. Explain why the auxiliary conditions at $S = Z$ and $S = W$ are prescribed by

$$\begin{cases} C(Z) = 0 \quad \text{and} \quad \frac{dC}{dS}(Z) = 0 \\ C(W) = W - X \quad \text{and} \quad \frac{dC}{dS}(W) = 1 \end{cases}.$$

Show that the solution to the perpetual American call model is given by

$$C(S) = \begin{cases} 0 & \text{if } S < Z \\ \alpha_1 S + \alpha_2 S^\lambda - \frac{H}{r} & \text{if } Z \leq S \leq W \\ S - X & \text{if } S > W \end{cases},$$

where $\lambda = -\frac{2r}{\sigma_S^2}$, $W = \frac{\lambda}{\lambda - 1}\left(X - \frac{H}{r}\right)\frac{1}{1-\alpha_1}$,

$$Z = \frac{\lambda}{\lambda - 1}\frac{H}{r}\left[1 - \left(1 - \frac{rX}{H}\right)^{(\lambda-1)/\lambda}\right],$$

$$\alpha_1 = \frac{1}{1 - \left(1 - \frac{rX}{H}\right)^{(\lambda-1)/\lambda}}, \qquad \alpha_2 = -\frac{\alpha_1}{\lambda Z^{\lambda-1}}.$$

This pricing model has two-sided free boundaries, one is associated with the right to abandon the land and the other with the right to build the structure.

5.39 Consider an American installment option in which the buyer pays a smaller upfront premium, while a constant stream of installments at a certain rate per unit time are paid subsequently throughout the whole life of the option. Let δ denote the above rate of installment flow. The holder has the right to exercise the option or stop the installment payment prematurely.

(a) Derive the linear complementarity formulation of an American installment option on a dividend yield paying asset with either a call or put payoff.

(b) Consider an American installment call option and let $q > 0$ denote the dividend yield. Show that the optimal stopping boundaries consist of two branches:

(i) The upper critical asset price $S^*_{up}(t)$ at which the option should be exercised prematurely.

(ii) The lower critical asset price $S^*_{low}(t)$ at which the option should be terminated prematurely by stopping the installment payment. Show that

$$\lim_{t \to T^-} S^*_{up}(t) = \max\left(\frac{rX - \delta}{q}, X\right) \quad \text{and} \quad \lim_{t \to T^-} S^*_{low}(t) = X,$$

where X is the strike price.

(c) Deduce similar results for an American installment option with the put payoff.

6

Numerical Schemes for Pricing Options

In previous chapters, we obtained closed form price formulas for a variety of option models. However, option models that lend themselves to analytic solutions are limited. In most cases, option valuation must be relegated to numerical procedures. The classes of numerical methods employed in option valuation include the *lattice tree methods, finite difference algorithms* and *Monte Carlo simulation.*

The finance community typically uses the binomial scheme for numerical valuation of a wide variety of option models, due primarily to its ease of implementation and pedagogical appeal. The primary essence of the binomial model is the simulation of the continuous asset price movement by a discrete random walk model. Interestingly, the concept of risk neutral valuation is embedded naturally in the binomial model. In Sect. 6.1, we revisit the binomial model and illustrate how to apply it to valuation of options on a discrete dividend paying asset and options with early exercise right and callable right. We examine the asymptotic limit of the discrete binomial model to the continuous Black–Scholes model. We also consider the extension of the binomial lattice tree to its trinomial counterpart. The trinomial lattice tree simulates the underlying asset price process using a discrete three-jump process. For numerical valuation of path dependent options, like Asian options and options with the Parisian feature of knock-out, we discuss the versatile forward shooting grid approach that allows us to keep track of the path dependence of the underlying state variables in a lattice tree.

The finite difference approach seeks the discretization of the differential operators in the Black–Scholes equation. The numerical schemes arising from the discretization procedure can be broadly classified as either implicit or explicit schemes. Each class of schemes has its merits and limitations. The explicit schemes have better computational efficiency, but they may be susceptible to numerical instabilities to round-off errors if the time steps in the numerical computation are not chosen to be sufficiently small. Interestingly, the lattice tree schemes are seen to have the same analytic forms as those of the explicit finite difference schemes, though the two classes of numerical schemes are derived using quite different approaches. Section 6.2 presents various versions of finite difference schemes for option valuation. In particular, we discuss the projected successive-over-relaxation scheme and the front-fixing method for numerical valuation of American options.

Nowadays, it is quite common to demand the computation of thousands of option values within a short duration of time, thus providing the impetus for developing numerical algorithms that compete favorably in terms of accuracy, efficiency and reliability. We discuss the theoretical concepts of order of accuracy and numerical stability in the analysis of a numerical scheme. We analyze the intricacies associated with the smoothing of the "kink" or "jump" in the terminal payoff function and the avoidance of spurious oscillations in the numerical solutions. Also, we consider the issues of implementing the boundary conditions in the barrier option and the look-back option.

The Monte Carlo method simulates the random movement of the asset price processes and provides a probabilistic solution to the option pricing models. Since most derivative pricing problems can be formulated as evaluation of the risk neutral expectation of the discounted terminal payoff function, the Monte Carlo simulation provides a direct numerical tool for pricing derivative securities, even without a full formulation of the pricing model. When faced with pricing a new derivative with complex payoffs, a market practitioner can always rely on the Monte Carlo simulation procedure to generate an estimate of the new derivative's price, though other more efficient numerical methods may be available when the analytic properties of the derivative model are better explored.

One main advantage of the Monte Carlo simulation is that it can accommodate complex payoff functions in option valuation without much additional effort. Also, the computational cost for Monte Carlo simulation increases linearly with the number of underlying state variables, so the method becomes more competitive in multi-state option models with a large number of risky assets. The most undesirable nature of Monte Carlo simulation is that a large number of simulation runs are generally required in order to achieve a desired level of accuracy, as the standard error of the estimate is inversely proportional to the square root of the number of simulation runs. To reduce the standard deviation of the estimate, there are several effective variance-reduction techniques, like the control variate technique and the antithetic variables technique. In Sect. 6.3, we examine how to apply these variance reduction techniques in the context of option pricing.

It had been commonly believed that the Monte Carlo simulation method cannot be used to handle the early exercise decision of an American option since one cannot predict whether the early exercise decision is optimal when the asset price reaches a certain level at a particular instant. Recently, several effective Monte Carlo simulation techniques have been proposed for the valuation of American options. These include the bundling and sorting algorithm, the method of parameterization of the optimal exercise boundary, the stochastic mesh method and the least squares regression method. An account of each of these techniques is presented in Sect. 6.3.

6.1 Lattice Tree Methods

We start the discussion on the lattice tree methods by revisiting the binomial model and consider its continuous limits. We then examine how to modify the binomial schemes so as to incorporate discrete dividends, early exercise and call features. Also, we illustrate how to construct the trinomial schemes where the asset price allows for trinomial jumps in each time step. At the end of this section, we consider the forward shooting grid approach of pricing path dependent options.

6.1.1 Binomial Model Revisited

In the discrete binomial pricing model, we simulate the stochastic asset price process by the discrete binomial process. In Sect. 2.1.4, we derive the risk neutral probability $p = \frac{R-d}{u-d}$ of the upward move in the discrete binomial process. Here, $R = e^{r\Delta t}$ is the growth factor of the risk free asset over one time period Δt, where r is the constant interest rate. However, the proportional upward jump u and downward jump d in the binomial asset price process have not yet been determined. We expect u and d to be directly related to the volatility of the continuous diffusion process of the asset price. We derive the relations that govern u, d and p by equating the mean and variance of the continuous process and its discrete binomial counterpart.

Let S_t and $S_{t+\Delta t}$ denote, respectively, the asset prices at the current time t and one period Δt later. In the Black–Scholes continuous model, the asset price dynamics are assumed to follow the Geometric Brownian process where $\frac{S_{t+\Delta t}}{S_t}$ is lognormally distributed. Under the risk neutral measure, $\ln \frac{S_{t+\Delta t}}{S_t}$ becomes normally distributed with mean $(r - \frac{\sigma^2}{2})\Delta t$ and variance $\sigma^2 \Delta t$ (see Sect. 2.4.2), where σ^2 is the variance rate. The mean and variance of $\frac{S_{t+\Delta t}}{S_t}$ are R and $R^2(e^{\sigma^2 \Delta t} - 1)$, respectively [see (2.3.20)–(2.3.21)]. On the other hand, for the one-period binomial option model under the risk neutral measure, the mean and variance of the asset price ratio $\frac{S_{t+\Delta t}}{S_t}$ are

$$pu + (1-p)d \quad \text{and} \quad pu^2 + (1-p)d^2 - [pu + (1-p)d]^2,$$

respectively. By equating the mean and variance of the asset price ratio in both continuous and discrete models, we obtain

$$pu + (1-p)d = R \tag{6.1.1a}$$

$$pu^2 + (1-p)d^2 - R^2 = R^2(e^{\sigma^2 \Delta t} - 1). \tag{6.1.1b}$$

Equation (6.1.1a) leads to $p = \frac{R-d}{u-d}$, the same risk neutral probability determined in Sect. 2.1.4. Equations (6.1.1a,b) provide only two equations for the three unknowns: u, d and p. The third condition can be chosen arbitrarily. A convenient choice is the tree-symmetry condition

$$u = \frac{1}{d}, \tag{6.1.1c}$$

so that the lattice nodes associated with the binomial tree are symmetrical. The asset price returns to the same value when the binomial process has realized one upward jump followed by one downward jump.

Writing $\widetilde{\sigma}^2 = R^2 e^{\sigma^2 \Delta t}$, the solution to (6.1.1a,b,c) is found to be

$$u = \frac{1}{d} = \frac{\widetilde{\sigma}^2 + 1 + \sqrt{(\widetilde{\sigma}^2 + 1)^2 - 4R^2}}{2R}, \qquad p = \frac{R - d}{u - d}. \tag{6.1.2}$$

The expression for u in the above formula appears to be quite cumbersome. It is tempting to seek a simpler formula for u, while not sacrificing the order of accuracy. By expanding u as defined in (6.1.2) in a Taylor series in powers of $\sqrt{\Delta t}$, we obtain

$$u = 1 + \sigma\sqrt{\Delta t} + \frac{\sigma^2}{2}\Delta t + \frac{4r^2 + 4\sigma^2 r + 3\sigma^4}{8\sigma}(\sqrt{\Delta t})^3 + O(\Delta t^2).$$

Observe that the first three terms in the above Taylor series agree with those of $e^{\sigma\sqrt{\Delta t}}$ up to $O(\Delta t)$ term. This suggests the judicious choice of the following set of parameter values (Cox, Ross and Rubinstein, 1979, Chap. 2)

$$u = e^{\sigma\sqrt{\Delta t}}, \quad d = e^{-\sigma\sqrt{\Delta t}}, \qquad p = \frac{R - d}{u - d}. \tag{6.1.3}$$

These parameter values appear to be in simpler analytic forms compared to those in formula (6.1.2). With this new set of parameters, the variance of the price ratio $\frac{S_{t+\Delta t}}{S_t}$ in the continuous and discrete models agree up to $O(\Delta t)^2$. That is, (6.1.1b) is now satisfied up to $O(\Delta t^2)$ since

$$pu^2 + (1 - p)d^2 - R^2 e^{\sigma^2 \Delta t} = -\frac{5\sigma^4 + 12r\sigma^2 + 12r^2}{12}\Delta t^2 + O(\Delta t^3).$$

Other choices of the set of parameter values in the binomial model have been proposed in the literature (see Problem 6.1). They all share the same order of accuracy in approximating (6.1.1b), but their analytic expressions are more cumbersome. This explains why the parameter values shown in (6.1.3) are most commonly used in binomial models.

6.1.2 Continuous Limits of the Binomial Model

Given the parameter values for u, d and p in (6.1.3), we consider the asymptotic limit $\Delta t \to 0$ of the binomial formula

$$c = [pc_u^{\Delta t} + (1 - p)c_d^{\Delta t}]\, e^{-r\Delta t}.$$

We would like to show that the Black–Scholes equation for the continuous option model is obtained as a result. Since the solution function of the binomial model is a grid function, so it is necessary to perform continuation of the grid function to its continuous extension such that the two functions agree with each other at the node points. The continuous analog of the binomial formula can be written as

$$c(S, t - \Delta t) = [pc(uS, t) + (1 - p)c(dS, t)]\, e^{-r\Delta t}. \tag{6.1.4}$$

For the convenience of presentation, we take the current time to be $t - \Delta t$ and S be the current asset value. Assuming sufficient continuity of $c(S, t)$, we perform the Taylor expansion of the binomial scheme at (S, t) as follows:

$$\begin{aligned} &-c(S, t - \Delta t) + [pc(uS, t) + (1 - p)c(dS, t)]e^{-r\Delta t} \\ &= \frac{\partial c}{\partial t}(S, t)\Delta t - \frac{1}{2}\frac{\partial^2 c}{\partial t^2}(S, t)\Delta t^2 + \cdots - (1 - e^{-r\Delta t})c(S, t) \\ &+ e^{-r\Delta t}\left\{[p(u-1) + (1-p)(d-1)]S\frac{\partial c}{\partial S}(S, t)\right. \\ &+ \frac{1}{2}[p(u-1)^2 + (1-p)(d-1)^2]S^2\frac{\partial^2 c}{\partial S^2}(S, t) \\ &\left. + \frac{1}{6}[p(u-1)^3 + (1-p)(d-1)^3]S^3\frac{\partial^3 c}{\partial S^3}(S, t) + \cdots\right\}. \end{aligned} \tag{6.1.5}$$

By observing thc rclation:

$$1 - e^{-r\Delta t} = r\Delta t + O(\Delta t^2),$$

it can be shown that

$$\begin{aligned} e^{-r\Delta t}\,[p(u-1) + (1-p)(d-1)] &= r\Delta t + O(\Delta t^2), \\ e^{-r\Delta t}\,[p(u-1)^2 + (1-p)(d-1)^2] &= \sigma^2\Delta t + O(\Delta t^2), \\ e^{-r\Delta t}\,[p(u-1)^3 + (1-p)(d-1)^3] &= O(\Delta t^2). \end{aligned}$$

Substituting the above results into (6.1.5), we obtain

$$\begin{aligned} &-c(S, t - \Delta t) + \big[pc(uS, t) + (1 - p)c(dS, t)\big]e^{-r\Delta t} \\ &= \left[\frac{\partial c}{\partial t}(S, t) + rS\frac{\partial c}{\partial S}(S, t) + \frac{\sigma^2}{2}S^2\frac{\partial^2 c}{\partial S^2}(S, t) - rc(S, t)\right]\Delta t + O(\Delta t^2). \end{aligned}$$

Since $c(S, t)$ satisfies the binomial formula (6.1.4), so we obtain

$$0 = \frac{\partial c}{\partial t}(S, t) + rS\frac{\partial c}{\partial S}(S, t) + \frac{\sigma^2}{2}S^2\frac{\partial^2 c}{\partial S^2}(S, t) - rc(S, t) + O(\Delta t).$$

In the limit $\Delta t \to 0$, the call value $c(S, t)$ obtained from the binomial model satisfies the Black–Scholes equation. We say that the binomial formula approximates the Black–Scholes equation to first-order accuracy in time.

Asymptotic Limit to the Black–Scholes Price Formula

We have seen that the continuous limit of the binomial formula tends to the Black–Scholes equation. One would expect that the call price formula for the n-period binomial model [see (2.2.25)] also tends to the Black–Scholes call price formula in the limit $n \to \infty$, or equivalently $\Delta t \to 0$ (since $n\Delta t$ is finite). Mathematically, we would like to show

$$\lim_{n\to\infty} [S\Phi(n,k,p') - XR^{-n}\Phi(n,k,p)] = SN(d_1) - Xe^{-r\tau}N(d_2), \tag{6.1.6}$$

where k is the minimum number of upward moves among the n binomial steps such that the call option expires in-the-money, and

$$d_1 = \frac{\ln\frac{S}{X} + \left(r + \frac{\sigma^2}{2}\right)\tau}{\sigma\sqrt{\tau}}, \quad d_2 = d_1 - \sigma\sqrt{\tau}.$$

The proof of the above asymptotic formula relies on the renowned result on the normal approximation to the binomial distribution. Let Y be the binomial random variable with parameters n and p, where n is the number of binomial trials and p is the probability of success. For large n, Y is approximately normal with mean np and variance $np(1-p)$.

To prove formula (6.1.6), it suffices to show

$$\lim_{n\to\infty} \Phi(n,k,p) = N\left(\frac{\ln\frac{S}{X} + (r - \frac{\sigma^2}{2})\tau}{\sigma\sqrt{\tau}}\right), \tag{6.1.7a}$$

and

$$\lim_{n\to\infty} \Phi(n,k,p') = N\left(\frac{\ln\frac{S}{X} + (r + \frac{\sigma^2}{2})\tau}{\sigma\sqrt{\tau}}\right), \quad \tau = T - t. \tag{6.1.7b}$$

The proof of (6.1.7a) will be presented below while that of (6.1.7b) is relegated to Problem 6.3.

Recall that $\Phi(n,k,p)$ is the risk neutral probability that the number of upward moves in the asset price is greater than or equal to k in the n-period binomial model, where p is the risk neutral probability of an upward move. Let j denote the random integer variable that gives the number of upward moves during the n periods. Consider

$$1 - \Phi(n,k,p) = P(j < k-1) = P\left(\frac{j - np}{\sqrt{np(1-p)}} < \frac{k-1-np}{\sqrt{np(1-p)}}\right), \tag{6.1.8}$$

where $\frac{j-np}{\sqrt{np(1-p)}}$ is the normalized binomial variable with zero mean and unit variance. Let S and S^* denote the known asset price at the current time and the random asset price at n periods later, respectively. Since S and S^* are related by $S^* = u^j d^{n-j} S$, the logarithm of the asset price ratio is also binomial variable and it is related linearly to j by the following relation:

$$\ln \frac{S^*}{S} = j \ln \frac{u}{d} + n \ln d. \tag{6.1.9}$$

For the binomial random variable j, its mean and variance are known to be $E[j] = np$ and $\text{var}(j) = np(1-p)$, respectively. Since $\ln \frac{S^*}{S}$ and j are linearly related, the mean and variance of $\ln \frac{S^*}{S}$ are given by

$$E\left[\ln \frac{S^*}{S}\right] = E[j] \ln \frac{u}{d} + n \ln d = n\left(p \ln \frac{u}{d} + \ln d\right)$$

$$\text{var}\left(\ln \frac{S^*}{S}\right) = \text{var}(j)\left(\ln \frac{u}{d}\right)^2 = np(1-p)\left(\ln \frac{u}{d}\right)^2.$$

In the limit $n \to \infty$, the mean and variance of the logarithm of the price ratio of the discrete binomial model and the continuous Black–Scholes model should agree with each other, that is,

$$\lim_{n\to\infty} n\left(p \ln \frac{u}{d} + \ln d\right) = \left(r - \frac{\sigma^2}{2}\right)(T-t) \tag{6.1.10a}$$

$$\lim_{n\to\infty} np(1-p)\left(\ln \frac{u}{d}\right)^2 = \sigma^2(T-t), \quad T = t + n\Delta t. \tag{6.1.10b}$$

Since k is the smallest nonnegative integer greater than or equal to $\dfrac{\ln \frac{X}{Sd^n}}{\ln \frac{u}{d}}$, we have

$$k - 1 = \frac{\ln \frac{X}{Sd^n}}{\ln \frac{u}{d}} - \alpha, \qquad \text{where} \quad 0 < \alpha \leq 1,$$

so that (6.1.8) can be rewritten as

$$\begin{aligned} 1 - \Phi(n,k,p) &= P(j < k-1) \\ &= P\left(\frac{j - np}{\sqrt{np(1-p)}} < \frac{\ln \frac{X}{S} - n(p \ln \frac{u}{d} + \ln d) - \alpha \ln \frac{u}{d}}{\sqrt{np(1-p)} \ln \frac{u}{d}}\right). \end{aligned} \tag{6.1.11}$$

In the limit $n \to \infty$, or equivalently $\Delta t \to 0$, the quantities $\sqrt{np(1-p)} \ln \frac{u}{d}$ and $n(p \ln \frac{u}{d} + \ln d)$ are finite [see (6.1.10a,b)] while $\alpha \ln \frac{u}{d}$ is $O(\sqrt{\Delta t})$. By virtue of the property of normal approximation to the binomial distribution, the normalized binomial variable $\frac{j-np}{\sqrt{np(1-p)}}$ becomes the standard normal random variable. Together with the asymptotic results in (6.1.10a,b), we obtain

$$\lim_{n\to\infty} \Phi(n,k,p) = 1 - N\left(\frac{\ln \frac{X}{S} - (r - \frac{\sigma^2}{2})\tau}{\sigma\sqrt{\tau}}\right) = N\left(\frac{\ln \frac{S}{X} + (r - \frac{\sigma^2}{2})\tau}{\sigma\sqrt{\tau}}\right), \tag{6.1.12}$$

where $\tau = T - t$.

6.1.3 Discrete Dividend Models

The binomial model can easily incorporate the effect of dividend yield paid by the underlying asset (see Problem 6.2). With some simplifying but reasonable assumptions, we can also incorporate discrete dividends into the discrete binomial model quite effectively.

First, we consider the naive construction of the binomial tree. Let S be the asset price at the current time which is $n\Delta t$ from expiry, and suppose a discrete dividend of amount D is paid at time between one time step and two time steps from the current time. The nodes in the binomial tree at two time steps from the current time would correspond to asset prices

$$u^2S - D, \quad S - D \quad \text{and} \quad d^2S - D,$$

since the asset price drops by the same amount as the dividend right after the dividend payment (see Fig. 6.1). Extending one time step further, there will be six nodes

$$(u^2S - D)u, (u^2S - D)d, (S - D)u, (S - D)d, (d^2S - D)u, (d^2S - D)d$$

instead of four nodes as in the usual binomial tree without the discrete dividend. This is because $(u^2S-D)d \neq (S-D)u$ and $(S-D)d \neq (d^2S-D)u$, so the interior nodes do not recombine. Extending one time step further, the number of nodes will grow to nine instead of five as in the usual binomial tree. In general, suppose a discrete dividend is paid in the future between k and $k+1$ time steps from the current time, then at $k+m$ time steps later from the current time, the number of nodes would be $(m+1)(k+1)$ rather than $k+m+1$ as in the usual reconnecting binomial tree. This is because each of the $k+1$ nodes at the time level right after the discrete dividend payment apparently serves as the tip of a binomial sub-tree. Each of these sub-trees generates $m+1$ nodes after m time steps. Hence, the total number of binomial nodes generated by all $k+1$ sub-trees is $(m+1)(k+1)$.

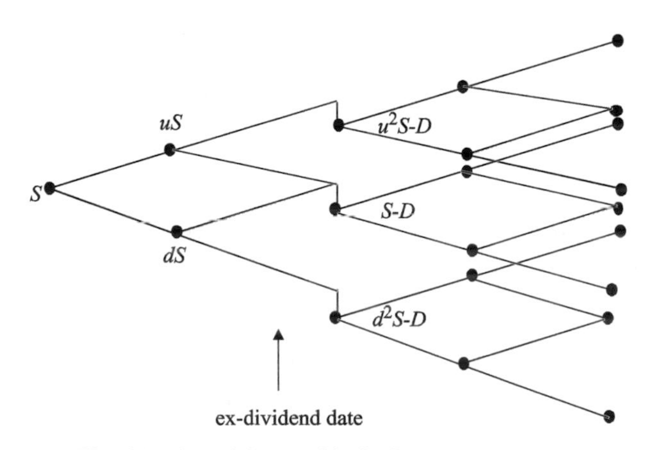

Fig. 6.1. Binomial tree with single discrete dividend.

The above difficulty of nodes exploding can be circumvented by splitting the asset price S_t into two parts: the risky component $\widetilde{S}_t$ that is stochastic and the remaining nonrisky component which is "reserved" for paying the discrete dividend (assumed to be deterministic) in the future. Suppose the dividend date is t^*, then at the current time t, the risky component $\widetilde{S}_t$ is given by (see Sect. 3.4)

$$\widetilde{S}_t = \begin{cases} S_t - De^{-r(t^*-t)}, & t \leq t^* \\ S_t, & t > t^*. \end{cases}$$

Let $\widetilde{\sigma}$ denote the volatility of $\widetilde{S}_t$ and assume $\widetilde{\sigma}$ to be constant rather than the volatility of S_t itself to be constant. Now, $\widetilde{\sigma}$ will be used instead of σ in the calculation of the binomial parameters: p, u and d, and a binomial tree is built to model the discrete jump process for $\widetilde{S}_t$. This assumption is similar in spirit as the common practice of using the Black–Scholes price formula with the asset price reduced by the present value of the sum of all future dividends. Now, the nodes in the tree for $\widetilde{S}_t$ become reconnected. To construct the reconnecting tree for S_t, at each node of the tree, the associated asset value is obtained by adding the sum of the present values of all future dividends to the risky component.

Let S and $\widetilde{S}$ denote the asset price and its risky component at the tip of the binomial tree, respectively, and let N denote the total number of time steps in the tree. Assume a discrete dividend D is paid at time t^*, which lies between the kth and $(k+1)$th time step. At the tip of the binomial tree, the risky component $\widetilde{S}$ is related to the asset price S by

$$S = \widetilde{S} + De^{-kr\Delta t}.$$

As an example, consider a binomial tree with four time steps and single discrete dividend D is paid between the second and third time step so that $N = 4$ and $k = 2$ (see Fig. 6.2). The present value of the dividend at the tip of the binomial tree is $De^{-2r\Delta t}$ so that the asset value at the tip (node P) is $\widetilde{S} + De^{-2r\Delta t}$. At node Q, which is one upward jump from the tip, the risky component becomes $\widetilde{S}u$ while the present value of the dividend is $De^{-r\Delta t}$. With one upward jump and two downward jumps, we reach node R. The node is three time steps from the tip and so the dividend

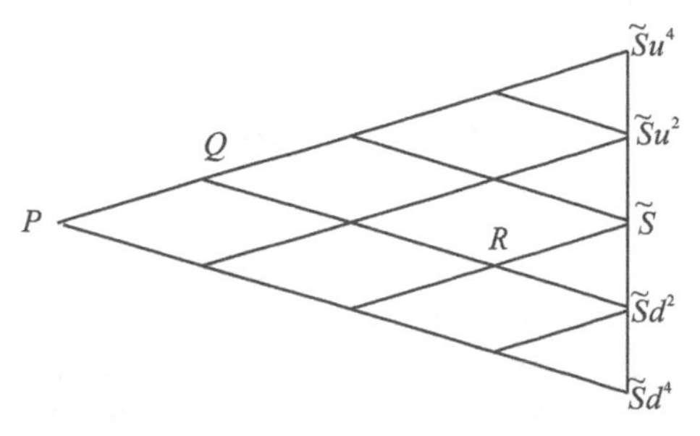

Fig. 6.2. Construction of a reconnecting binomial tree with single discrete dividend D, $N = 4$ and $k = 2$. The asset value at nodes P, Q and R are $\widetilde{S} + De^{-2r\Delta t}$, $\widetilde{S}u + De^{-r\Delta t}$ and $\widetilde{S}d$, respectively.

has been paid. Therefore, the asset value is simply $\widetilde{S}ud^2 = \widetilde{S}d$. In general, the asset value at the (n, j)th node (which corresponds to n time steps from the tip and j upward jumps among the n steps) is given by

$$\widetilde{S}u^j d^{n-j} + De^{-(k-n)r\Delta t}\mathbf{1}_{\{n\leq k\}}, \quad n = 0, 1, \cdots, N \text{ and } j = 0, 1, \cdots, n.$$

Once the reconnecting tree for the discrete asset price process is available, the option values at the nodes can be found using the binomial formula following the backward induction procedure. It is quite straightforward to generalize the above splitting approach to option models with several discrete dividends.

6.1.4 Early Exercise Feature and Callable Feature

Recall that an American option can be terminated prematurely due to possibility of early exercise by the holder. Without the early exercise privilege, risk neutral valuation leads to the usual binomial formula

$$V_{cont} = \frac{pV_u^{\Delta t} + (1-p)V_d^{\Delta t}}{R}.$$

Here, we use V_{cont} to represent the option value at the state of continuation when the option is kept alive. To incorporate the early exercise feature embedded in an American option, we compare at each binomial node the continuation value V_{cont} with the option's intrinsic value, which is the payoff upon early exercise. The following simple dynamic programming procedure is applied at each binomial node

$$V = \max(V_{cont}, h(S)), \tag{6.1.13}$$

where $h(S)$ is the exercise payoff function.

As an example, we consider the numerical valuation of an American put option. First, we construct the usual binomial tree and let N denote the total number of time steps in the tree. Let S_j^n and P_j^n denote the asset price and put value at the (n, j)th node, respectively. The intrinsic value of a put option is $X - S_j^n$ at the (n, j)th node, where X is the strike price. Hence, the dynamic programming procedure applied at each node is given by

$$P_j^n = \max\left(\frac{pP_{j+1}^{n+1} + (1-p)P_j^{n+1}}{R}, X - S_j^n\right), \tag{6.1.14}$$

where $n = 0, 1, \cdots, N-1$, and $j = 0, 1, \cdots, n$.

Also, the binomial scheme can be easily modified to incorporate additional embedded features in an American option contract. For example, the callable feature entitles the issuer to buy back the American option at any time at a predetermined call price. Upon issuer's call, the holder can choose either to exercise the option or receive the call price as cash. The interplay between the holder and issuer can be seen as a game option between the two counterparties. Consider a callable American put option with call price K. To price this callable put option, the dynamic programming procedure applied at each node is modified as follows (Kwok and Wu, 2000, Chap. 5)

$$P_j^n = \min\left(\max\left(\frac{pP_{j+1}^{n+1} + (1-p)P_j^{n+1}}{R}, X - S_j^n\right),\right.$$
$$\left.\max(K, X - S_j^n)\right). \tag{6.1.15}$$

The first term: $\max(\frac{pP_{j+1}^{n+1}+(1-p)P_j^{n+1}}{R}, X - S_j^n)$ represents the optimal strategy of the holder, given no call of the option by the issuer. Upon call by the issuer, the payoff is given by the second term: $\max(K, X - S_j^n)$ since the holder can either receive the call price of cash amount K or exercise the option to receive the exercise payoff $X - S_j^n$. From the perspective of the issuer, he or she chooses to call or restrain from calling so as to minimize the option value with reference to the possible actions of the holder. Hence, the value of the callable American put option at the node is given by taking the minimum value of the above two terms.

Other enhanced numerical schemes for valuation of American options with various embedded features have been proposed in the literature (Dempster and Hutton, 1999). A good survey of comparison of the performance of different numerical schemes can be found in Broadie and Detemple (1996).

6.1.5 Trinomial Schemes

In the binomial model, we assume a two-jump process for the asset price over each time step. One may query whether accuracy and reliability of option valuation can be improved by allowing a three-jump process for the stochastic asset price. In a trinomial model, the asset price S is assumed to jump to either uS, mS or dS after one time period Δt, where $u > m > d$. We consider a trinomial formula of option valuation of the form

$$V = \frac{p_1 V_u^{\Delta t} + p_2 V_m^{\Delta t} + p_3 V_d^{\Delta t}}{R}, \qquad R = e^{r\Delta t}. \tag{6.1.16}$$

Here, $V_u^{\Delta t}$ denotes the option price when the asset price takes the value uS at one period later, and there is a similar interpretation for $V_m^{\Delta t}$ and $V_d^{\Delta t}$. The new trinomial model may allow greater freedom in the selection of the parameters to achieve some desirable properties, like avoiding instability to roundoff errors, attaining a faster rate of convergence, etc. The tradeoff is a decrease of computational efficiency in general because a trinomial scheme involves more computational steps compared to that of a binomial scheme (see Problem 6.7). Cox, Ross and Rubinstein (1979) cautioned that the trinomial model (unlike the binomial model) will not lead to an option pricing formula based solely on arbitrage considerations. However, a direct link between the approximating process of the asset price and arbitrage strategy is not essential. In fact, any contingent claim can be valued by computing conditional expectation under an appropriate pricing measure. One may adopt various forms of a discrete stochastic process to approximate the underlying asset price process, and in turn these lead to different numerical schemes.

Recall that under the risk neutral measure, $\ln \frac{S_{t+\Delta t}}{S_t}$ is assumed to be normally distributed with mean $(r - \frac{\sigma^2}{2})\Delta t$ and variance $\sigma^2 \Delta t$. Alternatively, we may write

$$\ln S_{t+\Delta t} = \ln S_t + \zeta, \tag{6.1.17}$$

where ζ is a normal random variable with mean $(r - \frac{\sigma^2}{2})\Delta t$ and variance $\sigma^2 \Delta t$. Kamrad and Ritchken (1991) proposed to approximate ζ by an approximate discrete random variable ζ^a with the following distribution

$$\zeta^a = \begin{cases} v & \text{with probability } p_1 \\ 0 & \text{with probability } p_2 \\ -v & \text{with probability } p_3, \end{cases}$$

where $v = \lambda\sigma\sqrt{\Delta t}$ and $\lambda \geq 1$ (see an explanation below). Note that with the choice of $\lambda = 1$ and $p_2 = 0$, ζ^a reduces to the same form as that of the binomial model. The corresponding values for u, m and d in the trinomial scheme are: $u = e^v$, $m = 1$ and $d = e^{-v}$. To find the probability values p_1, p_2 and p_3, the mean and variance of ζ^a are chosen to be equal to those of ζ. These lead to

$$E[\zeta^a] = v(p_1 - p_3) = \left(r - \frac{\sigma^2}{2}\right)\Delta t \tag{6.1.18a}$$

$$\operatorname{var}(\zeta^a) = v^2(p_1 + p_3) - v^2(p_1 - p_3)^2 = \sigma^2 \Delta t. \tag{6.1.18b}$$

From (6.1.18a), we see that $v^2(p_1 - p_3)^2 = O(\Delta t^2)$. Suppose we seek an approximation up to $O(\Delta t)$, we may drop this term from (6.1.18b). We then have

$$v^2(p_1 + p_3) = \sigma^2 \Delta t, \tag{6.1.18c}$$

while accuracy of $O(\Delta t)$ is maintained. Without this simplication, the final expressions for p_1, p_2 and p_3 would become more cumbersome. Finally, the probabilities must be summed to one so that

$$p_1 + p_2 + p_3 = 1. \tag{6.1.18d}$$

We then solve (6.1.18a,c,d) together to obtain

$$p_1 = \frac{1}{2\lambda^2} + \frac{(r - \frac{\sigma^2}{2})\sqrt{\Delta t}}{2\lambda\sigma} \tag{6.1.19a}$$

$$p_2 = 1 - \frac{1}{\lambda^2} \tag{6.1.19b}$$

$$p_3 = \frac{1}{2\lambda^2} - \frac{(r - \frac{\sigma^2}{2})\sqrt{\Delta t}}{2\lambda\sigma}. \tag{6.1.19c}$$

It is now apparent to see why we require $\lambda \geq 1$. If otherwise, p_2 would become negative. By choosing different values for the free parameter λ, we can obtain a range of probability values. With the choice of $\lambda = 1$, we obtain $p_2 = 0$. It is quite desirable to see that the popular Cox–Ross–Rubinstein binomial scheme happens to be a special case of this class of trinomial schemes.

Though a trinomial scheme is seen to require more computational work than that of a binomial scheme, one can show easily that a trinomial scheme with n steps requires less computational work (measured in terms of number of multiplications and additions) than a binomial scheme with $2n$ steps (see Problem 6.7). The numerical tests performed by Kamrad and Ritchken (1991) reveal that the trinomial scheme with n steps invariably performs better in accuracy than the binomial scheme with $2n$ steps. In terms of order of accuracy, both the binomial scheme and trinomial scheme satisfy the Black–Scholes equation to first-order accuracy (see Problem 6.10).

Multistate Options

The extension of the above approach to two-state options is quite straightforward. First, we assume the joint density of the prices of the two underlying assets S_1 and S_2 to be bivariate lognormal. Let σ_i be the volatility of asset price S_i, $i = 1, 2$ and ρ be the correlation coefficient between the two lognormal diffusion processes. Let S_i and $S_i^{\Delta t}$ denote, respectively, the price of asset i at the current time and one period Δt later. Under the risk neutral measure, we have

$$\ln \frac{S_i^{\Delta t}}{S_i} = \zeta_i, \qquad i = 1, 2,$$

where ζ_i is a normal random variable with mean $(r - \frac{\sigma_i^2}{2})\Delta t$ and variance $\sigma_i^2 \Delta t$. The instantaneous correlation coefficient between ζ_1 and ζ_2 is ρ. The joint bivariate normal processes $\{\zeta_1, \zeta_2\}$ is approximated by a pair of joint discrete random variables $\{\zeta_1^a, \zeta_2^a\}$ with the following distribution

ζ_1^a	ζ_2^a	Probability
v_1	v_2	p_1
v_1	$-v_2$	p_2
$-v_1$	$-v_2$	p_3
$-v_1$	v_2	p_4
0	0	p_5

where $v_i = \lambda_i \sigma_i \sqrt{\Delta t}$, $i = 1, 2$. There are five probability values to be determined. In our approximation procedures, we set the first two moments of the approximating distribution (including the covariance) to the corresponding moments of the continuous distribution. Equating the corresponding means gives

$$E[\zeta_1^a] = v_1(p_1 + p_2 - p_3 - p_4) = \left(r - \frac{\sigma_1^2}{2}\right)\Delta t \tag{6.1.20a}$$

$$E[\zeta_2^a] = v_2(p_1 - p_2 - p_3 + p_4) = \left(r - \frac{\sigma_2^2}{2}\right)\Delta t. \tag{6.1.20b}$$

By equating the corresponding variances and covariance to $O(\Delta t)$ accuracy, we have

$$\text{var}(\zeta_1^a) = v_1^2(p_1 + p_2 + p_3 + p_4) = \sigma_1^2 \Delta t \tag{6.1.20c}$$

$$\text{var}(\zeta_2^a) = v_2^2(p_1 + p_2 + p_3 + p_4) = \sigma_2^2 \Delta t \tag{6.1.20d}$$

$$E[\zeta_1^a \zeta_2^a] = v_1 v_2(p_1 - p_2 + p_3 - p_4) = \sigma_1 \sigma_2 \rho \Delta t. \tag{6.1.20e}$$

So that (6.1.20c,d) are consistent, we must set $\lambda_1 = \lambda_2$. Writing $\lambda = \lambda_1 = \lambda_2$, we have the following four independent equations for the five probability values

$$p_1 + p_2 - p_3 - p_4 = \frac{(r - \frac{\sigma_1^2}{2})\sqrt{\Delta t}}{\lambda \sigma_1}$$

$$p_1 - p_2 - p_3 + p_4 = \frac{(r - \frac{\sigma_2^2}{2})\sqrt{\Delta t}}{\lambda \sigma_2}$$

$$p_1 + p_2 + p_3 + p_4 = \frac{1}{\lambda^2}$$

$$p_1 - p_2 + p_3 - p_4 = \frac{\rho}{\lambda^2}.$$

Noting that the probabilities must be summed to one, this gives the remaining condition as

$$p_1 + p_2 + p_3 + p_4 + p_5 = 1.$$

The solution of the above linear algebraic system of five equations gives

$$p_1 = \frac{1}{4}\left[\frac{1}{\lambda^2} + \frac{\sqrt{\Delta t}}{\lambda}\left(\frac{r - \frac{\sigma_1^2}{2}}{\sigma_1} + \frac{r - \frac{\sigma_2^2}{2}}{\sigma_2}\right) + \frac{\rho}{\lambda^2}\right] \tag{6.1.21a}$$

$$p_2 = \frac{1}{4}\left[\frac{1}{\lambda^2} + \frac{\sqrt{\Delta t}}{\lambda}\left(\frac{r - \frac{\sigma_1^2}{2}}{\sigma_1} - \frac{r - \frac{\sigma_2^2}{2}}{\sigma_2}\right) - \frac{\rho}{\lambda^2}\right] \tag{6.1.21b}$$

$$p_3 = \frac{1}{4}\left[\frac{1}{\lambda^2} + \frac{\sqrt{\Delta t}}{\lambda}\left(-\frac{r - \frac{\sigma_1^2}{2}}{\sigma_1} - \frac{r - \frac{\sigma_2^2}{2}}{\sigma_2}\right) + \frac{\rho}{\lambda^2}\right] \tag{6.1.21c}$$

$$p_4 = \frac{1}{4}\left[\frac{1}{\lambda^2} + \frac{\sqrt{\Delta t}}{\lambda}\left(-\frac{r - \frac{\sigma_1^2}{2}}{\sigma_1} + \frac{r - \frac{\sigma_2^2}{2}}{\sigma_2}\right) - \frac{\rho}{\lambda^2}\right] \tag{6.1.21d}$$

$$p_5 = 1 - \frac{1}{\lambda^2}, \qquad \lambda \geq 1 \text{ is a free parameter.} \tag{6.1.21e}$$

For convenience, we write $u_i = e^{v_i}$, $d_i = e^{-v_i}$, $i = 1, 2$. Let V denote the price of a two-state option with underlying asset prices S_1 and S_2. Also, let $V_{u_1 u_2}^{\Delta t}$ denote the option price at one time period later with asset prices $u_1 S_1$ and $u_2 S_2$, and similar meaning for $V_{u_1 d_2}^{\Delta t}$, $V_{d_1 u_2}^{\Delta t}$ and $V_{d_1 d_2}^{\Delta t}$. We let $V_{0,0}^{\Delta t}$ denote the option price one period later with no jumps in asset prices. The corresponding five-point formula for the two-state trinomial model can be expressed as (Kamrad and Ritchken, 1991)

$$V = (p_1 V_{u_1 u_2}^{\Delta t} + p_2 V_{u_1 d_2}^{\Delta t} + p_3 V_{d_1 d_2}^{\Delta t} + p_4 V_{d_1 u_2}^{\Delta t} + p_5 V_{0,0}^{\Delta t})/R. \tag{6.1.22}$$

In particular, when $\lambda = 1$, we have $p_5 = 0$ and the above five-point formula reduces to the four-point formula.

The presence of the free parameter λ in the five-point formula provides the flexibility to better explore convergence behavior of the discrete pricing formula. With a proper choice of λ, Kamrad and Ritchken (1991) observed from their numerical experiments that convergence of the numerical values obtained from the five-point formula to the continuous solution is invariably smoother and more rapid than those obtained from the four-point formula. The extension of the present approach to the three-state option models can be derived in a similar manner (see Problem 6.13).

6.1.6 Forward Shooting Grid Methods

For path dependent options, the option value also depends on the path function $F_t = F(S, t)$ defined specifically for the given nature of path dependence. For example, the path dependence may be defined by the minimum asset price realized over a specific time period. In order to reflect the impact of path dependence on the option value, it is necessary to find the corresponding option values at each node in the lattice tree for all possible values of F_t that can occur. In order that the numerical scheme competes well in terms of efficiency, it is desirable that the value $F_{t+\Delta t}$ can be computed easily from F_t and $S_{t+\Delta t}$ (that is, the path function is Markovian) and the number of alternative values for $F(S, t)$ cannot grow too large with an increasing number of binomial steps. The approach of appending an auxiliary state vector at each node in the lattice tree to model the correlated evolution of F_t with S_t is commonly called the *forward shooting grid* (FSG) *method*.

The FSG approach was pioneered by Hull and White (1993b) for pricing American and European Asian and lookback options. A systematic framework of constructing the FSG schemes for pricing path dependent options was presented by Barraquand and Pudet (1996). Forsyth, Vetzal and Zvan (2002) showed that convergence of the numerical solutions of the FSG schemes for pricing Asian arithmetic averaging options depend on the method of interpolation of the average asset values between neighboring lattice nodes. Jiang and Dai (2004) used the notion of viscosity solution to show uniform convergence of the FSG schemes for pricing American and European arithmetic Asian options.

For some exotic path dependent options, like an option with the window Parisian feature of knock-out (see Problem 6.14), the governing option pricing equation cannot be derived. However, by relating the correlated evolution of the path dependent state variable with the asset price process, it becomes feasible to devise the FSG schemes for pricing these exotic options.

Consider a trinomial tree whose probabilities of upward, zero and downward jump of the asset price are denoted by p_u, p_0 and p_d, respectively. Let $V_{j,k}^n$ denote the numerical option value of the exotic path dependent option at the nth-time level

(n time steps from the tip of the tree). The index j denotes j upward jumps among n moves from the initial asset value while k denotes the numbering index for the various possible values of the augmented state variable F_t at the (n, j)th node in the trinomial tree. Let G denote the function that describes the correlated evolution of F_t with S_t over the time interval Δt, that is,

$$F_{t+\Delta t} = G(t, F_t, S_{t+\Delta t}). \tag{6.1.23}$$

Let $g(k, j; n)$ denote the grid function which is considered as the discrete analog of the correlated evolution function G. The trinomial version of the FSG scheme can be represented as

$$V_{j,k}^n = \left[p_u V_{j+1,g(k,j+1;n)}^{n+1} + p_0 V_{j,g(k,j;n)}^{n+1} + p_d V_{j-1,g(k,j-1;n)}^{n+1}\right] e^{-r\Delta t}, \tag{6.1.24}$$

where $e^{-r\Delta t}$ is the discount factor over each time interval Δt. The numbering index changes from k to $g(k, j+1; n)$ when the asset value encounters an upward jump from the (n, j)th node to the $(n+1, j+1)$th node, and a similar adjustment on k when the asset value has zero or downward jump. To price a specific path dependent option, the design of the FSG algorithm requires the specification of the grid function $g(k, j; n)$. We illustrate how to find $g(k, j; n)$ for various types of path dependent options, which include the barrier options with the Parisian style of knock-out and the floating strike arithmetic averaging options.

Barrier Options with Parisian Style of Knock-Out

The one-touch breaching of barrier in barrier options has the undesirable effect of knocking out the option when the asset price spikes, no matter how briefly the spiking occurs. Hedging barrier options may become difficult when the asset price is very close to the barrier. In the foreign exchange markets, market volatility may increase around popular barrier levels due to plausible price manipulation aimed at activating knock-out.

To circumvent the spiking effect and short-period price manipulation, various forms of Parisian style of knock-out provision have been proposed in the literature. The Parisian style of knock-out is activated only when the underlying asset price breaches the barrier for a prespecified period of time. The breaching can be counted consecutively or cumulatively. In actual market practice, breaching is monitored at discrete time instants rather than continuously, so the number of breaching occurrences at the monitoring instants is counted. We would like to derive the FSG scheme for pricing barrier options with the cumulative Parisian style of knock-out. The construction of the FSG schemes for the moving window Parisian feature is relegated to Problems 6.14. The application of the FSG approach to price convertible bonds with the Parisian style of soft call requirement can be found in Lau and Kwok (2004).

Cumulative Parisian Feature

Let M denote the prespecified number of cumulative breaching occurrences that is required to activate knock-out in a barrier option, and let k be the integer index that counts the number of breaching so far. Let B denote the down barrier associated

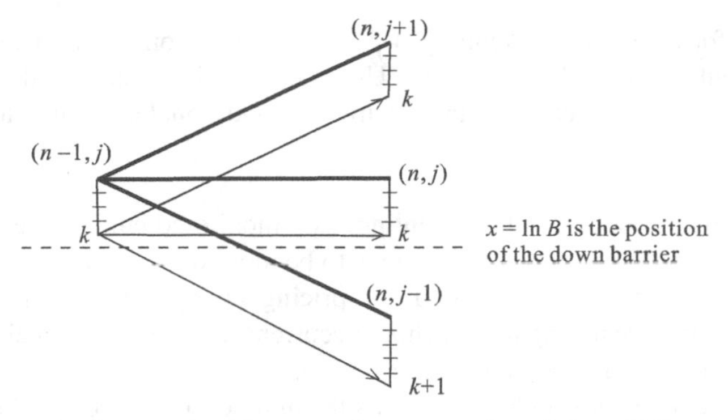

Fig. 6.3. Schematic diagram that illustrates the construction of the grid function $g_{cum}(k, j)$ that models the path dependence of the cumulative Parisian feature. The down barrier ln B is placed mid-way between two horizontal rows of trinomial nodes. Here, the nth-time level is a monitoring instant.

with the knock-out feature. Now, the augmented path dependent state variable at each node is the index k. The value of k is not changed except at a time step which corresponds to an instant at which breaching of the barrier is monitored. Let $V_{j,k}^n$ denote the value of the option with cumulative Parisian counting index k at the (n, j)th node in a trinomial tree. Let x_j denote the value of $x = \ln S$ that corresponds to j upward jumps in the trinomial tree. When $n\Delta t$ happens to be a monitoring instant, the index k increases its value by 1 if the asset price S falls on or below the barrier B, that is, $x_j \le \ln B$. To model the path dependence of the cumulative Parisian feature, the appropriate choice of the grid function $g_{cum}(k, j)$ is defined by

$$g_{cum}(k, j) = k + \mathbf{1}_{\{x_j \le \ln B\}}. \tag{6.1.25}$$

Note that $g_{cum}(k, j)$ has no dependence on n since the correlated evolution function G has no time dependence. The schematic diagram that illustrates the construction of the grid function $g_{cum}(k, j)$ is shown in Fig. 6.3.

When $n\Delta t$ is not a monitoring instant, the trinomial tree calculations proceed like those for usual options. The FSG algorithm for pricing an option with the cumulative Parisian feature can be summarized as

$$V_{j,k}^{n-1} = \begin{cases} p_u V_{j+1,k}^n + p_0 V_{j,k}^n + p_d V_{j-1,k}^n \\ \qquad \text{if } n\Delta t \text{ is not a monitoring instant} \\ p_u V_{j+1,g_{cum}(k,j+1)}^n + p_0 V_{j,g_{cum}(k,j)}^n + p_d V_{j-1,g_{cum}(k,j-1)}^n \\ \qquad \text{if } n\Delta t \text{ is a monitoring instant} \end{cases}. \tag{6.1.26}$$

Let M be the number of breaching occurrences counted cumulatively that is required to activate knock-out of the option. Assuming no rebate is paid upon knock-out, the value of the option becomes zero when $k = M$. In typical FSG calculations, it is necessary to start with $V_{j,M-1}^n$, then $V_{j,M-2}^n, \cdots$, and proceed down until the index k hits 0. We compute $V_{j,M-1}^n$ by setting $k = M - 1$ in (6.1.26) and observe that

$V_{j,M}^n = 0$ for all n and j. Actually, $V_{j,M-1}^n$ is the option value of the one-touch down-and-out option at the same node. Under cumulative counting, the option with $M-1$ breaching occurrences so far requires one additional breaching to knock out.

Remarks.

1. The pricing of options with the continuously monitored cumulative Parisian feature is obtained by setting all time steps to be monitoring instants.
2. The computational time required for pricing an option with the cumulative Parisian feature requiring M breaching occurrences to knock out is about M times that of a one-touch knock-out barrier option.
3. The consecutive Parisian feature counts the number of consecutive breaching occurrences that the asset price stays in the knock-out region. The count is reset to zero once the asset price moves out from the knock-out region. Assuming B to be the down barrier, the appropriate grid function $g_{con}(k, j)$ in the FSG algorithm is given by

$$g_{con}(k, j) = (k+1)\mathbf{1}_{\{x_j \leq \ln B\}}. \tag{6.1.27}$$

Floating Strike Arithmetic Averaging Call

To price an Asian option, we find the option value at each node for all possible values of the path function $F(S, t)$ that can occur at that node. Unfortunately, the number of possible values for the averaging value F at a binomial node for arithmetic averaging options grows exponentially at 2^n, where n is the number of time steps from the tip of the binomial tree. Therefore, the binomial schemes that place no constraint on the number of possible F values at a node become computationally infeasible. A possible remedy is to restrict the possible values for F to a certain set of predetermined values. The option value $V(S, F, t)$ for other values of F is obtained from the known values of V at predetermined F values by interpolation between nodal values (Barraquand and Pudet, 1996; Forsyth, Vetzal and Zvan, 2002). The methods of interpolation include the nearest node interpolation, linear and quadratic interpolation.

We illustrate the interpolation technique through the construction of the FSG algorithm for pricing the floating strike arithmetic averaging call option. First, we define the arithmetic averaging state variable by

$$A_t = \frac{1}{t}\int_0^t S_u \, du. \tag{6.1.28a}$$

The terminal payoff of the floating strike Asian call option is given by $\max(S_T - A_T, 0)$, where A_T is the arithmetic average of S_t over the time period $[0, T]$. For a given time step Δt, we fix the stepwidths to be

$$\Delta W = \sigma\sqrt{\Delta t} \text{ and } \Delta Y = \rho \Delta W, \quad \rho < 1,$$

and define the possible values for S_t and A_t at the nth time step by

$$S_j^n = S_0 e^{j\Delta W} \text{ and } A_k^n = S_0 e^{k\Delta Y},$$

where j and k are integers, and S_0 is the asset price at the tip of the binomial tree. We take $1/\rho$ to be an integer. The larger integer value chosen for $1/\rho$, the finer the quantification of the average asset value. By differentiating (6.1.28a) with respect to t, we obtain

$$d(tA_t) = S_t\, dt,$$

and whose discrete analog is given by

$$A_{t+\Delta t} = \frac{(t+\Delta t)A_t + \Delta t\; S_{t+\Delta t}}{t+2\Delta t}. \tag{6.1.28b}$$

Consider the binomial procedure at the (n, j)th node, suppose we have an upward move in asset price from S_j^n to S_{j+1}^{n+1} and let $A_{k^+(j)}^{n+1}$ be the corresponding new value of A_t changing from A_k^n. Setting $A_0^0 = S_0$, the equivalence of (6.1.28b) is given by

$$A_{k^+(j)}^{n+1} = \frac{(n+1)A_k^n + S_{j+1}^{n+1}}{n+2}. \tag{6.1.29a}$$

Similarly, for a downward move in asset price from S_j^n to S_{j-1}^{n+1}, A_k^n changes to $A_{k^-(j)}^{n+1}$ where

$$A_{k^-(j)}^{n+1} = \frac{(n+1)A_k^n + S_{j-1}^{n+1}}{n+2}. \tag{6.1.29b}$$

Note that $A_{k^\pm(j)}^{n+1}$ in general do not coincide with $A_{k'}^{n+1} = S_0 e^{k'\Delta Y}$, for some integer k'. We define the integers $k_{floor}^{\pm}$ such that $A_{k_{floor}^{\pm}}^{n+1}$ are the largest possible $A_{k'}^{n+1}$ values less than or equal to $A_{k^\pm(j)}^{n+1}$. Accordingly, we compute the indexes $k^{\pm}(j)$ by

$$k^{\pm}(j) = \frac{\ln \frac{(n+1)e^{k\Delta Y} + e^{(j\pm 1)\Delta W}}{n+2}}{\Delta Y}. \tag{6.1.30}$$

We then set $k_{floor}^{+} = \text{floor}(k^+(j))$ and $k_{floor}^{-} = \text{floor}(k^-(j))$, where floor$(x)$ denotes the largest integer less than or equal to x.

What would be the possible range of k at the nth time step? We observe that the average A_t must lie between the maximum asset value S_n^n and the minimum asset value S_{-n}^n, so k must lie between $-\frac{n}{\rho} \le k \le \frac{n}{\rho}$. Unless that ρ assumes a very small value, the number of predetermined values for A_t is in general manageable.

Consider A_ℓ^n, where ℓ is in general a real number. We write $\ell_{floor} = \text{floor}(\ell)$ and let $\ell_{ceil} = \ell_{floor} + 1$, then A_ℓ^n lies between $A_{\ell_{floor}}^n$ and $A_{\ell_{ceil}}^n$. Though the number of possible values of ℓ grows exponentially with the number of time steps in the binomial tree, both ℓ_{floor} and ℓ_{ceil} at the nth time level assume an integer value lying between $-\frac{n}{\rho}$ and $\frac{n}{\rho}$. Let $c_{j,\ell}^n$ denote the Asian call value at the (n, j)th node with the averaging state variable assuming the value A_ℓ^n, and similar notations for $c_{j,\ell_{floor}}^n$ and $c_{j,\ell_{ceil}}^n$. For a noninteger value ℓ, $c_{j,\ell}^n$ is approximated through interpolation using the call values at the neighboring nodes. We approximate $c_{j,\ell}^n$ in terms of $c_{j,\ell_{floor}}^n$ and $c_{j,\ell_{ceil}}^n$ by the following linear interpolation formula

$$c^n_{j,\ell} = \epsilon_\ell c^n_{j,\ell_{ceil}} + (1 - \epsilon_\ell) c^n_{j,\ell_{floor}}, \tag{6.1.31}$$

where

$$\epsilon_\ell = \frac{\ln A^n_\ell - \ln A^n_{\ell_{floor}}}{\Delta Y}.$$

By applying the above linear interpolation formula [taking ℓ to be $k^+(j)$ and $k^-(j)$ successively], the FSG algorithm with linear interpolation for pricing the floating strike arithmetic averaging call option is given by

$$\begin{aligned} c^n_{j,k} &= e^{-r\Delta t}\big[pc^{n+1}_{j+1,k^+(j)} + (1-p)c^{n+1}_{j-1,k^-(j)}\big] \\ &= e^{-r\Delta t}\big\{p\big[\epsilon_{k^+(j)}c^{n+1}_{j+1,k^+_{ceil}} + (1-\epsilon_{k^+(j)})c^{n+1}_{j+1,k^+_{floor}}\big] \\ &\quad + (1-p)\big[\epsilon_{k^-(j)}c^{n+1}_{j-1,k^-_{ceil}} + (1-\epsilon_{k^-(j)})c^{n+1}_{j-1,k^-_{floor}}\big]\big\}, \end{aligned} \tag{6.1.32}$$

$n = N-1, \cdots, 0$, $j = -n, \cdots, n$, k is an integer between $-\frac{n}{\rho}$ and $\frac{n}{\rho}$, $k^\pm(j)$ are given by (6.1.30), and

$$\epsilon_{k^\pm(j)} = \frac{\ln A^{n+1}_{k^\pm(j)} - \ln A^{n+1}_{k^\pm_{floor}}}{\Delta Y}. \tag{6.1.33}$$

The final condition is

$$\begin{aligned} c^N_{j,k} &= \max(S^N_j - A^N_k, 0) \\ &= \max(S_0 e^{j\Delta W} - S_0 e^{k\Delta Y}, 0), \quad j = -N, \cdots, N, \end{aligned} \tag{6.1.34}$$

and k is an integer between $-\frac{N}{\rho}$ and $\frac{N}{\rho}$. At each terminal node (N, j), we compute all possible payoff values of the Asian call option with varying values of k. To proceed with the backward induction procedure, at a typical (n, j)th node, we find all possible call values with varying integer values of k lying between $-\frac{n}{\rho}$ and $\frac{n}{\rho}$ using (6.1.32). For a given integer value k, we compute $k^\pm(j)$ and $\epsilon_{k^\pm(j)}$ using (6.1.30) and (6.1.33), respectively.

As a cautious remark, Forsyth, Vetzal and Zvan (2002) proved that the FSG algorithm using the nearest lattice point interpolation may exhibit large errors as the number of time steps becomes large. They also showed that when the linear interpolation method is used, the FSG scheme converges to the correct solution plus a constant error term which cannot be reduced by decreasing the size of the time step. Some of these shortcomings may be remedied by adopting a quadratic interpolation of nodal values.

6.2 Finite Difference Algorithms

Finite difference methods are popular numerical techniques for solving science and engineering problems modeled by differential equations. The earliest application of the finite difference methods to option valuation was performed by Brennan and Schwartz (1978). Tavella and Randall (2000) presented a comprehensive survey of finite difference methods applied to numerical pricing of financial instruments. In the construction of finite difference schemes, we approximate the differential operators in the governing differential equation of the option model by appropriate finite difference operators, hence the name of this approach.

In this section, we first show how to develop the family of explicit finite difference schemes for numerical valuation of options. Interestingly, the binomial and trinomial schemes can be shown to be members in the family of explicit schemes. In an explicit scheme, the option values at the computational nodes along the new time level can be calculated explicitly from known option values at the nodes along the old time level. However, if the discretization of the spatial differential operators involves option values at the nodes along the new time level, then the finite difference calculations involve solution of a system of linear equations at every time step. We discuss how to construct the implicit finite difference schemes and illustrate the method of their solution using the effective Thomas algorithm. We also consider how to apply the finite difference methods for solving American-style option models. In the front fixing method, we apply a transformation of variable so that the *front* or free boundary associated with the optimal exercise price is transformed to a *fixed* boundary of the solution domain. Unlike the binomial and trinomial schemes, the construction procedure of a finite difference scheme allows for direct incorporation of the boundary conditions associated with the option models. We illustrate the methods of implementing the Dirichlet condition in barrier options and Neumann condition in lookback options. To resolve the computational nuisance arising from nondifferentiability of the "initial" condition, we introduce several effective smoothing techniques that lessen the deterioration in accuracy due to a nonsmooth terminal payoff.

6.2.1 Construction of Explicit Schemes

By applying the transformed variable $x = \ln S$, the Black–Scholes equation for the price of a European option becomes

$$\frac{\partial V}{\partial \tau} = \frac{\sigma^2}{2}\frac{\partial^2 V}{\partial x^2} + \left(r - \frac{\sigma^2}{2}\right)\frac{\partial V}{\partial x} - rV, \quad -\infty < x < \infty, \tag{6.2.1a}$$

where $V = V(x, \tau)$ is the option value. Here, we adopt time to expiry τ as the temporal variable. Suppose we define $W(x, \tau) = e^{r\tau} V(x, \tau)$, then $W(x, \tau)$ satisfies

$$\frac{\partial W}{\partial \tau} = \frac{\sigma^2}{2}\frac{\partial^2 W}{\partial x^2} + \left(r - \frac{\sigma^2}{2}\right)\frac{\partial W}{\partial x}, \quad -\infty < x < \infty. \tag{6.2.1b}$$

To derive the finite difference algorithm, we first transform the domain of the continuous problem $\{(x, \tau) : -\infty < x < \infty, \tau \geq 0\}$ into a discretized domain. The infinite extent of $x = \ln S$ in the continuous problem is approximated by a finite truncated interval $[-M_1, M_2]$, where M_1 and M_2 are sufficiently large positive

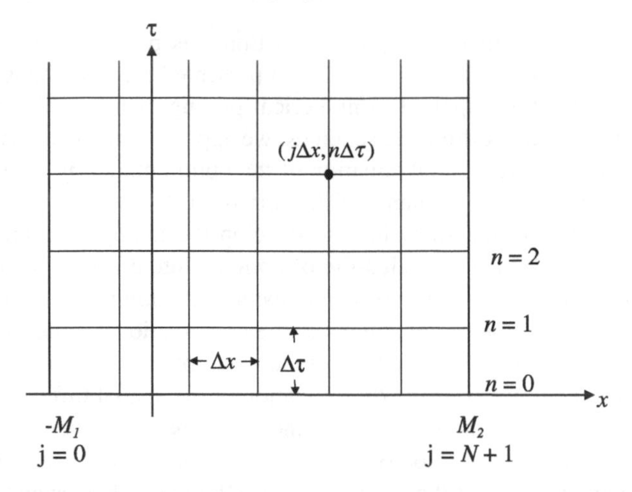

Fig. 6.4. Finite difference mesh with uniform stepwidth Δx and time step $\Delta\tau$. Numerical option values are computed at the node points $(j\Delta x, n\Delta\tau)$, $j = 1, 2, \cdots, N$, $n = 1, 2, \cdots$. Option values along the boundaries: $j = 0$ and $j = N + 1$ are prescribed by the boundary conditions of the option model. The "initial" values V_j^0 along the zeroth time level, $n = 0$, are given by the terminal payoff function.

constants so that the boundary conditions at the two ends of the infinite interval can be applied with sufficient accuracy. The discretized domain is overlaid with a uniform system of meshes or node points $(j\Delta x, n\Delta\tau)$, $j = 0, 1, \cdots, N + 1$, where $(N+1)\Delta x = M_1 + M_2$ and $n = 0, 1, 2, \cdots$ (see Fig. 6.4). The stepwidth Δx and time step $\Delta\tau$ are in general independent. In the discretized finite difference formulation, the option values are computed only at the node points.

We start with the discretization of (6.2.1b) and let W_j^n denote the numerical approximation of $W(j\Delta x, n\Delta\tau)$. The continuous temporal and spatial derivatives in (6.2.1b) are approximated by the following finite difference operators

$$\frac{\partial W}{\partial \tau}(j\Delta x, n\Delta\tau) \approx \frac{W_j^{n+1} - W_j^n}{\Delta\tau} \qquad \text{(forward difference)}$$

$$\frac{\partial W}{\partial x}(j\Delta x, n\Delta\tau) \approx \frac{W_{j+1}^n - W_{j-1}^n}{2\Delta x} \qquad \text{(centered difference)}$$

$$\frac{\partial^2 W}{\partial x^2}(j\Delta x, n\Delta\tau) \approx \frac{W_{j+1}^n - 2W_j^n + W_{j-1}^n}{\Delta x^2} \qquad \text{(centered difference).}$$

Similarly, we let V_j^n denote the numerical approximation of $V(j\Delta x, n\Delta\tau)$. By observing

$$W_j^{n+1} = e^{r(n+1)\Delta\tau} V_j^{n+1} \quad \text{and} \quad W_j^n = e^{rn\Delta\tau} V_j^n,$$

then canceling $e^{rn\Delta\tau}$, we obtain the following *explicit* Forward-Time-Centered-Space (FTCS) finite difference scheme for the Black–Scholes equation [see (6.2.1a)]:

$$V_j^{n+1} = \left[V_j^n + \frac{\sigma^2}{2}\frac{\Delta\tau}{\Delta x^2}\left(V_{j+1}^n - 2V_j^n + V_{j-1}^n\right)\right.$$
$$\left. + \left(r - \frac{\sigma^2}{2}\right)\frac{\Delta\tau}{2\Delta x}\left(V_{j+1}^n - V_{j-1}^n\right)\right]e^{-r\Delta\tau}. \tag{6.2.2}$$

Since V_j^{n+1} is expressed *explicitly* in terms of option values at the nth time level, one can compute V_j^{n+1} directly from known values of V_{j-1}^n, V_j^n and V_{j+1}^n. Suppose we are given "initial" values V_j^0, $j = 0, 1, \cdots, N+1$ along the zeroth time level, we can use scheme (6.2.2) to find values V_j^1, $j = 1, 2, \cdots, N$ along the first time level $\tau = \Delta\tau$. The values at the two ends V_0^1 and V_{N+1}^1 are given by the numerical boundary conditions specified for the option model. In this sense, the boundary conditions are *naturally incorporated* into the finite difference calculations. For example, the Dirichlet boundary conditions in barrier options and the Neumann boundary conditions in lookback options can be embedded into the finite difference algorithms (see Sect. 6.2.6 for details). The computational procedure then proceeds in a manner similar to successive time levels $\tau = 2\Delta\tau, 3\Delta\tau, \cdots$, through forward marching along the τ-direction. This is similar to the backward induction procedure (in terms of calendar time) in the lattice tree method.

We consider the class of two-level four-point explicit schemes of the form

$$V_j^{n+1} = b_1 V_{j+1}^n + b_0 V_j^n + b_{-1} V_{j-1}^n, \quad j = 1, 2, \cdots, N, \; n = 0, 1, 2, \cdots, \tag{6.2.3}$$

where b_1, b_0 and b_{-1} are coefficients specified for each individual scheme. For example, the above FTCS scheme corresponds to

$$b_1 = \left[\frac{\sigma^2}{2}\frac{\Delta\tau}{\Delta x^2} + \left(r - \frac{\sigma^2}{2}\right)\frac{\Delta\tau}{2\Delta x}\right]e^{-r\Delta\tau},$$
$$b_0 = \left[1 - \sigma^2\frac{\Delta\tau}{\Delta x^2}\right]e^{-r\Delta\tau},$$
$$b_{-1} = \left[\frac{\sigma^2}{2}\frac{\Delta\tau}{\Delta x^2} - \left(r - \frac{\sigma^2}{2}\right)\frac{\Delta\tau}{2\Delta x}\right]e^{-r\Delta\tau}.$$

An important observation is that both the binomial and trinomial schemes are members of the family specified in (6.2.3), when the reconnecting condition $ud = 1$ holds. Suppose we write $\Delta x = \ln u$, then $\ln d = -\Delta x$; the binomial scheme can be expressed as

$$V^{n+1}(x) = \frac{pV^n(x+\Delta x) + (1-p)V^n(x-\Delta x)}{R}, \quad x = \ln S, \text{ and } R = e^{r\Delta\tau}, \tag{6.2.4}$$

where $V^{n+1}(x)$, $V^n(x+\Delta x)$ and $V^n(x-\Delta x)$ are analogous to c, c_u and c_d, respectively [see (2.1.21)]. The above representation of the binomial scheme corresponds to the following specification of coefficients:

$$b_1 = p/R, \quad b_0 = 0 \quad \text{and} \quad b_{-1} = (1-p)/R$$

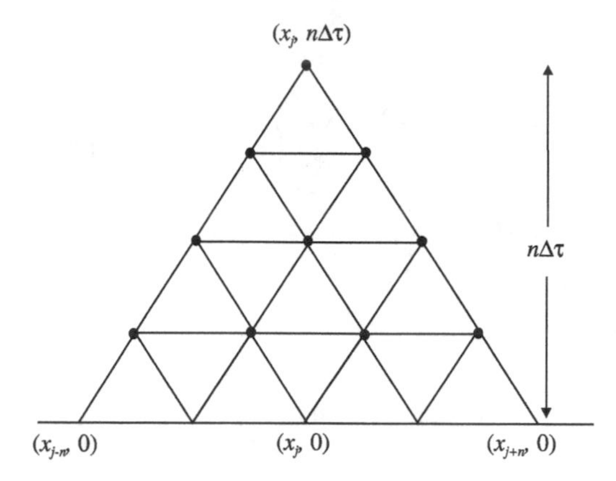

Fig. 6.5. The domain of dependence of a trinomial scheme with n time steps to expiry.

in (6.2.3). Similarly, suppose we choose $\Delta x = \ln u = -\ln d$ and $m = 1$, the trinomial scheme can be expressed as

$$V^{n+1}(x) = \frac{p_1 V^n(x + \Delta x) + p_2 V^n(x) + p_3 V^n(x - \Delta x)}{R}, \tag{6.2.5}$$

which also belongs to the family of explicit FTCS schemes.

While the usual finite difference calculations give option values at all node points along a given time level $\tau = n\Delta\tau$, we compute the option value at single asset value S at $\tau = n\Delta\tau$ in typical binomial/trinomial calculations. To illustrate, we relate the computational procedure for the trinomial scheme to finite difference calculations. The tip of the trinomial tree is referred as the jth-level node x_j (which assumes the value $\ln S$) and n time steps are taken to reach expiry $\tau = 0$ from the current time. Following the backward induction procedure in lattice tree calculations, the trinomial scheme computes $V^n(x_j)$ from known values of $V^{n-1}(x_{j-1})$, $V^{n-1}(x_j)$, $V^{n-1}(x_{j+1})$. Going down one time level, the computation of $V^n(x_j)$ requires the five values $V^{n-2}(x_{j-2})$, $V^{n-2}(x_{j-1})$, $\cdots$, $V^{n-2}(x_{j+2})$. Deductively, the $2n+1$ values $V^0(x_{j-n})$, $V^0(x_{j-n+1})$, $\cdots$, $V^0(x_{j+n})$ along $\tau = 0$ are involved to find $V^n(x_j)$. The triangular region in the computational domain with vertices $(x_j, n\Delta\tau)$, $(x_{j-n}, 0)$ and $(x_{j+n}, 0)$ is called the *domain of dependence* for the computation of $V^n(x_j)$ (see Fig. 6.5) since the option values at all node points inside the domain of dependence are required for finding $V^n(x_j)$. The lattice tree calculations confine computation of option values within a triangular domain of dependence. This may be seen to be more efficient when single option value at given values of S and τ is required.

Suppose boundary nodes are not included in the domain of dependence, then the boundary conditions of the option model do not have any effect on the numerical solution of the discrete model. This negligence of the boundary conditions does not reduce the accuracy of calculations when the boundary points are at infinity, as in vanilla option models where the domain of definition for $x = \ln S$ is infinite. This is no longer true when the domain of definition for x is truncated, as in barrier option models. To achieve a high level of numerical accuracy, it is important that the numerical scheme takes into account the effect of boundary conditions. The issues of numerical approximation of the auxiliary conditions in option models are examined in Sect. 6.2.6.

Note that the stepwidth Δx and time step $\Delta\tau$ in the binomial scheme are dependent. In the Cox–Ross–Rubinstein scheme, they are related by $\Delta x = \ln u = \sigma\sqrt{\Delta\tau}$ or $\sigma^2\Delta\tau = \Delta x^2$. However, in the trinomial scheme, their relation is given by $\lambda^2\sigma^2\Delta\tau = \Delta x^2$, where the free parameter λ can assume many possible values (though there are constraints on the choice of λ, like $\lambda \geq 1$).

The explicit schemes seem easy to implement. However, compared to the implicit schemes discussed in the next section, they exhibit lower order of accuracy. Also, the time step in explicit schemes cannot be chosen to be too large due to numerical stability considerations. The concepts of order of accuracy and stability will be explored in Sect. 6.2.5.

6.2.2 Implicit Schemes and Their Implementation Issues

Suppose the discount term $-rV$ and the spatial derivatives are approximated by the average of the centered difference operators at the nth and $(n+1)$th time levels

$$-rV\left(j\Delta x, \left(n+\frac{1}{2}\right)\Delta\tau\right) \approx -\frac{r}{2}(V_j^n + V_j^{n+1})$$

$$\frac{\partial V}{\partial x}\left(j\Delta x, \left(n+\frac{1}{2}\right)\Delta\tau\right) \approx \frac{1}{2}\left(\frac{V_{j+1}^n - V_{j-1}^n}{2\Delta x} + \frac{V_{j+1}^{n+1} - V_{j-1}^{n+1}}{2\Delta x}\right)$$

$$\frac{\partial^2 V}{\partial x^2}\left(j\Delta x, \left(n+\frac{1}{2}\right)\Delta\tau\right) \approx \frac{1}{2}\left(\frac{V_{j+1}^n - 2V_j^n + V_{j-1}^n}{\Delta x^2} + \frac{V_{j+1}^{n+1} - 2V_j^{n+1} + V_{j-1}^{n+1}}{\Delta x^2}\right),$$

and the temporal derivative by

$$\frac{\partial V}{\partial \tau}\left(j\Delta x, \left(n+\frac{1}{2}\right)\Delta\tau\right) \approx \frac{V_j^{n+1} - V_j^n}{\Delta\tau},$$

we then obtain the following two-level implicit finite difference scheme

$$\begin{aligned} V_j^{n+1} = V_j^n &+ \frac{\sigma^2}{2}\frac{\Delta\tau}{\Delta x^2}\left(\frac{V_{j+1}^n - 2V_j^n + V_{j-1}^n + V_{j+1}^{n+1} - 2V_j^{n+1} + V_{j-1}^{n+1}}{2}\right) \\ &+ \left(r - \frac{\sigma^2}{2}\right)\frac{\Delta\tau}{2\Delta x}\left(\frac{V_{j+1}^n - V_{j-1}^n + V_{j+1}^{n+1} - V_{j-1}^{n+1}}{2}\right) \\ &- r\Delta\tau\left(\frac{V_j^n + V_j^{n+1}}{2}\right), \end{aligned} \tag{6.2.6}$$

which is commonly known as the *Crank–Nicolson scheme*.

The above Crank–Nicolson scheme is seen to be a member of the general class of two-level six-point schemes of the form

$$a_1 V_{j+1}^{n+1} + a_0 V_j^{n+1} + a_{-1} \; V_{j-1}^{n+1} = b_1 V_{j+1}^n + b_0 V_j^n + b_{-1} V_{j-1}^n,$$
$$j = 1, 2, \cdots, N, \quad n = 0, 1, \cdots. \tag{6.2.7}$$

One can observe easily that the Crank–Nicolson scheme corresponds to

$$\begin{aligned}
a_1 &= -\frac{\sigma^2}{4} \frac{\Delta\tau}{\Delta x^2} - \left(r - \frac{\sigma^2}{2}\right) \frac{\Delta\tau}{4\Delta x}, \\
a_0 &= 1 + \frac{\sigma^2}{2} \frac{\Delta\tau}{\Delta x^2} + \frac{r}{2}\Delta\tau, \\
a_{-1} &= -\frac{\sigma^2}{4} \frac{\Delta\tau}{\Delta x^2} + \left(r - \frac{\sigma^2}{2}\right) \frac{\Delta\tau}{4\Delta x},
\end{aligned}$$

and

$$\begin{aligned}
b_1 &= \frac{\sigma^2}{4} \frac{\Delta\tau}{\Delta x^2} + \left(r - \frac{\sigma^2}{2}\right) \frac{\Delta\tau}{4\Delta x}, \\
b_0 &= 1 - \frac{\sigma^2}{2} \frac{\Delta\tau}{\Delta x^2} - \frac{r}{2}\Delta\tau, \\
b_{-1} &= \frac{\sigma^2}{4} \frac{\Delta\tau}{\Delta x^2} - \left(r - \frac{\sigma^2}{2}\right) \frac{\Delta\tau}{4\Delta x}.
\end{aligned}$$

A wide variety of implicit finite difference schemes of the class depicted in (6.2.7) can be derived in a systematic manner (Kwok and Lau, 2001b).

Suppose the values for V_j^n are all known along the nth time level, the solution for V_j^{n+1} requires the inversion of a tridiagonal system of equations. This explains the use of the term *implicit* for this class of schemes. In matrix form, the two-level six-point scheme can be represented as

$$\begin{pmatrix}
a_0 & a_1 & 0 & \cdots & \cdots & 0 \\
a_{-1} & a_0 & a_1 & 0 & \cdots & 0 \\
 & \cdots & & & & \\
 & & \cdots & & & \\
 & & & \cdots & & \\
0 & \cdots & \cdots & 0 & a_{-1} & a_0
\end{pmatrix}
\begin{pmatrix}
V_1^{n+1} \\ V_2^{n+1} \\ \cdot \\ \cdot \\ \cdot \\ V_N^{n+1}
\end{pmatrix}
=
\begin{pmatrix}
c_1 \\ c_2 \\ \cdot \\ \cdot \\ \cdot \\ c_N
\end{pmatrix}, \tag{6.2.8}$$

where

$$\begin{aligned}
c_1 &= b_1 V_2^n + b_0 V_1^n + b_{-1} V_0^n - a_{-1} V_0^{n+1}, \\
c_N &= b_1 V_{N+1}^n + b_0 V_N^n + b_{-1} V_{N-1}^n - a_1 V_{N+1}^{n+1}, \\
c_j &= b_1 V_{j+1}^n + b_0 V_j^n + b_{-1} V_{j-1}^n, \quad j = 2, \cdots, N-1.
\end{aligned}$$

The solution of the above tridiagonal system can be effected by the well-known *Thomas algorithm*. The algorithm is an efficient implementation of the Gaussian elimination procedure, the details of which are outlined as follows.

Thomas Algorithm
Consider the solution of the following tridiagonal system of the form

$$-a_j V_{j-1} + b_j V_j - c_j V_{j+1} = d_j, \quad j = 1, 2, \cdots N, \tag{6.2.9}$$

with $V_0 = V_{N+1} = 0$. This form is more general in the sense that the coefficients can differ among equations. In the first step of elimination, we reduce the system to the upper triangular form by eliminating V_{j-1} in each of the equations. Starting from the first equation, we can express V_1 in terms of V_2 and other known quantities. This relation is then substituted into the second equation giving a new equation involving V_2 and V_3 only. Again, we express V_2 in terms of V_3 and some known quantities. We then substitute into the third equation, . . . , and so on.

Suppose the first k equations have been reduced to the form

$$V_j - e_j V_{j+1} = f_j, \quad j = 1, 2, \cdots, k.$$

We use the kth reduced equation to transform the original $(k+1)$th equation to the same form, namely

$$V_{k+1} - e_{k+1} V_{k+2} = f_{k+1}. \tag{6.2.10a}$$

Now, we consider

$$V_k - e_k V_{k+1} = f_k,$$

and

$$-a_{k+1} V_k + b_{k+1} V_{k+1} - c_{k+1} V_{k+2} = d_{k+1},$$

the elimination of V_k from these two equations gives a new equation involving V_{k+1} and V_{k+2}, namely,

$$V_{k+1} - \frac{c_{k+1}}{b_{k+1} - a_{k+1} e_k} V_{k+2} = \frac{d_{k+1} + a_{k+1} f_k}{b_{k+1} - a_{k+1} e_k}. \tag{6.2.10b}$$

Comparing (6.2.10a) and (6.2.10b), and replacing the dummy variable $k+1$ by j, we can deduce the following recurrence relations for e_j and f_j:

$$e_j = \frac{c_j}{b_j - a_j e_{j-1}}, \quad f_j = \frac{d_j + a_j f_{j-1}}{b_j - a_j e_{j-1}}, \quad j = 1, 2, \cdots N. \tag{6.2.11a}$$

Corresponding to the boundary value $V_0 = 0$, we must have

$$e_0 = f_0 = 0. \tag{6.2.11b}$$

Starting from the above initial values, the recurrence relations (6.2.11a) can be used to find all values e_j and f_j, $j = 1, 2, \cdots, N$. Once the system is in an upper triangular form, we can solve for $V_N, V_{N-1}, \cdots V_1$, successively by backward substitution, starting from $V_{N+1} = 0$.

The Thomas algorithm is a very efficient algorithm where the tridiagonal system (6.2.8) can be solved with four (add/subtract) and six (multiply/divide) operations per node point. Compared to the explicit schemes, it takes about twice the number of operations per time step. Using the Thomas algorithm, the solution of a tridiagonal system required by an implicit scheme does not add much computational complexity.

On the control of the growth of roundoff errors, we observe that the calculations would be numerically stable provided that $|e_j| < 1$ so that error in V_{j+1} will not be magnified and propagated to V_j [see (6.2.10a)]. A set of sufficient conditions to guarantee $|e_j| < 1$ is given by

$$a_j > 0, b_j > 0, c_j > 0 \quad \text{and} \quad b_j > a_j + c_j.$$

Fortunately, the above conditions can be satisfied easily by appropriate choices of $\Delta\tau$ and Δx in the Crank–Nicolson scheme.

6.2.3 Front Fixing Method and Point Relaxation Technique

In this section, we consider several numerical approaches for solving American option models using the finite difference methods. The difficulties in the construction of numerical algorithms for solving American style option models arise from the unknown optimal exercise boundary (which has to be obtained as part of the solution). First, we discuss the front fixing method, where a transformation of the independent variable is applied so that the free boundary associated with the optimal exercise boundary is converted into a fixed boundary. The extension of the front fixing method to pricing of convertible bonds is reported by Zhu and Sun (1999).

Recall that in the binomial/trinomial algorithm for pricing an American option, a dynamic programming procedure is applied at each node to determine whether the continuation value is less than the intrinsic value or otherwise. If this is the case, the intrinsic value is taken as the option value (signifying the early exercise of the American option). Difficulties in implementing the above dynamic programming procedure are encountered when an implicit scheme is employed since option values are obtained implicitly. We examine how the difficulty can be resolved by a point relaxation scheme.

The third approach is called the penalty method, where we append an extra penalty term into the governing equation. In the limit where the penalty parameter becomes infinite, the resulting solution is guaranteed to satisfy the constraint that the option value cannot be below the exercise payoff. The construction of the penalty approximation scheme is relegated to an exercise (see Problem 6.26).

Front Fixing Method

We consider the construction of the front fixing algorithm for finding the option value and the associated optimal exercise boundary $S^*(\tau)$ of an American put. For simplicity, we take the strike price to be unity. This is equivalent to normalizing the underlying asset price and option value by the strike price. In the continuation region, the put value $P(S, \tau)$ satisfies the Black–Scholes equation

$$\frac{\partial P}{\partial \tau} - \frac{\sigma^2}{2}S^2\frac{\partial^2 P}{\partial S^2} - rS\frac{\partial P}{\partial S} + rP = 0, \quad \tau > 0, S^*(\tau) < S < \infty, \tag{6.2.12}$$

subject to the boundary conditions:

$$P(S^*(\tau), \tau) = 1 - S^*(\tau), \frac{\partial P}{\partial S}(S^*(\tau), \tau) = -1, \lim_{S\to\infty} P(S, \tau) = 0,$$

and the initial condition

$$P(S, 0) = 0 \quad \text{for} \quad S^*(0) < S < \infty,$$

with $S^*(0) = 1$. We apply the transformation of the state variable $y = \ln \frac{S}{S^*(\tau)}$ so that $y = 0$ at $S = S^*(\tau)$. Now, the free boundary $S = S^*(\tau)$ becomes the fixed boundary $y = 0$, hence the name of this method. In terms of the new independent variable y, the above governing equation becomes

$$\frac{\partial P}{\partial \tau} - \frac{\sigma^2}{2}\frac{\partial^2 P}{\partial y^2} - \left(r - \frac{\sigma^2}{2}\right)\frac{\partial P}{\partial y} + rP = \frac{S^{*'}(\tau)}{S^*(\tau)}\frac{\partial P}{\partial y}, \tag{6.2.13}$$

subject to the new set of auxiliary conditions

$$P(0, \tau) = 1 - S^*(\tau), \frac{\partial P}{\partial y}(0, \tau) = -S^*(\tau), P(\infty, \tau) = 0, \tag{6.2.14a}$$

$$P(y, 0) = 0 \quad \text{for } 0 < y < \infty. \tag{6.2.14b}$$

The nonlinearity in the American put model is revealed by the nonlinear term $\frac{S^{*'}(\tau)}{S^*(\tau)}\frac{\partial P}{\partial y}$. Along the boundary $y = 0$, we apply the continuity properties of P, $\frac{\partial P}{\partial y}$ and $\frac{\partial P}{\partial \tau}$ so that $\frac{\partial^2 P}{\partial y^2}(0^+, \tau)$ observes the relation

$$\begin{aligned}\frac{\sigma^2}{2}\frac{\partial^2 P}{\partial y^2}(0^+, \tau) &= \frac{\partial}{\partial \tau}[1 - S^*(\tau)] - \left(r - \frac{\sigma^2}{2}\right)[-S^*(\tau)] \\ &\quad + r[1 - S^*(\tau)] - \frac{S^{*'}(\tau)}{S^*(\tau)}[-S^*(\tau)] \\ &= r - \frac{\sigma^2}{2}S^*(\tau).\end{aligned} \tag{6.2.15}$$

This derived relation is used to determine $S^*(\tau)$ once we have obtained $\frac{\partial^2 P}{\partial y^2}(0^+, \tau)$.

The direct Crank–Nicolson discretization of (6.2.13) would result in a nonlinear algebraic system of equations for the determination of V_j^{n+1} due to the presence of the nonlinear term $\frac{S^{*'}(\tau)}{S^*(\tau)}\frac{\partial P}{\partial y}$. To circumvent the difficulties while maintaining the same order of accuracy as that of the Crank–Nicholson scheme, we adopt a three-level scheme of the form

$$\begin{aligned}&\frac{P_j^{n+1} - P_j^{n-1}}{2\Delta\tau} - \left[\frac{\sigma^2}{2}D_+D_- + \left(r - \frac{\sigma^2}{2}\right)D_0 - r\right]\frac{P_j^{n+1} + P_j^{n-1}}{2} \\ &= \frac{S_{n+1}^* - S_{n-1}^*}{2\Delta\tau S_n^*}D_0P_j^n,\end{aligned} \tag{6.2.16}$$

where S_n^* denotes the numerical approximation to $S^*(n\Delta\tau)$, while D_+, D_- and D_0 are discrete difference operators defined by

$$D_+ = \frac{1}{\Delta y}(E^1 - I),\ D_- = \frac{1}{\Delta y}(I - E^{-1}), \quad D_0 = \frac{1}{2\Delta y}(E^1 - E^{-1}).$$

Here, I denotes the identity operator and E^i, $i = -1, 1$, denotes the spatial shifting operator on a discrete function P_j, defined by $E^i P_j = P_{j+i}$.

The discretization of the value matching condition, smooth pasting condition and the boundary equation (6.2.15) lead to the following system of equations that relate P_{-1}^n, P_0^n, P_1^n and S_n^*:

$$P_0^n = 1 - S_n^* \tag{6.2.17a}$$

$$\frac{P_1^n - P_{-1}^n}{2\Delta y} = -S_n^* \tag{6.2.17b}$$

$$\frac{\sigma^2}{2}\left[\frac{P_1^n - 2P_0^n + P_{-1}^n}{\Delta y^2}\right] + \frac{\sigma^2}{2}S_n^* - r = 0. \tag{6.2.17c}$$

Here, P_{-1}^n is a fictitious value outside the computational domain. By eliminating P_{-1}^n, we obtain

$$P_1^n = \alpha - \beta S_n^*, \quad n \geq 1, \tag{6.2.18}$$

where

$$\alpha = 1 + \frac{\Delta y^2}{\sigma^2} r \quad \text{and} \quad \beta = \frac{1 + (1 + \Delta y)^2}{2}.$$

Once P_1^n is known, we can find S_n^* using (6.2.18) and P_0^n using (6.2.17a). For the boundary condition at the right end of the computational domain, we observe that the American put value tends to zero when S is sufficiently large. Therefore, we choose M to be sufficiently large such that we set $P_{M+1}^n = 0$ with sufficient accuracy.

Let $\mathbf{P}^n = (P_1^n \quad P_2^n \quad \cdots \quad P_M^n)^T$ and $\mathbf{e}_1 = (1 \quad 0 \quad \cdots \quad 0)^T$. By putting all the auxiliary conditions into the finite difference scheme (6.2.16), we would like to show how to calculate $\mathbf{P}^{n+1}$ from known values of $\mathbf{P}^n$ and $\mathbf{P}^{n-1}$. First, we define the following parameters

$$a = \mu\sigma^2 + r\Delta\tau, \quad b = \frac{\mu}{2}\left[\sigma^2 - \Delta y\left(r - \frac{\sigma^2}{2}\right)\right],$$
$$c = \frac{\mu}{2}\left[\sigma^2 + \Delta y\left(r - \frac{\sigma^2}{2}\right)\right],$$

where $\mu = \frac{\Delta\tau}{\Delta y^2}$. Also, we define the tridiagonal matrix

$$A = \begin{pmatrix} a & -c & 0 & \cdots & \cdots & 0 \\ -b & a & -c & 0 & \cdots & 0 \\ 0 & -b & a & -c & 0 & \cdots \\ \vdots & \ddots & \ddots & \ddots & \ddots & \vdots \\ 0 & \cdots & \cdots & -b & a & -c \\ 0 & 0 & \cdots & 0 & -b & a \end{pmatrix}.$$

In terms of A, the finite difference scheme (6.2.16) can be expressed as

$$(I + A)\mathbf{P}^{n+1} = (I - A)\mathbf{P}^{n-1} + bP_0^{n-1}\mathbf{e}_1 + bP_0^{n+1}\mathbf{e}_1 + g^n D_0 \mathbf{P}^n, \quad n > 1, \tag{6.2.19}$$

where $g^n = \frac{S_{n+1}^* - S_{n-1}^*}{S_n^*}$. By inverting the matrix $(I + A)$, (6.2.19) can be expressed as

$$\mathbf{P}^{n+1} = \mathbf{f}_1 + bP_0^{n+1}\mathbf{f}_2 + g^n \mathbf{f}_3, \tag{6.2.20}$$

where

$$\mathbf{f}_1 = (I + A)^{-1}[(I - A)\mathbf{P}^{n-1} + bP_0^{n-1}\mathbf{e}_1],$$
$$\mathbf{f}_2 = (I + A)^{-1}\mathbf{e}_1,$$
$$\mathbf{f}_3 = (I + A)^{-1} D_0 \mathbf{P}^n.$$

Note that P_0^{n+1} and S_{n+1}^* can be expressed in terms of P_1^{n+1} using (6.2.17a) and (6.2.18).

Since (6.2.20) is a three-level scheme, we need $\mathbf{P}^1$ in addition to $\mathbf{P}^0$ to initialize the computation. To maintain an overall second order accuracy, we employ the following two-step predictor–corrector technique to obtain $\mathbf{P}^1$:

$$\left(I + \frac{A}{2}\right)\widetilde{\mathbf{P}} = \left(I - \frac{A}{2}\right)\mathbf{P}^0 + \frac{b}{2}\widetilde{P}_0\mathbf{e}_1 + \widetilde{g} D_0 \mathbf{P}^0,$$
$$\left(I + \frac{A}{2}\right)\mathbf{P}^1 = \left(I - \frac{A}{2}\right)\mathbf{P}^0 + \frac{b}{2}P_0^1\mathbf{e}_1 + g^1 D_0\left(\frac{\widetilde{\mathbf{P}} + \mathbf{P}^0}{2}\right),$$

where the first equation gives the predictor value $\widetilde{\mathbf{P}}$ and the corrector value $\mathbf{P}_1$ is finally obtained using the second equation. The provisional values and g^1 are given by

$$\widetilde{P}_0 = 1 - \widetilde{S}_0^*, \quad \widetilde{S}_0^* = \frac{\alpha - \widetilde{P}_1}{\beta},$$
$$\widetilde{g} = \frac{\widetilde{S}_0^* - S_0^*}{S_0^*} \quad \text{and} \quad g^1 = \frac{S_1^* - S_0^*}{\frac{\widetilde{S}_0^* + S_0^*}{2}}.$$

Further details on the implementation procedures and numerical performance of the front fixing scheme can be found in Wu and Kwok (1997).

Projected Successive-Over-Relaxation Method

Consider an implicit finite difference scheme in the form [see (6.2.7)]

$$a_{-1}V_{j-1} + a_0V_j + a_1V_{j+1} = d_j, \quad j = 1, 2, \cdots, N, \tag{6.2.21}$$

where the superscript "$n + 1$" is omitted for brevity, and d_j represents the known quantities. Instead of solving the tridiagonal system by direct elimination (Thomas algorithm), one may choose to use an iteration method. The *Gauss–Seidel* iterative procedure produces the kth iterate of V_j by

$$\begin{aligned} V_j^{(k)} &= \frac{1}{a_0}\left(d_j - a_{-1}V_{j-1}^{(k)} - a_1 V_{j+1}^{(k-1)}\right) \\ &= V_j^{(k-1)} + \frac{1}{a_0}\left(d_j - a_{-1}V_{j-1}^{(k)} - a_0 V_j^{(k-1)} - a_1 V_{j+1}^{(k-1)}\right), \end{aligned} \tag{6.2.22}$$

where the last term in the above equation represents the correction made on the $(k-1)$th iterate of V_j. We start from $j = 1$ and proceed sequentially with increasing value of j. Hence, when we compute $V_j^{(k)}$ in the kth iteration, the new kth iterate $V_{j-1}^{(k)}$ is already available while only the old $(k-1)$th iterate $V_{j+1}^{(k-1)}$ is known. To accelerate the rate of convergence of the iteration, we multiply the correction term by a relaxation parameter ω. The corresponding iterative procedure becomes

$$V_j^{(k)} = V_j^{(k-1)} + \frac{\omega}{a_0}\left(d_j - a_{-1}V_{j-1}^{(k)} - a_0 V_j^{(k-1)} - a_1 V_{j+1}^{(k-1)}\right), \quad 0 < \omega < 2. \tag{6.2.23}$$

This procedure is called the *successive-over-relaxation*. As a necessary condition for convergence, the relaxation parameter ω must be chosen between 0 and 2.

Let h_j denote the intrinsic value of the American option at the jth node. To incorporate the constraint that the option value must be above the intrinsic value, the dynamic programming procedure in combination with the above relaxation procedure is given by

$$V_j^{(k)} = \max\left(V_j^{(k-1)} + \frac{\omega}{a_0}\left(d_j - a_{-1}V_{j-1}^{(k)} - a_0 V_j^{(k-1)} - a_1 V_{j+1}^{(k-1)}\right), h_j\right). \tag{6.2.24}$$

A sufficient number of iterations are performed until the following termination criterion is met:

$$\sqrt{\sum_{j=1}^{N}\left(V_j^{(k)} - V_j^{(k-1)}\right)^2} < \epsilon, \qquad \epsilon \text{ is some small tolerance value.}$$

The convergent value $V_j^{(k)}$ is then taken to be the numerical solution for V_j. The above iterative scheme is called the *projected successive-over-relaxation method.*

6.2.4 Truncation Errors and Order of Convergence

The local truncation error measures the discrepancy that the continuous solution does not satisfy the numerical scheme at the node point. The local truncation error of a given numerical scheme is obtained by substituting the exact solution of the continuous problem into the numerical scheme. Let $V(j\Delta x, n\Delta\tau)$ denote the exact solution of the continuous Black–Scholes equation. We illustrate the procedure of finding the local truncation error of the explicit FTCS scheme by substituting the exact solution into the numerical scheme. The local truncation error at the node point $(j\Delta x, n\Delta\tau)$ is given by

$$\begin{aligned}
&T(j\Delta x, n\Delta\tau) \\
&= \frac{V(j\Delta x, (n+1)\Delta\tau) - V(j\Delta x, n\Delta\tau)}{\Delta\tau} \\
&\quad - \frac{\sigma^2}{2}\frac{V((j+1)\Delta x, n\Delta\tau) - 2V(j\Delta x, n\Delta\tau) + V((j-1)\Delta x, n\Delta\tau)}{\Delta x^2} \\
&\quad - \left(r - \frac{\sigma^2}{2}\right)\frac{V((j+1)\Delta x, n\Delta\tau) - V((j-1)\Delta x, n\Delta\tau)}{2\Delta x} \\
&\quad + rV(j\Delta x, n\Delta\tau). \qquad (6.2.25)
\end{aligned}$$

We then expand each term by performing the Taylor expansion at the node point $(j\Delta x, n\Delta\tau)$. After some cancellation of terms, we obtain

$$\begin{aligned}
&T(j\Delta x, n\Delta\tau) \\
&= \frac{\partial V}{\partial\tau}(j\Delta x, n\Delta\tau) + \frac{\Delta\tau}{2}\frac{\partial^2 V}{\partial\tau^2}(j\Delta x, n\Delta\tau) + O(\Delta\tau^2) \\
&\quad - \frac{\sigma^2}{2}\left[\frac{\partial^2 V}{\partial x^2}(j\Delta x, n\Delta\tau) + \frac{\Delta x^2}{12}\frac{\partial^4 V}{\partial x^4}(j\Delta x, n\Delta\tau) + O(\Delta x^4)\right] \\
&\quad - \left(r - \frac{\sigma^2}{2}\right)\left[\frac{\partial V}{\partial x}(j\Delta x, n\Delta\tau) + \frac{\Delta x^2}{3}\frac{\partial^3 V}{\partial x^3}(j\Delta x, n\Delta\tau) + O(\Delta x^4)\right] \\
&\quad + rV(j\Delta x, n\Delta\tau).
\end{aligned}$$

Since $V(j\Delta x, n\Delta\tau)$ satisfies the Black–Scholes equation, this leads to

$$\begin{aligned}
T(j\Delta x, n\Delta\tau) &= \frac{\Delta\tau}{2}\frac{\partial^2 V}{\partial\tau^2}(j\Delta x, n\Delta\tau) - \frac{\sigma^2}{24}\Delta x^2\frac{\partial^4 V}{\partial x^4}(j\Delta x, n\Delta\tau) \\
&\quad - \left(r - \frac{\sigma^2}{2}\right)\frac{\Delta x^2}{3}\frac{\partial^3 V}{\partial x^3}(j\Delta x, n\Delta\tau) + O(\Delta\tau^2) \\
&\quad + O(\Delta x^4). \qquad (6.2.26)
\end{aligned}$$

A necessary condition for the convergence of the numerical solution to the continuous solution is that the local truncation error of the numerical scheme must tend to zero for vanishing stepwidth and time step. In this case, the numerical scheme is said to be *consistent* . The *order of accuracy* of a scheme is defined to be the order in powers of Δx and $\Delta\tau$ in the leading truncation error terms. Suppose the leading truncation terms are $O(\Delta\tau^k, \Delta x^m)$, then the numerical scheme is said to be kth order time accurate and mth order space accurate. From (6.2.26), we observe that the explicit FTCS scheme is first-order time accurate and second-order space accurate. Suppose we choose $\Delta\tau$ to be the same order as Δx^2, that is, $\Delta\tau = \lambda\Delta x^2$ for some finite constant λ (recall that the same relation between $\Delta\tau$ and Δx has been adopted by the trinomial scheme), then the leading truncation error terms in (6.2.26) become $O(\Delta\tau)$.

By performing the corresponding Taylor expansion, one can show that all versions of the binomial scheme are first-order time accurate (recall that $\Delta\tau$ and Δx are dependent in binomial schemes). This is not surprising since this is consistent with a similar error analysis presented in Sect. 6.1. The earlier error analysis attempts to find the order of approximation that the numerical solution from the binomial scheme satisfies the continuous Black–Scholes equation. Both approaches give the same conclusion on the order of accuracy.

For the implicit Crank–Nicolson scheme, it can be shown that it is second-order time accurate and second-order space accurate (see Problem 6.19). The highest order of accuracy that can be achieved for a two-level six-point scheme is known to be $O(\Delta\tau^2, \Delta x^4)$ (see the compact scheme given in Problem 6.20). With regard to accuracy consideration, higher order schemes should be preferred over lower order schemes.

Suppose the leading truncation error terms of a numerical scheme are $O(\Delta\tau^m)$, m is some positive integer, one can show from more advanced theoretical analysis that the numerical solution $V_j^n(\Delta\tau)$ using time step $\Delta\tau$ has the asymptotic expansion of the form

$$V_j^n(\Delta\tau) = V_j^n(0) + K\Delta\tau^m + O(\Delta\tau^{m+1}), \tag{6.2.27}$$

where $V_j^n(0)$ is visualized as the exact continuous solution obtained in the limit $\Delta\tau \to 0$, and K is some constant independent of $\Delta\tau$. Suppose we perform two numerical calculations using time step $\Delta\tau$ and $\frac{\Delta\tau}{2}$ successively, it is easily seen that

$$V_j^n(0) - V_j^n(\Delta\tau) \approx 2^m\left[V_j^n(0) - V_j^n\left(\frac{\Delta\tau}{2}\right)\right]. \tag{6.2.28}$$

That is, the error in the numerical solution of a mth-order time accurate scheme is reduced by a factor of $\frac{1}{2^m}$ when we reduce the time step by a factor of $\frac{1}{2}$.

6.2.5 Numerical Stability and Oscillation Phenomena

A numerical scheme must be consistent so that the numerical solution converges to the exact solution of the underlying differential equation. However, consistency is only a necessary but not sufficient condition for convergence. The roundoff errors incurred during numerical calculations may lead to the blow up of the solution and erode the whole computation. Besides the analysis of the truncation error, it is necessary to analyze the stability properties of a numerical scheme. Loosely speaking, a scheme is said to be stable if roundoff errors are not amplified in numerical computation. For a linear evolutionary differential equation, like the Black–Scholes equation, the *Lax Equivalence Theorem* states that numerical stability is the necessary and sufficient condition for the convergence of a consistent difference scheme.

Another undesirable feature in the behaviors of the finite difference solution is the occurrence of spurious oscillations. It is possible to generate negative option values even if the scheme is stable (see Fig. 6.6). The oscillation phenomena in the

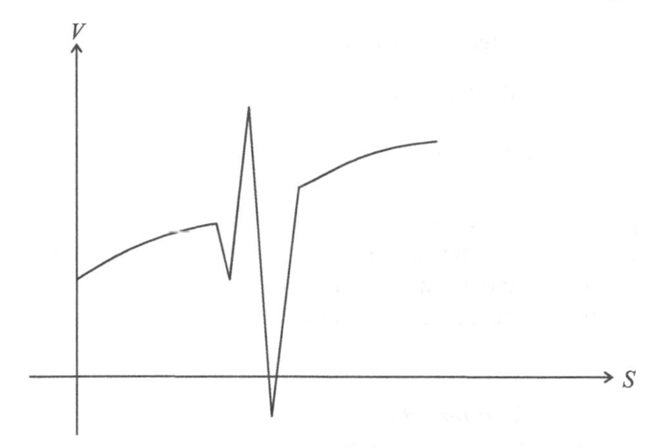

Fig. 6.6. Spurious oscillations in numerical solution of an option price.

numerical calculations of the barrier and Asian option models were discussed in detail by Zvan, Forsyth and Vetzal (1998) and Zvan, Vetzal and Forsyth (2000).

Fourier Method of Stability Analysis

There is a huge body of literature on the stability analysis of numerical schemes, and different notions of stability have been defined. Here, we discuss only the *Fourier method of stability analysis*. The Fourier method is based on the assumption that the numerical scheme admits a solution of the form

$$V_j^n = A^n(k)e^{ikj\Delta x}, \tag{6.2.29}$$

where k is the wavenumber and $i = \sqrt{-1}$. The von Neumann stability criterion examines the growth of the above Fourier component. Substituting (6.2.29) into the two-level six-point scheme (6.2.7), we obtain

$$G(k) = \frac{A^{n+1}(k)}{A^n(k)} = \frac{b_1 e^{ik\Delta x} + b_0 + b_1 e^{-ik\Delta x}}{a_1 e^{ik\Delta x} + a_0 + a_{-1} e^{-ik\Delta x}}, \tag{6.2.30}$$

where $G(k)$ is the amplification factor that governs the growth of the Fourier component over one time period. The strict von Neumann stability condition is given by

$$|G(k)| \leq 1, \tag{6.2.31}$$

for $0 \leq k\Delta x \leq \pi$. Henceforth, we write $\beta = k\Delta x$.

We now apply the Fourier stability analysis to study the stability properties of the Cox–Ross–Rubinstein binomial scheme and the implicit Crank–Nicolson scheme.

Stability of the Cox–Ross–Rubinstein Binomial Scheme

The corresponding amplification factor of the Cox–Ross–Rubinstein binomial scheme is

$$G(\beta) = (\cos\beta + iq\sin\beta)e^{-r\Delta\tau}, \quad q = 2p - 1. \tag{6.2.32}$$

The von Neumann stability condition requires

$$|G(\beta)|^2 = \left[1 + (q^2 - 1)\sin^2\beta\right]e^{-2r\Delta\tau} \le 1, \quad 0 \le \beta \le \pi. \tag{6.2.33}$$

When $0 \le p \le 1$, we have $|q| \le 1$ so that $|G(\beta)| \le 1$ for all β. Under this condition, the scheme is guaranteed to be stable in the von Neumann sense. We then obtain the following sufficient condition for von Neumann stability of the Cox–Ross–Rubinstein scheme: nonoccurrence of negative probability values in the binomial scheme.

Stability of the Crank–Nicolson Scheme

The corresponding amplification factor of the Crank–Nicolson scheme is found to be

$$G(\beta) = \frac{1 - \sigma^2\frac{\Delta\tau}{\Delta x^2}\sin^2\frac{\beta}{2} + \left(r - \frac{\sigma^2}{2}\right)\frac{\Delta\tau}{2\Delta x}i\sin\beta - \frac{r}{2}\Delta\tau}{1 + \sigma^2\frac{\Delta\tau}{\Delta x^2}\sin^2\frac{\beta}{2} - \left(r - \frac{\sigma^2}{2}\right)\frac{\Delta\tau}{2\Delta x}i\sin\beta + \frac{r}{2}\Delta\tau}. \tag{6.2.34}$$

The von Neumann stability condition requires

$$|G(\beta)|^2 = \frac{\left(1 - \sigma^2\frac{\Delta\tau}{\Delta x^2}\sin^2\frac{\beta}{2} - \frac{r}{2}\Delta\tau\right)^2 + \left(r - \frac{\sigma^2}{2}\right)^2\frac{\Delta\tau^2}{4\Delta x^2}\sin^2\beta}{\left(1 + \sigma^2\frac{\Delta\tau}{\Delta x^2}\sin^2\frac{\beta}{2} + \frac{r}{2}\Delta\tau\right)^2 + \left(r - \frac{\sigma^2}{2}\right)^2\frac{\Delta\tau^2}{4\Delta x^2}\sin^2\beta} \le 1, \quad 0 \le \beta \le \pi. \tag{6.2.35}$$

It is easily seen that the above condition is satisfied for any choices of $\Delta\tau$ and Δx. Hence, the Crank–Nicolson scheme is unconditionally stable. In other words, numerical stability (in the von Neumann sense) is ensured without any constraint on the choice of $\Delta\tau$.

The implicit Crank–Nicolson scheme is observed to have second-order temporal accuracy and unconditional stability. Also, the implementation of the numerical computation can be quite efficient with the use of the Thomas algorithm. Apparently, practitioners should favor the Crank–Nicolson scheme over other conditionally stable and first-order time accurate explicit schemes. Unfortunately, the numerical accuracy of the Crank–Nicholson solution can be much deteriorated due to the nonsmooth property of the terminal payoff function in most option models. The issues of implementation of the auxiliary conditions in option pricing using finite difference schemes are discussed in Sect. 6.2.6.

Spurious Oscillations of Numerical Solution

It is relatively easy to find the sufficient conditions for non-appearance of spurious oscillations in the numerical solution of a two-level explicit scheme. The following theorem reveals the relation between the signs of the coefficients in the explicit scheme and spurious oscillations of the computed solution (Kwok and Lau, 2001b).

Theorem 6.1. *Suppose the coefficients in the two-level explicit scheme* (6.2.3) *are all nonnegative, and the initial values are bounded, that is,* $\max_j |V_j^0| \le M$ *for some constant* M*, then*

$$\max_j |V_j^n| \le M \quad \textit{for all} \quad n \ge 1. \tag{6.2.36}$$

The proof of the above theorem is quite straightforward. From the explicit scheme, we deduce that

$$|V_j^{n+1}| \le |b_{-1}|\,|V_{j-1}^n| + |b_0|\,|V_j^n| + |b_1|\,|V_{j+1}^n|,$$

so

$$\max_j |V_j^{n+1}| \le b_{-1} \max_j |V_{j-1}^n| + b_0 \max_j |V_j^n| + b_1 \max_j |V_{j+1}^n|$$

since b_{-1}, b_0 and b_1 are nonnegative. Let E^n denote $\max_j |V_j^n|$, the above inequality can be expressed as

$$E^{n+1} \le b_{-1}E^n + b_0E^n + b_1E^n = E^n$$

since $b_{-1} + b_0 + b_1 = 1$. Deductively, we obtain

$$E^n \le E^{n-1} \le \cdots \le E^0 = \max_j |V_j^0| = M.$$

What happens when one or more of the coefficients of the explicit scheme become negative? For example, we take $b_0 < 0, b_{-1} > 0$ and $b_1 > 0$, and let $V_0^0 = \varepsilon > 0$ and $V_j^0 = 0, j \ne 0$. At the next time level, $V_{-1}^1 = b_1\varepsilon, V_0^1 = b_0\varepsilon$ and $V_1^1 = b_{-1}\varepsilon$, where the sign of V_j^1 alternates with j. This alternating sign property can be shown to persist at all later time levels. In this way, we deduce that

$$|V_j^{n+1}| = b_{-1}|V_{j-1}^n| - b_0|V_j^n| + b_1|V_{j+1}^n|.$$

We sum over all values of j of the above equation and let $\mathcal{S}^n = \sum_j |V_j^n|$. As a result, we obtain

$$\mathcal{S}^{n+1} = (b_{-1} - b_0 + b_1)\mathcal{S}^n = (1 - 2b_0)\mathcal{S}^n.$$

Note that $1 - 2b_0 > 1$ since $b_0 < 0$. Deductively, we obtain

$$\mathcal{S}^n = (1 - 2b_0)^n \mathcal{S}_0 = (1 - 2b_0)^n \varepsilon,$$

and as $n \to \infty, \mathcal{S}^n \to \infty$. The solution values oscillate in signs at neighboring nodes, and the oscillation amplitudes grow with increasing number of time steps.

6.2.6 Numerical Approximation of Auxiliary Conditions

The errors observed in the finite difference solution may arise from various sources. The major source is the truncation error, which stems from the difference approximation of the differential operators. Another source comes from the numerical approximation of the auxiliary conditions in the option models. Auxiliary conditions refer to the terminal payoff function plus (possibly) additional boundary conditions due to the embedded path dependent features in the option contract. It is commonly observed that numerical option values obtained from trinomial or finite difference calculations exhibit wavy or erratic pattern of convergence to the continuous solutions. Heston and Zhou (2000) illustrated from their numerical experiments that the rate of convergence of binomial calculations fluctuate between $O(\sqrt{\Delta t})$ and $O(\Delta t)$. Due to the lack of smooth convergence, an extrapolation technique for the enhancement of the rate of convergence cannot be routinely applied to numerical option values. In this section, we present several smoothness-enhancement techniques for dealing with discontinuity and nondifferentiability of the terminal payoff function. We also consider the proper treatment of numerical boundary conditions that are associated with the barrier and lookback features. Interestingly, the optimal positioning of the boundary grids depends on whether the path dependent feature is continuously or discretely monitored.

Smoothing of Discontinuities in the Terminal Payoff Functions

Most terminal payoff function of options have some form of discontinuity (like binary payoff) or nondifferentiability (like call or put payoff). Quantization error arises because the payoff function is sampled at discrete node points. Several smoothing techniques have been proposed in the literature. Heston and Zhou (2000) proposed to set the payoff value at a node point in the computational mesh by the average of the payoff function over the surrounding node cells rather than sampled at the node point. Let $V_T(S)$ denote the terminal payoff function. The payoff value at node S_j is given by

$$V_j^0 = \frac{1}{\Delta S} \int_{S_j - \frac{\Delta S}{2}}^{S_j + \frac{\Delta S}{2}} V_T(S)\, dS \tag{6.2.37}$$

instead of simply taking the value $V_T(S_j)$. Take the call payoff $\max(S - X, 0)$ as an example. If the strike price X falls exactly on a node point, then $V_T(S_j) = 0$ while the cell-averaged value is $\Delta S/8$. In their binomial calculations, Heston and Zhou (2000) found that averaging the payoff for vanilla European and American calls provides a more smooth convergence. The smoothed numerical solutions then allow the application of extrapolation for convergence enhancement. Another simple technique is the method of node positioning. Tavella and Randall (2000) proposed placing the strike price halfway between two neighboring node points. The third technique is called the Black–Scholes approximation, which is useful for pricing American options and exotic options for which the Black–Scholes solution is a good approximation at time close to expiry. The trick is to use the Black–Scholes values along the first time level and proceed with usual finite difference calculations for subsequent time levels.

More advanced methods for minimizing the quantization errors in higher order schemes have also been studied. Pooley, Vetzal and Forsyth (2003) showed that if discontinuous terminal payoff is present, the expected second-order convergence of the Crank–Nicolson scheme cannot be realized. They managed to develop elaborate techniques that can be used to recover the quadratic rate of convergence of the Crank–Nicolson scheme.

Barrier Options

The two major factors that lead to deterioration of numerical accuracy in barrier option calculations are (i) positioning of the nodes relative to the barrier and (ii) proximity of the initial asset price to the barrier.

Several papers have reported that better numerical accuracy can be achieved if the barrier is placed to pass through a layer of nodes for the continuously monitored barrier, and located halfway between two layers of nodes for the discretely monitored barrier. Heuristic arguments that explain why these choices of positioning achieve better numerical accuracy can be found in Kwok and Lau (2001b). To remedy the proximity problem, Figlewski and Gao (1999) suggested constructing fine meshes near the barrier to improve the level of accuracy. Boyle and Tian (1998) showed that the application of spline interpolation of option values at three adjacent nodes is a simple approach to resolve the problem of dealing with the proximity issue. For implicit schemes, "initial asset price close to the barrier" is not an issue because the response to boundary conditions are felt almost instantaneously across the entire solution in implicit scheme calculations (Zvan, Vetzal and Forsyth, 2000).

Lookback Options

It is relatively straightforward to price lookback options using the forward shooting grid approach (see Problem 6.15). For floating strike lookback options, by applying appropriate choices of similarity variables, the pricing formulation reduces to the form similar to that of usual one-asset option models, except that the Neumann boundary condition appears at one end of the domain of the lookback option model. Let $c(S, m, t)$ denote the price of a continuously monitored European floating strike lookback call option, where m is the realized minimum asset price from T_0 to t, $T_0 < t$. The terminal payoff at time T of the lookback call is given by

$$c(S, m, T) = S - m. \tag{6.2.38a}$$

Recall that $S \geq m$ and the boundary condition at $S = m$ is given by

$$\frac{\partial c}{\partial m} = 0 \quad \text{at} \quad S = m. \tag{6.2.38b}$$

We choose the following set of similarity variables:

$$x = \ln \frac{S}{m} \quad \text{and} \quad V(x, \tau) = \frac{c(S, m, t)}{S} e^{-q\tau}, \tag{6.2.39}$$

where $\tau = T - t$, then the Black–Scholes equation for $c(S, m, t)$ is transformed into the following equation for V.

$$\frac{\partial V}{\partial \tau} = \frac{\sigma^2}{2} \frac{\partial^2 V}{\partial x^2} + \left(r - q + \frac{\sigma^2}{2}\right) \frac{\partial V}{\partial x}, \quad x > 0, \tau > 0. \tag{6.2.40}$$

Note that $S > m$ corresponds to $x > 0$. The terminal payoff condition becomes the following initial condition

$$V(x,0) = 1 - e^{-x}, \quad x > 0. \tag{6.2.41}$$

The boundary condition at $S = m$ becomes the Neumann condition

$$\frac{\partial V}{\partial x}(0,\tau) = 0. \tag{6.2.42}$$

Suppose we discretize the governing equation using the FTCS scheme, we obtain

$$V_j^{n+1} = \left[\frac{\alpha + \mu}{2} V_{j+1}^n + (1-\alpha)V_j^n + \frac{\alpha - \mu}{2} V_{j-1}^n\right], \; j = 1, 2, \cdots, \tag{6.2.43}$$

where $\alpha = (r - q + \frac{\sigma^2}{2})\frac{\Delta\tau}{\Delta x}$ and $\mu = \sigma^2 \frac{\Delta\tau}{\Delta x^2}$. Consider the continuously monitored lookback option model, we place the reflecting boundary $x = 0$ (corresponding to the Neumann boundary condition) along a layer of nodes, where the node $j = 0$ corresponds to $x = 0$. To approximate the Neumann boundary condition at $x = 0$, we use the centered difference

$$\left.\frac{\partial V}{\partial x}\right|_{x=0} \approx \frac{V_1^n - V_{-1}^n}{2\Delta x}, \tag{6.2.44}$$

where V_{-1}^n is the option value at a fictitious node one cell to the left of node $j = 0$. By setting $j = 0$ in (6.2.43) and applying the approximation of the Neumann condition: $V_1^n = V_{-1}^n$, we obtain

$$V_0^{n+1} = \alpha V_1^n + (1-\alpha)V_0^n. \tag{6.2.45}$$

Numerical results obtained from the above scheme demonstrate $O(\Delta t)$ rate of convergence (Kwok and Lau, 2001b). However, suppose forward difference is used to approximate $\frac{\partial V}{\partial x}|_{x=0}$ so that the Neumann boundary condition is approximated by $V_0^n = V_{-1}^n$ (Cheuk and Vorst, 1997), then the order of convergence reduces to $O(\sqrt{\Delta t})$ only. Also, when the nodes are not chosen to align along the reflecting boundary, we observe erratic convergence behavior of the numerical results. Problem 6.22 illustrates the failure of a naive treatment of the reflecting boundary condition of a lookback put option while Problem 6.23 demonstrates another approach to constructing a numerical boundary condition that approximates the Neumann boundary condition.

It is quite tricky to price discretely sampled lookback options since the Neumann condition is applied only on those time steps that correspond to monitoring instants. Discussion of the construction of effective algorithms for pricing these lookback options can be found in Andreasen (1998) and Kwok and Lau (2001b).

6.3 Monte Carlo Simulation

We have observed that a wide class of derivative pricing problems come down to the evaluation of the following expectation functional

$$E[f(Z(T; t, z))].$$

Here, Z denotes the stochastic process that describes the price evolution of one or more underlying financial variables, such as asset prices and interest rates, under the respective risk neutral probability measure. The process Z has the initial value z at time t, and the function f specifies the value of the derivative at the expiration time T.

As the third alternative other than the lattice tree algorithms and finite difference methods for the numerical valuation of derivative pricing problems, the Monte Carlo simulation has been proven to be a powerful and versatile technique. The Monte Carlo method is basically a numerical procedure for estimating the expected value of a random variable, so it leads itself naturally to derivative pricing problems represented as expectations. The simulation procedure involves generating random variables with a given probability density and using the law of large numbers to take the sample mean of these values as an estimate of the expected value of the random variable. In the context of derivative pricing, the Monte Carlo procedure involves the following steps:

(i) Simulate sample paths of the underlying state variables in the derivative model such as asset prices and interest rates over the life of the derivative, according to the risk neutral probability distributions.
(ii) For each simulated sample path, evaluate the discounted cash flows of the derivative.
(iii) Take the sample mean of the discounted cash flows over all sample paths.

As an example, we consider the valuation of a European vanilla call option to illustrate the Monte Carlo procedure. The numerical procedure requires the computation of the expected payoff of the call option at expiry, $E_t[\max(S_T - X, 0)]$, and discounted to the present value at time t, namely $e^{-r(T-t)}E_t[\max(S_T - X, 0)]$. Here, S_T is the asset price at expiration time T and X is the strike price. Assuming a lognormal distribution for the asset price process, the price dynamics under the risk neutral measure is given by [see (2.4.16)]

$$\frac{S_{t+\Delta t}}{S_t} = e^{\left(r-\frac{\sigma^2}{2}\right)\Delta t+\sigma\epsilon\sqrt{\Delta t}}, \tag{6.3.1}$$

where Δt is the time step, σ is the volatility and r is the riskless interest rate. Here, ϵ denotes a normally distributed random variable with zero mean and unit variance, so $\sigma\epsilon\sqrt{\Delta t}$ represents a discrete approximation to an increment in the Wiener process of the asset price with volatility σ over time increment Δt. The random number ϵ can be generated in most computer programming languages, and by virtue of its randomness, it assumes a different value in each generation run. Suppose there are N time steps between the current time t and expiration time T, then $\Delta t = (T - t)/N$. The numerical procedure given in (6.3.1) is repeated N times to simulate the price path from S_t to $S_T = S_{t+N\Delta t}$. The call price resulted from this particular simulated asset price path is then computed using the formula of discounted payoff:

$$c = e^{-r(T-t)}\max(S_T - X, 0). \tag{6.3.2}$$

This completes one-sample iteration of the Monte Carlo simulation for the European call option model.

After repeating the above simulation for a sufficiently large number of runs, the expected call value is obtained by computing the sample mean of the estimates of the call value found in the sample simulation. Also, the standard deviation of the estimate of the call value can be found. Let c_i denote the estimate of the call value obtained in the ith simulation and let M be the total number of simulation runs. The expected call value is given by

$$\hat{c} = \frac{1}{M} \sum_{i=1}^{M} c_i, \tag{6.3.3}$$

and the sample variance of the estimate is computed by

$$\hat{s}^2 = \frac{1}{M-1} \sum_{i=1}^{M} (c_i - \hat{c})^2. \tag{6.3.4}$$

For a sufficiently large value of M, the distribution

$$\frac{\hat{c} - c}{\sqrt{\frac{\hat{s}^2}{M}}}, \qquad c \text{ is the true call value,}$$

tends to the standard normal distribution. Note that the standard deviation of $\hat{c}$ is equal to $\hat{s}/\sqrt{M}$ and so the confidence limit of estimation can be reduced by increasing the number of simulation runs M. The appearance of M as the factor $1/\sqrt{M}$ implies that the reduction of the standard deviation by a factor of 10 will require an increase in the number of simulation runs by 100 times.

One major advantage of the Monte Carlo method is that the error is independent of the dimension of the option problem. Another advantage is its ease of accommodating complicated payoff in an option model. For example, the terminal payoff of an Asian option depends on the average of the asset price over a certain time interval while that of a lookback option depends on the extremum value of the asset price over some period of time. It is quite straightforward to obtain the average or extremum value in the price path in each simulated path. The main drawback of the Monte Carlo simulation is its demand for a large number of simulation trials in order to achieve a sufficiently high level of accuracy. This makes the simulation method less competitive in computational efficiency when compared to the lattice tree algorithms and finite difference methods. However, viewed from another perspective, practitioners dealing with a newly structured derivative product may obtain an estimate of its price using the Monte Carlo approach through routine simulation, rather than risking themselves in the construction of an analytic pricing model for the new derivative.

The efficiency of a Monte Carlo simulation can be greatly enhanced through the use of various variance reduction techniques (Hull and White, 1988; Boyle, Broadie and Glasserman, 1997), some of which are discussed below.

6.3.1 Variance Reduction Techniques

It is desirable to reduce the sample variance $\hat{s}^2$ of the estimate so that a significant reduction in the number of simulation trials M may result. The two most common techniques of variance reduction are the *antithetic variates method* and the *control variate method*.

First, we describe how to assess the effectiveness of a variance reduction technique from the perspective of computational efficiency. Suppose W_T is the total amount of computational work units available to generate an estimate of the value of an option V. Assume that there are two methods for generating the Monte Carlo estimates for the option value, requiring W_1 and W_2 units of computation work, respectively, for each simulation run. For simplicity, we assume W_T to be divisible by both W_1 and W_2. Let $V_i^{(1)}$ and $V_i^{(2)}$ denote the estimator of V in the ith simulation using Methods 1 and 2, respectively, and their respective standard deviations are σ_1 and σ_2. The sample means for estimating V from the two methods using W_T amount of work are, respectively,

$$\frac{W_1}{W_T}\sum_{i=1}^{W_T/W_1} V_i^{(1)} \quad \text{and} \quad \frac{W_2}{W_T}\sum_{i=1}^{W_T/W_2} V_i^{(2)}.$$

By the law of large numbers, the above two estimators are approximately normally distributed with mean V and their respective standard deviation are

$$\sigma_1\sqrt{\frac{W_1}{W_T}} \quad \text{and} \quad \sigma_2\sqrt{\frac{W_2}{W_T}}.$$

Hence, the first method would be preferred over the second one in terms of computational efficiency provided that

$$\sigma_1^2 W_1 < \sigma_2^2 W_2. \tag{6.3.5}$$

Alternatively speaking, a lower variance estimator is preferred only if the variance ratio σ_1^2/σ_2^2 is less than the work ratio W_2/W_1 when the aspect of computational efficiency is taken into account.

Antithetic Variates Method

Suppose $\{\epsilon^{(i)}\}$ denotes the independent samples from the standard normal distribution for the ith simulation run of the asset price path so that

$$S_T^{(i)} = S_t\, e^{\left(r-\frac{\sigma^2}{2}\right)(T-t)+\sigma\sqrt{\Delta t}\sum_{j=1}^{N}\epsilon_j^{(i)}}, \quad i = 1, 2, \cdots, M, \tag{6.3.6}$$

where $\Delta t = \frac{T-t}{N}$ and M is the total number of simulation runs. Note that $\epsilon_j^{(i)}$ is randomly sampled from the standard normal distribution. From (6.3.2)–(6.3.3), an unbiased estimator of the price of a European call option with strike price X is given by

$$\hat{c} = \frac{1}{M}\sum_{i=1}^{M} c_i = \frac{1}{M}\sum_{i=1}^{M} e^{-r(T-t)} \max(S_T^{(i)} - X, 0). \tag{6.3.7a}$$

We observe that if $\{\epsilon^{(i)}\}$ has a standard normal distribution, so does $\{-\epsilon^{(i)}\}$, and the simulated price $\tilde{S}_T{}^{(i)}$ obtained from (6.3.7a) using $\{-\epsilon^{(i)}\}$ is also a valid sample from the terminal asset price distribution. A new unbiased estimator of the call price can be obtained from

$$\tilde{c} = \frac{1}{M}\sum_{i=1}^{M} \tilde{c}_i = \frac{1}{M}\sum_{i=1}^{M} e^{-r(T-t)} \max(\tilde{S}_T^{(i)} - X, 0). \tag{6.3.7b}$$

Normally we would expect c_i and $\tilde{c}_i$ to be negatively correlated, that is, if one estimate overshoots the true value, the other estimate downshoots the true value. It seems sensible to take the average of these two estimates. Indeed, we take the antithetic variates estimate to be

$$\bar{c}_{AV} = \frac{\hat{c} + \tilde{c}}{2}. \tag{6.3.8}$$

Considering the aspect of computational efficiency as governed by inequality (6.3.5), it can be shown that the antithetic variates method improves efficiency provided that $\text{cov}(c_i, \tilde{c}_i) \leq 0$ (see Problem 6.27).

Control Variate Method

The control variate method is applicable when there are two similar options, A and B. Option A is the one whose price is desired, while option B is similar to option A in nature but its analytic price formula is available. Let V_A and V_B denote the true value of option A and option B, respectively, and let $\hat{V}_A$ and $\hat{V}_B$ denote the respective estimated value of option A and option B using the Monte Carlo simulation. How does the knowledge of V_B and $\hat{V}_B$ help improve the estimate of the value of option A beyond the available estimate $\hat{V}_A$?

The control variate method aims to provide a better estimate of the value of option A using the formula

$$\hat{V}_A^{cv} = \hat{V}_A + (V_B - \hat{V}_B), \tag{6.3.9}$$

where the error $V_B - \hat{V}_B$ is used as a control in the estimation of V_A. To justify the method, we consider the following relation between the variances of the above quantities

$$\text{var}\big(\hat{V}_A^{cv}\big) = \text{var}(\hat{V}_A) + \text{var}(\hat{V}_B) - 2\,\text{cov}(\hat{V}_A, \hat{V}_B),$$

so that

$$\text{var}(\hat{V}_A^{cv}) < \text{var}(\hat{V}_A) \text{ provided that } \text{var}(\hat{V}_B) < 2\,\text{cov}(\hat{V}_A, \hat{V}_B).$$

Hence, the control variate technique reduces the variance of the estimator of V_A when the covariance between $\hat{V}_A$ and $\hat{V}_B$ is large. This is true when the two options are strongly correlated. In terms of computational efforts, we need to compute two estimates $\hat{V}_A$ and $\hat{V}_B$. However, if the underlying asset price paths of the two options are identical, then there is only slight additional work to evaluate $\hat{V}_B$ along with $\hat{V}_A$ on the same set of simulated price paths.

To facilitate the more optimal use of the control $V_B - \hat{V}_B$, we define the control variate estimate to be

$$\hat{V}_A^\beta = \hat{V}_A + \beta(V_B - \hat{V}_B), \tag{6.3.10}$$

where β is a parameter with value other than 1. The new relation between the variances is now given by

$$\text{var}\left(\hat{V}_A^\beta\right) = \text{var}\left(\hat{V}_A\right) + \beta^2 \text{var}\left(\hat{V}_B\right) - 2\beta \text{ cov}\left(\hat{V}_A, \hat{V}_B\right). \tag{6.3.11}$$

The particular choice of β which minimizes var $(\hat{V}_A^\beta)$ is found to be

$$\beta^* = \frac{\text{cov}(\hat{V}_A, \hat{V}_B)}{\text{var}(\hat{V}_B)}. \tag{6.3.12}$$

Unlike the choice of $\beta = 1$ used in (6.3.9), the control variate estimate based on β^* is guaranteed to decrease variance. Unfortunately, the determination of β^* requires the knowledge of $\text{cov}(\hat{V}_A, \hat{V}_B)$, which is in general not available. However, one may estimate β^* using the regression technique from the simulated option values $V_A^{(i)}$ and $V_B^{(i)}$, $i = 1, 2, \cdots, M$, obtained from the simulation runs.

Valuation of Asian Options

A nice example of applying the control variate method is the estimation of the value of an arithmetic averaging Asian option. We base this estimation on the knowledge of the exact analytic formula of the corresponding geometric averaging Asian option. The two types of Asian options are very similar in nature except that the terminal payoff function depends on either arithmetic averaging or geometric averaging of the asset prices.

The averaging feature in the Asian options does not pose any difficulty in Monte Carlo simulation because the average of the asset prices at different observational instants in a given simulated path can be easily computed. Since option price formulas are readily available for the majority of geometrically averaged Asian options, we can use this knowledge to include a variance reduction procedure to reduce the confidence interval in the Monte Carlo simulation performed for valuation of the corresponding arithmetically averaged Asian options (Kemna and Vorst, 1990).

Let V_A denote the price of an option whose payoff depends on the arithmetic averaging of the underlying asset price and V_G be the price of an option similar to the above option except that geometric averaging is taken. Let $\hat{V}_A$ and $\hat{V}_G$ denote the discounted option payoff for a single simulated path of the asset price with respect to arithmetic and geometric averaging, respectively, so that

$$V_A = E[\widehat{V}_A] \text{ and } V_G = E[\widehat{V}_G]. \tag{6.3.13}$$

We then have

$$V_A = V_G + E[\widehat{V}_A - \widehat{V}_G],$$

so that an unbiased estimator of V_A is given by

$$\hat{V}_A^{cv} = \hat{V}_A + (V_G - \hat{V}_G). \tag{6.3.14}$$

6.3.2 Low Discrepancy Sequences

The crude Monte Carlo method uses random (more precisely pseudo-random) points and the rate of convergence is known to be $O(\frac{1}{\sqrt{M}})$, where M is the number of simulation trials. The order of convergence of $O(\frac{1}{\sqrt{M}})$ implies that $O(\frac{1}{\epsilon^2})$ simulations are required to achieve $O(\epsilon)$ level of accuracy. Such a low rate of convergence is certainly not quite desirable. Also, it is quite common to have the accuracy of simulation to be sensitive to the initial seed.

It is commonly observed that the pseudo-random points may not be quite uniformly dispersed throughout the domain of the problem. It seems reasonable to postulate that convergence may be improved if these points are more uniformly distributed. A notion in number theory called *discrepancy* measures the deviation of a set of points in d dimensions from uniformity. Lower discrepancy means the points are more evenly dispersed. There have been a few well-tested sequences, called the quasi-random sequences (though they are deterministic in nature), which demonstrate a low level of discrepancy. Some of these examples are the Sobol points and Halton points (Paskov and Traub, 1995). These low discrepancy sequences have the nice property that when successive points are added, the entire sequence of points still remains at a similar level of discrepancy. The routines for generating these sequences are readily available in many software texts (e.g., Press et al., 1992).

The rate of convergence of simulation with respect to the use of different sequences can be assessed through the numerical approximation of an integral by a discrete average. If we use equally spaced points, which is simply the trapezoidal rule of numerical integration, the error is $O(M^{-2/d})$ where d is the dimension of the integral. For the Sobol points or Halton points, the rate of convergence is $O(\frac{(\ln M)^d}{M})$. This is still in favor of $O(\frac{1}{\sqrt{M}})$ rate of convergence of the Monte Carlo method when d is modest.

Various numerical studies on the use of low discrepancy sequences in finance applications reveal that the errors produced are substantially lower than the corresponding errors using the crude random sequences. Paskov and Traub (1995) employed both Sobol sequences and Halton sequences to evaluate mortgage-backed security prices, which involves the evaluation of integrals with d up to 360. They showed that the Sobol sequences outperform the Halton sequences which in turn perform better than the standard Monte Carlo method. The reason for the better performance may be attributed to the smoothness of the integrand functions. Strong research interests still persist in the continual search for better low discrepancy sequences in finance applications.

6.3.3 Valuation of American Options

There had been a general belief that the Monte Carlo approach can be used only for European style derivatives. The apparent difficulties of using simulation to price American options stem from the *backward* nature of the early exercise feature since there is no way of knowing whether early exercise is optimal when a particular asset value is reached at a given time. The estimated option value with respect to a given simulated path can be determined only with a prespecified exercise policy.

A variety of simulation algorithms have been proposed in the literature to tackle the above difficulties. The earliest simulation algorithm is the "bundling and sorting" algorithm proposed by Tilley (1993). The algorithm computes an estimate for the option's continuation value by using backward induction and a bundling technique. At each time instant, simulation path with similar asset prices are grouped together to obtain an estimate of the one-period-ahead option value. Another approach (Grant, Vora and Weeks, 1996) attempts to approximate the exercise boundary at each early exercise point using backward induction as the first step, then estimates the option price in a forward simulation based on the known exercise policies. The other approach (Broadie and Glasserman, 1997) attempts to find the efficient upper and lower bounds on option value from simulated paths, the upper bound is based on a nonrecombining tree and the lower bound is based on a stochastic mesh. These high and low estimates for the option value converge asymptotically to the true option value.

The more recent and possibly most popular approach is the linear regression method via basis functions. The conditional expectations in the dynamic programming procedure are approximated by projections on a finite set of basis functions. Monte Carlo simulations and least squares regression techniques are used to compute the above approximated value function. Longstaff and Schwartz (2001) chose the Laguerre polynomials as the basis functions. The guidelines on the choice of the basis functions were discussed by Tsitsiklis and Van Roy (2001), Lai and Wong (2004). Clément, Lamberton and Protter (2002) proved the almost sure convergence of the algorithm. Glasserman and Yu (2004) analyze the convergence of the algorithm when both the number of basis functions and the number of simulated paths increase.

Four classes of algorithms are presented below, namely, the "bundling and sorting" algorithm, method of parameterization of the early exercise boundary, stochastic mesh method and linear regression method via basis functions. A comparison of performance of various Monte Carlo simulation approaches for pricing American style options was reported by Fu et al. (2001). A comprehensive review of Monte Carlo methods in financial engineering can be found in Glasserman (2004).

Tilley's Bundling and Sorting Algorithm

Tilley (1993) proposes a "bundling and sorting" algorithm which computes an estimate for the American option's continuation value using backward induction. At each time step in the simulation procedure, simulated asset price paths are ordered by asset price and bundled into groups. The method rests on the belief that the price paths within a given bundle are sufficiently alike so that they can be considered to have the same expected one-period-ahead option value. The boundary between the exercise-or-hold decisions is determined for each time step.

The options are assumed to be exercisable at specified instants $t = 1, 2, \cdots, N$. Actually, this discretization assumption transforms the American options with continuous early exercise right to the Bermudan options with discrete exercise opportunities (see Problem 6.28). The simulation procedure generates a finite sample of R asset price paths from $t = 0$ to $t = N$, where the realization of the asset price of the kth price path is represented by the sequence $\{S_0(k), \cdots, S_N(k)\}$. Let d_t denote the discount factor from t to $t+1$ and D_t be the discount factor from 0 to time t, so that $D_t = d_0 d_1 \cdots d_{t-1}$. Let X be the strike price of the option. The backward induction procedure starts at $t = N - 1$. At each t, we proceed inductively according to the following steps.

1. Sort the price paths by order of the asset price by partitioning the ordered paths in Q distinct bundles of P paths in each bundle ($R = QP$). We write $B_t(k)$ as the set of price paths in the bundle containing path k at time t. For each path k, compute the intrinsic value $I_t(k)$ of the option.
2. Compute the option's continuation value $H_t(k)$, defined as the present value of the expected one-period-ahead option value:
$$H_t(k) = \frac{d_t(k)}{P} \sum_{j \in B_t(k)} V_{k+1}(j), \tag{6.3.15}$$
where $V_{t+1}(j)$ has been computed in the previous time step. In particular, $V_N(j) = I_N(j)$ for all j.
3. For each path k, compare $H_t(k)$ to $I_t(k)$ and decide "tentatively" whether to exercise the option or continue holding it. Define $x_t(k)$ as the "tentative" exercise-or-hold indicator variable, where
$$x_t(k) = \begin{cases} 1 & \text{when } I_t(k) \geq H_t(k) \\ 0 & \text{when } I_t(k) < H_t(k) \end{cases}.$$
Here, "1" and "0" represent "exercise" and "hold", respectively.
4. In general, there may be more than one bundle in which $x_t(k) = 1$ for some $k \in B_t(k)$ but 0 for other paths within the same bundle. These bundles have a "transition zone" in asset price from "hold" to "exercise" decision. The algorithm has to be refined by creating a sharp boundary between the "hold" and "exercise" decisions. To achieve this goal, we examine the sequence $\{x_t(k) : k = 1, \cdots, R\}$, and determine the sharp boundary as the start of the first string of "1"'s, the length of which exceeds the length of every subsequent string of "0"'s. For example, consider the following string

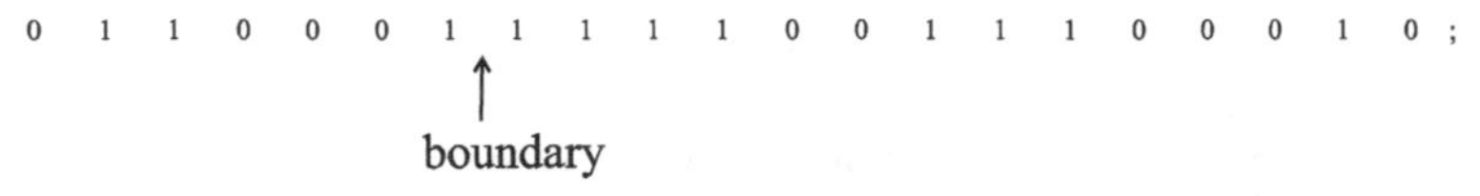

The length of the first string of 1's is 5 while the lengths of subsequent strings of 0's are 2, 3 and 1. The path index of the leading "1" is called k_t^*. Next, we define the "update" exercise-or-hold indicator variable $y_t(k)$ by

$$y_t(k) = \begin{cases} 1 & \text{when } k \geq k_t^* \\ 0 & \text{when } k < k_t^* \end{cases}.$$

5. For each path k, define the current value $V_t(k)$ of the option by

$$V_t(k) = \begin{cases} I_t(k) & \text{when } y_t(k) = 1 \\ H_t(k) & \text{when } y_t(k) = 0 \end{cases}.$$

The above procedure proceeds backwards from $t = N - 1$ to $t = 0$. Finally, we define the exercise-or-hold indicator variable by

$$Z_t(k) = \begin{cases} 1 & \text{if } y_t(k) = 1 \text{ and } y_s(t) = 0 \text{ for all } s < t \\ 0 & \text{otherwise} \end{cases}.$$

Once the exercise policy of each price path is established, the option price estimator is given by

$$\frac{1}{R} \sum_{k=1}^{R} \sum_{t=1}^{N} Z_t(k) D_t(k) I_t(k).$$

For each path k, $Z_t(k)$ equals one at only one time instant and $D_t(k) I_t(k)$ gives the discount value of the option payoff of the path.

There are several major weaknesses in Tilley's algorithm. The algorithm is not computationally efficient since it requires storage of all simulated asset price paths at all time steps. The bundling and sorting of all price paths pose stringent requirement on storage and computation even when the number of simulated paths is moderate. As shown by Tilley's own numerical experiments, there is no guarantee on the convergence of the algorithm to the true option value. Also, the extension of the algorithm to multi-asset option models can be very tedious (see Problem 6.29).

Grant–Vora–Weeks Algorithm

The simulation algorithm proposed by Grant, Vora and Weeks (1996) first identifies the optimal exercise price $S_{t_i}^*$ at selected time instants t_i, $i = 1, 2, \cdots, N-1$ between the current t and expiration time T. The determination of the optimal exercise prices is done by simulation at successive time steps proceeding backwards in time. Once the exercise boundary is identified, the option value can be estimated by the usual simulation procedure, respecting the early exercise strategy as dictated by the known exercise boundary.

We illustrate the procedure by considering the valuation of an American put option and choosing only three time steps between the current time t and the expiration time T, where $t_0 = t$ and $t_3 = T$. Assuming a constant dividend yield q, the optimal exercise price at T is equal to $\min(\frac{r}{q}X, X)$, where X is the strike price of the option and r is the riskless interest rate. At time t_2 which is one time period prior to expiration, the put value is $X - S_{t_2}$ when $S_{t_2} \leq S_{t_2}^*$, and $E[P_T]e^{-r(T-t_2)}$ when $S_{t_2} > S_{t_2}^*$. Here, $P_T = \max(X - S_T, 0)$ denotes the put option value at expiration time T. Obviously, $E[P_T]$ is dependent on S_{t_2}. For a given value of S_{t_2}, one can perform a sufficient number of simulations to estimate $E[P_T]$. The optimal exercise price $S_{t_2}^*$ is identified by finding the appropriate value of S_{t_2} such that

$$X - S_{t_2}^* = e^{-r(T-t_2)} E[P_T | S_{t_2}^*]. \tag{6.3.16}$$

We find the simulation estimate of $e^{-r(T-t_2)} E[P_T]$ as a function of S_{t_2} by starting with S_{t_2} close to but smaller than $S_{t_3}^*$ (remark: $S_{t_3}^*$ is known and $S_{t_2}^*$ must be less that $S_{t_3}^*$) and repeat the simulation process for a series of S_{t_2} which decreases systematically. Once the functional dependence of the discounted expectation value $e^{-r(T-t_2)} E[P_T]$ in S_{t_2} is available, one can find a good estimate of $S_{t_2}^*$ such that (6.3.16) is satisfied.

Proceeding backwards in time, we continue to estimate the optimal exercise price at time t_1. The simulation now starts at t_1. The initial asset value S_{t_1} is first chosen with a value slightly less than $S_{t_2}^*$ and simulation is repeated with decreasing S_{t_1}. Again, we would like to find the estimate of the discounted expectation value of holding the put, and this expectation value is a function of S_{t_1}. In a typical simulation run, an asset value S_{t_2} is generated at t_2 with an initial asset value S_{t_1}. Using the estimate of $S_{t_2}^*$ obtained in the previous step, we can determine whether S_{t_2} falls in the stopping region or otherwise. If the answer is yes, the estimated put value for that simulated path is the present value of the early exercise value. Otherwise, the simulation continues by generating an asset value at expiration T. The put value for this simulation path then equals the present value of the corresponding terminal payoff. This simulation procedure is repeated a sufficient number of times so that an estimate of the discounted expectation value can be obtained. In a similar manner, we determine the critical value $S_{t_1}^*$ such that when S_{t_1} is chosen to be $S_{t_1}^*$, the intrinsic value $X - S_{t_1}^*$ equals the estimate of the discounted expectation value of holding the put.

Once the optimal exercise prices at t_1 and t_2 are available, one can mimic the above numerical procedure to find the estimate of the discounted expectation value of holding the put at time t_0 by performing simulation runs with an initial asset value S_{t_0}. The put value at time t_0 for a given S_{t_0} is the maximum of the estimate of the discounted expectation value obtained from simulation (taking into account the early exercise strategy as already determined at t_1 and t_2) and the intrinsic value $X - S_{t_0}$ from early exercise.

Broadie–Glasserman Algorithm

The stochastic mesh algorithm of Broadie and Glasserman (1997) produces two estimators for the true option value, one biased high and the other biased low, but both asymptotically unbiased as the number of simulations tends to infinity. The two estimates together provide a conservative confidence interval for the option value.

First, a random tree with b branches per node is constructed (see Fig. 6.7 for $b = 3$) and the asset values at the nodes at time t_j are denoted by

$$S_j^{i_1 i_2 \cdots i_j}, \quad j = 1, 2, \cdots, N, \quad 1 \le i_1, \cdots, i_j \le b,$$

where N is the total number of time steps. The total number of nodes at time t_j will be b^j. Here, S_0 is the fixed initial state and each sequence S_0, $S_1^{i_1}$, $S_2^{i_1 i_2}$, $\cdots$, $S_N^{i_1 i_2 \cdots i_N}$ is a realization of the Markov process for the asset price, and two such sequences evolve independently of each other once they differ in some i_j.

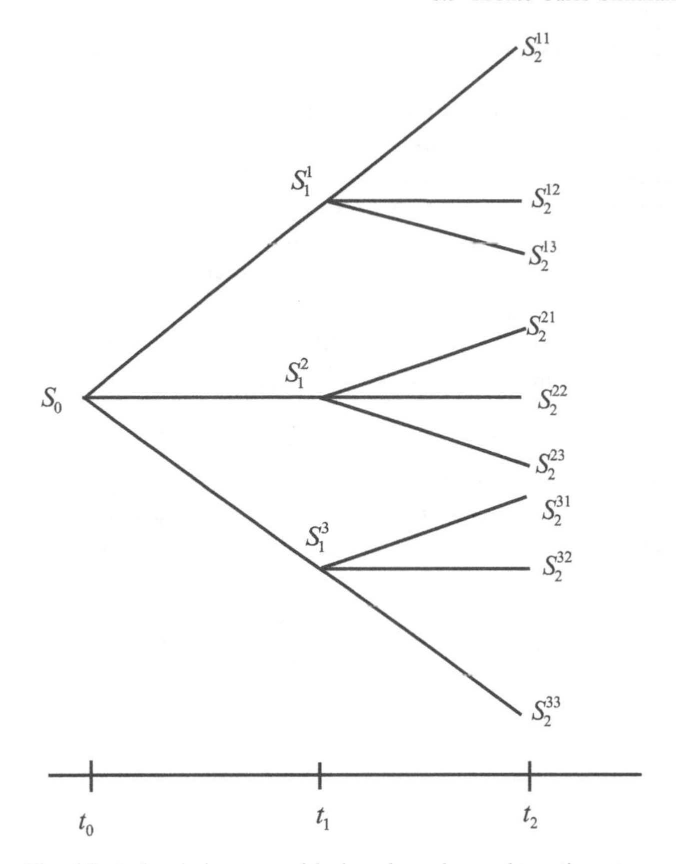

Fig. 6.7. A simulation tree with three branches and two time steps.

Let $\theta_{\mathrm{high},j}^{i_1\cdots i_j}$ and $\theta_{\mathrm{low},j}^{i_1\cdots i_j}$ denote, respectively, the high and low estimators of the option value at the $(i_1,\cdots,i_j)$th node at time t_j. Also, let $h_j(s)$ be the payoff from exercise at time t_j in state s and $1/R_{j+1}$ be the discount factor from t_j to t_{j+1}. Broadie and Glasserman defined the high estimator for the option value at the $(i_1,\cdots,i_j)$ node at time t_j to be the maximum of the early exercise payoff and the estimate of the continuation value from the b successor nodes, namely,

$$\theta_{\mathrm{high},j}^{i_1\cdots i_j} = \max\left(h_j(S_j^{i_1\cdots i_j}), \frac{1}{b}\sum_{i_{j+1}=1}^{b}\frac{1}{R_{j+1}}\theta_{\mathrm{high},j+1}^{i_1\cdots i_j i_{j+1}}\right). \tag{6.3.17}$$

Simple arguments can be used to explain why the above estimate is biased high. If the asset prices at the nodes at time t_{j+1} turn out to be too high in the simulation process, the above dynamic programming procedure will choose not to exercise and take a value higher than the optimal decision to exercise. On the other hand, if the simulated asset prices at t_{j+1} turn out to be too low, the dynamic programming procedure will choose to exercise even when the optimal decision is not to exercise. The option value is over-estimated since we have taken advantage of knowledge of the future.

The numerical algorithm for the low estimator is slightly more complicated. At each node, one branch is used to estimate the continuation value and the other $b-1$ branches are used to estimate the exercise decision. The same procedure is repeated b times, where each branch is chosen in turn. To explain the procedure in more detail, suppose the kth branch is chosen to estimate the continuation value while the other $b-1$ branches are used to estimate the exercise decision. Early exercise is chosen if the payoff $h_j(S_j^{i_1\cdots i_j})$ is greater than or equal to the expectation of the continuation value. This expectation is computed by taking the average among $b-1$ branches of the discounted values $\frac{1}{R_{j+1}}\theta_{\text{low},j+1}^{i_1\cdots i_{j+1}}, i_{j+1}=1,\cdots,b, i_{j+1}\neq k$. If early exercise is chosen, then the estimate $\eta_j^{i_1\cdots i_jk}$ takes the payoff value $h_j(S_j^{i_1\cdots i_j})$, otherwise, it takes the continuation value $\frac{1}{R_{j+1}}\theta_{\text{low},j+1}^{i_1\cdots i_jk}$. Thus b estimates are obtained in these b steps of calculations and they are then averaged to determine the option value estimate at the node. The procedure can be succinctly described as follows. Write

$$\eta_j^{i_1\cdots i_jk}=\begin{cases} h_j(S_j^{i_1\cdots i_j}), & \text{if } h_j(S_j^{i_1\cdots i_j})\geq \frac{1}{b-1}\sum_{\substack{i_{j+1}=1\\ i_{j+1}\neq k}}^{b}\frac{1}{R_{j+1}}\theta_{\text{low},j+1}^{i_1\cdots i_ji_{j+1}}\\ \frac{1}{R_{j+1}}\theta_{\text{low},j+1}^{i_1\cdots i_jk}, & \text{if } h_j(S_j^{i_1\cdots i_j})< \frac{1}{b-1}\sum_{\substack{i_{j+1}=1\\ i_{j+1}\neq k}}^{b}\frac{1}{R_{j+1}}\theta_{\text{low},j+1}^{i_1\cdots i_ji_{j+1}}, \end{cases}$$
$$k=1,\cdots,b, \tag{6.3.18}$$

then

$$\theta_{\text{low},j}^{i_1\cdots i_j}=\frac{1}{b}\sum_{k=1}^{b}\eta_j^{i_1\cdots i_jk}. \tag{6.3.19}$$

The explanation as to why the above procedure gives a biased low estimator is relegated to an exercise.

Both algorithms (6.3.17) and (6.3.18)–(6.3.19) are backward induction, that is, knowing estimates at time t_{j+1}, we compute estimates at t_j one period earlier. For both high- and low-biased estimators, the starting iterates at expiration time $T=t_N$ are both given by the following terminal payoff function:

$$\theta_N^{i_1\cdots i_N}=h_N\left(S_N^{i_1\cdots i_N}\right). \tag{6.3.20}$$

The Broadie–Glasserman algorithm can be extended to deal with multi-asset options, and the computation can be made parallelized to work on a cluster of workstations. Variance reduction techniques can also be employed to fasten the rate of convergence. The algorithm can allow multiple decisions other than the two-fold decision: exercise or hold.

Linear Regression Method via Basis Functions

Under the discrete assumption of exercise opportunities, the option values satisfy the following dynamic programming equations

$$V_n = \max(h_n(S), H_n(S)), \quad n = 0, 1, \cdots, N-1, \tag{6.3.21}$$

where $S = S(t_n)$, $H_n(S)$ is the continuation value at time t_n and $h_n(S)$ is the exercise payoff. At maturity date $t_N = T$, we have $V_N(S) = h_N(S)$ [for notational convenience, we set $H_N(S) = 0$]. The continuation values at different time instants are given by the following recursive scheme

$$H_n(S) = E\big[\max(h_{n+1}(S(t_{n+1})), H_{n+1}(S(t_{n+1}))|S(t_n) = S\big]. \tag{6.3.22}$$

The difficulty of estimating the above conditional expectations may be resolved by considering an approximation of $H_n(S)$ in the form

$$H_n(S) \approx \sum_{m=0}^{M} \alpha_{nm}\phi_{nm}(S), \tag{6.3.23}$$

for some choice of the basis functions $\phi_{nm}(S)$. Longstaff and Schwartz (2001) proposed determining the coefficients α_{nm} through the least squares projection onto the span of basis functions. Their chosen basis functions are the Laguerre polynomials defined by

$$L_m(S) = e^{-S/2}\frac{e^S}{m!}\frac{d^m}{dS^m}\left(S^m e^{-S}\right), \quad m = 0, 1, 2, \cdots. \tag{6.3.24}$$

The first few members are:

$$L_0(S) = e^{-S/2}, L_1(S) = e^{-S/2}(1 - S), L_2(S) = e^{-S/2}\left(1 - 2S + \frac{S^2}{2}\right).$$

Following the description of the algorithm by Longstaff and Schwartz (2001), we use $C(\omega, s; t, T)$ to denote the path of cash flows generated by the option, conditional on the option not being exercised at or prior to time t. Here, ω represents a sample path and T is option's maturity date. The holder is assumed to follow the optimal stopping strategy for all subsequent times s, where $t < s \leq T$. Recall that the value of an American option is given by maximizing the discounted cash flows from the option, where the maximum is taken over all stopping times. We seek a pathwise approximation to the optimal stopping rule associated with the early exercise right in the American option. Like other simulation algorithms, the key is to identity the conditional expected value of continuation.

Let $H_n(\omega; t_n)$ denote the continuation value at time t_n. By the no arbitrage principle, $H_n(\omega)$ is given by the expectation of the remaining discounted cash flows under the risk neutral measure. At time t_n, $H_n(\omega)$ is given by

$$H_n(\omega; t_n) = E\left[\sum_{j=n+1}^{N} e^{-r(t_j - t_n)} C(\omega, t_j; t_n, T)\right], \tag{6.3.25}$$

where the expectation is taken under the risk neutral measure conditional on the filtration at time t_n. Suppose we have chosen M basis functions, then $H_n(\omega)$ is estimated by regressing the discounted cash flow onto the basis functions for the paths where the option is in-the-money at time t_n. Longstaff and Schwartz (2001) proposed that only the in-the-money paths be used in the estimation since the exercise decision is relevant only in the in-the-money regime. Once the functional form of the estimated continuation value $\widehat{H}_n(\omega)$ is obtained from linear regression, we can calculate the estimated continuation value from the known asset price at time t_n for that path ω.

Our goal is to solve for the stopping rule that maximizes the option value at every time point along each asset price path. We start from the maturity date t_N, and proceed backward in time. At t_N, the cash flows are given by the terminal payoff function and thus they are readily known. At one time step backward, we search for those paths that are in-the-money at t_{N-1}. Only in-the-money paths are used since one can better estimate the conditional expectation in the region where exercise is relevant. From these paths, we compute the discounted cash flow received at time t_N given that the option remains alive at time t_{N-1}. Consider path k, its asset price at t_{N-1} and t_N are denoted by $S_{N-1}^{(k)}$ and $S_N^{(k)}$, respectively, $k = 1, \cdots, K$, where K is the total number of paths that are in-the-money at t_{N-1}. The discounted cash flow at t_{N-1} for path k is given by $e^{-r(t_N - t_{N-1})} h_N(S_N^{(k)})$, where h_N is the terminal payoff function of the option. Using the information of these K data points and choosing M basis functions, we estimate the continuation value $\widehat{H}_{N-1}^{(k)}$ by regressing the discounted cash flow at t_{N-1} with respect to the asset price at t_{N-1}. Early exercise at time t_{N-1} is optimal for an in-the-money path ω if the immediate exercise value is greater than or equal to the estimated continuation value. In this case, the cash flow at t_{N-1} is set equal the exercise value.

Once the cash flow paths and stopping rule at t_{N-1} have been determined, we then proceed recursively in the same manner to the earlier time points $t_{N-2}, \cdots, t_1$. As a result, we obtain the optimal stopping rule at all time points for every path. Once the cash flows generated by the option for all paths are identified, we can compute an estimate of the option value by discounting each cash flow back to the issue date and averaging over all sample price paths.

To illustrate the above numerical procedures, we adopt the numerical example presented by Longstaff and Schwartz (2001). Consider a three-year American (actually Bermudan) put option on a nondividend paying asset with strike price 1.1. The put can be exercised only at $t = 1, 2, 3$. We take the riskless interest rate to be 0.06. Only eight sample paths of the asset price are generated under the risk neutral measure. These paths are shown in the following table.

Path	Asset price paths				Cash flow at $t = 3$
	$t = 0$	$t = 1$	$t = 2$	$t = 3$	
1	1.00	1.09	1.08*	1.34	0.00
2	1.00	1.16	1.26	1.54	0.00
3	1.00	1.22	1.07*	1.03	0.07
4	1.00	0.93	0.97*	0.92	0.18
5	1.00	1.11	1.56	1.52	0.00
6	1.00	0.76	0.77*	0.90	0.20
7	1.00	0.92	0.84*	1.01	0.09
8	1.00	0.88	1.22	1.34	0.00

* Sample path for which the put is in-the-money at $t = 2$.

Note that there are five paths for which the put is in-the-money at $t = 2$. How can we solve for the optimal stopping rule that maximizes the value of the put at each time point along each path?

For the five paths that are in-the-money at $t = 2$, we compute the corresponding discounted cash flows received at $t = 2$ if the put is not exercised at $t = 2$. Let X and Y denote, respectively, the asset price at $t = 2$ and the discounted cash flow at $t = 3$ conditional on no exercise at $t = 2$. The values of X and Y for those in-the-money asset price paths are listed below:

Path	Y	X	Exercise value	Continuation value
1	0.00×0.94176	1.08	0.02	0.0369
3	0.07×0.94176	1.07	0.03	0.0461
4	0.18×0.94176	0.97	0.13	0.1176
6	0.20×0.94176	0.77	0.33	0.1520
7	0.09×0.94176	0.84	0.26	0.1565

* The discount factor is $e^{-0.06} = 0.94176$.

For simplicity, suppose the basis functions: 1, X and X^2 are chosen, we regress Y on these basis functions. Longstaff and Schwartz (2001) obtained the following conditional expectation function:

$$E[Y|X] = -1.070 + 2.983X - 1.813X^2.$$

Now, we compare the value of immediate exercise at $t = 2$ and the value from continuation (calculated using the above conditional expectation function). For example, for Path 1 where $X = 1.08$, the immediate exercise value equals $1.10 - 1.08 = 0.02$ while the continuation value is $-1.070 + 2.983 \times 1.08 - 1.813 \times 1.08^2 = 0.0369$. Since the continuation value is higher, it is not optimal to exercise the put at $t = 2$ for the first path. The corresponding cash flow received by the option holder for Path 1 conditional on not exercising prior to $t = 2$ is zero. For Path 4, since the exercise value is higher than the continuation value, the cash flow for this path at $t = 2$ is set equal to the exercise value. One can check that it is also optimal to exercise at $t = 2$ for Paths 6 and 7.

The cash flow received at $t = 2$ and $t = 3$ for the eight simulated price paths are summarized in the following table:

Path	$t = 1$	$t = 2$	$t = 3$
1	–	0.00	0.00
2	–	0.00	0.00
3	–	0.00	0.07
4	–	0.13	0.00
5	–	0.00	0.00
6	–	0.33	0.00
7	–	0.26	0.00
8	–	0.00	0.00

Note that it is optimal to exercise the put at $t = 2$ for Paths 4, 6 and 7. Once the option has been exercised at $t = 2$, the cash flow at $t = 3$ becomes zero.

Next, we proceed recursively to determine the stopping rule at $t = 1$. There are five paths (Paths $1, 4, 6, 7$ and 8) for which the put is in-the-money at $t = 1$. Similarly, we solve for the estimated expectation function at $t = 1$ by regressing the discounted value of subsequent option cash flow at $t = 1$ on a constant, X, and X^2, where X is the asset price at $t = 1$. Again, we can compute the estimated continuation values and immediate exercise values at $t = 1$ (see table below).

Path	Y	X	Exercise value	Continuation value
1	0.00×0.94176	1.09	0.01	0.0139
4	0.13×0.94176	0.93	0.17	0.1092
6	0.33×0.94176	0.76	0.34	0.2866
7	0.26×0.94176	0.92	0.18	0.1175
8	0.00×0.94176	0.88	0.22	0.1533

* The estimated conditional expectation function at $t = 1$ is
$E[Y|X] = 2.038 - 3.335X + 1.356X^2$.

Note that exercise at $t = 1$ is optimal for Paths 4, 6, 7 and 8. The optimal stopping rules at all times are now identified.

Path	Stopping rule			Option cash flow matrix		
	$t = 1$	$t = 2$	$t = 3$	$t = 1$	$t = 2$	$t = 3$
1	0	0	0	0.00	0.00	0.00
2	0	0	0	0.00	0.00	0.00
3	0	0	1	0.00	0.00	0.07
4	1	0	0	0.17	0.00	0.00
5	0	0	0	0.00	0.00	0.00
6	1	0	0	0.34	0.00	0.00
7	1	0	0	0.18	0.00	0.00
8	1	0	0	0.22	0.00	0.00

* "1" represents exercise optimally at the exercise date.

Once optimal exercise for a given path has been chosen at an earlier time, the stopping rules that have been obtained for later times in the backward induction procedure becomes immaterial. When the cash flows generated by the put option at each time along each path have been identified, the put option value can be computed by discounting each cash flow back to current time, and taking average value over all paths.

6.4 Problems

6.1 Instead of the tree-symmetry condition: $u = 1/d$ [see (6.1.1c)], Jarrow and Rudd (1983) chose the third condition to be $p = 1/2$. By solving together with (6.1.1a,b), show that

$$u = R(1+\sqrt{e^{\sigma^2\Delta t}-1}), d = R(1-\sqrt{e^{\sigma^2\Delta t}-1}) \quad \text{and} \quad p = \frac{1}{2}.$$

6.2 Suppose the underlying asset is paying a continuous dividend yield at the rate q, the two governing equations for u, d and p are modified as

$$pu + (1-p)d = e^{(r-q)\Delta t}$$
$$pu^2 + (1-p)d^2 = e^{2(r-q)\Delta t}e^{\sigma^2\Delta t}.$$

Show that the parameter values in the binomial model are modified by replacing the growth factor of the asset price $e^{r\Delta t}$ (under the risk neutral measure) by the new factor $e^{(r-q)\Delta t}$ while the discount factor in the binomial formula remains to be $e^{-r\Delta t}$.

6.3 Show that

$$\lim_{n\to\infty} \Phi(n, k, p') = N(d_1)$$

where $p' = ue^{-r\Delta t}p$ and $d_1 = \dfrac{\ln\frac{S}{X} + \left(r+\frac{\sigma^2}{2}\right)\tau}{\sigma\sqrt{\tau}}$.

Hint: Note that

$$1 - \Phi(n, j, p')$$
$$= P\left(\frac{j-np'}{\sqrt{np'(1-p')}} < \frac{\ln\frac{X}{S} - n\left(p'\ln\frac{u}{d} + \ln d\right) - \alpha\ln\frac{u}{d}}{\sqrt{np'(1-p')}\ln\frac{u}{d}}\right),$$
$$0 < \alpha \leq 1.$$

By considering the Taylor expansion of $n(p'\ln\frac{u}{d} + \ln d)$ and $np'(1-p')(\ln\frac{u}{d})^2$ in powers of Δt, show that

$$\lim_{n\to\infty} n\left(p'\ln\frac{u}{d}+\ln d\right)=\left(r+\frac{\sigma^2}{2}\right)\tau$$

$$\lim_{n\to\infty} np'(1-p')\left(\ln\frac{u}{d}\right)^2=\sigma^2\tau,$$

where $n\Delta t=\tau$.

6.4 Consider the modified binomial formula employed for the numerical valuation of an American put on a nondividend paying asset [see (6.1.14)], deduce the optimal exercise price at time close to expiry from the binomial formula. Compare the result with that of the continuous model by taking the limit $\Delta t\to 0$.

6.5 Consider the nodes in the binomial tree employed for the numerical valuation of an American put option on a nondividend paying asset. The (n,j)th node corresponds to the node which is n time steps from the current time and has j upward jumps among n moves. The put value at the (n,j)th node is denoted by P_j^n. Similar to the continuous models, we define the stopping region $\mathcal{S}$ and continuation region $\mathcal{C}$ by

$$\mathcal{S}=\left\{(n,j)|P_j^n=X-Su^jd^{n-j}\right\}$$
$$\mathcal{C}=\left\{(n,j)|P_j^n>X-Su^jd^{n-j}\right\}.$$

That is, $\mathcal{S}$ ($\mathcal{C}$) represents the set of nodes where the put is exercised (alive). Let N be the total number of time steps in the tree. Prove the following properties of $\mathcal{S}$ and $\mathcal{C}$ (Kim and Byun, 1994):

(i) Suppose both $(n+1,j)$ and $(n+1,j+1)$ belong to $\mathcal{S}$, then $(n,j)\in\mathcal{S}$ for $0\le n\le N-1, 0\le j\le n$.

(ii) Suppose $(n+2,j+1)\in\mathcal{C}$, then $(n,j)\in\mathcal{C}$ for $0\le n\le N-2, 0\le j\le n$.

(iii) Suppose $(n,j)\in\mathcal{S}$, then both $(n,j-1)$ and $(n-1,j-1)\in\mathcal{S}$; also, suppose $(n,j)\in\mathcal{C}$, then $(n,j+1)\in\mathcal{C}$ and $(n-1,j)\in\mathcal{C}$, for $1\le n\le N, 1\le j\le n-1$.

6.6 Consider the pricing of the callable American put option by binomial calculations, let us write

$$P_{cont}=\frac{pP_{j+1}^{n+1}+(1-p)P_j^{n+1}}{R}.$$

Show that binomial scheme (6.1.15) can be modified to become

$$P_j^n=\max(\min(P_{cont},K),X-S_j^n).$$

Give the financial interpretation of the modified scheme.

6.7 Show that the total number of multiplications and additions in performing n steps of numerical calculations using the trinomial and binomial schemes are given by

Scheme	*Number of multiplications*	*Number of additions*
trinomial	$3n^2$	$2n^2$
binomial	$n^2 + n$	$\frac{1}{2}(n^2 + n)$

6.8 Suppose we let $p_2 = 0$ and write $p_1 = -p_3 = p$ in the trinomial scheme. By matching the mean and variance of $\zeta(t)$ and $\zeta^a(t)$ accordingly

$$E[\zeta^a(t)] = 2pv - v = \left(r - \frac{\sigma^2}{2}\right)\Delta t$$
$$\operatorname{var}(\zeta^a(t)) = v^2 - E[\zeta^a(t)]^2 = \sigma^2 \Delta t,$$

show that the parameters v and p obtained by solving the above pair of equations are found to be

$$v = \sqrt{\left(r - \frac{\sigma^2}{2}\right)^2 \Delta t^2 + \sigma^2 \Delta t}$$
$$p = \frac{1}{2}\left[1 + \frac{\left(r - \frac{\sigma^2}{2}\right)\Delta t}{\sqrt{\sigma^2 \Delta t + \left(r - \frac{\sigma^2}{2}\right)^2 \Delta t^2}}\right].$$

6.9 Boyle (1988) proposed the following three-jump process for the approximation of the asset price process over one period:

Nature of jump	*Probability*	*Asset price*
up	p_1	uS
horizontal	p_2	S
down	p_3	dS

where S is the current asset price. The middle jump ratio m is chosen to be 1. There are five parameters in Boyle's trinomial model: u, d and the probability values. Thc governing equations for the parameters can be obtained by:

(i) Setting the sum of probabilities to be 1; and

$$p_1 + p_2 + p_3 = 1,$$

(ii) equating the first two moments of the approximating discrete distribution and the corresponding continuous lognormal distribution

$$p_1 u + p_2 + p_3 d = e^{r\Delta t} = R$$

$$p_1 u^2 + p_2 + p_3 d^2 - (p_1 u + p_2 + p_3 d)^2 = e^{2r\Delta t}\,(e^{\sigma^2 \Delta t} - 1).$$

The last equation can be simplified as

$$p_1 u^2 + p_2 + p_3 d^2 = e^{2r\Delta t}\, e^{\sigma^2 \Delta t}.$$

The remaining two conditions can be chosen freely. They were chosen by Boyle (1988) to be

$$ud = 1$$

and

$$u = e^{\lambda\sigma\sqrt{\Delta t}}, \qquad \lambda \text{ is a free parameter.}$$

By solving the five equations together, show that

$$p_1 = \frac{(W-R)u - (R-1)}{(u-1)(u^2-1)}, \qquad p_3 = \frac{(W-R)u^2 - (R-1)u^3}{(u-1)(u^2-1)},$$

where $W = R^2 e^{\sigma^2 \Delta t}$. Also show that Boyle's trinomial model reduces to the Cox–Ross–Rubinstein binomial scheme when $\lambda = 1$.

6.10 Suppose we let $y = \ln S$, the Kamrad–Ritchken trinomial scheme can be expressed as

$$c(y, t - \Delta t) = [p_1 c(y + v, t) + p_2 c(y, t) + p_3 (y - v, t)]\, e^{-r\Delta t}.$$

Show that the Taylor expansion of the above trinomial scheme is given by

$$\begin{aligned}
&-c(y, t - \Delta t) + [p_1 c(y + v, t) + p_2 c(y, t) + p_3 (y - v, t)]\, e^{-r\Delta t} \\
&= \Delta t \frac{\partial c}{\partial t}(y, t) - \frac{\Delta t^2}{2} \frac{\partial^2 c}{\partial t^2}(y, t) + \cdots + (1 - e^{-r\Delta t}) c(y, t) \\
&\quad + e^{-r\Delta t}\left[(p_1 - p_3) v \frac{\partial c}{\partial y} + \frac{1}{2}(p_1 + p_3) v^2 \frac{\partial^2 c}{\partial y^2}\right. \\
&\quad \left. + \frac{1}{6}(p_1 - p_3) v^3 \frac{\partial^3 c}{\partial y^3} + \cdots\right].
\end{aligned}$$

Given the probability values stated in (6.1.19a,b,c), show that the numerical solution $c(y, t)$ of the trinomial scheme satisfies

$$0 = \frac{\partial c}{\partial t}(y, t) + \left(r - \frac{\sigma^2}{2}\right) \frac{\partial c}{\partial y}(y, t) + \frac{\sigma^2}{2} \frac{\partial^2 c}{\partial y^2}(y, t) - rc(y, t) + O(\Delta t).$$

6.11 Show that the width of the domain of dependence of the trinomial scheme (see Fig. 6.5) increases as $\sqrt{n}$, where n is the number of time steps to expiry.

6.12 Consider the five-point multinomial scheme defined in (6.1.22) and the corresponding four-point scheme (obtained by setting $\lambda = 1$), show that the total number of multiplications and additions in performing n steps of the schemes are given by (Kamrad and Ritchken, 1991)

Scheme	*Number of multiplications*	*Number of additions*
5-point	$\frac{5}{3}(2n^3+n)$	$\frac{4}{3}(2n^3+n)$
4-point	$\frac{2}{3}(2n^3+3n^2+n)$	$\frac{1}{2}(2n^3+3n^2+n)$

6.13 Consider a three-state option model where the logarithmic return processes of the underlying assets are given by

$$\ln \frac{S_i^{\Delta t}}{S_i} = \zeta_i, \quad i = 1, 2, 3.$$

Here, ζ_i denotes the normal random variable with mean $(r - \frac{\sigma_i^2}{2})\Delta t$ and variance $\sigma_i^2 \Delta t$, $i = 1, 2, 3$. Let ρ_{ij} denote the instantaneous correlation coefficient between ζ_i and ζ_j, $i, j = 1, 2, 3$, $i \neq j$. Suppose the approximating multivariate distribution ξ_i^a, $i = 1, 2, 3$, is taken to be

ζ_1^a	ζ_2^a	ζ_3^a	Probability
v_1	v_2	v_3	p_1
v_1	v_2	$-v_3$	p_2
v_1	$-v_2$	v_3	p_3
v_1	$-v_2$	$-v_3$	p_4
$-v_1$	v_2	v_3	p_5
$-v_1$	v_2	$-v_3$	p_6
$-v_1$	$-v_2$	v_3	p_7
$-v_1$	$-v_2$	$-v_3$	p_8
0	0	0	p_9

where $v_i = \lambda \sigma_i \sqrt{\Delta t}$, $i = 1, 2, 3$. Following the Kamrad–Ritchken approach, find the probability values so that the approximating discrete distribution converges to the continuous multivariate distribution as $\Delta t \to 0$.

Hint: The first and last probability values are given by

$$p_1 = \frac{1}{8}\left\{\frac{1}{\lambda^2} + \frac{\sqrt{\Delta t}}{\lambda}\left(\frac{r - \frac{\sigma_1^2}{2}}{\sigma_1} + \frac{r - \frac{\sigma_2^2}{2}}{\sigma_2} + \frac{r - \frac{\sigma_3^2}{2}}{\sigma_3}\right) + \frac{\rho_{12} + \rho_{13} + \rho_{23}}{\lambda^2}\right\}$$

$$p_9 = 1 - \frac{1}{\lambda^2}.$$

6.14 Consider the window Parisian feature. Associated with each time point, a moving window is defined with $\widehat{m}$ consecutive monitoring instants before and including that time point. The option is knocked out at a given time when the asset price has already stayed within the knock-out region exactly m times, $m \leq \widehat{m}$, within the moving window. Under what condition does the window Parisian feature reduce to the consecutive Parisian feature? How can we construct the corresponding discrete grid function g_{win} in the forward shooting grid (FSG) algorithm?

Hint: We define a binary string $A = a_1 a_2 \cdots a_{\widehat{m}}$ to represent the history of the asset price path falling inside or outside the knock-out region within the moving window. The augmented path dependence state vector has binary strings as elements (Kwok and Lau, 2001a).

6.15 Construct the FSG scheme for pricing the continuously monitored European style floating strike lookback call option. In particular, describe how to define the terminal payoff values. How can we modify the FSG scheme in order to incorporate the American early exercise feature?

6.16 Consider the European put option with the automatic strike reset feature, where the strike price is reset to the prevailing asset price on a prespecified reset date if the option is out-of-the-money on that date. The strike price at expiry is not known a priori, rather it depends on the actual realization of the asset price on those prespecified reset dates. Construct the FSG scheme that prices the strike reset put option (Kwok and Lau, 2001a).

Hint: Let t_ℓ, $\ell = 1, 2, \cdots, m$ be the prespecified reset dates, and let X_ℓ denote the strike price reset at t_ℓ. Explain why

$$X_\ell = \max(X, X_{\ell-1}, S(t_\ell)),$$

where X is the original strike price and $S(t_\ell)$ denotes the asset price at t_ℓ.

6.17 Suppose we would like to approximate $\frac{df}{dx}$ at x_0 up to $O(\Delta x^2)$ using function values at $x_0, x_0 - \Delta x$ and $x - 2\Delta x$, that is,

$$\left.\frac{df}{dx}\right|_{x_0} = \alpha_{-2} f(x_0 - 2\Delta x) + \alpha_{-1} f(x_0 - \Delta x) + \alpha_0 f(x_0) + O(\Delta x^2),$$

where α_{-2}, α_{-1} and α_0 are unknown coefficients to be determined. Show that these coefficients are obtained by solving

$$\begin{pmatrix} 1 & 1 & 1 \\ -2 & -1 & 0 \\ 4 & 1 & 0 \end{pmatrix} \begin{pmatrix} \alpha_{-2} \\ \alpha_{-1} \\ \alpha_0 \end{pmatrix} = \begin{pmatrix} 0 \\ 1 \\ 0 \end{pmatrix}.$$

6.18 Consider the following difference operators, show that they approximate the corresponding differential operator up to second-order accuracy

(i) $$\left.\frac{d^2 f}{dx^2}\right|_{x_0} = \frac{2f(x_0) - 5f(x_0 - \Delta x) + 4f(x - 2\Delta x) - f(x_0 - 3\Delta x)}{\Delta x^2} + O(\Delta x^2)$$

(ii) $$\frac{\partial^2 f}{\partial x \partial y} = [f(x_0+\Delta x, y_0+\Delta y) - f(x_0+\Delta x, y_0-\Delta y) - f(x_0-\Delta x), y_0+\Delta y) + f(x_0-\Delta x, y_0-\Delta y)]/(4\Delta x \Delta y) + O(\Delta x^2) + O(\Delta y^2).$$

6.19 Show that the leading local truncation error terms of the following Crank–Nicolson scheme

$$\frac{V_j^{n+1} - V_j^n}{\Delta\tau} = \frac{\sigma^2}{4}\left(\frac{V_{j+1}^n - 2V_j^n + V_{j-1}^n}{\Delta x^2} + \frac{V_{j+1}^{n+1} - 2V_j^{n+1} + V_{j-1}^{n+1}}{\Delta x^2}\right) + \frac{1}{2}\left(r - \frac{\sigma^2}{2}\right)\left(\frac{V_{j+1}^n - V_{j-1}^n}{2\Delta x} + \frac{V_{j+1}^{n+1} - V_{j-1}^{n+1}}{2\Delta x}\right) - \frac{r}{2}(V_j^n + V_j^{n+1})$$

are $O(\Delta\tau^2, \Delta x^2)$.
Hint: Perform the Taylor expansion at $(j\Delta x, (n+\frac{1}{2})\Delta\tau)$.

6.20 Consider the following form of the Black–Scholes equation:

$$\frac{\partial W}{\partial \tau} = \frac{\sigma^2}{2}\frac{\partial^2 W}{\partial x^2} + \left(r - q - \frac{\sigma^2}{2}\right)\frac{\partial W}{\partial x}, \qquad W = e^{-r\tau}V \text{ and } x = \ln S,$$

where $V(S, \tau)$ is the option price and S is the asset price. The two-level six-point implicit *compact scheme* takes the form:

$$a_1 W_{j+1}^{n+1} + a_0 W_j^{n+1} + a_{-1} W_{j-1}^{n+1} = b_1 W_{j+1}^n + b_0 W_j^n + b_{-1} W_{j-1}^n,$$

where

$$c = \left(r - q - \frac{\sigma^2}{2}\right)\frac{\Delta\tau}{\Delta x}, \qquad \mu = \sigma^2 \frac{\Delta\tau}{\Delta x^2},$$

$$a_1 = 1 - 3\mu - 3c - \frac{c^2}{\mu} + \frac{c}{\mu}, \qquad a_0 = 10 + 6\mu + \frac{2c^2}{\mu},$$

$$a_{-1} = 1 - 3\mu + 3c - \frac{c^2}{\mu} - \frac{c}{\mu}, \qquad b_1 = 1 + 3\mu + 3c + \frac{c^2}{\mu} + \frac{c}{\mu},$$

$$b_0 = 10 - 6\mu - \frac{2c^2}{\mu}, \qquad b_{-1} = 1 + 3\mu - 3c + \frac{c^2}{\mu} - \frac{c}{\mu}.$$

Show that the compact scheme is second-order time accurate and fourth order space accurate.

6.21 Use the Fourier method to deduce the von Neumann stability condition for (i) the Jarrow–Rudd binomial scheme (see Problem 6.1), (ii) the Kamrad–Ritchken trinomial scheme, and (iii) the explicit FTCS scheme [see (6.2.2)].

6.22 Let $p(S, M, t)$ denote the price function of the European floating strike lookback put option. Define $x = \ln \frac{M}{S}$ and $V(x,t) = \frac{p(S,M,t)}{S}$. The pricing formulation of $V(x,t)$ is given by

$$\frac{\partial V}{\partial t} + \frac{\sigma^2}{2}\frac{\partial^2 V}{\partial x^2} + \left(q - r - \frac{\sigma^2}{2}\right)\frac{\partial V}{\partial x} - qV = 0, \quad x > 0,\ 0 < t < T.$$

The final and boundary conditions are

$$V(x,T) = e^x - 1 \quad \text{and} \quad \frac{\partial V}{\partial x}(0,t) = 0,$$

respectively. By writing $\alpha = \frac{1}{2} + \frac{\Delta x}{2}(\frac{r-q}{\sigma^2} + \frac{1}{2})$ and setting $\Delta x = \sigma\sqrt{\Delta t}$, the binomial scheme takes the form

$$V_j^n = \frac{1}{1+q\Delta t}\left[\alpha V_{j-1}^{n+1} + (1-\alpha)V_{j+1}^{n+1}\right], \quad j \geq 0.$$

Suppose the boundary condition at $x = 0$ is approximated by

$$V_{-1}^{n+1} = V_0^{n+1},$$

then the numerical boundary value is given by

$$V_0^n = \frac{1}{1+q\Delta t}\left[\alpha V_0^{n+1} + (1-\alpha)V_1^n\right].$$

Let $\mathcal{T}_0^n$ denote the local truncation error at $j = 0$ of the above binomial scheme, show that

$$\mathcal{T}_0^n = -\frac{1}{1+q\Delta t}\frac{\sigma^2}{4}\frac{\partial^2 V}{\partial x^2}\bigg|_{x=0} + O(\Delta x).$$

Therefore, the proposed binomial scheme is not consistent.

6.23 To obtain a consistent binomial scheme for the floating strike lookback put option, we derive the binomial discretization at $j = 0$ using the finite volume approach. First, we integrate the governing differential equation from $x = 0$ to $x = \frac{\Delta x}{2}$ to obtain

$$0 = \int_0^{\frac{\Delta x}{2}} \left(\frac{\partial V}{\partial t} - qV\right)dx + \frac{\sigma^2}{2}\left[\frac{\partial V}{\partial x}\bigg|_{\frac{\Delta x}{2}} - \frac{\partial V}{\partial x}\bigg|_0\right]$$
$$+ \left(q - r - \frac{\sigma^2}{2}\right)(V_{\frac{\Delta x}{2}} - V_0).$$

Suppose we adopt the following approximations:

$$\int_0^{\frac{\Delta x}{2}} \left(\frac{\partial V}{\partial t} - qV\right)dx \approx \left(\frac{V_0^{n+1} - V_0^n}{\Delta t} - qV_0^n\right)\Delta x$$

$$\frac{\partial V}{\partial x}\bigg|_{\frac{\Delta x}{2}} \approx \frac{V_1^{n+1} - V_0^{n+1}}{\Delta x}, \quad V_{\frac{\Delta x}{2}} \approx \frac{V_1^{n+1} + V_0^{n+1}}{2}.$$

Show that the binomial approximation at $j = 0$ is given by

$$V_0^n = \frac{1}{1 + q\Delta t}\left[(2\alpha - 1)V_0^{n+1} + 2(1-\alpha)V_1^{n+1}\right].$$

Examine the consistency of the above binomial approximation.

6.24 Suppose we use the FTCS scheme to solve the Black–Scholes equation so that

$$\frac{V_j^{n+1} - V_j^n}{\Delta\tau} = \frac{\sigma^2}{2} S_j^2 \frac{V_{j+1}^n - 2V_j^n + V_{j-1}^n}{\Delta S^2} + rS_j \frac{V_{j+1}^n - V_{j-1}^n}{2\Delta S} - rV_j^n.$$

Show that the sufficient conditions for nonappearance of spurious oscillations in the numerical scheme are given by (Zvan, Forsyth and Vetzal, 1998)

$$\Delta S < \frac{\sigma^2 S_i}{r} \quad \text{and} \quad \frac{1}{\Delta\tau} > \frac{\sigma^2 S_i^2}{\Delta S^2} + r.$$

6.25 A sequential barrier option has two-sided barriers. Unlike the usual double-barrier options, the order of breaching of the barrier is specified. The second barrier is activated only after the first barrier has been hit earlier, and the option is knocked out only if both barriers have been hit in the prespecified order. Construct the explicit finite difference scheme for pricing this sequential barrier option under the Black–Scholes pricing framework (Kwok, Wong and Lau, 2001).

6.26 The penalty method is characterized by the replacement of the linear complementarity formulation of the American option model by appending a nonlinear penalty term in the Black–Scholes equation. Let $h(S)$ denote the exercise payoff of an American option. The nonlinear penalty term takes the form $\rho \max(h - V, 0)$, where ρ is the positive penalty parameter and $V(S, \tau)$ is the option price function. It can be shown that when $\rho \to \infty$, the solution of the following equation

$$\frac{\partial V}{\partial \tau} = \frac{\sigma^2}{2} S^2 \frac{\partial^2 V}{\partial S^2} + (r - q)S\frac{\partial V}{\partial S} - rV + \rho \max(h - V, 0)$$

gives the solution of the American option price function. Discuss the construction of the Crank–Nicolson scheme for solving the above nonlinear differential equation, paying special attention to the solution of the resulting nonlinear algebraic system of equations. Note that the nonlinearity stems from the penalty term (Forsyth and Vetzal, 2002).

6.27 Considering the antithetic variates method [see (6.3.7a,b)], explain why

$$\text{var}\left(\frac{c_i + \widetilde{c}_i}{2}\right) = \frac{1}{2}\left[\text{var}(c_i) + \text{cov}(c_i, \widetilde{c}_i)\right].$$

Note that the amount of computational work to generate $\bar{c}_{AV}$ [see (6.3.8)] is about twice the work to generate $\hat{c}$. By applying (6.3.5), show that the antithetic variates method improves computational efficiency provided that

$$\text{cov}(c_i, \widetilde{c}_i) \leq 0.$$

Give a statistical justification why the above negative correlation property is in general valid (Boyle, Broadie and Glasserman, 1997).

6.28 Consider the Bermudan option pricing problem, where the Bermudan option has d exercise opportunities at times $t_1 < t_2 < \cdots < t_d = T$, with $t_1 \geq 0$. Here, the issue date and maturity date of the Bermudan option are taken to be 0 and T, respectively. Let M_t denote the value at time t of \$1 invested in the riskless money market account at time 0. Let h_t denote the payoff from exercise at time t and τ^* be a stopping time taking values in $\{t_1, t_2, \cdots, t_d\}$. The value of the Bermudan option at time 0 is given by

$$V_0 = \sup_{\tau^*} E_0\left[\frac{h_\tau}{M_\tau}\right].$$

Consider the quantity defined by (Andersen and Broadie, 2004)

$$Q_{t_i} = \max\left(h_{t_i}, E_{t_i}\left[\frac{M_{t_i}}{M_{t_{i+1}}} Q_{t_{i+1}}\right]\right), \quad i = 1, 2, \cdots, d-1,$$

explain why Q_{t_i} gives the value of a Bermudan option that is newly issued at time t_i. Is it the same as the value at t_i of a Bermudan option issued at time 0? If not, explain why?

6.29 It has been generally believed that the extension of the Tilley algorithm to multiasset American options is not straightforward. Discuss the modifications to the bundling and sorting procedure required in the path grouping of all the asset price paths of n assets, $n > 1$. Also, think about how to determine the exercise-or-hold indicator variables when the exercise boundary is defined by a high-dimensional surface (Fu et al., 2001).

6.30 Discuss how to implement the secant method in the root-finding procedure of solving the optimal exercise price $S^*_{t_i}$ from the following algebraic equation

$$X - S^*_{t_i} = e^{-r(T-t_i)} E\left[P_{i+1} \middle| S_{t_i} = S^*_{t_i}\right]$$

in the Grant–Vora–Weeks algorithm (Fu et al., 2001).

6.31 Judge whether the simulation estimator on the option price given by the Grant–Vora–Weeks algorithm is biased high or low or unbiased.

6.32 Explain why the estimator $\theta_{\text{low},j}^{i_1 \cdots i_j}$ defined by (6.3.18)–(6.3.19) is biased low.
Hint: Upward bias is eliminated since the continuation value and the early exercise decision are determined from independent information sets. The early exercise decision is always suboptimal with a finite sample.

7

Interest Rate Models and Bond Pricing

The riskless interest rate has been assumed to be constant in most of the option pricing models discussed in earlier chapters. Such an assumption is acceptable for short-lived options when the interest rate appears only in the discount factor. In recent decades, we have witnessed a proliferation of fixed income derivatives and exotic interest rate products whose payoffs are strongly dependent on the interest rates. In these products, interest rates are used for discounting as well as for defining the payoff of the derivative. The values of these interest rate derivative products are highly sensitive to the level of interest rates. The correct modeling of the stochastic behavior of the term structure of interest rates is important for the construction of reliable pricing models of interest rate derivatives.

The trading of interest rate derivatives provides the market information on how the interest rates and cost of raising capital are set. Bonds, swaps and swaptions are traded securities and their prices are directly observable in the market. The bond price depends crucially on the random fluctuation of the interest rates over the term of bond's life. Unlike bonds, interest rates themselves are not "tradeable" securities. We only trade bonds and other fixed income instruments that depend on interest rates. In this chapter, we consider various stochastic models of interest rate dynamics and derive the mathematical relations that govern the related dynamics of interest rates and bond prices. We relegate the discussion of various pricing models of interest rate derivatives to the next chapter.

In Sect. 7.1, we discuss the relations between discount bond prices and yield curves, and illustrate various approaches to defining different types of interest rates that are derived from discount bond prices. We examine the product nature of a forward rate agreement, bond forward and vanilla interest rate swap. Several theories on the evolution of the term structures of interest rates are briefly discussed.

Various versions of the one-factor short rate models are considered in Sect. 7.2. We apply the no arbitrage argument to derive the governing differential equation for the price of a bond when the short rate is modeled by an Ito stochastic process. We show how to express the bond price as the expectation of a stochastic integral. We deduce the conditions on the parameter functions in the short rate models under which the bond price function admits an affine term structure. The Vasicek mean-reversion model and Cox–Ross–Ingersoll square root diffusion model are discussed in details. We examine and analyze the term structure of interest rates obtained from these two prototype affine term structure models. We comment on the empirical studies of the applicability of the class of generalized one-factor short rate models. It is commonly observed that the interest rate term structure derived from the interest rate models do not fit with the observed initial term structures. We consider interest rate models with parameters that are functions of time and show how these parameter functions can be calibrated from the current term structures of traded bond prices.

Under the one-factor interest rate models, the instantaneous returns on bonds of varying maturities are perfectly correlated. Multifactor models overcome this major drawback of the one-factor models. In Sect. 7.3, we consider multifactor interest rate models, including the two-factor long rate and short rate models, stochastic volatility models and multifactor affine term structure models. However, the analytic tractability of most multifactor models is very limited. We address the merits and drawbacks of these multifactor models.

In Sect. 7.4, we consider the Heath–Jarrow–Morton (HJM) approach to modeling the stochastic movement of forward rates. Most popular short rate models can be visualized as special cases within the HJM framework. The HJM methodologies provide a unified approach to the modeling of instantaneous interest rates. The HJM type models are in general non-Markovian. The numerical implementation of these models can become quite cumbersome, thus limiting their practical use. We consider the conditions under which an HJM model becomes Markovian, and present the characterization of various classes of the Markovian HJM models. The chapter concludes with the formulation of the forward LIBOR processes under the Gaussian HJM framework.

7.1 Bond Prices and Interest Rates

A bond is a financial contract under which the issuer promises to pay the bondholder a stream of coupon payments (usually periodic) on specified coupon dates and principal on the maturity date. If there is no coupon payment, the bond is said to be a *discount bond or zero coupon bond*. The upfront premium paid by the bondholder can be considered as a loan to the issuer. The face value of the bond is usually called the *par value*. A natural question: How much premium should be paid by the bond investor at the initiation of the contract so that it is fair to both the issuer and investor? The amount of premium is the value of the bond. The value of a bond can be interpreted as the present value of the cash flows that the bondholder expects to realize throughout the life of the bond. The discount factors employed in the calculation of the present value of the cash flows from the bond are stochastic due to the random fluctuations in the interest rates.

After the bond has been launched, the value of the bond changes over the bond's life due to the change in its life span (remaining coupon payments outstanding), fluctuations in interest rates, and factors like change in the credit quality of the bond issuer. Throughout this chapter and the next chapter, we consider only bonds that are default free. Although no corporate bonds are absolutely free from default, the U.S. Treasury bonds are generally considered default free. The interest rates that are derived from the default free bond prices are termed the *riskless* rates. The pricing of a defaultable bond and modeling of the credit process of the bond issuer are huge subjects of their own and a substantial literature on credit risk models has been developed. These issues are not taken up in this book.

7.1.1 Bond Prices and Yield Curves

The discount bond price $B(t, T)$ is a function of both the current time t and maturity T. Therefore, the plot of $B(t, T)$ is indeed a two-dimensional surface with varying values of t and T. At the current time t_0, the plot of $B(t_0, T)$ against T represents the whole spectrum of bond prices of different maturities (see Fig. 7.1).

For notational convenience, we take the par value to be unity, unless otherwise stated. When the interest rates stay positive, $B(t, T)$ is always a decreasing function of T since a higher discount factor results when the time horizon is longer. The market bond prices indicate the market expectation of the interest rates at future times. We expect that prices of bonds with maturity dates that are close to each other should exhibit strong correlation. To understand and model bond price dynamics, one should explore these correlation relations. On the other hand, we can plot $B(t, T_0)$ against t, $t < T_0$, for a discount bond with fixed maturity date T_0 (see Fig. 7.2). The evolution of the bond price as a function of time t can be considered a stochastic process, tending toward its par value as maturity T_0 is approached. This is known as the *pull-to-par phenomenon*.

Yield to Maturity and Yield Curve

The yield to maturity $R(t, T)$ of a bond is defined in terms of traded discount bond price, $B(t, T)$:

$$R(t, T) = -\frac{1}{T - t} \ln B(t, T). \tag{7.1.1}$$

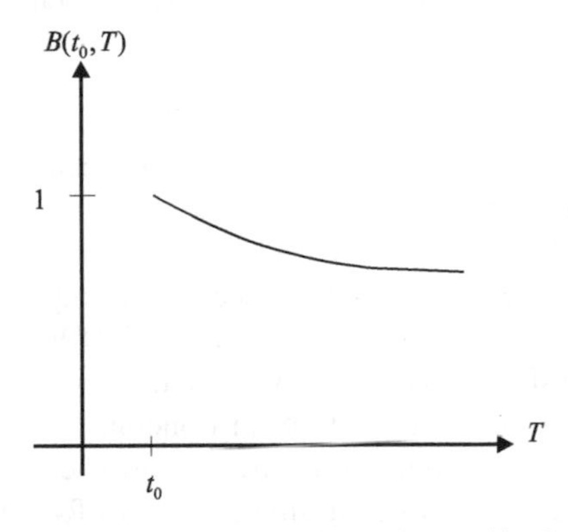

Fig. 7.1. Plot of the spectrum of discount bond prices of maturities beyond t_0. The bond prices $B(t_0, T)$ decrease monotonically with maturity T.

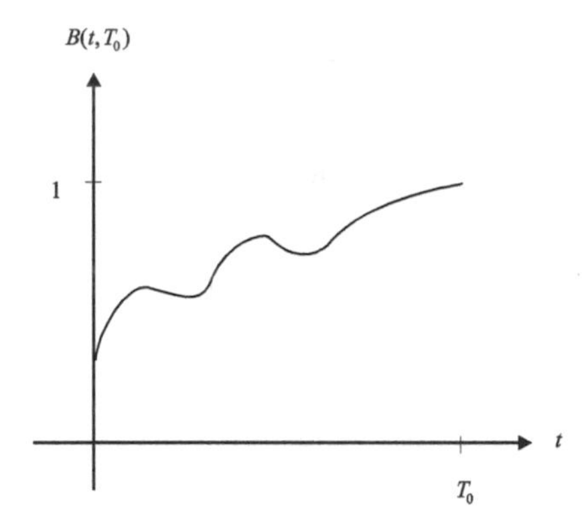

Fig. 7.2. Evolution of the price of a discount bond with fixed maturity T_0. Observe that $B(t, T_0)|_{t=T_0} = 1$ since unit par is paid at $t = T_0$.

This is precisely the *internal rate of return* at time t on the bond with time to maturity $T - t$. The *yield curve* is the plot of $R(t, T)$ against $T - t$ and the dependence of the yield curve on the time to maturity $T - t$ is called its *term structure*. The term structure reveals market beliefs on the bond yields at different maturities. Normally, the yield increases with maturity due to higher uncertainties associated with a longer time horizon. However, if the short-term interest rates are already high, the longer-term bond yield may be lower than its shorter-term counterpart.

7.1.2 Forward Rate Agreement, Bond Forward and Vanilla Swap

The most basic interest rate instrument is the forward rate agreement, which involves a single exchange of floating and fixed interest payments on a preset future date. A bond forward is a forward contract whose underlying asset is a bond. A vanilla interest rate swap can be considered as a portfolio of forward rate agreements, since it involves a stream of scheduled exchange of floating and fixed interest payments. We discuss the structures of these interest rate instruments, then derive their no arbitrage prices in terms of traded bond prices.

Forward Rate Agreement

A forward rate agreement (FRA) is an agreement between two counterparties to exchange floating and fixed interest payments at the future settlement date S. The floating rate is the London Interbank Offered Rate (LIBOR), denoted by $L[R, S]$, that is observed at the future reset date R for the accrual period $[R, S]$. The length of the accrual period is usually three or six months. The LIBORs are rates at which highly credit-rated financial institutions can borrow U.S. dollars in the interbank market for a series of possible maturities, fixed daily in London. LIBOR is always quoted as an annual rate of interest though it applies over a period commonly shorter than one year. Note that there are three time parameters in an FRA, namely, current time t, reset date R and settlement date S. Let $\mathcal{N}$ denote the notional of the contract and α_R^S denote the accrual factor (in years) of the period $[R, S]$. Depending on the day

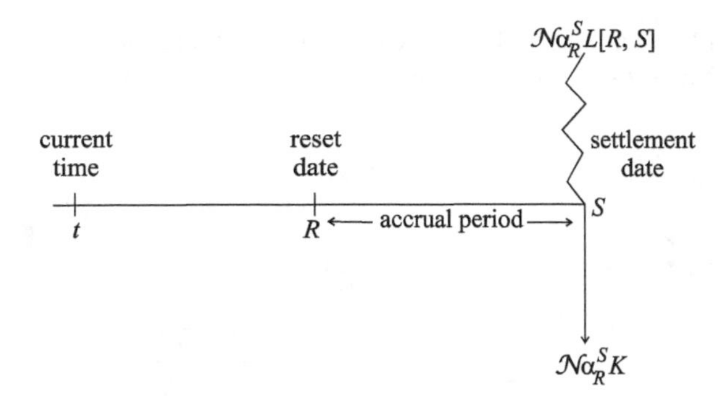

Fig. 7.3. Timing of the cash flows of a forward rate agreement.

count convention used (see Problem 7.1), α_R^S may be slightly different from the actual length of the time period $S - R$. The floating payment will be $\mathcal{N}\alpha_R^S L[R, S]$ while the fixed payment will be $\mathcal{N}\alpha_R^S K$, where K is the fixed rate. Figure 7.3 shows the timing of the cash flows of an FRA.

To find the value of the FRA to the fixed rate receiver at time t earlier than R, we replicate the cash settlement at time S of the fixed rate receiver by

(i) long holding of $\mathcal{N}(1 + \alpha_R^S K)$ units of S-maturity unit par discount bond and
(ii) short holding of $\mathcal{N}$ units of R-maturity unit par discount bond.

To see how the replication works, the floating rate payment at time S is financed by the payout of $\mathcal{N}$ dollars at time R due to the short position of the R-maturity bonds. The $\mathcal{N}$ dollars at time R will accumulate to $N(1 + \alpha_R^S L[R, S])$ at time S. The net cash settlement at time S from these bond positions is $\mathcal{N}\alpha_R^S(K - L[R, S])$, which replicates exactly the net cash settlement of the fixed rate receiver. The value of the replicating portfolio at time t is

$$\mathcal{N}[(1 + \alpha_R^S K)B(t, S) - B(t, R)],$$

which gives the time-t value of the FRA to the fixed rate receiver.

One can solve for the fixed rate K such that the value of the FRA at time t is zero. The breakeven value for K is called the *forward LIBOR*, denoted by $L_t[R, S]$, which is the time-t forward price of the LIBOR over the future period $[R, S]$. In terms of the market prices of discount bonds observed at time t, this forward LIBOR is given by

$$L_t[R, S] = \frac{1}{\alpha_R^S}\left[\frac{B(t, R)}{B(t, S)} - 1\right]. \tag{7.1.2}$$

Bond Forward

Consider a forward contract maturing at T_F whose underlying asset is a bond with maturity T_B, where $T_B > T_F$. Suppose the multiple coupons received after T_F from the bond are C_i at T_i, $i = 1, 2, \cdots, n$, and par value P is paid at T_B. Let F denote the time-t bond forward price. The value of the time-t bond forward at time t is given by the sum of the present values of the future cash flows, so we obtain

$$V = \sum_{i=1}^{n} C_i B(t, T_i) + P B(t, T_B) - F B(t, T_F).$$

To find the time-t forward price, we set $V = 0$ so that

$$F = \sum_{i=1}^{n} C_i \frac{B(t, T_i)}{B(t, T_F)} + P \frac{B(t, T_B)}{B(t, T_F)}. \tag{7.1.3}$$

Vanilla Interest Rate Swap

In Sect. 1.4.1, we saw that the cash flows of the fixed rate receiver of a vanilla interest rate swap can be replicated by long holding a fixed rate bond and short holding a floating rate bond. Alternatively, an interest rate swap can be visualized as a series of FRAs. Let $T_1, \cdots, T_n$ be the preassigned payment dates where floating rate interest payments are exchanged for fixed rate interest payments, and T_n is the maturity date of the swap. Let α_i denote the accrual factor over the time interval $[T_{i-1}, T_i], i = 1, 2, \cdots, n$. At time T_i, the fixed rate receiver receives the fixed interest payment $\mathcal{N}\alpha_i K$, where $\mathcal{N}$ is the notional and K is the fixed interest rate. The floating rate receiver receives the floating interest payment $\mathcal{N}\alpha_i L[T_{i-1}, T_i]$, where $L[T_{i-1}, T_i]$ is the LIBOR reset at T_{i-1} that is applied over the period (T_{i-1}, T_i). The net cash flow received by the fixed rate receiver at T_i is $N\alpha_i(K - L[T_{i-1}, T_i])$. For $t < T_0$, the time-t value of the net cash flow at T_i is given by

$$\mathcal{N}[(1 + \alpha_i K)B(t, T_i) - B(t, T_{i-1})].$$

By summing all these discounted cash flows at $T_i, i = 1, 2, \cdots, n$, the time-t value of the interest rate swap to the fixed rate receiver (or called the floating rate payer) is given by

$$\begin{aligned} V(t; \mathcal{N}, K) &= \sum_{i=1}^{n} \mathcal{N}[B(t, T_i) - B(t, T_{i-1}) + \alpha_i K B(t, T_i)] \\ &= \left[\mathcal{N} B(t, T_n) + \sum_{i=1}^{n} \mathcal{N}\alpha_i K B(t, T_i)\right] - \mathcal{N} B(t, T_0). \end{aligned}$$

The sum of all positive terms in the above equation gives the value of the long position of a fixed rate T_n-maturity bond which pays interest payment $\mathcal{N}\alpha_i K$ at $T_i, 1, 2, \cdots, n$, and par payment $\mathcal{N}$ at T_n. For the floating rate bond with par value $\mathcal{N}$ paying floating rate interest at $T_i = i = 1, 2, \cdots, n$, and par payment $\mathcal{N}$ at T_n, its value at time T_0 is $\mathcal{N}$. This is because the same cash amount $\mathcal{N}$ at time T_0 placed in a floating LIBOR-earning deposit generates the same stream of cash flows as the floating rate bond. Hence, the value of the floating rate bond at time t is $\mathcal{N} B(t, T_0)$.

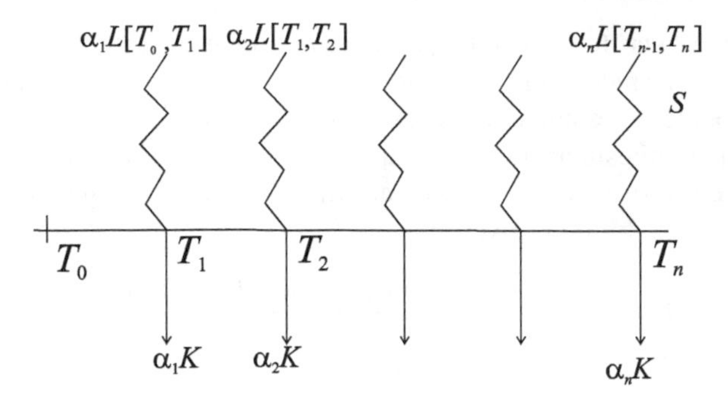

Fig. 7.4. Cash flows of an interest rate swap.

The *forward swap rate* $K_t[T_0, T_n]$ is defined to be the value of K that sets the time-t value $V(t; \mathcal{N}, K)$ of the interest rate swap to both counterparties be zero. The forward swap rate expressed in terms of the traded discount bond prices can be obtained as follows:

$$K_t[T_0, T_n] = \frac{B(t, T_0) - B(t, T_n)}{\sum_{i=1}^{n} \alpha_i B(t, T_i)}. \tag{7.1.4}$$

7.1.3 Forward Rates and Short Rates

The forward LIBOR $L_t[R, S]$ at time t assumes discrete compounding of interest over the accrual period $[R, S]$. Under the setting of continuous interest rate models, it is more convenient to define the time-t *continuous forward rate* $f(t, T_1, T_2)$ for the future period $[T_1, T_2]$, where $t < T_1 < T_2$. To relate $f(t, T_1, T_2)$ with the traded discount bond prices $B(t, T_1)$ and $B(t, T_2)$, we consider the time-t price of a bond forward contract where the buyer agrees to purchase at time T_1 a unit par discount bond with maturity date T_2. By setting the coupons to be zero and par to be unity in (7.1.3), the forward price of a T_2-maturity forward on a unit par T_1-maturity discount bond is seen to be $B(t, T_2)/B(t, T_1)$. One pays this forward price $B(t, T_2)/B(t, T_1)$ at time T_1 and he is entitled to receive \$1 at time T_2. By definition, the time-t value of the continuous forward rate $f(t, T_1, T_2)$ is related to this bond forward price via the relation

$$\frac{B(t, T_2)}{B(t, T_1)} e^{f(t,T_1,T_2)(T_2-T_1)} = 1.$$

That is, the amount $B(t, T_2)/B(t, T_1)$ grows to \$1 at the forward rate $f(t, T_1, T_2)$ compounded continuously over the period $[T_1, T_2]$. Under continuous compounding at the rate $f(t, T_1, T_2)$ over the finite time interval $[T_1, T_2]$, the growth factor is $e^{f(t,T_1,T_2)(T_2-T_1)}$. We obtain

$$f(t, T_1, T_2) = -\frac{1}{T_2 - T_1} \ln \frac{B(t, T_2)}{B(t, T_1)}. \tag{7.1.5}$$

In financial markets, forward rates are always applied over a finite time interval. However, for convenience in the mathematical formulation of bond pricing and interest rate models, we commonly deal with instantaneous forward rates that are applied over an infinitesimal time interval. By taking $T_1 = T$ and $T_2 = T + \Delta T$, the *instantaneous forward rate* as seen at time t over the infinitesimal time interval $[T, T + \Delta T]$ in the future is given by

$$\begin{aligned} F(t,T) &= -\lim_{\Delta T \to 0} \frac{\ln B(t, T+\Delta T) - \ln B(t,T)}{\Delta T} \\ &= -\frac{1}{B(t,T)} \frac{\partial B}{\partial T}(t,T), \quad t < T. \end{aligned} \tag{7.1.6}$$

Here, $F(t, T)$ can be interpreted as the marginal rate of return from committing a bond investment maturing at time T for an additional infinitesimal time interval. Conversely, by integrating (7.1.6) with respect to T, the bond price $B(t, T)$ can be expressed in terms of the instantaneous forward rate as follows:

$$B(t,T) = \exp\left(-\int_t^T F(t,u)\,du\right). \tag{7.1.7}$$

Furthermore, by combining (7.1.1) and (7.1.7), $F(t, T)$ can be expressed as

$$R(t,T) = \frac{1}{T-t} \int_t^T F(t,u)\,du, \tag{7.1.8a}$$

or equivalently,

$$F(t,T) = \frac{\partial}{\partial T}[R(t,T)(T-t)] = R(t,T) + (T-t)\frac{\partial R}{\partial T}(t,T). \tag{7.1.8b}$$

In (7.1.8a), $F(t, u)$ gives the internal rate of return as seen at time t over the future infinitesimal time interval $(u, u + du)$, and its average over the time period (t, T) gives the yield to maturity. The above equations indicate that the bond price and bond yield can be recovered from the knowledge of the term structure of the instantaneous forward rates. On the other hand, the instantaneous forward rate provides the sense of instantaneity as dictated by the nature of its definition. The *short rate* $r(t)$ (also called the *instantaneous spot rate*) is simply

$$r(t) = \lim_{T \to t} R(t,T) = R(t,t) = F(t,t). \tag{7.1.9}$$

The plot of $B(t, T)$ against T is inevitably a downward sloping curve since bonds with longer maturity always have lower prices under positive interest rates (see Fig. 7.1). However, the yield curve [plot of $R(t, T)$ against T] reveals the average rate of return of the bonds over varying maturities, so it can be an increasing or decreasing curve. Therefore, the yield curves provide more visual information compared to the bond price curves. As deduced from (7.1.8b), the forward rate curve [plot of $F(t, T)$ against T] will be above the yield curve if the yield curve is increasing or below the yield curve if otherwise.

Money Market Account Process

The wealth accumulation process of a money market account is obtained by investing in a self-financing "rolling over" trading strategy with rate of return $r(t)$ over $[t, t + \Delta t]$. Assuming that the money market account starts with one dollar at time zero, the money market account process $M(t)$ is then governed by

$$\frac{dM(t)}{dt} = r(t)M(t), \quad M(0) = 1,$$

the solution of which is seen to be

$$M(t) = \exp\left(\int_0^t r(u)\,du\right). \tag{7.1.10}$$

Theories of Term Structures

Several theories of term structures have been proposed to explain the shape of a yield curve. One of them is the *expectation theory*, which states that the forward rates reflect the expected future short-term interest rates. Let $E_t[r(u)]$ denote the expectation of $r(u)$, $u > t$, conditional on the information $\mathcal{F}_t$. The yield to maturity under the expectation theory can be expressed as [comparing (7.1.8a)]

$$R(t, T) = \frac{1}{T - t} \int_t^T E_t[r(u)]\,du. \tag{7.1.11}$$

The second theory is the *market segmentation theory*, which states that each borrower or lender has a preferred maturity so that the slope of the yield curve will depend on the supply-and-demand conditions for funds in the long-term market relative to the short-term market. The third theory is the *liquidity preference theory*. It conjectures that lenders prefer to make short-term loans rather than long-term loans since liquidity of capital is in general preferred. Hence, long-term bonds normally have a better yield than short-term bonds. The representation equations of the term structures for the market segmentation theory and the liquidity preference theory are similar, namely,

$$R(t, T) = \frac{1}{T - t}\left[\int_t^T E_t[r(u)]\,du + \int_t^T L(u, T)\,du\right], \tag{7.1.12}$$

where $L(u, T)$ is interpreted as the instantaneous term premium at time u of a bond maturing at time T (Langetieg, 1980). The premium represents the deviation from the expectation theory, which could be irregular as implied by the market segmentation theory or monotonically increasing as implied by the liquidity preference theory.

7.1.4 Bond Prices under Deterministic Interest Rates

Before we consider bond pricing under stochastic interest rates, it may be worthwhile to derive the bond price formula under deterministic time dependent interest rates.

Under the deterministic setting, the short rate $r(t)$ becomes a deterministic function of time. Normally, the bond price is a function of the interest rate and time. When the interest rate is not an independent state variable but itself is a known function of time, the bond price is a function of time only. Let $B(t)$ and $k(t)$ denote the bond price and the deterministic coupon rate, respectively. The final condition at bond's maturity T is given by $B(T) = P$, where P is the par value.

The derivation of the governing equation for $B(t), t < T$, leads to a first-order linear ordinary differential equation. Over an infinitesimal time increment dt from the current time t, the change in value of the bond is $\frac{dB}{dt}dt$ and the coupon received is $k(t)\,dt$. By no arbitrage, the above sum must equal the riskless interest return $r(t)B(t)\,dt$. This gives

$$\frac{dB}{dt} + k(t) = r(t)B, \quad t < T. \tag{7.1.13}$$

To solve the differential equation, we multiply both sides by the integrating factor $e^{\int_t^T r(s)\,ds}$ to obtain

$$\frac{d}{dt}\left[B(t)e^{\int_t^T r(s)\,ds}\right] = -k(t)e^{\int_t^T r(s)\,ds}.$$

Together with the final condition, $B(T) = P$, the bond price function is found to be

$$B(t) = e^{-\int_t^T r(s)\,ds}\left[P + \int_t^T k(u)e^{\int_u^T r(s)\,ds}du\right]. \tag{7.1.14}$$

The above bond price formula has a nice financial interpretation. The coupon amount $k(u)\,du$ received over the period $[u, u + du]$ will grow to the amount $k(u)e^{\int_u^T r(s)\,ds}\,du$ at maturity T. The future value at T of all coupons received during the bond life is given by $\int_t^T k(u)e^{\int_u^T r(s)\,ds}\,du$. The present value of the par value and coupon stream is obtained by discounting the sum received at T by the discount factor $e^{-\int_t^T r(s)\,ds}$. This is precisely the current bond value at time t, which agrees with the price function given in (7.1.14). Depending on the relative magnitude of $r(t)B(t)$ and $k(t)$ [see (7.1.13)] the bond price function can be an increasing or decreasing function of time.

7.2 One-Factor Short Rate Models

From the fundamental result in the martingale theory of option pricing, the existence of a risk neutral measure Q implies that the no arbitrage price at time t of a contingent claim with payoff $H(T)$ at time T is given by [see (3.2.16)]

$$V(t) = E_Q^t\left[e^{-\int_t^T r(u)\,du}H(T)\right], \tag{7.2.1}$$

where E_Q^t is the expectation under Q conditional on the filtration $\mathcal{F}_t$. For the case of a discount bond whose terminal payoff is $H(T) = 1$, we have

$$B(t,T) = E_Q^t\left[e^{-\int_t^T r(u)\,du}\right]. \tag{7.2.2}$$

Once the dynamics of the short rate $r(t)$ is specified, we are able to compute the bond price. This is why most earlier interest rate models are based on the characterization of the dynamics of the short rate.

7.2.1 Short Rate Models and Bond Prices

Assume that the short rate r_t follows the Ito process as described by the following stochastic differential equation

$$dr_t = \mu(r_t, t)\,dt + \rho(r_t, t)\,dZ_t, \tag{7.2.3}$$

where dZ_t is the differential of the standard Brownian process, $\mu(r_t, t)$ and $\rho(r_t, t)^2$ are the instantaneous drift and variance. We would like to derive the governing differential equation for the bond price using the no arbitrage argument. Since the short rate is not a traded security, the differential equation is expected to involve the market price of risk of r_t. The prices of bonds with varying maturities are shown to satisfy certain consistency relations in order to ensure absence of arbitrage opportunities. We express the bond price in terms of the expectation under the physical measure, and from which we deduce the Radon–Nikodym derivative for the change of measure from the physical measure to the risk neutral measure.

Throughout this section, we assume the bond price to be dependent on r_t only, independent of default risk, liquidity and other factors. If we write the bond price as $B(r,t)$ (suppressing T when there is no ambiguity and dropping the time index t in stochastic processes of r_t, Z_t, etc.), then the use of Ito's lemma gives the dynamics of the bond price as

$$dB = \left(\frac{\partial B}{\partial t} + \mu\frac{\partial B}{\partial r} + \frac{\rho^2}{2}\frac{\partial^2 B}{\partial r^2}\right)dt + \rho\frac{\partial B}{\partial r}dZ. \tag{7.2.4}$$

When the above dynamics of $B(r,t)$ is expressed in the following lognormal form

$$\frac{dB}{B} = \mu_B(r,t)\,dt + \sigma_B(r,t)\,dZ,$$

the drift rate $\mu_B(r,t)$ and volatility $\sigma_B(r,t)$ of the bond price process are found to be

$$\mu_B(r,t) = \frac{1}{B}\left(\frac{\partial B}{\partial t} + \mu\frac{\partial B}{\partial r} + \frac{\rho^2}{2}\frac{\partial^2 B}{\partial r^2}\right) \tag{7.2.5a}$$

$$\sigma_B(r,t) = \frac{\rho}{B}\frac{\partial B}{\partial r}. \tag{7.2.5b}$$

Since the short rate is not a traded security, it cannot be used to hedge with the bond, like the role of the underlying asset in an equity option. Instead, we try to hedge bonds of different maturities. This is possible because the instantaneous returns on bonds of varying maturities are correlated as there exists the common underlying stochastic short rate that drives the bond prices. The following portfolio is constructed: we buy a bond of dollar value V_1 with maturity T_1 and sell another bond of dollar value V_2 with maturity T_2. The portfolio value Π is given by

$$\Pi = V_1 - V_2.$$

According to the bond price dynamics defined by (7.2.4), the change in portfolio value in time dt is

$$\begin{aligned} d\Pi = &\left[V_1\mu_B(r,t;T_1) - V_2\mu_B(r,t;T_2)\right]dt \\ &+ \left[V_1\sigma_B(r,t;T_1) - V_2\sigma_B(r,t;T_2)\right]dZ. \end{aligned}$$

Suppose V_1 and V_2 are chosen such that

$$V_1 = \frac{\sigma_B(r,t;T_2)}{\sigma_B(r,t;T_2) - \sigma_B(r,t;T_1)}\Pi \quad \text{and} \quad V_2 = \frac{\sigma_B(r,t;T_1)}{\sigma_B(r,t;T_2) - \sigma_B(r,t;T_1)}\Pi,$$

then the stochastic term in $d\Pi$ vanishes and the equation becomes

$$\frac{d\Pi}{\Pi} = \frac{\mu_B(r,t;T_1)\sigma_B(r,t;T_2) - \mu_B(r,t;T_2)\sigma_B(r,t;T_1)}{\sigma_B(r,t;T_2) - \sigma_B(r,t;T_1)}dt. \tag{7.2.6a}$$

Since the portfolio is instantaneously riskless, in order to avoid arbitrage opportunities, it must earn the riskless short rate so that

$$d\Pi = r\Pi\, dt. \tag{7.2.6b}$$

Combining (7.2.6a,b), we obtain

$$\frac{\mu_B(r,t;T_1) - r}{\sigma_B(r,t;T_1)} = \frac{\mu_B(r,t;T_2) - r}{\sigma_B(r,t;T_2)}.$$

The above relation is valid for arbitrary maturity dates T_1 and T_2, so the ratio $\frac{\mu_B(r,t)-r}{\sigma_B(r,t)}$ should be independent of maturity T. Let the common ratio be defined by $\lambda(r,t)$, that is,

$$\frac{\mu_B(r,t) - r}{\sigma_B(r,t)} = \lambda(r,t). \tag{7.2.7}$$

The quantity $\lambda(r,t)$ is called the *market price of risk* of the short rate (see Problem 7.4), since it gives the extra increase in expected instantaneous rate of return on a bond per an additional unit of risk. In a market that admits no arbitrage opportunity, bonds that are hedgeable among themselves should have the same market price of risk, regardless of maturity. If we substitute $\mu_B(r,t)$ and $\sigma_B(r,t)$ into (7.2.7), we obtain the following governing differential equation for the price of a zero-coupon bond

$$\frac{\partial B}{\partial t} + \frac{\rho^2}{2}\frac{\partial^2 B}{\partial r^2} + (\mu - \lambda\rho)\frac{\partial B}{\partial r} - rB = 0, \quad t < T, \tag{7.2.8}$$

with final condition: $B(T, T) = 1$. Once the diffusion process for the short rate r and the market price of risk $\lambda(r, t)$ are specified, the bond value can be obtained by solving (7.2.8). Since the short rate is not a traded asset, we are unable to eliminate the dependence of $B(r, t)$ on preferences, as what has been done in stock/option hedge.

Market Price of Risk

The drift $\mu(r, t)$ and volatility $\rho(r, t)$ in the bond price equation may be obtained by statistical analysis of the observable process of the short rate. Once $\mu(r, t)$ and $\rho(r, t)$ are available, the market price of risk $\lambda(r, t)$ can be estimated using the following relation (see Problem 7.6)

$$\left.\frac{\partial R}{\partial T}\right|_{T=t} = \frac{1}{2}\left[\mu(r, t) - \rho(r, t)\lambda(r, t)\right], \tag{7.2.9}$$

where $\frac{\partial R}{\partial T}|_{T=t}$ is the slope of the yield curve $R(t, T)$ at immediate maturity.

Bond Price Function as Expectation under Physical Measure

The solution of the bond price can be formally represented in an integral form as an expectation under the physical measure P:

$$\begin{aligned} B(r, t; T) = E_P^t\Bigg[\exp\Bigg(&-\int_t^T \left[r(u) - \frac{\lambda^2(r(u), u)}{2}\right] du \\ &+ \int_t^T \lambda(r(u), u)\, dZ(u)\Bigg)\Bigg], \quad t \leq T, \end{aligned} \tag{7.2.10}$$

where E_P^t denotes the expectation under P conditional on filtration $\mathcal{F}_t$. To show the result, we define the following auxiliary function:

$$V(r, t; \xi) = \exp\left(-\int_t^\xi \left[r(u) - \frac{\lambda^2(r(u), u)}{2}\right] du + \int_t^\xi \lambda(r(u), u)\, dZ(u)\right), \quad t \leq \xi,$$

and apply Ito's differential rule to compute the differential of $B(r, \xi; T)V(r, t; \xi)$. By observing

$$\begin{aligned} \frac{dV}{V} &= \left(-r - \frac{\lambda^2}{2}\right) d\xi + \lambda\, dZ + \frac{\lambda^2}{2}\, d\xi = -r\, d\xi + \lambda\, dZ \\ dB\, dV &= -\lambda V \rho \frac{\partial B}{\partial r}\, d\xi \end{aligned}$$

and the relation in (7.2.8), we obtain

$$\begin{aligned} d(BV) &= V\, dB + B\, dV + dB\, dV \\ &= V\left(\frac{\partial B}{\partial \xi} + u\frac{\partial B}{\partial r} + \frac{\rho^2}{2}\frac{\partial^2 B}{\partial r^2}\right) d\xi + V\rho\frac{\partial B}{\partial r}\, dZ \end{aligned}$$

$$
\begin{aligned}
&+ BV(-r\,d\xi + \lambda\,dZ) - \lambda V\rho \frac{\partial B}{\partial r} d\xi \\
&= V\left[\frac{\partial B}{\partial \xi} + (u - \lambda\rho)\frac{\partial B}{\partial r} + \frac{\rho^2}{2}\frac{\partial^2 B}{\partial r^2} - rB\right] d\xi \\
&\quad + BV\lambda\,dZ + V\rho\frac{\partial B}{\partial r} dZ \\
&= BV\lambda\,dZ + V\rho\frac{\partial B}{\partial r} dZ.
\end{aligned}
$$

Next, we integrate the above equation from t to T and take the expectation with respect to P. Since the expectation of the stochastic integral is zero, we have

$$
E_P^t[B(r,T;T)V(r,t;T) - B(r,t;T)V(r,t;t)] = 0.
$$

Applying the terminal conditions $B(r,T;T) = 1$ and $V(r,t;t) = 1$, we finally obtain

$$
B(r,t;T) = E_P^t[V(r,t;T)],
$$

which agrees with the result in (7.2.10).

We would like to apply the change of measure from the physical measure P to the risk neutral measure Q such that the bond price is a martingale under Q satisfying (7.2.2). Assuming that $\lambda(r,t)$ satisfies the Novikov condition

$$
E_Q\left[\exp\left(\int_0^T \frac{\lambda^2(r(u),u)}{2}\,du\right)\right] < \infty,
$$

we define

$$
\widetilde{Z}(t) = Z(t) + \int_0^t \lambda(r(u),u)\,du,
$$

then there exists an equivalent measure Q under which $\widetilde{Z}(t)$ is Q-Brownian. The corresponding Radon–Nikodym derivative is given by

$$
\frac{dQ}{dP} = \exp\left(-\int_0^T \lambda(r(u),u)\,dZ(u) - \int_0^T \frac{\lambda^2(r(u),u)}{2}\,du\right).
$$

By virtue of the representation in (7.2.10) and applying the above change of measure, we obtain

$$
B(r,t;T) = E_Q^t\left[\exp\left(-\int_t^T r(u)\,du\right)\right], \tag{7.2.11}
$$

hence Q is a martingale measure [see (7.2.2)]. The bond price is given by the expectation of the stochastic discount factor under the risk neutral measure Q. By observing the relations

$$
d\widetilde{Z}_t = dZ_t + \lambda(r_t,t)\,dt
$$

and (7.2.3), the dynamics of the short rate r_t under Q becomes

$$
dr_t = \left[\mu(r_t,t) - \lambda(r_t,t)\rho(r_t,t)\right]dt + \rho(r_t,t)\,d\widetilde{Z}_t. \tag{7.2.12}
$$

Suppose the bond price function $B(r, t; T)$ satisfies (7.2.8) and the dynamics of r_t are governed by (7.2.12), by virtue of the Feynman–Kac representation formula, $B(r, t; T)$ admits the expectation representation given in (7.2.11). Under Q, the stochastic differential equation for $B(r, t)$ becomes (see Problem 7.7)

$$\frac{dB}{B} = r_t\,dt + \sigma_B(t, T)\,d\widetilde{Z}_t, \tag{7.2.13a}$$

where

$$\sigma_B(t, T) = -\frac{\rho}{B}\frac{\partial B}{\partial r}. \tag{7.2.13b}$$

Affine Term Structure Models

A short rate model that generates the bond price solution of the form

$$B(t, T) = e^{a(t,T)-b(t,T)r} \tag{7.2.14}$$

is called an affine term structure model. Suppose the dynamics of the short rate r_t under the risk neutral measure Q is governed by

$$dr_t = \mu(r_t, t)\,dt + \rho(r_t, t)\,dZ_t, \tag{7.2.15}$$

then the governing equation for $B(t, T)$ is given by

$$\frac{\partial B}{\partial t} + \frac{\rho^2}{2}\frac{\partial^2 B}{\partial r^2} + \mu\frac{\partial B}{\partial r} - rB = 0, \quad t < T, \tag{7.2.16}$$

with terminal condition $B(T, T) = 1$. Substituting the assumed affine solution of bond price into (7.2.16), we obtain

$$\frac{\partial a}{\partial t}(t, T) - \left[1 + \frac{\partial b}{\partial t}(t, T)\right]r - \mu(r, t)b(t, T) + \frac{\rho^2(r, t)}{2}b^2(t, T) = 0, \tag{7.2.17}$$

with $a(T, T) = 0$ and $b(T, T) = 0$. Given an arbitrary set of functions of $\mu(r, t)$ and $\rho(r, t)$, there will be no solution to $a(t, T)$ and $b(t, T)$. However, when $\mu(r, t)$ and $\rho(r, t)$ are both an affine function of r, where

$$\mu(r, t) = \mu_0(t) + \mu_1(t)r \quad \text{and} \quad \rho^2(r, t) = \alpha_0(t) + \alpha_1(t)r, \tag{7.2.18}$$

then (7.2.17) becomes

$$\begin{aligned} &\frac{\partial a}{\partial t}(t, T) - \mu_0(t)b(t, T) + \frac{\alpha_0^2(t)}{2}b^2(t, T) \\ &- \left[\frac{\partial b}{\partial t}(t, T) + \mu_1(t)b(t, T) - \frac{\alpha_1(t)}{2}b^2(t, T) + 1\right]r = 0. \end{aligned}$$

Since the above equation is valid for all values of r, we then deduce that $a(t, T)$ and $b(t, T)$ satisfy the following pair of equations:

$$\frac{\partial b}{\partial t}(t,T) + \mu_1(t)b(t,T) - \frac{\alpha_1(t)}{2}b^2(t,T) + 1 = 0, \quad b(T,T) = 0, \quad (7.2.19a)$$

$$\frac{\partial a}{\partial t}(t,T) - \mu_0(t)b(t,T) + \frac{\alpha_0^2(t)}{2}b^2(t,T) = 0, \quad a(T,T) = 0. \quad (7.2.19b)$$

The nonlinear differential equation for $b(t, T)$ is called the *Ricatti equation*. For some special cases of $\mu_1(t)$ and $\alpha_1(t)$, it is possible to derive a closed form solution to $b(t, T)$. Once the analytic solution to $b(t, T)$ is available, we can obtain $a(t, T)$ by direct integration of (7.2.19b). In the next two sections, we consider two renowned short rate models that admit the bond price solution in an affine form.

7.2.2 Vasicek Mean Reversion Model

Vasicek (1977) proposed the stochastic process for the short rate r_t under the physical measure to be governed by the Ornstein–Uhlenbeck process:

$$dr_t = \alpha(\gamma - r_t)\,dt + \rho\,dZ_t, \quad \alpha > 0. \quad (7.2.20)$$

The above process is sometimes called the elastic random walk or *mean reversion process*. The instantaneous drift $\alpha(\gamma - r_t)$ represents the effect of pulling the process toward its long-term mean γ with magnitude proportional to the deviation of the process from the mean. The mean reversion assumption agrees with the economic phenomenon that interest rates appear over time to be pulled back to some long-run average value. To explain the mean reversion phenomenon, we argue that when interest rates increase, the economy slows down and there is less demand for loans; this leads to the tendency for rates to fall. The stochastic differential equation (7.2.20) can be integrated to give

$$r(T) = \gamma + [r(t) - \gamma]e^{-\alpha(T-t)} + \rho\int_t^T e^{-\alpha(T-t)}\,dZ(t). \quad (7.2.21)$$

Due to the Brownian term in the stochastic integral, it is possible that the short rate may become negative under the Vasicek model. Conditional on the current level of short rate $r(t)$, the mean of the short rate at T is found to be

$$E[r(T)|r(t)] = \gamma + [r(t) - \gamma]e^{-\alpha(T-t)}. \quad (7.2.22)$$

The variance of the mean reversion process is governed by

$$\frac{d}{dt}\,\mathrm{var}(r(t)) = -2\alpha\,\mathrm{var}(r(t)) + \rho^2.$$

By observing the initial condition that the variance at the current time is zero (see Problem 7.11), we obtain

$$\mathrm{var}(r(T)|r(t)) = \frac{\rho^2}{2\alpha}\left[1 - e^{-2\alpha(T-t)}\right], \quad t < T. \quad (7.2.23)$$

Analytic Bond Price Formula
Suppose we assume the market price of risk λ to be constant, independent of r and t, then it is possible to derive an analytic formula for the bond price under the Vasicek model. The Vasicek mean reversion model corresponds to $\mu_0 = \alpha\gamma - \lambda\rho, \mu_1 = -\alpha, \alpha_0 = \rho$ and $\alpha_1 = 0$ in (7.2.18). We obtain the following pair of differential equations for $a(t, T)$ and $b(t, T)$:

$$\frac{da}{dt} + (\lambda\rho - \alpha\gamma)b + \frac{\rho^2}{2}b^2 = 0, \quad t < T$$
$$\frac{db}{dt} - \alpha b + 1 = 0, \qquad t < T,$$

with final conditions: $a(T, T) = 0$ and $b(T, T) = 0$. Solving the coupled system of differential equations, we obtain

$$B(r, t; T) = \exp\Bigg(\frac{1}{\alpha}\big[1 - e^{-\alpha(T-t)}\big](R_\infty - r) - R_\infty(T-t) - \frac{\rho^2}{4\alpha^3}\big[1 - e^{-\alpha(T-t)}\big]^2\Bigg), \quad t < T, \quad (7.2.24)$$

where $R_\infty = \gamma - \frac{\rho\lambda}{\alpha} - \frac{\rho^2}{2\alpha^2}$ [R_∞ is actually equal to $\lim_{T\to\infty} R(t, T)$, see (7.2.26)]. Using (7.2.5a,b), the mean and standard deviation of the instantaneous rate of return of a bond maturing at time T are found to be

$$\mu_B(r, t; T) = r(t) + \frac{\rho\lambda}{\alpha}\big[1 - e^{-\alpha(T-t)}\big] \quad (7.2.25a)$$

$$\sigma_B(r, t; T) = \frac{\rho}{\alpha}\big[1 - e^{-\alpha(T-t)}\big]. \quad (7.2.25b)$$

The yield to maturity is found to be

$$R(t, T) = R_\infty + \frac{[r(t) - R_\infty][1 - e^{-\alpha(T-t)}]}{\alpha(T-t)} + \frac{\rho^2}{4\alpha^3(T-t)}[1 - e^{-\alpha(T-t)}]^2. \quad (7.2.26)$$

By taking $T \to \infty$, the last two terms in (7.2.26) vanish so that the long-term internal rate of return is seen to be constant. Note that $R(t, T)$ and $\ln B(r, t; T)$ are linear functions of $r(t)$. Since $r(t)$ is normally distributed, it then follows that $R(t, T)$ is also normally distributed and $B(r, t; T)$ is lognormally distributed. Suppose we set $T = T_1$ and $T = T_2$ in (7.2.26), and subsequently eliminate $r(t)$, we obtain a relation between $R(t, T_1)$ and $R(t, T_2)$ that is dependent only on the parameter values.

Readers are invited to explore additional properties of the term structures of the yield curve associated with the Vasicek model in Problem 7.12. Also, a discrete version of the Vasicek model is presented in Problem 7.13.

7.2.3 Cox–Ingersoll–Ross Square Root Diffusion Model

Recall that the short rate may become negative under the Vasicek model due to its Gaussian nature. To rectify the problem, Cox, Ingersoll and Ross (1985) proposed the following square root diffusion process for the short rate:

$$dr_t = \alpha(\gamma - r_t)\,dt + \rho\sqrt{r_t}\,dZ_t, \quad \alpha, \gamma > 0. \tag{7.2.27}$$

With an initially nonnegative interest rate, r_t will never be negative. This is attributed to the mean-reverting drift rate that tends to pull r_t towards the long-run average γ and the diminishing volatility as r_t declines to zero (recall that volatility is constant in the Vasicek model). It can be shown that r_t can reach zero only if $\rho^2 > 2\alpha\gamma$; while the upward drift is sufficiently strong to make $r_t = 0$ impossible when $2\alpha\gamma \geq \rho^2$ [for a rigorous proof, see Cairns, 2004]. A heuristic argument is presented below. Define $L_t = \ln r_t$, then by Ito's lemma, the differential of L_t is found to be

$$dL = \left[\left(\alpha\gamma - \frac{\rho^2}{2}\right)e^{-L} - \alpha\right]dt + \rho e^{-L/2}\,dZ. \tag{7.2.28}$$

The drift and volatility coefficients are well behaved for positive L but they may blow up for large negative L. If $2\alpha\gamma < \rho^2$, the drift becomes negative for large negative L, pulling L further toward $-\infty$. This indicates that $2\alpha\gamma \geq \rho^2$ is a necessary condition for the short rate process to remain strictly positive.

The probability density of the short rate at time T, conditional on its value at the current time t, is given by

$$g(r(T); r(t)) = ce^{-u-v}\left(\frac{v}{u}\right)^{q/2} I_q\big(2(uv)^{1/2}\big), \tag{7.2.29}$$

where

$$c = \frac{2\alpha}{\rho^2\left[1 - e^{-\alpha(T-t)}\right]}, \quad u = cr(t)e^{-\alpha(T-t)}, \quad v = cr(T), \quad q = \frac{2\alpha\gamma}{\rho^2} - 1,$$

and I_q is the modified Bessel function of the first kind of order q [see Feller, 1951 for details]. The mean and variance of $r(T)$ conditional on $r(t)$ are given by (see Problem 7.11)

$$E[r(T)|r(t)] = r(t)e^{-\alpha(T-t)} + \gamma\left[1 - e^{-\alpha(T-t)}\right] \tag{7.2.30a}$$

$$\text{var}(r(T)|r(t)) = r(t)\left(\frac{\rho^2}{\alpha}\right)\left[e^{-\alpha(T-t)} - e^{-2\alpha(T-t)}\right] + \frac{\gamma\rho^2}{2\alpha}\left[1 - e^{-\alpha(T-t)}\right]^2. \tag{7.2.30b}$$

The distribution of the future short rates has the following properties:

(i) as $\alpha \to \infty$, the mean tends to γ and the variance to zero;
(ii) as $\alpha \to 0^+$, the mean tends to $r(t)$ and the variance to $\rho^2(T-t)r(t)$.

The Cox–Ingersoll–Ross model falls within the class of affine term structure models, so the price of the discount bond assumes the same form as in (7.2.14).

The corresponding pair of differential equations for $a(t, T)$ and $b(t, T)$ are given by

$$\frac{da}{dt} - \alpha\gamma b = 0, \qquad t < T, \tag{7.2.31a}$$

$$\frac{db}{dt} - (\alpha + \lambda\rho)b - \frac{\rho^2}{2}b^2 + 1 = 0, \quad t < T, \tag{7.2.31b}$$

where the market price of risk is taken to be $\lambda\sqrt{r}$, and λ is assumed to be constant. The final conditions are

$$a(T, T) = 0 \quad \text{and} \quad b(T, T) = 0.$$

The solutions to the above equations are found to be (Cox, Ingersoll and Ross, 1985)

$$a(t, T) = \frac{2\alpha\gamma}{\rho^2} \ln\left(\frac{2\theta e^{(\theta+\psi)(T-t)/2}}{(\theta + \psi)\left[e^{\theta(T-t)} - 1\right] + 2\theta}\right) \tag{7.2.32a}$$

$$b(t, T) = \frac{2\left[e^{\theta(T-t)} - 1\right]}{(\theta + \psi)\left[e^{\theta(T-t)} - 1\right] + 2\theta}, \tag{7.2.32b}$$

where

$$\psi = \alpha + \lambda\rho, \quad \theta = \sqrt{\psi^2 + 2\rho^2}.$$

Note that the market price of risk λ appears only through the sum ψ in the above solution. The properties of the comparative statics for the bond price and the yield to maturity of the Cox–Ingersoll–Ross model are addressed in Problems 7.15–7.17.

7.2.4 Generalized One-Factor Short Rate Models

Besides the Vasicek and Cox–Ingersoll–Ross models, several other one-factor short rate models have also been proposed in the literature. Many of these models can be nested within the stochastic process represented by

$$dr_t = (\alpha + \beta r_t)dt + \rho r_t^\gamma \, dZ_t, \tag{7.2.33}$$

where the parameters α, β, γ and ρ are constants. For example, the Vasicek and Cox–Ingersoll–Ross models correspond to $\gamma = 0$ and $\gamma = 1/2$, respectively, and the Geometric Brownian model corresponds to $\alpha = 0$ and $\gamma = 1$. The stochastic interest rate model used by Merton (1973, Chap. 1) can be nested within the Vasicek model with $\beta = 0$ and $\gamma = 0$. Other examples of one-factor interest rate models nested within the stochastic process of (7.2.33) are:

Dothan model (1978)	$dr = \rho r \, dZr$
Brennan–Schwartz model (1980)	$dr = (\alpha + \beta r)dt + \rho r \, dZ$
Cox–Ingersoll–Ross variable rate model (1980)	$dr = \rho r^{3/2} dZ$
Constant elasticity of variance model	$dr = \beta r \, dt + \rho r^\gamma \, dZ$

Note that when $\gamma > 0$, the volatility increases with the level of interest rate. This is called the *level effect*.

Chan ct al. (1992) performed an empirical analysis on the above list of one-factor short rate models. They found that the most successful models which capture the dynamics of the short-term interest rate are those that allow changes of volatility to be highly sensitive to the level of the short rate. The findings confirm the financial intuition that the term structure of volatility is an important factor governing the value of contingent claims. Using data from one-month Treasury bill yields, they discovered that those models with $\gamma \geq 1$ can capture the dynamics of the short-term interest rate better than those models with $\gamma < 1$. The relation between the short rate volatility and the level of r is more important than the mean reversion feature in the characterization of the dynamic short rate models. The incorporation of mean reversion feature usually causes complexity in the analysis of the term structure. They argued that since mean reversion plays the lesser role, the additional generality of adding mean reversion in a model may not be well justified. The Vasicek and Merton models have always been criticized for allowing negative interest rate values. However, their far more serious deficiency is the assumption of $\gamma = 0$ in the models. This assumption implies the conditional volatility of changes in the interest rate to be constant, independent on the interest rate level. Another disquieting conclusion deduced from their empirical studies is that the range of possible call option values varies significantly across various models. Also, the term structures derived from these models provide only a limited family which cannot correctly price many traded bonds. This stems from the inherent shortcomings that these models price interest rate derivatives with reference to a theoretical yield curve rather than the actually observed curve. Once the short rate process is fully defined, everything about the initial term structure and how it can evolve at future times is then fully defined. The initial term structure is an output from the model rather than an input to the model. All these indicate that the present framework of the one-factor diffusion process for the short rate may not be adequate to describe the true term structure of interest rates over time.

In the next section we consider how to extend the short rate models to allow for time dependent parameter functions. These parameter functions are determined through calibration to the current term structures of bond prices.

7.2.5 Calibration to Current Term Structures of Bond Prices

An interest rate model should take the current term structure of bond prices as an input rather than an output. Arbitrage exists if the theoretical bond prices solved from the model do not agree with the observed bond prices. To resolve the above shortcomings, we consider the class of term structure fitting models that allow for time dependent parameter functions and they are calibrated in such a way that the current bond prices obtained from the model coincide with the observed term structure of bond prices or forward rates. The two popular models with the short rate r_t as the underlying state variable are the Ho–Lee model and Hull–White model. To remedy the possibility of the short rate assuming a negative value, the Black–Derman–Toy model and Black–Karasinski model use $\ln r_t$ instead of r_t as the underlying state variable.

Ho–Lee (HL) Model

This is the first term structure fitting model proposed in the literature (Ho and Lee, 1986), whose initial formulation is in the form of a binomial tree. The continuous time limit of the model takes the form

$$dr_t = \phi(t)\,dt + \sigma\,dZ_t, \tag{7.2.34}$$

where r_t is the short rate and σ is the constant instantaneous standard deviation of the short rate. The time dependent drift function $\phi(t)$ is chosen to ensure that the model fits the initial term structure (see Problem 7.19).

Hull–White (HW) Model

The Ho–Lee model assumes a constant volatility structure and incorporates no mean reversion. Hull and White (1990) proposed the following model for the short rate:

$$dr_t = [\phi(t) - \alpha(t)r_t]\,dt + \sigma(t)r_t^\beta\,dZ_t. \tag{7.2.35}$$

The mean reversion level is given by $\frac{\phi(t)}{\alpha(t)}$. The model can be considered as an extended Vasicek model when $\beta = 0$ and an extended Cox–Ingersoll–Ross model when $\beta = 1/2$.

Fitting Term Structures of Bond Prices

We consider a special form of the Hull–White model, where $\phi(t)$ in the drift term is the only time dependent function in the model. Under the risk neutral measure Q, the short rate is assumed to follow

$$dr_t = [\phi(t) - \alpha r_t]\,dt + \sigma\,dZ_t, \tag{7.2.36}$$

where α and σ are constant parameters. The model possesses the mean reversion property and exhibits nice analytic tractability. We illustrate the analytic procedure for the determination of $\phi(t)$ using the information of the current term structure of bond prices.

The governing equation for the bond price $B(r, t; T)$ is given by

$$\frac{\partial B}{\partial t} + \frac{\sigma^2}{2}\frac{\partial^2 B}{\partial r^2} + [\phi(t) - \alpha r]\frac{\partial B}{\partial r} - rB = 0. \tag{7.2.37}$$

The bond price function assumes the affine form shown in (7.2.14). Solving the pair of ordinary differential equations for $a(t, T)$ and $b(t, T)$, we obtain

$$b(t, T) = \frac{1}{\alpha}\left[1 - e^{-\alpha(T-t)}\right] \tag{7.2.38a}$$

$$a(t, T) = \frac{\sigma^2}{2}\int_t^T b^2(u, T)\,du - \int_t^T \phi(u)b(u, T)\,du. \tag{7.2.38b}$$

Our goal is to determine $\phi(T)$ in terms of the current term structure of bond prices $B(r,t;T)$. From (7.2.38b) and applying the relation

$$\ln B(r,t;T) + rb(t,T) = a(t,T),$$

we have

$$\int_t^T \phi(u)b(u,T)\,du = \frac{\sigma^2}{2}\int_t^T b^2(u,T)\,du - \ln B(r,t;T) - rb(t,T). \quad (7.2.39)$$

To solve for $\phi(u)$, the first step is to obtain an explicit expression for $\int_t^T \phi(u)\,du$. This can be achieved by differentiating $\int_t^T \phi(u)b(u,T)\,du$ with respect to T and subtracting the terms involving $\int_t^T \phi(u)e^{-\alpha(T-t)}\,du$. The derivative of the left-hand side of (7.2.39) with respect to T gives

$$\begin{aligned}\frac{\partial}{\partial T}\int_t^T \phi(u)b(u,T)\,du &= \phi(u)b(u,T)\Big|_{u=T} + \int_t^T \phi(u)\frac{\partial}{\partial T}b(u,T)\,du \\ &= \int_t^T \phi(u)e^{-\alpha(T-u)}\,du. \end{aligned} \quad (7.2.40)$$

Next, we equate the derivatives on both sides to obtain

$$\begin{aligned}\int_t^T \phi(u)e^{-\alpha(T-u)}\,du = {}& \frac{\sigma^2}{\alpha}\int_t^T [1-e^{-\alpha(T-u)}]e^{-\alpha(T-u)}\,du \\ & - \frac{\partial}{\partial T}\ln B(r,t;T) - re^{-\alpha(T-t)}. \end{aligned} \quad (7.2.41)$$

We multiply (7.2.39) by α and add it to (7.2.41) to obtain

$$\begin{aligned}\int_t^T \phi(u)\,du = {}& \frac{\sigma^2}{2\alpha}\int_t^T [1-e^{-2\alpha(T-u)}]\,du - r \\ & - \frac{\partial}{\partial T}\ln B(r,t;T) - \alpha\ln B(r,t;T). \end{aligned}$$

Finally, by differentiating the above equation with respect to T again, we obtain $\phi(T)$ in terms of the current term structure of bond prices $B(r,t;T)$ as follows:

$$\begin{aligned}\phi(T) = {}& \frac{\sigma^2}{2\alpha}[1-e^{-2\alpha(T-t)}] - \frac{\partial^2}{\partial T^2}\ln B(r,t;T) \\ & - \alpha\frac{\partial}{\partial T}\ln B(r,t;T). \end{aligned} \quad (7.2.42a)$$

Alternatively, one may express $\phi(T)$ in terms of the current term structure of forward rates $F(t,T)$. Recall that $-\frac{\partial}{\partial T}\ln B(r,t;T) = F(t,T)$ so that we may rewrite $\phi(T)$ in the form

$$\phi(T) = \frac{\sigma^2}{2\alpha}[1-e^{-2\alpha(T-t)}] + \frac{\partial}{\partial T}F(t,T) + \alpha F(t,T). \quad (7.2.42b)$$

An analytic representation of the drift function $\phi(T)$ in terms of the current yield curve $R(t, T)$ can also be derived [see Problem 7.22].

Using (7.2.13b), the bond price volatility $\sigma_B(t, T)$ is given by (note that absolute value is enforced)

$$\sigma_B(t, T) = \left|\frac{\sigma}{B}\frac{\partial B}{\partial r}\right| = |-\sigma b(t, T)| = \frac{\sigma}{\alpha}[1 - e^{-\alpha(T-t)}].$$

Substituting into (7.2.13a), the dynamics of the bond price process under the risk neutral measure Q is given by

$$\frac{dB}{B} = r\,dt + \frac{\sigma}{\alpha}[1 - e^{-\alpha(T-t)}]\,dZ. \tag{7.2.43}$$

As the bond volatility is independent of r, so the distribution of the bond price at any given time conditional on its price at an earlier time is lognormal.

A similar procedure of calibrating the time dependent drift term by matching data of the initial term structure of bond prices or forward rates can be applied to the following generalized Vasicek mean reversion short rate model of the form (Hull and White, 1990)

$$dr_t = [\theta(t) + \alpha(t)(d - r_t)]\,dt + \sigma_r(t)\,dZ_t. \tag{7.2.44}$$

The details of calibration are documented in Problem 7.23.

Black–Derman–Toy Model and Black–Karasinski Model

Similar to the Ho–Lee model, the original formulation of the Black–Derman–Toy (BDT) model (Black, Derman and Toy, 1990) is in the form of a binomial tree. The continuous time equivalent of the model can be shown to be

$$d\ln r_t = \left[\theta(t) - \frac{\sigma_r'(t)}{\sigma_r(t)}\ln r_t\right]dt + \sigma_r(t)\,dZ_t. \tag{7.2.45}$$

In this model, the changes in the short rate in the model are lognormally distributed so that the short rates are always nonnegative. The first parameter function $\theta(t)$ is chosen so that the model fits the term structure of short rates while the second parameter function $\sigma_r(t)$ is chosen to fit the term structure of short rate volatilities. When the volatility function $\sigma_r(t)$ is taken to be constant, the BDT model reduces to a lognormal version of the HL model.

Suppose the reversion rate and volatility in the BDT model are decoupled, we then have

$$d\ln r_t = [\theta(t) - \alpha(t)\ln r_t]\,dt + \sigma_r(t)\,dZ_t. \tag{7.2.46}$$

This modified version is called the Black–Karasinski (BK) model (Black and Karasinski, 1991).

7.3 Multifactor Interest Rate Models

In the one-factor interest rate models discussed in Sect. 7.2, the short rate is assumed to follow a specific parametric one-factor continuous time model. Most of these models offer nice analytic tractability where a closed form solution for the bond price can be found. However, under the one-factor assumption, the possible forms of term structure of interest rates that can be generated are limited. In other words, a one-factor model tends to oversimplify the true stochastic behavior of the interest rate movement. To rectify these shortcomings, one may consider the construction of multifactor interest rate models that involve the short rate together with other factors. The higher degree of freedom used to model the behavior of the term structure of interest rates offered by the multifactor models do come with a price: the analytic tractability of the pricing model is usually much reduced. In most cases, one has to resort to numerical methods for valuation of bond prices. Also, it is extremely difficult to calibrate the parameters in the models. These aspects have been criticized by some practitioners in the market as counterproductive.

We discuss several popular classes of multifactor models. The first class of models use the short rate and the long rate as the underlying state variables. The state variables in the second class are the short rate and the variance of the short rate. The general formulation of the multifactor affine term structure models is discussed in Sect. 7.3.3.

7.3.1 Short Rate/Long Rate Models

Market practitioners generally classify interest rates into the short-term and long-term categories. The short-term rates include rates on Treasury bills, interbank trading of deposits and certificates of deposit. These rates normally have a life span less than one year. The long-term rates are implicitly implied in the prices of long-term bonds, some of these bonds may have maturity up to 30 years. The two classes of interest rates are certainly not locked with each other in any fixed manner.

Brennan and Schwartz (1979) chose the two underlying stochastic factors which govern the term structure of interest rates to be the short-term interest rate r_t and the long-term interest rate ℓ_t (for simplicity of notation, we drop the time index t in r_t, ℓ_t and Brownian processes etc. in subsequent exposition). Their model assumes that r and ℓ follow the stochastic processes of the form

$$dr = \beta_r(r, \ell, t)\, dt + \eta_r(r, \ell, t)\, dZ_r \tag{7.3.1a}$$

$$d\ell = \beta_\ell(r, \ell, t)\, dt + \eta_\ell(r, \ell, t)\, dZ_\ell, \tag{7.3.1b}$$

where dZ_r and dZ_ℓ are differentials of the standard Brownian processes. Let ρ denote the constant correlation coefficient between dZ_r and dZ_ℓ, where $\rho\, dt = dZ_r\, dZ_\ell$. The expected change and the variance of the change in each interest rate are assumed to be functions of both r and ℓ. Let $B(r, \ell, t; T)$ be the price of the T-maturity discount bond with unit par. By Ito's lemma, the stochastic process of the price of the discount bond is given by

$$\frac{dB}{B} = \mu(r, \ell, t)\, dt + \sigma_r(r, \ell, t)\, dZ_r + \sigma_\ell(r, \ell, t)\, dZ_\ell, \tag{7.3.2}$$

where

$$\mu(r, \ell, t) = \frac{1}{B}\left(\frac{\partial B}{\partial t} + \beta_r \frac{\partial B}{\partial r} + \beta_\ell \frac{\partial B}{\partial \ell} + \frac{\eta_r^2}{2}\frac{\partial^2 B}{\partial r^2} + \rho\eta_r\eta_\ell \frac{\partial^2 B}{\partial r \partial \ell} + \frac{\eta_\ell^2}{2}\frac{\partial^2 B}{\partial \ell^2}\right),$$

$$\sigma_r(r, \ell, t) = \frac{\eta_r}{B}\frac{\partial B}{\partial r} \quad \text{and} \quad \sigma_\ell(r, \ell, t) = \frac{\eta_\ell}{B}\frac{\partial B}{\partial \ell}. \tag{7.3.3}$$

Since there are two stochastic factors in the model, we need bonds of three different maturities to form a riskless hedging portfolio. Suppose we construct a portfolio which contains V_1, V_2, V_3 units of bonds with maturity dates T_1, T_2, T_3, respectively. Let Π denote the value of the portfolio. As usual, we follow the Black–Scholes approach of keeping the portfolio composition to be instantaneously "frozen". By virtue of (7.3.2), the rate of return on the portfolio over time dt is given by

$$\begin{aligned} d\Pi &= [V_1\mu(T_1) + V_2\mu(T_2) + V_3\mu(T_3)]\,dt \\ &\quad + [V_1\sigma_r(T_1) + V_2\sigma_r(T_2) + V_3\sigma_r(T_3)]\,dZ_r \\ &\quad + [V_1\sigma_\ell(T_1) + V_2\sigma_\ell(T_2) + V_3\sigma_\ell(T_3)]\,dZ_\ell. \end{aligned} \tag{7.3.4}$$

Here, $\mu(T_i) = \mu(r, \ell, t; T_i)$ denotes the drift rate of the bond with maturity $T_i, i = 1, 2, 3$; and similar notational interpretation for $\sigma_r(T_i)$ and $\sigma_\ell(T_i)$. Suppose we choose V_1, V_2, V_3 such that the coefficients of the stochastic terms in (7.3.4) are zero, thus making the portfolio value to be instantaneously riskless. This leads to two equations for V_1, V_2 and V_3, namely,

$$\begin{aligned} V_1\sigma_r(T_1) + V_2\sigma_r(T_2) + V_3\sigma_r(T_3) &= 0 \\ V_1\sigma_\ell(T_1) + V_2\sigma_\ell(T_2) + V_3\sigma_\ell(T_3) &= 0. \end{aligned} \tag{7.3.5}$$

Since the portfolio is now instantaneously riskless, it must earn the riskless short rate to avoid arbitrage, that is,

$$d\Pi = [V_1\mu(T_1) + V_2\mu(T_2) + V_3\mu(T_3)]dt = r(V_1 + V_2 + V_3)\,dt,$$

so that

$$V_1\,[\mu(T_1) - r] + V_2\,[\mu(T_2) - r] + V_3\,[\mu(T_3) - r] = 0. \tag{7.3.6}$$

Putting the results together, we obtain a system of homogeneous equations for V_1, V_2, V_3, namely,

$$\begin{pmatrix} \sigma_r(T_1) & \sigma_r(T_2) & \sigma_r(T_3) \\ \sigma_\ell(T_1) & \sigma_\ell(T_2) & \sigma_\ell(T_3) \\ \mu(T_1) - r & \mu(T_2) - r & \mu(T_3) - r \end{pmatrix} \begin{pmatrix} V_1 \\ V_2 \\ V_3 \end{pmatrix} = \begin{pmatrix} 0 \\ 0 \\ 0 \end{pmatrix}. \tag{7.3.7}$$

The above system possesses nontrivial solutions provided that the last row is a linear combination of the first two rows. Since the maturity dates T_1, T_2 and T_3 are arbitrary, this dictates the following relation between the drift rate and volatility functions

$$\mu(r, \ell, t) - r = \lambda_r(r, \ell, t)\sigma_r(r, \ell, t) + \lambda_\ell(r, \ell, t)\sigma_\ell(r, \ell, t). \tag{7.3.8}$$

In general, the multipliers λ_r and λ_ℓ should have dependence on r, ℓ and t.

Here, λ_r and λ_ℓ are recognized as the respective market price of risk of the short-term and long-term rates. Substituting the expressions for $\mu(r, \ell, t)$, $\sigma_r(r, \ell, t)$ and $\sigma_\ell(r, \ell, t)$ from (7.3.3) into (7.3.8), we obtain the following governing equation for the bond price

$$\begin{aligned}\frac{\partial B}{\partial t} &+ \frac{\eta_r^2}{2}\frac{\partial^2 B}{\partial r^2} + \rho\eta_r\eta_\ell\frac{\partial^2 B}{\partial r\partial \ell} + \frac{\eta_\ell^2}{2}\frac{\partial^2 B}{\partial \ell^2} \\ &+ (\beta_r - \lambda_r\eta_r)\frac{\partial B}{\partial r} + (\beta_\ell - \lambda_\ell\eta_\ell)\frac{\partial B}{\partial \ell} - rB = 0. \qquad (7.3.9)\end{aligned}$$

Note that the market prices of risks are present in the above governing equation.

Consol Bond and Market Price of Interest Rate Risk
It occurs that the long rate can be related directly to a traded security, the consol bond. A *consol bond* is a perpetual bond (with infinite maturity) which promises to pay a continuous constant coupon rate c. Let $G(\ell)$ denote the value of the consol bond, then

$$G(\ell) = \frac{c}{\ell}. \qquad (7.3.10)$$

Since the long rate is a function of a traded asset, the market price of risk of the long rate λ_ℓ can be expressed in terms of r, ℓ and other parameters defining the stochastic process for ℓ. First, we apply Ito's lemma to find the differential of G and obtain

$$dG = \frac{\partial G}{\partial \ell}\,d\ell + \frac{\eta_\ell^2}{2}\frac{\partial^2 G}{\partial \ell^2}\,dt = c\left(-\frac{\beta_\ell}{\ell^2} + \frac{\eta_\ell^2}{\ell^3}\right)dt - c\frac{\eta_\ell}{\ell^2}\,dZ_\ell$$

so that

$$\frac{dG}{G} = \left(\frac{\eta_\ell^2}{\ell^2} - \frac{\beta_\ell}{\ell}\right)dt - \frac{\eta_\ell}{\ell}\,dZ_\ell. \qquad (7.3.11)$$

The instantaneous rate of return μ_c on the consol bond is the sum of coupon rate ℓ and drift rate of G, that is,

$$\mu_c = \frac{\eta_\ell^2}{\ell^2} - \frac{\beta_\ell}{\ell} + \ell. \qquad (7.3.12)$$

We then obtain (Brennan and Schwartz, 1979)

$$\lambda_\ell(r, \ell, t) = \frac{\mu_c - r}{\sigma_\ell(r, \ell, t)} = \frac{\beta_\ell - \ell^2 + r\ell}{\eta_\ell} - \frac{\eta_\ell}{\ell}, \qquad (7.3.13)$$

where

$$\sigma_\ell = \frac{\eta_\ell}{G}\frac{\partial G}{\partial \ell} = -\eta_\ell/\ell. \qquad (7.3.14)$$

What are the intricacies faced in the implementation of the model? It is quite cumbersome to determine the parameters in the stochastic processes for the two rates. Also, the reliability of the estimation is quite difficult to be assessed. In the numerical solution of the two-factor interest rate model by finite difference calculations, one requires the prescription of the full set of boundary conditions for the governing bond price equation. It is anticipated that the bond price would tend to zero as either one of the interest rates goes to infinity. However, the boundary condition at the limiting case of zero interest rate can be quite tricky to implement.

A similar version of the long rate and short rate model was proposed by Schaefer and Schwartz (1984). They chose the long rate and the spread (defined as short rate minus long rate) as the underlying state variables. By arguing that the long rate and spread are almost uncorrelated and using the technique of frozen coefficients, they managed to obtain an analytic approximate solution to the bond price. Details of the Schaefer–Schwartz model are presented in Problem 7.27.

7.3.2 Stochastic Volatility Models

Other forms of multifactor interest rate models have also been proposed in the literature. Fong and Vasicek (1991) postulated that the volatility of the short rate should also be a stochastic state variable in order to better characterize the term structure of interest rates. Using the interest rate volatility as the second state variable is intuitively appealing since volatility is always a dominant factor in determining the prices of bonds and options. They proposed that the diffusion processes for the short rate r and the instantaneous variance of the short rate v are governed by the following mean reversion processes:

$$dr = \alpha(\overline{r} - r)\, dt + \sqrt{v}\, dZ_r \tag{7.3.15a}$$

$$dv = \gamma(\overline{v} - v)\, dt + \xi\sqrt{v}\, dZ_v, \tag{7.3.15b}$$

where $\overline{r}$ and $\overline{v}$ are the long-term mean of r and v, respectively. The parameters α, γ and ξ are taken to be constant. Further, the market prices of risk of r and v are assumed to be proportional to $\sqrt{v}$. Let ρ denote the constant correlation coefficient between dZ_r and dZ_v. Using a similar argument of no arbitrage pricing, the governing equation for the price of a discount bond can be shown to be

$$\begin{aligned}\frac{\partial B}{\partial \tau} &= \frac{v}{2}\frac{\partial^2 B}{\partial r^2} + \rho\xi v\frac{\partial^2 B}{\partial r \partial v} + \frac{\xi^2 v}{2}\frac{\partial^2 B}{\partial v^2} \\ &\quad + (\alpha\overline{r} - \alpha r + \lambda v)\frac{\partial B}{\partial r} + \left[\gamma\overline{v} - (\gamma + \xi\eta)v\right]\frac{\partial B}{\partial v} - rB,\end{aligned} \tag{7.3.16}$$

where $\lambda\sqrt{r}$ and $\eta\sqrt{v}$ are assumed to be the respective market price of risk of r and v, λ and η are taken to be constant. Analytic bond price formula under the Fong–Vasicek model can be derived, though the final form involves the confluent hypergeometric functions with complex arguments (Selby and Strickland, 1995).

Balduzzi et al. (1996) extended the Fong–Vasicek model by adding an additional state variable: the stochastic mean level of the short rate $\overline{r}$. They assumed the following dynamics for r, $\overline{r}$ and v:

$$\begin{aligned}dr &= \alpha(\overline{r} - r)\, dt + \sqrt{v}\, dZ_r \\ d\overline{r} &= \beta(\theta - \overline{r})\, dt + \eta\, dZ_{\overline{r}} \\ dv &= \gamma(\overline{v} - v)\, dt + \xi\sqrt{v}\, dZ_v,\end{aligned} \tag{7.3.17}$$

where $dZ_r\,dZ_v = \rho\,dt$ while other correlation coefficients are taken to be zero. This three-factor model embeds the Fong–Vasicek model (which assumes constant mean). It can be shown that the discount bond price $B(t,T)$ admits an exponential affine term structure, where

$$B(t,T) = a(\tau)\exp(-b(\tau)r - c(\tau)\overline{r} - d(\tau)v), \quad \tau = T - t. \tag{7.3.18}$$

The details of the analytic calculations of $a(\tau), b(\tau), c(\tau)$ and $d(\tau)$ are relegated to Problem 7.30.

7.3.3 Affine Term Structure Models

Affine term structure models are popular due to their nice analytic tractability. In fact, all one-factor short rate models considered in Sect. 7.2 are affine models. It was shown earlier that a one-factor short rate model admits the affine form only if the drift term and volatility term in the stochastic dynamics of the short rate are an affine function of r [see (7.2.18)]. In this section, we extend a similar result to multifactor interest rate models. In particular, we discuss the characterization of admissible affine term structure models. We show that the stochastic volatility models presented in the last section also belong to the class of affine models.

Consider a multifactor interest rate model with dependence on n stochastic state variables $X_1(t), \cdots, X_n(t)$ [preferably written in vector form $\mathbf{X}(t) = (X_1(t)\cdots X_n(t))^T$], the model is said to be in the affine form if the discount bond prices admit the following exponential affine form in $X(t)$:

$$B(t,T) = \exp(a(t,T) + \mathbf{b}(t,T)^{\mathrm{T}}X(t)), \tag{7.3.19}$$

where $\mathbf{b}(t,T) = (b_1(t,T)\cdots b_n(t,T))^{\mathrm{T}}$. The model is time homogeneous if the state variables $\mathbf{X}(t)$ are time homogeneous and the functions $a(t,T)$ and $\mathbf{b}(t,T)$ are functions of $T-t$ only. In this case, the yield $R(t,t+\tau)$ with time to maturity τ is found to be

$$R(t,t+\tau) = -\frac{a(\tau)}{\tau} - \frac{\mathbf{b}(\tau)^{\mathrm{T}}\mathbf{X}(t)}{\tau}. \tag{7.3.20}$$

Taking the limit $\tau \to 0^+$ and observing

$$\lim_{\tau\to 0^+}\frac{a(\tau)}{\tau} = \frac{da}{d\tau}(0) \quad \text{and} \quad \lim_{\tau\to 0^+}\frac{\mathbf{b}(\tau)}{\tau} = \frac{d\mathbf{b}}{d\tau}(0),$$

the short rate is given by

$$r(t) = -\frac{da}{d\tau}(0) - \frac{d\mathbf{b}^{\mathrm{T}}}{d\tau}(0)\mathbf{X}(t). \tag{7.3.21}$$

Obviously, certain conditions must be set on $\mathbf{X}(t)$ in order that the bond price $B(t,t+\tau)$ admits a solution of the form:

$$B(t,t+\tau) = \exp(a(\tau) + \mathbf{b}^{\mathrm{T}}(\tau)\mathbf{X}(t)). \tag{7.3.22}$$

To address the above issue, Duffie and Kan (1996) proved that the stochastic differential equation for $\mathbf{X}(t)$ under the risk neutral measure Q must take the form:

$$d\mathbf{X}(t) = [\boldsymbol{\gamma} + \delta\mathbf{X}(t)]\,dt + \Sigma \begin{pmatrix} \sqrt{\alpha_1 + \boldsymbol{\beta}_1^T\mathbf{X}(t)} & \cdots & 0 \\ \vdots & \ddots & \vdots \\ 0 & \cdots & \sqrt{\alpha_n + \boldsymbol{\beta}_n^T\mathbf{X}(t)} \end{pmatrix} d\mathbf{Z}(t), \quad (7.3.23)$$

where $\alpha_i, i = 1, \cdots, n$, are constant scalars, $\boldsymbol{\gamma}$ and $\boldsymbol{\beta}_i, i = 1, \cdots, n$, are constant n-dimensional vectors, δ and Σ are $n \times n$ constant matrices, $\mathbf{Z}(t)$ is an n-dimensional vector Brownian process. Similar to the one-factor CIR process, certain conditions on the parameters α_i and $\boldsymbol{\beta}_i$ are required in order that the volatility $\alpha_i + \boldsymbol{\beta}_i^T\mathbf{X}$ remains positive. The coefficient vectors $\boldsymbol{\beta}_1, \cdots, \boldsymbol{\beta}_n$ generate stochastic volatility unless they are set to be identically zero. In this case, the model reduces to the Gaussian model.

Analytic Solution to the Multifactor Gaussian Model

Let the short rate $r(t)$ be defined

$$r(t) = f(t) + \mathbf{g}(t)^{\mathrm{T}}\mathbf{X}(t). \quad (7.3.24)$$

The dynamics of $\mathbf{X}(t)$ under the risk neutral measure Q is assumed to follow the Gaussian process

$$d\mathbf{X}(t) = [\boldsymbol{\gamma} + \delta\mathbf{X}(t)]\,dt + \Sigma\, d\mathbf{Z}(t), \quad (7.3.25)$$

where all parameters are constant. Under these assumptions, it is possible to derive an analytic representation of the bond price formula. First, we consider the following system of equations

$$\frac{d\mathbf{y}}{dt} = \delta\mathbf{y} \quad (7.3.26)$$

with initial condition: $\mathbf{y}(0) = \mathbf{y}_0$. It is known that the solution to the above system is given by

$$\mathbf{y}(t) = \Phi(t)\mathbf{y}_0. \quad (7.3.27)$$

Here, $\Phi(t)$ is called the fundamental matrix of the differential equation system. It satisfies the system of equations

$$\frac{d\Phi}{dt} = \delta\Phi(t) \quad (7.3.28)$$

with initial condition $\Phi(0) = I$, where I is the identity matrix. The solution to the system of stochastic differential equation (7.3.25) can be deduced to be

$$\mathbf{X}(t) = \Phi(t)\left[\mathbf{X}(0) + \int_0^t \Phi^{-1}(u)\boldsymbol{\gamma}\,du + \int_0^t \Phi^{-1}(u)\Sigma\,d\mathbf{Z}(u)\right]. \quad (7.3.29)$$

Define the auxiliary function

$$d(t,T) = -\int_t^T r(u)\,du = -\int_t^T \left[f(u) + \mathbf{g}(u)^T\mathbf{X}(u)\right]du. \tag{7.3.30}$$

As the short rate is affine in $\mathbf{X}(t)$, so both $r(t)$ and $d(t,T)$ remain to be Gaussian. Recall that

$$B(t,T) = E_Q^t[\exp(d(t,T))],$$

and since $d(t,T)$ is Gaussian, so

$$B(t,T) = \exp\left(E_t[d(t,T)] + \frac{1}{2}\mathrm{var}_t(d(t,T))\right). \tag{7.3.31}$$

It is relatively straightforward to show that

$$E_t[d(t,T)]$$
$$= -\int_t^T \left[f(u) + \mathbf{g}^T(u)\Phi(u-t)\mathbf{X}(t) + \mathbf{g}^T(u)\int_t^u \Phi(u-s)\boldsymbol{\gamma}\,ds\right]du \tag{7.3.32a}$$
$$\mathrm{var}_t(d(t,T))$$
$$= \int_t^T\int_t^T\int_t^{\min(u,u^*)} \mathbf{g}^T(u)\Phi(u-s)\Sigma\Phi^T(u^*-s)\mathbf{g}(u^*)\,ds\,du\,du^*. \tag{7.3.32b}$$

Equivalent Classes of Affine Models

Even with a small number of stochastic state variables in an affine term structure model, the number of admissible forms of affine diffusion as defined by (7.3.23) can be quite numerous due to the large number of parameters involved in characterizing $\mathbf{X}(t)$. Dai and Singleton (2000) claimed that two models are considered equivalent if they generate identical prices for all contingent claims. In their paper, they presented a canonical representative of each equivalent class of affine models.

For the class of one-factor affine models, there are only two equivalent classes, namely, the Vasicek model and the CIR model. When there are two stochastic state variables, the number of equivalent classes becomes three. Besides the two-factor Vasicek model and the two-factor CIR model, the third class is the stochastic volatility model. The canonical forms of these equivalent classes are listed below.

1. Two-factor Vasicek model

$$\begin{aligned} dX_1(t) &= -\delta_{11}X_1(t)\,dt + dZ_1(t) \\ dX_2(t) &= -[\delta_{21}X_1(t) + \delta_{22}X_2(t)]\,dt + dZ_2(t). \end{aligned} \tag{7.3.33}$$

2. Two-factor CIR model (Longstaff–Schwartz model as a typical example)

$$\begin{aligned} dX_1(t) &= \{\delta_{11}[\alpha_1 - X_1(t)]\,dt + \delta_{12}[\alpha_2 - X_2(t)]\}\,dt + \sqrt{X_1(t)}\,dZ_1(t) \\ dX_2(t) &= \{\delta_{21}[\alpha_1 - X_1(t)] + \delta_{22}[\alpha_2 - X_2(t)]\}\,dt + \sqrt{X_2(t)}\,dZ_2(t). \end{aligned} \tag{7.3.34}$$

3. Two-factor stochastic volatility model (Fong–Vasicek model as a typical example)

$$\begin{aligned} dX_1(t) &= \delta_{11}[\alpha_1 - X_1(t)]\,dt + \sqrt{X_1(t)}\,dZ_1(t) \\ dX_2(t) &= \{\delta_{21}[\alpha_1 - X_1(t)] - \delta_{22}X_2(t)\}\,dt \\ &\quad + [1 + \beta_{21}\sqrt{X_1(t)}]\,dZ_2(t). \end{aligned} \tag{7.3.35}$$

For the class of three-factor models, the number of equivalent classes increases to four. Besides the three-factor Vasicek model and CIR model, the other two equivalent classes are stochastic volatility models. Their canonical forms are given below.

4. Three-factor type-one stochastic volatility model (Balduzzi et al.'s model as a typical example)

$$\begin{aligned} dX_1(t) &= \delta_{11}[\alpha_1 - X_1(t)]\,dt + \sqrt{X_1(t)}\,dZ_1(t) \\ dX_2(t) &= \{\delta_{21}[\alpha_1 - X_1(t)] - \delta_{22}X_2(t) - \delta_{23}X_3(t)\}\,dt \\ &\quad + [1 + \beta_{21}\sqrt{X_1(t)}]\,dZ_2(t) \\ dX_3(t) &= \{\delta_{31}[\alpha_1 - X_1(t)] - \delta_{32}X_2(t) - \delta_{33}X_3(t)\}\,dt \\ &\quad + [1 + \beta_{31}\sqrt{X_1(t)}]\,dZ_3(t). \end{aligned} \tag{7.3.36}$$

5. Three-factor type-two stochastic volatility model

$$\begin{aligned} dX_1(t) &= \{\delta_{11}[\alpha_1 - X_1(t)] + \delta_{12}[\alpha_2 - X_2(t)]\}\,dt + \sqrt{X_1(t)}\,dZ_1(t) \\ dX_2(t) &= \{\delta_{21}[\alpha_1 - X_1(t)] + \delta_{22}[\alpha_2 - X_2(t)]\}\,dt + \sqrt{X_2(t)}\,dZ_2(t) \\ dX_3(t) &= \{\delta_{31}[\alpha_1 - X_1(t)] + \delta_{32}[\alpha_2 - X_2(t)] - \delta_{33}X_3(t)\}\,dt \\ &\quad + [1 + \beta_{31}\sqrt{X_1(t)} + \beta_{32}\sqrt{X_2(t)}]\,dZ_3(t). \end{aligned} \tag{7.3.37}$$

Extension Beyond Affine Forms—Quadratic Term Structure Models

In the quadratic term structure models, the short rate is given by a quadratic function of a vector of stochastic state variables, say, $\mathbf{X}(t) = (X_1(t) \cdots X_n(t))^{\mathrm{T}}$. For example, the short rate $r(t)$ is defined by (see Problem 7.32).

$$r(t) = f(t) + \mathbf{g}^{\mathrm{T}}(t)\mathbf{X}(t) + \mathbf{X}(t)^{\mathrm{T}}Q(t)\mathbf{X}(t), \tag{7.3.38}$$

where $f(t)$ is a scalar function, $\mathbf{g}(t)$ is an n-dimensional vector function and $Q(t)$ is an $n \times n$ matrix function. Under certain restrictions imposed on $\mathbf{X}(t)$, the bond prices become an exponential quadratic form. This class of quadratic term structure models have been fully discussed in the literature [see Ahn, Dittmar and Gallant (2002) and Leippold and Wu (2002)].

7.4 Heath–Jarrow–Morton Framework

The information on the term structure of interest rates can be provided either by the bond prices $B(t, T)$, yield curve $R(t, T)$ or instantaneous forward rates $F(t, T)$.

The Heath–Jarrow–Morton (HJM) framework (Heath, Jarrow and Morton, 1992) attempts to construct a family of continuous time stochastic processes for the term structure. It is based on an exogenous specification of the dynamics of $F(t, T)$. The driving state variable of the model is chosen to be the entire forward rate curve $F(t, T)$. In the most general multistate version of the model, the stochastic process for $F(t, T)$ is assumed to be governed by

$$F(t, T) = F(0, T) + \int_0^t \alpha_F(u, T)\, du + \sum_{i=1}^n \int_0^t \sigma_F^i(u, T)\, dZ_i(u), \tag{7.4.1a}$$

or in differential form,

$$dF(t, T) = \alpha_F(t, T)\, dt + \sum_{i=1}^n \sigma_F^i(t, T)\, dZ_i(t), \quad 0 \leq t \leq T. \tag{7.4.1b}$$

The last equation represents the stochastic differential equation for $F(t, T)$ in the t-variable indexed by the parameter T. Apparently, we are involved with infinitely many processes, one process for each maturity T. However, if the HJM model depends on n random sources, we only need to know the changes over an infinitesimal interval in the forward rate curve at n maturity dates in order to specify the changes over the same interval at all maturities. In (7.4.1a), $F(0, T)$ is the known market information of initial forward rate curve, $\alpha_F(t, T)$ is the drift of the instantaneous forward rate, $\sigma_F^i(t, T)$ is the ith volatility function of the forward rate, and Z_i is the ith Brownian process. There are n independent Brownian processes determining the stochastic fluctuation of the forward rate curve. The drift function $\alpha_F(t, T)$ and volatility functions $\sigma_F^i(t, T)$ are adapted processes. The forward rate process starts with the initial value $F(0, T)$, then evolves under a drift specification and influences of several Brownian processes. Such specification of the initial condition gives an automatic fit between the observed and theoretical forward rates at $t = 0$.

For an arbitrary set of drift and volatility structures, the forward rate dynamics as posed in (7.4.1a) are not necessarily arbitrage free. In Sect. 7.4.1, we show that the drift $\alpha_F(t, T)$ must be related to the volatility functions $\sigma_F^i(t, T), i = 1, \cdots, n$, in order that the derived system of bond prices admit no arbitrage opportunities. It is important to understand that the HJM approach is a framework for analyzing interest rate dynamics, and is not a specified model itself. In Sect. 7.4.2, we demonstrate how some of the common short rate models can be formulated under the HJM framework. The evolution of the forward rate under the HJM dynamics is in general non-Markovian. This non-Markovian nature poses the "bushy tree" phenomena in numerical implementation using the binomial tree approach since the discrete HJM tree is nonrecombining thus causing the number of nodes to grow exponentially. We consider the conditions under which the short rate model derived under the HJM framework becomes Markovian. In Sect. 7.4.3, we examine the dynamics of the forward LIBOR process under the HJM framework. We derive the relation between the volatility functions of the instantaneous forward process and the forward LIBOR process.

7.4.1 Forward Rate Drift Condition

We would like to show how the HJM approach exploits the arbitrage relation between the forward rates and bond prices. As a consequence, this imposes restrictions on the drift $\alpha_F(t, T)$ of the instantaneous forward rate $F(t, T)$. With absence of arbitrage opportunities, the drift depends only on the volatility functions of $F(t, T)$. We start with the stochastic differential equation for the bond price, then derive the corresponding stochastic differential equation for the instantaneous forward rate.

Under the risk neutral measure Q, the drift rate of the discount bond price $B(t, T)$ must be the short rate $r(t)$, given that the discount bond does not pay coupons. Assuming n risk factors as modeled by n uncorrelated Q-Brownian processes $Z_i(t), i = 1, 2, \cdots, n$, that drive the bond prices, the dynamics of $B(t, T)$ under Q is governed by the following stochastic differential equation:

$$\frac{dB(t, T)}{B(t, T)} = r(t)\, dt - \sum_{i=1}^{n} \sigma_B^i(t, T)\, dZ_i(t), \tag{7.4.2}$$

where $\sigma_B^i(t, T), i = 1, 2, \cdots, n$, are adapted processes with the terminal condition $\sigma_B^i(T, T) = 0$. Since $-Z_i$ is distributed like Z_i, it causes no confusion to put a negative sign in the stochastic term in (7.4.2). According to the definition of continuous forward rate $f(t, T_1, T_2)$ defined in (7.1.5), its differential is given by

$$df(t, T_1, T_2) = \frac{d \ln B(t, T_1) - d \ln B(t, T_2)}{T_2 - T_1}. \tag{7.4.3}$$

By Ito's lemma, the logarithm derivative of the bond price with maturity T_j is given by

$$d \ln B(t, T_j) = \left[r(t) - \frac{1}{2} \sum_{i=1}^{n} \sigma_B^i(t, T_j)^2 \right] dt - \sum_{i=1}^{n} \sigma_B^i(t, T_j)\, dZ_i(t). \tag{7.4.4}$$

Recall that the instantaneous forward rate is defined by

$$F(t, T) = \lim_{\Delta T \to 0} f(t, T, T + \Delta T), \quad \Delta T > 0.$$

By substituting (7.4.4) into (7.4.3), setting $T_1 = T$ and $T_2 = T + \Delta T$ and taking the limit $\Delta T \to 0$, we obtain the following stochastic differential equation for the instantaneous forward rate:

$$dF(t, T) = \sum_{i=1}^{n} \frac{\partial \sigma_B^i}{\partial T}(t, T) \sigma_B^i(t, T)\, dt + \sum_{i=1}^{n} \frac{\partial \sigma_B^i}{\partial T}(t, T)\, dZ_i(t).$$

Suppose we write $\sigma_F^i(t, T) = \frac{\partial \sigma_B^i}{\partial T}(t, T)$, we obtain

$$dF(t,T) = \left(\sum_{i=1}^{n} \sigma_F^i(t,T) \int_t^T \sigma_F^i(t,u)\,du \right) dt + \sum_{i=1}^{n} \sigma_F^i(t,T)\,dZ_i(t). \tag{7.4.5}$$

Hence, under the risk neutral measure Q, the drift of $F(t,T)$ is related to the volatility function $\sigma_F^i(t,T), i = 1, 2, \cdots, n$, by the following *forward rate drift condition*:

$$\alpha_F(t,T) = \sum_{i=1}^{n} \sigma_F^i(t,T) \int_t^T \sigma_F^i(t,u)\,du. \tag{7.4.6}$$

The class of HJM interest rate models specify the volatility functions $\sigma_F^i(t,T)$ of all instantaneous forward rates at all future times. Once the *volatility structure* $\sigma_F^i(t,T)$ is specified, the drift $\alpha_F(t,T)$ can be found using (7.4.6). As an analogy to the Black–Scholes equity option model, the inputs to HJM involve the specification of an underlying and a measure of its volatility. The underlying is the entire term structure and the volatility structure describes how this term structure evolves over time. The initial term structure plays a similar role as the asset price in the Black–Scholes model.

Suppose the volatility functions $\sigma_F^i(t,T)$ are taken to be deterministic, then the forward rate $F(t,T)$ and bond price $B(t,T)$ have Gaussian probability laws under the risk neutral measure Q. This class of the HJM models are said to be *Gaussian*.

7.4.2 Short Rate Processes and Their Markovian Characterization

Recall that under the risk neutral measure Q, the stochastic process for $F(t,T)$ takes the form

$$F(t,T) = F(0,T) + \sum_{i=1}^{n} \int_0^t \alpha_F^i(u,T)\,du + \int_0^t \sigma_F^i(u,T)\,dZ_i(u), \tag{7.4.7}$$

where

$$\alpha_F^i(t,T) = \sigma_F^i(t,T) \int_t^T \sigma_F^i(t,u)\,du.$$

The short rate process under Q is then given by

$$r(t) = F(0,t) + \sum_{i=1}^{n} \left[\int_0^t \alpha_F^i(u,t)\,du + \int_0^t \sigma_F^i(u,t)\,dZ_i(u) \right]. \tag{7.4.8a}$$

By direct differentiation with respect to t, we obtain

$$dr(t) = \left\{ \frac{\partial}{\partial t} F(0,t) + \sum_{i=1}^{n} \left[\int_0^t \frac{\partial}{\partial t} \alpha_F^i(u,t)\,du + \int_0^t \frac{\partial}{\partial t} \sigma_F^i(u,t)\,dZ_i(u) \right] \right\} dt + \sum_{i=1}^{n} \sigma_F^i(t,t)\,dZ_i(t). \tag{7.4.8b}$$

The non-Markovian nature of the short rate stems from the integrals involving the stochastic quantities in the drift term. This leads to path dependence since these stochastic integrals represent weighted sum of Brownian increments realized over the time interval $(0, t)$.

For some special choices of volatility structures, it is possible to obtain Markovian short rate process. We now discuss how some popular Markovian short rate models can be formulated under the HJM framework. We also explore the form of the volatility function under which the short rate process becomes Markovian. In subsequent discussion, we concentrate on the class of one-factor models with $n = 1$. Extension and generalization to multifactor models can be readily made, some of which are relegated to exercises.

First, by taking the simple choice of $\sigma_F(t,T) = \sigma =$ constant, we obtain

$$r(t) = F(0,t) + \frac{\sigma^2 t^2}{2} + \sigma Z(t); \tag{7.4.9a}$$

and in differential form

$$dr(t) = \left[\frac{\partial F}{\partial t}(0,t) + \sigma^2 t\right] dt + \sigma\, dZ(t). \tag{7.4.9b}$$

This is just the Ho–Lee model (see Problem 7.19). The stochastic differential equation of $F(t,T)$ under Q is found to be

$$dF(t,T) = \sigma^2(T-t)\,dt + \sigma\, dZ(t). \tag{7.4.10}$$

Another volatility structure that admits nice analytic tractability is furnished by the exponential function

$$\sigma_F(t,T) = \sigma e^{-\alpha(T-t)}. \tag{7.4.11}$$

The corresponding drift coefficient of the forward rate under Q is given by

$$\alpha_F(t,T) = \sigma_F(t,T)\int_t^T \sigma_F(t,u)\,du = \frac{\sigma^2}{\alpha} e^{-\alpha(T-t)}[1 - e^{-\alpha(T-t)}].$$

The forward rate and its stochastic differential equation are found to be

$$F(t,T) = F(0,T) + \sigma \int_0^t e^{-\alpha(T-s)}\,dZ(s) - \frac{\sigma^2}{2\alpha^2}\left[e^{-2\alpha T}(e^{2\alpha t} - 1) - 2e^{-\alpha T}(e^{\alpha t} - 1)\right] \tag{7.4.12a}$$

$$dF(t,T) = \frac{\sigma^2}{2}\left[e^{-\alpha(T-t)} - e^{-2\alpha(T-t)}\right] dt + \sigma e^{-\alpha(T-t)}\,dZ(t). \tag{7.4.12b}$$

By taking the limit $T \to t$ in $F(t,T)$, we obtain

$$r(t) = F(0,t) + \sigma \int_0^t e^{-\alpha(t-u)}\,dZ(u) - \frac{\sigma^2}{2\alpha^2}(2e^{-\alpha t} - e^{-2\alpha t} - 1).$$

Subsequently, we differentiate to obtain

$$\begin{aligned} &dr(t) \\ &= \left[\frac{\partial F}{\partial t}(0,t) + \alpha F(0,t) + \frac{\sigma^2}{2\alpha}(1 - e^{-2\alpha t}) - \alpha r(t)\right] dt + \sigma\, dZ(t). \end{aligned} \quad (7.4.13)$$

The choice of the exponential volatility function leads to the Hull–White model as defined by (7.2.36), with the same time dependent parameter function [comparing $\phi(t)$ given in (7.2.42)]. Furthermore, using the analytic form of $F(t,T)$ in (7.4.12a), it becomes straightforward to obtain the discount bond price $B(t,T)$ via the following calculations:

$$\begin{aligned} &B(t,T) \\ &= \exp\left(-\int_t^T F(t,u)\,du\right) \\ &= \exp\left(-\int_t^T F(0,u)\,du + \frac{\sigma^2}{2\alpha^2}\int_t^T \left[e^{-2\alpha u}(e^{2\alpha t} - 1) - 2e^{-\alpha u}(e^{\alpha t} - 1)\right] du \right. \\ &\qquad \left. - \sigma \int_t^T \int_0^t e^{-\lambda(u-s)}\, dZ(s)\, du\right) \\ &= \exp\left(-\int_t^T F(0,u)\,du - [r(t) - F(0,t)]X(t,T) \right. \\ &\qquad \left. - \frac{\sigma^2}{4\alpha} X^2(t,T)(1 - e^{-2\alpha t})\right), \end{aligned} \quad (7.4.14)$$

where

$$X(t,T) = \int_t^T \sigma_F(t,u)\,du = \frac{1 - e^{-\alpha(T-t)}}{\alpha}.$$

Under the HJM framework, the initial term structure of the forward rate $F(0,T)$ is automatically incorporated as input (initial condition) in the bond price solution.

Both the Ho–Lee and Vasicek models are one-factor Markovian short rate models. It would be natural to ask whether there exist other Markovian short rate models under the HJM framework. This issue has been well explored in a series of papers (Ritchken and Sankarasubramanian, 1995; Inui and Kijima, 1998; Chiarella and Kwon, 2003).

Consider the one-factor HJM model, it can be shown easily that if the volatility function $\sigma_F(t,T)$ admits the separable form, that is,

$$\sigma_F(t,T) = \gamma(t)\beta(T), \quad (7.4.15)$$

then the HJM model reduces to the Markovian short rate model. Both the Ho–Lee and Vasicek models are seen to be embedded within this separable form. From (7.4.8b), we have

$$r(t) = F(0,t) + \int_0^t \alpha_F(u,t)\,du + \beta(t)\int_0^t r(u)\,dZ(u), \quad (7.4.16a)$$

and

$$dr(t) = \left[\frac{\partial}{\partial t}F(0,t) + \int_0^t \frac{\partial}{\partial t}\alpha_F(u,t)\,du + \beta'(t)\int_0^t \gamma(u)\,dZ(u)\right]dt$$
$$+\gamma(t)\beta(t)\,dZ(t), \tag{7.4.16b}$$

where

$$\alpha_F(t,T) = \sigma_F(t,T)\int_t^T \sigma_F(t,u)\,du = \gamma^2(t)\beta(T)\int_t^T \beta(u)\,du.$$

One may eliminate the stochastic integral terms in $r(t)$ and $dr(t)$ and obtain the following stochastic differential equation of $r(t)$:

$$dr(t) = [a(t) + b(t)r(t)]\,dt + c(t)\,dZ(t), \tag{7.4.17}$$

where

$$a(t) = g'(t) - \frac{g(t)\beta'(t)}{\beta(t)}, \quad b(t) = \frac{\beta'(t)}{\beta(t)},$$
$$c(t) = \gamma(t)\beta(t) \quad \text{and} \quad g(t) = F(0,t) + \int_0^t \frac{\partial}{\partial t}\alpha_F(u,t)\,du.$$

The coefficient functions are deterministic, so the short rate model is Markovian. Indeed, it falls within the class of one-factor affine term structure models.

Inui and Kijima (1998) proposed an elegant approach to deriving a sufficient condition on the volatility structures of the forward rates in order for the short rate process to be Markovian. Consider the one-factor HJM model, the drift of the dynamics of $r(t)$ is given by

$$\mu_r(t) = \frac{\partial}{\partial t}F(0,t) + \int_0^t \frac{\partial}{\partial t}\alpha_F(u,t)\,du + \int_0^t \frac{\partial}{\partial t}\sigma_F(u,t)\,dZ(u). \tag{7.4.18}$$

Suppose $\sigma_F(t,T)$ satisfies the differential equation

$$\frac{\partial}{\partial T}\sigma_F(t,T) = -k(T)\sigma_F(t,T), \tag{7.4.19a}$$

for some deterministic function $k(T)$, with initial condition:

$$\sigma_F(t,t) = \sigma_0(r(t),t). \tag{7.4.19b}$$

Here, $\sigma_0(r(t),t)$ can be interpreted as the short rate volatility [see (7.4.8b)], with possible dependence on the short rate $r(t)$. It is straightforward to solve for $\sigma_F(t,T)$ from (7.4.19a,b):

$$\sigma_F(t,T) = \sigma_0(r(t),t)e^{-\int_t^T k(u)\,du}. \tag{7.4.20}$$

By observing that

$$\begin{aligned}
&\frac{\partial}{\partial T}\alpha_F(t,T) \\
&= \frac{\partial \sigma_F}{\partial T}(t,T)\int_t^T \sigma_F(t,u)\,du + \sigma_F^2(t,T) \\
&= -k(T)\sigma_F(t,T)\int_t^T \sigma_F(t,u)\,du + \sigma_F^2(t,T) \\
&= -k(T)\alpha_F(t,T) + \sigma_F(t,T)^2,
\end{aligned}$$

and

$$r(t) = F(0,t) + \int_0^t \alpha_F(u,t)\,du + \int_0^t \sigma_F(u,t)\,dZ(u),$$

the drift of the short rate can be expressed as

$$\begin{aligned}
\mu_r(t) &= \frac{\partial}{\partial t}F(0,t) - k(t)\left[\int_0^t \alpha_F(u,t)\,du + \int_0^t \sigma_F(u,t)\,dZ(u)\right] \\
&\quad + \int_0^t \sigma_F(u,t)^2\,du \\
&= \frac{\partial}{\partial t}F(0,t) + k(t)\big[F(0,t) - r(t)\big] + \int_0^t \sigma_F(u,t)^2\,du. \qquad (7.4.21)
\end{aligned}$$

Suppose σ_0 is set to be time dependent only, $\sigma_F(t,T)$ then becomes deterministic since there is no stochastic term in the drift of the short rate. Hence, the condition stated in (7.4.19a) becomes sufficient for the short rate process to be Markovian. The resulting short rate process may be identified with the Hull–White model. By setting $k(t) = 0$ and $\sigma_0(r,t) = \sigma$, we recover the Ho–Lee model.

For the more general case where σ_0 is a function of $r(t)$ and t, the short rate process depends on the two stochastic state variables, $r(t)$ and $\phi(t)$, where

$$\phi(t) = \int_0^t \sigma_F(u,t)^2\,du, \qquad (7.4.22a)$$

and

$$d\phi(t) = \big[\sigma_0(r(t),t)^2 - 2k(t)\phi(t)\big]dt. \qquad (7.4.22b)$$

The dynamics of the short rate process is governed by

$$dr(t) = \left\{\frac{\partial}{\partial t}F(0,t) + k(t)[F(0,t) - r(t)] + \phi(t)\right\}dt + \sigma_0(r(t),t)\,dZ(t). \qquad (7.4.23)$$

Here, $\{r(t), \phi(t)\}$ forms a two-dimensional Markov process.

7.4.3 Forward LIBOR Processes under Gaussian HJM Framework

The term structure models considered so far are based on the diffusion type behavior of the *instantaneous* short rate or forward rate. However, it may not be too effective to use the instantaneous rates as state variables for pricing models of interest rate derivatives with payoff functions that are explicitly expressed in terms of market rates (LIBOR or swap rates). Recall that the forward LIBOR is the market observable discrete forward rate as implied by the tradeable forward rate agreement. Suppose the forward rate dynamics follows the Gaussian HJM model under the risk neutral measure Q as governed by (7.4.7), we would like to find the dynamics of the forward LIBOR under Q. Let $L_t[T, T+\delta]$ denote the forward LIBOR process applied over the future period $(T, T+\delta]$ as observed at time t, $t < T$ and $\delta > 0$. From (7.1.2) and (7.1.7), the LIBOR process is related to the instantaneous forward rate $F(t, T)$ via

$$1 + \alpha L_t[T, T+\delta] = \frac{B(t,T)}{B(t,T+\delta)} = \exp\left(\int_T^{T+\delta} F(t,u)\,du\right), \tag{7.4.24}$$

where α is the accrual factor of the interest compounding period $(T, T+\delta]$. For notational convenience, we define

$$h(t; T, T+\delta) = \int_T^{T+\delta} F(t,u)\,du$$

and recall that the ith component in the volatility of the bond price process $\sigma_B^i(t,T)$ is given by

$$\sigma_B^i(t,T) = \int_t^T \sigma_F^i(t,u)\,du.$$

To compute the differential of the LIBOR process L_t, we use Ito's lemma to obtain

$$dL_t = \frac{e^{h(t)}}{\alpha}\left[dh(t) + \frac{(dh(t))^2}{2}\right].$$

It is straightforward to show that

$$\begin{aligned}
dh(t) &= \int_T^{T+\delta} dF(t,u)\,du \\
&= \int_T^{T+\delta} \frac{1}{2}\sum_{i=1}^n \frac{\partial}{\partial u}[\sigma_B^i(t,u)^2]\,du\,dt \\
&\quad + \int_T^{T+\delta} \sum_{i=1}^n \frac{\partial \sigma_B^i}{\partial u}(t,u)\,du\,dZ_i(t) \\
&= \frac{1}{2}\sum_{i=1}^n \left[\sigma_B^i(t,T+\delta)^2 - \sigma_B^i(t,T)^2\right] dt \\
&\quad + \sum_{i=1}^n \left[\sigma_B^i(t,T+\delta) - \sigma_B^i(t,T)\right] dZ_i(t),
\end{aligned}$$

and

$$(dh(t))^2 = \sum_{i=1}^{n} \left[\sigma_B^i(t, T+\delta) - \sigma_B^i(t, T)\right]^2 dt.$$

We then obtain

$$\begin{aligned} dL_t = \frac{1+\alpha L_t}{\alpha} \sum_{i=1}^{n} & [\sigma_B^i(t, T+\delta) - \sigma_B^i(t, T)]\sigma_B^i(t, T+\delta)\, dt \\ & + [\sigma_B^i(t, T+\delta) - \sigma_B^i(t, T)]\, dZ_i(t). \end{aligned} \tag{7.4.25}$$

We write the stochastic differential equation for $L_t[T, T+\delta]$ under Q in the lognormal form:

$$dL_t = \mu_L(t; T, T+\delta)L_t\, dt + L_t \sum_{i=1}^{n} \sigma_L^i(t; T, T+\delta)\, dZ_i(t). \tag{7.4.26}$$

How are σ_B^i and μ_L related to σ_L^i? By comparing the stochastic terms in (7.4.25) and (7.4.26), we deduce that the LIBOR processes are related to the forward rate volatilities by

$$\sigma_B^i(t, T+\delta) - \sigma_B^i(t, T) = \frac{\alpha L_t}{1+\alpha L_t} \sigma_L^i(t). \tag{7.4.27}$$

Subsequently, the drift rate $\mu_L(t)$ can be expressed either as

$$\mu_L(t) = \sum_{i=1}^{n} \sigma_B^i(t, T+\delta)\sigma_L^i(t), \tag{7.4.28a}$$

or

$$\mu_L(t) = \sum_{i=1}^{n} \sigma_B^i(t, T)\sigma_L^i(t) + \frac{\alpha L_t}{1+\alpha L_t} \sum_{i=1}^{n} \sigma_L^i(t)^2. \tag{7.4.28b}$$

Note that $L_t[T, T+\delta]$ is not a martingale under the risk neutral measure Q. Under Q, the dependence of the drift rate $\mu_L(t; T, T+\delta)$ on the forward rate and LIBOR volatilities is seen to be quite complicated. In Sect. 8.1, it will be shown that $L_t[T, T+\delta]$ is a martingale under the so-called forward measure under which the price of the $(T+\delta)$-maturity bond is used as the numeraire. The discussion of these market rate models and their uses on pricing different types of caps, swaptions and exotic LIBOR products is relegated to the next chapter.

7.5 Problems

7.1 For the 30/360 day count convention of the time period $[D_1, D_2)$, with D_1 included but D_2 excluded, the year fraction is given

$$\frac{\max(30 - d_1, 0) + \min(d_2, 30) + 360 \times (y_2 - y_1) + 30 \times (m_2 - m_1 - 1)}{360},$$

where d_i, m_i and y_i represent the day, month and year of date $D_i, i = 1, 2$. Compute the year fraction between February 27, 2006 and July 31, 2008.

7.2 Let $F_t(T_0, T_i)$ denote the forward price at time t for buying at time T_0 a unit-par zero-coupon bond with maturity T_i. Show that the forward swap rate [see (7.1.4)] can be expressed as

$$K_t[T_0, T_n] = \frac{1 - F_t(T_0, T_n)}{\sum_{i=1}^{n} \alpha_i F_t(T_i, T_n)},$$

where α_i is the accrual factor of the period $[T_{i-1}, T_i]$. Alternatively, suppose we write the forward LIBOR as $L_i(t) = L_t[T_{i-1}, T_i]$. By observing that

$$\frac{B(t, T_k)}{B(t, T_0)} = \prod_{i=1}^{k} \frac{B(t, T_i)}{B(t, T_{i-1})} = \prod_{i=1}^{k} \frac{1}{1 + \alpha_i L_i(t)}, \qquad k \geq 1,$$

show that the forward swap rate can be expressed as

$$K_t[T_0, T_n] = \frac{1 - \displaystyle\sum_{i=1}^{n} \frac{1}{1 + \alpha_i L_i(t)}}{\displaystyle\sum_{i=1}^{n} \alpha_i \prod_{j=1}^{i} \frac{1}{1 + \alpha_j L_j(t)}}.$$

7.3 Define the instantaneous forward rate to be $\lim_{S \to R^+} L_t[R, S]$. Show that

$$\lim_{S \to R^+} L_t[R, S] = -\frac{\partial}{\partial R} \ln B(t, R).$$

7.4 (Market price of risk.) Consider two securities, both of them are dependent on the interest rate. Suppose security A has an expected return of 4% per annum and a volatility of 10% per annum, while security B has a volatility of 20% per annum. Suppose the riskless interest rate is 7% per annum. Find the market price of interest rate risk and the expected returns from security B per annum. Give a financial argument why the market price of the interest rate risk is usually negative.

Hint: The returns on the stocks and bonds are negatively correlated to changes in interest rates.

7.5 Suppose the price of a bond is dependent on the price of a commodity, denoted by Q_t. Let the stochastic process followed by Q_t be governed by

$$\frac{dQ_t}{Q_t} = \alpha\, dt + \sigma\, dZ_t.$$

By hedging bonds of different maturities, show that the governing equation for the bond price $B(Q, t)$ is given by [see (7.2.8)]

$$\frac{\partial B}{\partial t} + \frac{\sigma^2}{2} Q^2 \frac{\partial^2 B}{\partial Q^2} + (\alpha - \lambda\sigma) Q \frac{\partial B}{\partial Q} - rB = 0,$$

where λ is the market price of risk and r is the riskless interest rate. Since the commodity is a *traded security* (unlike the interest rate), the price of the commodity should also satisfy the same governing differential equation. Substituting Q for B into the differential equation, show that

$$\alpha - \lambda\sigma = r.$$

Argue why the governing equation for the bond price now takes the same form as the Black–Scholes equation, which has no dependence on the risk preference.

7.6 From the bond price representation formula (7.2.10), use Ito's differentiation to show

$$\left. \frac{\partial^2 B}{\partial T^2} \right|_{T=t} = r^2(t) - 2 \left. \frac{\partial R}{\partial T} \right|_{T=t},$$

where $R(t, T)$ is the yield to maturity. Also, try to relate the market price of interest rate risk $\lambda(r, t)$ to $\frac{\partial R}{\partial T}|_{T=t}$.

7.7 Suppose the dynamics of the short rate $r(t)$ is governed by

$$dr(t) = \mu(r, t)\, dt + \rho(r, t)\, dZ(t),$$

the governing differential equation for the price of a zero coupon bond $B(r, t)$ is given by [see (7.2.8)]

$$\frac{\partial B}{\partial t} + \frac{\rho^2}{2} \frac{\partial^2 B}{\partial r^2} + (\mu - \lambda\rho) \frac{\partial B}{\partial r} - rB = 0.$$

For any noncoupon bearing paying claim whose payoff depends on $r(T)$, its price function $U(r, t)$ is governed by the same differential equation as above. Now, suppose we relate the price of the claim to the bond price by defining

$$V(B(r, t), t) = U(r, t),$$

show that $V(B, t)$ satisfies

$$\frac{\partial V}{\partial t} + \frac{\sigma_B^2}{2} B^2 \frac{\partial^2 V}{\partial B^2} + rB \frac{\partial V}{\partial B} - rV = 0,$$

where the volatility of bond returns σ_B is given by

$$\sigma_B(r, t) = -\frac{\rho(r, t)}{B(r, t)} \frac{\partial B}{\partial r}(r, t).$$

Suppose the claim's payoff is $f(B_T)$ at maturity T, by applying the Feynman–Kac Theorem, show that $V(B, t)$ admits the following representation

$$V(B,t) = E_Q\left[e^{-\int_t^T r(u)\,du} f(B_T)\middle| B_t = B\right],$$

where the measure Q is defined so that

$$\frac{dB(r,t)}{B(r,t)} = r(t)\,dt + \sigma_B(r,t)\,dZ(t).$$

7.8 Suppose the *duration* D of a coupon-bearing coupon bond B at the current time t is defined by

$$D(B,t) = \left[\sum_{i=1}^{n} c_i(t_i - t)e^{-R(t_i-t)} + (t_n - t)Fe^{-R(t_n-t)}\right]\Big/ B,$$

where $c_i, i = 1, 2, \cdots, n$, is the ith coupon on the bond paid at time t_i, F is the face value. Here, R is the yield to maturity on the bond, which is given by the solution to

$$B = \sum_{i=1}^{n} c_i e^{-R(t_i-t)} + Fe^{-R(t_n-t)}.$$

Show that

$$D(B,t) = -\frac{1}{B}\frac{\partial B}{\partial R}.$$

Give a financial interpretation of duration.

7.9 Recall that

$$B(r,t) = \exp\left(-\int_t^T F(t,u)\,du\right) = E_Q^t\left[\exp\left(-\int_t^T r(u)\,du\right)\right],$$

show that the forward rate is given by

$$F(t,T) = \frac{E_Q^t[r(T)d(t,T)]}{E_Q^t[d(t,T)]},$$

where the stochastic discount factor is defined by

$$d(t,T) = \exp\left(-\int_t^T r(u)\,du\right).$$

7.10 Suppose the forward rate as a function of time t evolves as

$$dF(t,T) = \mu(t,T)\,dt + \sigma\,dZ_t,$$

where $\mu(t,T)$ is a deterministic function of t and T. Show that the forward rate is normally distributed, where

$$F(t, T) = F(0, T) + \int_0^t \mu(u, T)\, du + \sigma Z_t.$$

Explain why $F(t, T) - F(t, S)$ is purely deterministic. Deduce that the forward rates at different maturities are perfectly correlated.

7.11 Consider the linear stochastic differential equation

$$dr(t) = [a(t)r(t) + b(t)]\, dt + \rho(t)\, dZ(t).$$

Show that the mean $E[r(t)]$ is governed by the following deterministic linear differential equation:

$$\frac{d}{dt} E[r(t)] = a(t)E[r(t)] + b(t),$$

while the variance $\text{var}(r(t))$ is governed by

$$\frac{d}{dt} \text{var}(r(t)) = 2a(t)\text{var}(r(t)) + \rho(t)^2.$$

When the results are applied to the CIR interest rate model:

$$dr(t) = \alpha[\gamma - r(t)]\, dt + \rho\sqrt{r(t)}\, dZ(t),$$

show that $E[r(T)|r(t)]$ and $\text{var}(r(T)|r(t))$ are given by (7.2.30a,b).

7.12 Consider the yield curve associated with the Vasicek model [see (7.2.26)]. Show that the yield curve is monotonically increasing when $r(t) \leq R_\infty - \frac{\rho^2}{4\alpha^2}$, monotonically decreasing when $r(t) \geq R_\infty + \frac{\rho^2}{2\alpha^2}$, and it is a bumped curve when $R_\infty - \frac{\rho^2}{4\alpha^2} < r(t) < R_\infty + \frac{\rho^2}{2\alpha^2}$. Further, suppose we define the *liquidity premium* of the term structure as

$$\pi(\tau) = F(t, t + \tau) - E_Q^t[r(t + \tau)], \quad \tau \geq 0,$$

where $F(t, t + \tau)$ is the forward rate and E_Q^t is the expectation under Q conditional on the filtration $\mathcal{F}_t$. Show that the liquidity premium for the Vasicek model is given by (Vasicek, 1977)

$$\pi(\tau) = \left(R_\infty - \gamma + \frac{\rho^2}{2\alpha^2} e^{-\alpha\tau}\right)\left(1 - e^{-\alpha\tau}\right), \quad \tau \geq 0.$$

7.13 Consider the following discrete version of the Vasicek model when the short rate $r(t)$ follows the discrete mean reversion binary random walk

$$r(t + 1) = \rho r(t) + \alpha \pm \sigma.$$

Let $V(t)$ denote the value of an interest rate contingent claim at current time t, $V_u(t+1)$ and $V_d(t+1)$ be the corresponding values of the contingent claim at time $t+1$ when the short rate moves up and down, respectively. Let $B(t,T)$ be the price of a zero-coupon bond that pays one dollar at time T and observe that $B(t,t+1) = e^{-r(t)}$. By adopting a similar approach as the Cox–Ross–Rubinstein binomial pricing model, show that the binomial formula that relates $V(t)$, $V_u(t+1)$ and $V_d(t+1)$ is given by (Heston, 1995)

$$V(t) = \frac{pV_u(t+1) + (1-p)V_d(t+1)}{e^{r(t)}},$$

where

$$p = \frac{e^{r(t)} - d}{u-d}, \qquad u = \frac{e^{-[\alpha+\rho r(t)+\sigma]}}{B(t,t+2)}, \qquad d = \frac{e^{-[\alpha+\rho r(t)-\sigma]}}{B(t,t+2)}.$$

7.14 Consider the pricing of a futures contract on a discount bond, where the short rate r_t is assumed to follow the Vasicek process defined by (7.2.20). On the expiration date T_F of the futures, a bond of unit par with maturity date T_B is delivered. Let $B(r,t;T_B)$ and $V(r,t;T_F,T_B)$ denote, respectively, the bond price and futures price at the current time t. Show that the governing equation for the futures price is given by

$$\frac{\partial V}{\partial t} + \frac{\rho^2}{2}\frac{\partial^2 V}{\partial r^2} + \big[\alpha(\gamma - r) - \lambda\rho\big]\frac{\partial V}{\partial r} = 0, \quad t < T_F,$$

with terminal payoff condition $V(r,T_F;T_F,T_B) = B(r,T_F;T_B)$. By assuming the solution of the futures price to be the form:

$$V(r,t;T_F,T_B) = e^{-rX(t)-Y(t)},$$

show that (Chen, 1992)

$$X(t) = E(t,T_B) - E(t,T_F)$$
$$Y(t) = D[T_B - T_F - X(t)] - \frac{\rho^2}{2\alpha^2}\Big[X(t) - \frac{\alpha}{2}X(t)^2 - E(T_F,T_B)\Big],$$

where

$$D = \gamma - \frac{\rho\lambda}{\alpha} - \frac{\rho^2}{2\alpha^2} \quad \text{and} \quad E(t,T) = \frac{1-e^{-\alpha(T-t)}}{\alpha}.$$

7.15 Show that the steady state density function of the short rate at time T in the Cox–Ingersoll–Ross model is given by (Cox, Ingersoll and Ross, 1985)

$$\lim_{T\to\infty} g(r(T);r(t)) = \frac{\omega^\nu}{\Gamma(\nu)} r^{\nu-1} e^{-\omega r},$$

where $\omega = 2\alpha/\rho^2$ and $\nu = 2\alpha\gamma/\rho^2$. Show that the corresponding steady state mean and variance of $r(T)$ are γ and $\frac{\gamma\rho^2}{2\alpha}$, respectively.

7.16 Show that the bond price for the Cox–Ingersoll–Ross model [see (7.2.32a,b)] is a decreasing convex function of the short rate and a decreasing function of time to maturity. Further, show that the bond price is a decreasing convex function of the mean short rate level γ and an increasing concave function of the speed of adjustment α if $r(t) > \gamma$. What would be the effects on the bond price when the short rate variance ρ^2 and the market price of risk λ increase?

7.17 Consider the yield to maturity $R(t, T)$ corresponding to the Cox–Ingersoll–Ross model. Show that [see (7.2.32a,b)]

$$\lim_{T\to\infty} R(t, T) = \frac{2\alpha\gamma}{\theta + \psi}.$$

Explain why an increase in the current short rate increases yields for all maturities, but the effect is more significant for shorter maturities, while an increase in the steady state mean γ increases all yields but the effect is more significant for longer maturities. What would be the effect on the yields when the short rate variance ρ^2 and the market price of risk λ increase?

7.18 Consider the extended CIR model where the short rate, $r_t \geq 0$, follows the process

$$dr_t = [\alpha(t) - \beta(t)r_t]\,dt + \sigma(t)\sqrt{r_t}\,dZ_t$$

for some smooth deterministic functions $\alpha(t), \beta(t)$ and $\sigma(t) > 0$, and Z_t is a Brownian process under the risk neutral measure. Show that the bond price function is given by (Jamshidian, 1995)

$$B(t, T) = \exp\left(-\int_t^T A_T(u)\alpha(u)\,du - A_T(t)r_t\right), \quad t < T,$$

where $A_T(t)$ satisfies the Ricatti equation

$$A_T'(t) = \beta(t)A_T(t) + \frac{\sigma^2(t)}{2}A_T^2(t) - 1, \quad A_T(T) = 0.$$

Also, show that the instantaneous T-maturity forward rate $f(t, T)$ is given by

$$f(t, T) = \int_t^T a_T(u)\,du + a_T(t)r_t, \quad f(T, T) = r,$$

where

$$a_T'(t) = [\beta(t) + \sigma^2(t)A_T(t)]a_T(t), \quad a_T(T) = 1.$$

7.19 Consider the continuous time equivalent of the Ho–Lee model as a degenerate case of the Hull–White model, where the diffusion process for the short rate r_t under the risk neutral measure Q is given by

$$dr_t = \theta(t)\,dt + \sigma\,dZ_t.$$

Show that the parameter $\theta(t)$ is related to the slope of the initial forward rate curve through the following formula

$$\theta(t) = \frac{\partial F}{\partial T}(0, T)\big|_{T=t} + \sigma^2 t.$$

The bond price function can be shown to admit the affine form where

$$B(t, T) = e^{a(t,T)-b(t,T)r}.$$

Show that

$$b(t, T) = T - t$$
$$a(t, T) = \ln \frac{B(0, T)}{B(0, t)} + (T - t)F(0, t) - \frac{\sigma^2}{2} t(T - t)^2.$$

7.20 Consider the Hull–White model where the short rate process follows

$$dr_t = [\phi(t) - \alpha r_t]\, dt + \sigma\, dZ_t,$$

where Z_t is a Brownian process under the risk neutral measure Q. Using the relation

$$d[r(t)e^{\alpha t}] = \phi(t)e^{\alpha t}\, dt + \sigma e^{\alpha t}\, dZ(t),$$

show that

$$\begin{aligned}\int_t^T r(u)\, du &= r(t)\frac{1 - e^{-\alpha(T-t)}}{\alpha} + \int_t^T \int_t^u \phi(u)e^{-\alpha(u-s)} ds\, du \\ &\quad + \int_t^T \int_t^u \sigma e^{-\alpha(u-s)} dZ(s)\, du \\ &= r(t)\frac{1 - e^{-\alpha(T-t)}}{\alpha} + \int_t^T \phi(u)\frac{1 - e^{-\alpha(T-u)}}{\alpha}\, du \\ &\quad + \int_t^T \frac{\sigma}{\alpha}[1 - e^{-\alpha(T-s)}]\, dZ(s).\end{aligned}$$

Accordingly, the mean and variance of $\int_t^T r(u)\, du$ are found to be

$$E_Q^t\left[\int_t^T r(u)\, du\right] = r(t)\frac{1 - e^{-\alpha(T-t)}}{\alpha} + \int_t^T \phi(u)\frac{1 - e^{-\alpha(T-u)}}{\alpha}\, du$$
$$\operatorname{var}\left(\int_t^T r(u)\, du\right) = \int_t^T \frac{\sigma^2}{\alpha^2}[1 - e^{-\alpha(T-u)}]^2\, du.$$

Compute the bond price $B(r, t; T)$.

7.21 The expression for $a(t, T)$ derived from in (7.2.38b) involves $\phi(t)$. It may be desirable to express $a(t, T)$ solely in terms of the initial bond prices $B(0, T)$

for all maturities. Show that (Hull and White, 1994)

$$a(t,T) = \ln \frac{B(0,T)}{B(0,t)} - b(t,T)\frac{\partial}{\partial t} \ln B(0,t) \\ - \frac{\sigma^2}{4\alpha^3}(e^{-\alpha T} - e^{-\alpha t})^2(e^{2\alpha t} - 1).$$

7.22 Consider the Hull–White model where the short rate is defined by

$$dr(t) = [\phi(t) - \alpha r(t)]\, dt + \sigma\, dZ(t).$$

Suppose we define a new variable $x(t)$ where

$$dx(t) = -\alpha x(t)\, dt + \sigma\, dZ(t),$$

and let $\psi(t) = r(t) - x(t)$. Show that $\phi(t)$ and $\psi(t)$ are related by

$$\psi'(t) + \alpha\psi(t) = \phi(t), \quad \psi(0) = r(0).$$

Let $Y(t) = R(0,t)$ where $R(t,T)$ is the yield to maturity. Show that

$$\psi(t) = \frac{d}{dt}[tY(t)] + \frac{\sigma^2}{2\alpha^2}(1 - e^{-\alpha t})^2, \quad t \geq 0.$$

Also, show that the bond price $B(t,T)$ can be expressed as (Kijima and Nagayama, 1994)

$$\ln B(t,T) = \ln \frac{B(0,T)}{B(0,t)} + \frac{1}{\alpha}[e^{-\alpha(T-t)} - 1][r(t) - \psi(t)] \\ + \frac{\sigma^2}{4\alpha^3}\{1 - [2 - e^{-\alpha(T-t)}]^2 + (2 - e^{-\alpha T})^2 - (2 - e^{-\alpha t})^2\}.$$

7.23 An extended version of the Vasicek model takes the form (Hull and White, 1990)

$$dr_t = [\theta(t) + \alpha(t)(d - r_t)]\, dt + \sigma(t)\, dZ_t.$$

Let $\lambda(t)$ denote the time dependent market price of risk. Show that the bond price equation is given by

$$\frac{\partial B}{\partial t} + [\phi(t) - \alpha(t)r]\frac{\partial B}{\partial r} + \frac{\sigma(t)^2}{2}\frac{\partial^2 B}{\partial r^2} - rB = 0,$$

where

$$\phi(t) = \alpha(t)d + \theta(t) - \lambda(t)\sigma(t).$$

Suppose we write the bond price $B(r,t;T)$ in the form

$$B(r,t;T) = e^{a(t,T) - b(t,T)r}.$$

Show that $a(t, T)$ and $b(t, T)$ are governed by

$$\frac{\partial a}{\partial t} - \phi(t)b + \frac{\sigma(t)^2}{2}b^2 = 0$$
$$\frac{\partial b}{\partial t} - \alpha(t)b + 1 = 0,$$

with auxiliary conditions:

$$a(T, T) = 1 \quad \text{and} \quad b(T, T) = 0.$$

Solve for $a(t, T)$ and $b(t, T)$ in terms of $\alpha(t)$, $\phi(t)$ and $\sigma(t)$. It is desirable to express $a(t, T)$ and $b(t, T)$ in terms of $a(0, t)$ and $b(0, t)$ instead of $\alpha(t)$ and $\phi(t)$. Show that the new set of governing equations for $a(t, T)$ and $b(t, T)$, independent of $\alpha(t)$ and $\phi(t)$, are given by

$$\frac{\partial b}{\partial t}\frac{\partial b}{\partial T} - b\frac{\partial^2 b}{\partial t \partial T} + \frac{\partial b}{\partial T} = 0$$
$$ab\frac{\partial^2 a}{\partial t \partial T} - b\frac{\partial a}{\partial t}\frac{\partial a}{\partial t} - a\frac{\partial a}{\partial t}\frac{\partial b}{\partial T} + \frac{\sigma(t)^2}{2}a^2 b^2 \frac{\partial b}{\partial T} = 0.$$

The auxiliary conditions are the known values of $a(0, T)$ and $b(0, T)$, $a(T, T) = 1$ and $b(T, T) = 0$. Finally, show that the solutions for $b(t, T)$ and $a(t, T)$, expressed in terms of $b(0, T)$ and $a(0, T)$, are given by

$$b(t, T) = \frac{b(0, T) - b(0, t)}{\frac{\partial b}{\partial T}(0, T)\big|_{T=t}}$$
$$a(t, T) = \frac{a(0, T)}{a(0, t)} - b(t, T)\frac{\partial}{\partial T}[a(0, T)]\big|_{T=t}$$
$$- \frac{1}{2}\left[b(t, T)\frac{\partial b}{\partial T}(0, T)\big|_{T=t}\right]^2 \int_0^t \left[\frac{\sigma(u)}{\frac{\partial b}{\partial T}(0, T)\big|_{T=u}}\right]^2 du.$$

7.24 Hull and White (1994) proposed the following two-factor short rate model whose dynamics under the risk neutral measure are governed by

$$dr(t) = [\phi(t) + u(t) - ar(t)]\,dt + \sigma_1\,dZ_1(t),$$

where u has an initial value of zero and follows the process

$$du(t) = -bu(t)\,dt + \sigma_2\,dZ_2(t).$$

The parameters a, b, σ_1 and σ_2 are constants and $dZ_1\,dZ_2 = \rho\,dt$, where ρ is the instantaneous correlation coefficient. Show that the zero-coupon bond price $B(t, T)$ takes the form

$$B(t, T) = \alpha(t, T)\exp(-\beta(t, T)r - \gamma(t, T)u).$$

Find the governing equations for $\alpha(t, T)$, $\beta(t, T)$ and $\gamma(t, T)$.

Hint: $\beta(t, T)$ and $\gamma(t, T)$ are readily found to be

$$\beta(t, T) = \frac{1}{a}\left[1 - e^{-a(T-t)}\right]$$
$$\gamma(t, T) = \frac{1}{a(a-b)}e^{-a(T-t)} - \frac{1}{b(a-b)}e^{-b(T-t)} + \frac{1}{ab}.$$

7.25 Suppose the dynamics of the short rate $r(t)$ are governed by

$$dr(t) = k_r[\theta(t) - r(t)]\,dt + \sigma_r\,dZ_r(t)$$
$$d\theta(t) = k_\theta[\overline{\theta} - \theta(t)]\,dt + \sigma_\theta\,dZ_\theta(t),$$

where the short rate mean reverts to a drift rate $\theta(t)$, which itself reverts to a fixed mean rate $\overline{\theta}$, $dZ_r\,dZ_\theta = \rho\,dt$, and all other parameters are constant (k_r and k_θ are both positive). Show that the expected value of $r(t)$ is given by (Beaglehole and Tenney, 1991)

$$E[r(t)] = r(0)e^{-k_r t} + \theta(0)\frac{k_r}{k_r - k_\theta}(e^{-k_\theta t} - e^{-k_r t})$$
$$+ \frac{k_r k_\theta}{k_r - k_\theta}\overline{\theta}\left(\frac{1 - e^{-k_\theta t}}{k_\theta} - \frac{1 - e^{-k_r t}}{k_r}\right).$$

7.26 Assume that the dynamics of the short rate process under the risk neutral measure is governed by

$$r(t) = x_1(t) + x_2(t) + \phi(t), \quad r(0) = r_0,$$

and

$$dx_1 = -\alpha_1 x_1(t)\,dt + \sigma_1\,dZ_1(t), \quad x_1(0) = 0,$$
$$dx_2 = -\alpha_2 x_2(t)\,dt + \sigma_2\,dZ_2(t), \quad x_2(0) = 0,$$

with $dZ_1(t)\,dZ_2(t) = \rho\,dt$. Show that the time-$t$ price of a unit par discount bond is given by

$$B(r, t) = \exp\left(-\int_t^T \phi(u)\,du - \frac{1 - e^{-\alpha_1(T-t)}}{\alpha_1}x_1(t)\right.$$
$$\left. - \frac{1 - e^{-\alpha_2(T-t)}}{\alpha_2}x_2(t) + \frac{V(t, T)}{2}\right),$$

where

$$V(t, T) = \frac{\sigma_1^2}{\alpha_1^2}\left[T - t + \frac{2}{\alpha_1}e^{-\alpha_1(T-t)} - \frac{e^{-2\alpha_1(T-t)}}{2\alpha_1} - \frac{3}{2\alpha_1}\right]$$

$$+\frac{\sigma_2^2}{\alpha_2^2}\left[T-t+\frac{2}{\alpha_2}e^{-\alpha_2(T-t)}-\frac{e^{-\alpha_2(T-t)}}{2\alpha_2}-\frac{3}{2\alpha_2}\right]$$

$$+\frac{2\rho\sigma_1\sigma_2}{\alpha_1\alpha_2}\left[T-t+\frac{e^{-\alpha_1(T-t)}-1}{\alpha_1}+\frac{e^{-\alpha_2(T-t)}-1}{\alpha_2}\right.$$

$$\left.-\frac{e^{-(\alpha_1+\alpha_2)(T-t)}-1}{\alpha_1+\alpha_2}\right].$$

Let $f_m(0,T)$ denote the term structure of the forward rates at time 0 for maturity T as implied by the market bond prices. Show that the parameter function $\phi(t)$ can be calibrated to $f_m(0,T)$ via the relation

$$\phi(T)=f_m(0,T)+\frac{\sigma_1^2}{2\alpha_1^2}(1-e^{-\alpha_1 T})^2+\frac{\sigma_2^2}{2\alpha_2^2}(1-e^{-\alpha_2 T})^2$$

$$+\frac{\rho\sigma_2\sigma_2}{\alpha_1\alpha_2}(1-e^{-\alpha_1 T})(1-e^{-\alpha_2 T}).$$

7.27 Empirical evidence reveals that the long rate and the spread (short rate minus long rate) are almost uncorrelated. Suppose we choose the stochastic state variables in the two-factor interest rate model to be the spread s and the long rate ℓ, where

$$ds=\beta_s(s,\ell,t)\,dt+\eta_s(s,\ell,t)\,dZ_s,\qquad s=r-\ell,$$

$$d\ell=\beta_\ell(s,\ell,t)\,dt+\eta_\ell(s,\ell,t)\,dZ_\ell,$$

where r is the short rate. Assuming zero correlation between the above processes, show that the price of a default free bond $B(s,\ell,\tau)$ is governed by

$$\frac{\partial B}{\partial\tau}=\frac{\eta_s^2}{2}\frac{\partial^2 B}{\partial s^2}+\frac{\eta_\ell^2}{2}\frac{\partial^2 B}{\partial\ell^2}+(\beta_s-\lambda_s\eta_s)\frac{\partial B}{\partial s}$$

$$+\left(\frac{\eta_\ell^2}{\ell}-s\ell\right)\frac{\partial B}{\partial\ell}-(s+\ell)B,$$

where λ_s is the market price of spread risk and the market price of long rate risk is given by (7.3.14). Schaefer and Schwartz (1984) proposed the following specified stochastic processes for s and ℓ

$$ds=m(\mu-s)\,dt+\gamma\,dZ_s$$

$$d\ell=\beta_\ell(s,\ell,t)\,dt+\sigma\sqrt{\ell}\,dZ_\ell.$$

Show that the above bond price equation becomes

$$\frac{\partial B}{\partial\tau}=\frac{\gamma^2}{2}\frac{\partial^2 B}{\partial s^2}+\frac{\sigma^2\ell}{2}\frac{\partial^2 B}{\partial\ell^2}+(m\mu-\lambda\gamma-ms)\frac{\partial B}{\partial s}$$

$$+(\sigma^2-\ell s)\frac{\partial B}{\partial\ell}-(s+\ell)B.$$

The payoff function at maturity is $B(s, \ell, 0) = 1$. The following analytic approximation procedure is proposed to solve the above equation. They take the coefficient of $\frac{\partial B}{\partial \ell}$ to be constant by treating s as a frozen constant $\widehat{s}$. Now, we write the bond price as the product of two functions, namely,

$$B(s, \ell, \tau) = X(s, \tau)Y(\ell, \tau).$$

Show that the bond price equation can be split into the following pair of equations:

$$\frac{\partial X}{\partial \tau} = \frac{\gamma^2}{2}\frac{\partial^2 X}{\partial s^2} + (m\mu - \lambda\gamma - ms)\frac{\partial X}{\partial s} - sX, \qquad X(s, 0) = 1,$$

and

$$\frac{\partial Y}{\partial \tau} = \frac{\sigma^2 \ell}{2}\frac{\partial^2 Y}{\partial \ell^2} + (\sigma^2 - \ell\widehat{s})\frac{\partial Y}{\partial \ell} - \ell Y, \qquad Y(\ell, 0) = 1.$$

Assuming that all parameters are constant, solve the above two equations for $X(s, \tau)$ and $Y(\ell, \tau)$.

7.28 Consider the multifactor extension of the CIR model, where the short rate $r(t)$ is defined by

$$r(t) = \sum_{i=1}^{n} X_i(t).$$

Here, $X_i(t), i = 1, \cdots, n$, are uncorrelated processes of the one-factor CIR type as governed by

$$dX_i(t) = \alpha_i[\gamma_i - X_i(t)]\,dt + \rho_i\sqrt{X_i(t)}\,dZ_i(t)$$

under the risk neutral measure. Show that the bond price function $B(t, T)$ is given by

$$B(t, T) = \exp\left(\sum_{i=1}^{n} a_i(T - t) - \sum_{i=1}^{n} b_i(T - t)X_i(t)\right),$$

where $\tau = T - t$ and

$$a_i(\tau) = \frac{2\alpha_i\gamma_i}{\rho_i^2}\ln\left(\frac{2\theta_i e^{(\theta_i+\alpha_i)\tau/2}}{(\theta_i + \alpha_i)(e^{\theta_i\tau} - 1) + 2\theta_i}\right),$$

$$b_i(\tau) = \frac{2(e^{\theta_i\tau} - 1)}{(\theta_i + \alpha_i)(e^{\theta_i\tau} - 1) + 2\theta_i},$$

$$\theta_i = \sqrt{\alpha_i^2 + 2\rho_i^2}, \quad i = 1, 2, \cdots, n.$$

7.29 For the two-factor CIR model proposed by Longstaff and Schwartz (1992), the short rate $r(t)$ is defined by

$$r = \alpha x + \beta y,$$

where α and β are positive constants, and $\alpha \neq \beta$. Under the risk neutral measure, the risk factors x and y are governed by

$$dx = (\gamma - \delta x)\,dt + \sqrt{x}\,dZ_1$$
$$dy = (\eta - \xi y)\,dt + \sqrt{y}\,dZ_2,$$

where Z_1 and Z_2 are uncorrelated standard Brownian processes. Let V denote the instantaneous variance of changes in the short rate.

(a) Show that

$$V = \alpha^2 x + \beta^2 y.$$

(b) Using Ito's lemma, show that the dynamics of r and V are governed by

$$dr = \left(\alpha\gamma + \beta\eta - \frac{\beta\delta - \alpha\xi}{\beta - \alpha} r - \frac{\xi - \delta}{\beta - \alpha} V\right) dt + \alpha\sqrt{\frac{\beta r - V}{\alpha(\beta - \alpha)}}\,dZ_1 + \beta\sqrt{\frac{V - \alpha r}{\beta(\beta - \alpha)}}\,dZ_2$$

$$dV = \left(\alpha^2\gamma + \beta^2\eta - \frac{\alpha\beta(\delta - \xi)}{\beta - \alpha} r - \frac{\beta\xi - \alpha\delta}{\beta - \alpha} V\right) dt + \alpha^2\sqrt{\frac{\beta r - V}{\alpha(\beta - \alpha)}}\,dZ_1 + \beta^2\sqrt{\frac{V - \alpha\gamma}{\beta(\beta - \alpha)}}\,dZ_2.$$

Note that r and V together form a joint Markov process.

(c) Show that r has a long-run stationary (unconditional) distribution with mean

$$E[r] = \frac{\alpha\gamma}{\delta} + \frac{\beta\eta}{\xi}$$

and variance

$$\text{var}(r) = \frac{\alpha^2\gamma}{2\delta^2} + \frac{\beta^2\eta}{2\xi^2}.$$

Similarly, show that V also has a stationary distribution with mean

$$E[V] = \frac{\alpha^2\gamma}{\delta} + \frac{\beta^2\eta}{\xi}$$

and variance

$$\text{var}(V) = \frac{\alpha^4\gamma}{2\delta^2} + \frac{\beta^4\eta}{2\xi^2}.$$

(d) Let $B(r, V, \tau)$ denote the price function of a unit discount bond with τ periods until maturity. Show that

$$B(r, V, \tau) = A^{2\gamma}(\tau) B^{2\eta}(\tau) \exp(\kappa\tau + C(\tau)r + D(\tau)V),$$

where

$$A(\tau) = \frac{2\phi}{(\delta + \phi)(\exp(\phi\tau) - 1) + 2\phi},$$

$$B(\tau) = \frac{2\psi}{(\upsilon + \psi)(\exp(\psi\tau) - 1) + 2\psi},$$

$$C(\tau) = \frac{\alpha\phi(\exp(\psi\tau) - 1)B(\tau) - \beta\psi(\exp(\phi\tau) - 1)A(\tau)}{\phi\psi(\beta - \alpha)},$$

$$D(\tau) = \frac{\psi(\exp(\phi\tau) - 1)A(\tau) - \phi(\exp(\psi\tau) - 1)B(\tau)}{\phi\psi(\beta - \alpha)},$$

and

$$\upsilon = \xi + \lambda, \qquad \phi = \sqrt{2\alpha + \delta^2},$$

$$\psi = \sqrt{2\beta + \upsilon^2}, \quad \kappa = \gamma(\delta + \phi) + \eta(\upsilon + \psi).$$

(e) Suppose $\alpha < \beta$, show that $V(t)$ is limited to the range $(\alpha r(t), \beta r(t))$ at any time t.

7.30 Consider the three-factor stochastic volatility model [see (7.3.17)], by assuming constant market prices of risk λ_r, $\lambda_{\overline{r}}$ and λ_v, show that the bond price function $B(t, T)$ satisfies the partial differential equation

$$\begin{aligned}\frac{\partial B}{\partial \tau} &= \frac{\xi^2 v}{2}\frac{\partial^2 B}{\partial v^2} + \frac{\eta^2}{2}\frac{\partial^2 B}{\partial \overline{r}^2} + \frac{v}{2}\frac{\partial^2 B}{\partial r^2} + \rho\xi v \frac{\partial^2 B}{\partial r \partial v} \\ &\quad + [\alpha(\overline{r} - r) - \lambda_r\sqrt{v}]\frac{\partial B}{\partial r} + [\beta(\theta - \overline{r}) - \lambda_{\overline{r}}\eta]\frac{\partial B}{\partial \overline{r}} \\ &\quad + [\gamma(\overline{v} - v) - \lambda_v\xi\sqrt{v}]\frac{\partial B}{\partial v} - rB, \quad \tau = T - t.\end{aligned}$$

Suppose the discount bond price function admits the following exponential affine term structure

$$B(t, T) = a(\tau)\exp(-b(\tau)r - c(\tau)\overline{r} - d(\tau)v),$$

show that

$$b(\tau) = \frac{1 - e^{-\alpha\tau}}{\alpha}, \quad c(\tau) = \frac{1 - e^{-\alpha\tau} + \frac{\alpha}{\beta}e^{-\beta\tau}(1 - e^{-\beta\tau})}{\beta - \alpha},$$

$$a'(\tau) = \frac{\eta^2}{2}c(\tau)^2 - \gamma\overline{v}a(\tau)^2 d(\tau) - \beta\theta\tau a(\tau)c(\tau) + \lambda_{\overline{r}}a(\tau)c(\tau),$$

$$\begin{aligned}&d'(\tau) + \frac{\xi^2}{2}d(\tau)^2 + \rho\xi b(\tau)d(\tau) + \gamma d(\tau) + \frac{b(\tau)^2}{2} \\ &\quad + \lambda_r b(\tau) + \lambda_v d(\tau) = 0.\end{aligned}$$

7.31 Consider the generalized multifactor Vasicek model with constant parameters (Babbs and Nowman, 1999), where the short rate is characterized by

$$r(t) = \mu - \sum_{i=1}^{I} X_i(t).$$

Here, $X_i(t)$ are stochastic state variables whose dynamics under the risk neutral measure are governed by

$$dX_i(t) = -\xi_i X_i(t)\,dt + c_i\,dW_i$$

with $dW_i\,dW_j = \rho_{ij}\,dt$. The parameters $\mu, \xi_i, c_i, \rho_{ij}$ are all constant. Suppose we rewrite the stochastic differential equation of $dX_i(t)$ as

$$dX_i(t) = -\xi_i X_i(t)\,dt + \sum_{j=1}^{J} \sigma_{ij}\,dZ_j(t),$$

where $Z_1, \cdots, Z_J$ are independent standard Brownian processes.

(a) Show that

$$\sum_{j=1}^{J} \sigma_{ij}\sigma_{kj} = \rho_{ik}c_i c_k.$$

(b) Show that the discount bond price $B(t,T)$ can be found to be

$$B(t,T) = \exp\left(-\tau\left[R(\infty) - w(\tau) - \sum_{i=1}^{I} H(\xi_i\tau)X_i(t)\right]\right),$$

where $\tau = T - t$, $H(x) = (1 - e^{-x})/x$, and

$$R(\infty) = \mu - \frac{1}{2}\sum_{j=1}^{J}\left(\sum_{i=1}^{I}\frac{\sigma_{ij}}{\xi_i}\right)^2,$$

$$\begin{aligned} w(\tau) = &- \sum_{i=1}^{I} H(\xi_i\tau)\sum_{j=1}^{J}\sum_{k=1}^{I}\frac{\sigma_{ij}\sigma_{kj}}{\xi_i\xi_k} \\ &+ \frac{1}{2}\sum_{k=1}^{I}\sum_{i=1}^{I} H((\xi_i + \xi_k)\tau)\sum_{j=1}^{J}\frac{\sigma_{ij}\sigma_{kj}}{\xi_i\xi_k}. \end{aligned}$$

7.32 Consider the following general formulation of the quadratic term structure model (Jamshidian, 1996), where the short rate is defined by

$$r(t) = \frac{1}{2}\mathbf{x}(t)^T Q(t)\mathbf{x}(t) + \mathbf{g}(t)^T\mathbf{x}(t) + f(t),$$

where $\mathbf{x}(t) = (x_1(t) \cdots x_m(t))^T$ is an m-component vector, $Q(t)$ is a symmetric matrix, $\mathbf{g}(t)$ is a vector function and $f(t)$ is a scalar function. All parameters $Q(t)$, $\mathbf{g}(t)$ and $f(t)$ are smooth and deterministic functions of t. Under the risk neutral measure, the stochastic state vector $\mathbf{x}(t)$ follows the Gaussian process as defined by

$$d\mathbf{x} = [\boldsymbol{\alpha}(t) - \beta(t)\mathbf{x}]\, dt + \sigma(t)\, d\mathbf{Z},$$

for some smooth deterministic vector $\boldsymbol{\alpha}(t)$, matrices $\beta(t)$ and $\sigma(t)$.

(a) Show that the governing partial differential equation for the price of a contingent claim $C(\mathbf{x}, t)$ is given by

$$\frac{\partial C}{\partial t} + (\boldsymbol{\alpha} - \beta\mathbf{x})^T \frac{\partial C}{\partial \mathbf{x}} + \frac{1}{2} tr\left(\sigma^T \frac{\partial^2 C}{\partial \mathbf{x}^2} \sigma\right) - rC = 0.$$

(b) Show that the price of a T-maturity discount bond admits the following exponential affine form

$$B(T, t) = \exp\left(-\frac{1}{2}\mathbf{x}(t)^T B_T(t)\mathbf{x}(t) - \mathbf{b}_T^T(t)\mathbf{x}(t) - c_T(t)\right),$$

where the matrix $B_T(t)$, vector $\mathbf{b}_T(t)$ and scalar $c_T(t)$ are governed by the following coupled system of ordinary differential equations:

$$\frac{dB_T}{dt} = \beta^T B_T + B_T^T \beta + B_T^T \sigma\sigma^T B_T - Q, \qquad B_T(T) = 0,$$
$$\frac{d\mathbf{b}_T}{dt} - (\beta + \sigma\sigma^T B_T)^T \mathbf{b}_T + B_T^T \boldsymbol{\alpha} + \mathbf{g} = 0, \qquad \mathbf{b}_T(T) = 0,$$
$$\frac{dc_T}{dt} + \boldsymbol{\alpha}^T \mathbf{b}_T + \frac{1}{2} tr(\sigma^T B_T \sigma) - \frac{1}{2}\mathbf{b}_T^T \sigma\sigma^T \mathbf{b}_T + f = 0, \qquad c_T(T) = 0.$$

7.33 Suppose the forward rate $\widehat{F}(t, \tau)$ is defined in terms of running time t and time to maturity τ (instead of maturity date T), where

$$\widehat{F}(t, \tau) = F(t, t + \tau).$$

Under the one-factor HJM framework, we write $\sigma_{\widehat{F}}(t, \tau) = \sigma_F(t, t+\tau)$. Show that the dynamics of $\widehat{F}(t, \tau)$ under the risk neutral measure Q are given by (Brace and Musiela, 1994)

$$d\widehat{F}(t, \tau) = \left[\frac{\partial}{\partial \tau}\widehat{F}(t, \tau) + \sigma_{\widehat{F}}(t, \tau) \int_0^\tau \sigma_{\widehat{F}}(t, u)\, du\right] dt + \sigma_{\widehat{F}}(t, \tau)\, dZ(t).$$

7.34 Under the one-factor HJM framework, show that

$$E_Q^t[r(T)] = F(t,T) + E_Q^t\left[\int_t^T \left(\sigma_F(u,T)\int_u^T \sigma_F(u,s)\,ds\right)du\right],$$

where E_Q^t denotes the expectation under the risk neutral measure Q conditional on information $\mathcal{F}_t$. Explain why the forward rate $F(t,T)$ is a biased estimator of $r(T)$ under Q.

7.35 Following the n-factor HJM framework, show that the dynamics of $r(t)$ under the risk neutral measure [see (7.4.8b)] can be expressed as

$$dr(t) = \frac{\partial}{\partial T}F(t,T)\bigg|_{T=t} + \sum_{i=1}^n \sigma_F^i(t,T)\,dZ_i(t),$$

where $F(t,T)$ is defined in (7.4.7). Can you provide a financial interpretation of the result?

7.36 Following the n-factor HJM framework, let the forward rate $F(t,T)$ be governed by the following dynamics

$$F(t,T) = F(0,T) + \int_0^t \alpha(u,T)\,du + \sum_{i=1}^n \int_0^t \sigma_i(u,T)\,dZ_i(u).$$

Show that the covariance of the increments of $F(t,T_1)$ and $F(t,T_2)$ is given by

$$\sum_{i=1}^n \sigma_i(t,T_1)\sigma_i(t,T_2).$$

Deduce that under the one-factor model, which corresponds to $n = 1$, the changes in $F(t,T_1)$ and $F(t,T_2)$ are fully correlated.

7.37 Using the analytic representation of $F(t,T)$ in (7.4.7), show that the discount bond price can be expressed as

$$B(t,T) = \frac{B(0,T)}{B(0,t)}\exp\left(-\sum_{i=1}^n\left[\int_t^T\int_0^t \alpha_F^i(u,s)\,du\,ds \right.\right.$$
$$\left.\left. + \int_t^T\int_0^t \sigma_F^i(u,s)\,dZ_i(u)\,ds\right]\right).$$

7.38 Consider the two-factor Gaussian model, which is a combination of the Ho–Lee and Vasicek models. Let the volatility structure in the HJM framework be given by

$$\sigma_F^1(t,T) = \sigma_2 \quad \text{and} \quad \sigma_F^2(t,T) = \sigma_2 e^{-k(T-t)}.$$

Show that the bond price $B(t,T)$ is given by (Heath, Jarrow and Morton, 1992)

$$B(t,T) = \frac{B(0,T)}{B(0,t)} \exp\bigg(-M_1(t,T) - M_2(t,T) - \sigma_1(T-t)Z_1(t) - \frac{\sigma_2}{k}[1-e^{-k(T-t)}]X_2(t)\bigg),$$

where

$$X_2(t) = \int_0^t e^{-k(t-u)}\,dZ_2(u)$$

$$M_1(t,T) = \frac{\sigma_1^2}{2} tT(T-t)$$

$$M_2(t,T) = \frac{\sigma_2^2}{4k^3}\{[1-e^{-k(T-t)}]^2(1-e^{-2kt}) + 2[1-e^{-k(T-t)}(1-e^{-kt})^2\}.$$

Also, show that the yield to maturity $R(t,T)$ is normally distributed with mean $\mu_R(t,T)$ and variance $\sigma_R(t,T)^2$:

$$\mu_R(t,T) = -\frac{\ln\frac{B(0,T)}{B(0,t)}}{T-t} + \frac{M_1(t,T)}{T-t} + \frac{M_2(t,T)}{T-t}$$

$$\sigma_R(t,T)^2 = \sigma_1^2 t + \frac{\sigma_2^2}{2k^3}\left[\frac{1-e^{-k(T-t)}}{T-t}\right]^2(1-e^{-2kt}).$$

7.39 Let $\sigma_B(t,T)$ denote the volatility structure of the return of a discount bond. The Gaussian term structure models are characterized by (i) deterministic $\sigma_B(t,T)$ and (ii) a Markov short rate process. Show that a necessary and sufficient condition for a one-factor HJM model to be Gaussian is given by (Hull and White, 1993a)

$$\sigma_B(t,T) = \alpha(t)\beta(t,T),$$

where

$$\beta(t,T) = \frac{y(T)-y(t)}{y'(t)}.$$

7.40 Under the one-factor Inui–Kijima model [see (7.4.23)], we would like to solve for $B(t,T)$ in terms of $r(t)$, $\phi(t)$, $F(0,t)$ and other parameter functions. Define

$$\beta(t,T) = \int_t^T e^{-\int_t^u k(s)\,ds}\,du, \quad t \le T,$$

show that

$$B(t,T)=\frac{B(0,T)}{B(0,t)}\exp\left(-\frac{\beta(t,T)^2}{2}\phi(t)+\beta(t,T)[F(0,t)-r(t)]\right),$$
$$0\le t\le T.$$

7.41 For the one-factor Inui–Kijima model, suppose the short rate volatility depends on its level and

$$k(t)F(0,t)+\frac{\partial}{\partial t}F(0,t)>0,$$

show that the forward rates are positive with probability one.

7.42 Consider the multifactor extension of the Inui–Kijima model. Let $\sigma_F^i(t,T)$, $i=1,2,\cdots,n$, satisfy

$$\frac{\partial}{\partial T}\sigma_F^i(t,T)=-k_i(T)\sigma_F^i(t,T),$$

for some deterministic function $k_i(T)$ and initial condition

$$\sigma_F^i(t,t)=\sigma_i(r(t),t).$$

Define

$$\phi_i(r(t),t)=\int_0^t\sigma_F^i(u,t)^2\,du$$
$$\psi_i(r(t),t)=\int_0^t\alpha_F^i(u,t)\,du+\int_0^t\sigma_F^i(u,t)\,dZ_i(u),$$

where

$$\alpha_F^i(t,T)=\sigma_F^i(t,T)\int_t^T\sigma_F^i(t,u)\,du.$$

Show that

$$dr(t)=\left\{\frac{\partial F}{\partial t}(0,t)+\sum_{i=1}^n[\phi_i(t)-k_i(t)\psi_i(t)]\right\}dt$$
$$+\sum_{i=1}^n\psi_i(r(t),t)\,dZ_i(t),$$
$$d\phi_i(t)=[\sigma_F^i(t,t)^2-2k_i(t)\phi_i(t)]\,dt,$$
$$d\psi_i(t)=[\phi_i(t)-k_i(t)\psi_i(t)]\,dt+\sigma_F^i(t,t)\,dZ_i(t).$$

Also, show that

(i) it is possible to express one of $\psi_i(t)$ in terms of $r(t)$ and remaining $\psi_i(t)$, that is, the process

$$\{r(t),\phi_i(t),\psi_i(t),i=1,2,\cdots,n\}$$

forms a $2n$-dimensional Markov system; and

(ii) $r(t)$ is mean-reverting.

7.43 Let $L(t, T)$ denote the time-t LIBOR process $L_t(T, T + \delta]$ over the period $(T, T + \delta]$, and $\sigma_L^i(t, T)$ be its ith component of volatility function [see (7.4.26)]. From the relation

$$\sigma_B^i(t, T + \delta) - \sigma_B^i(t, T) = \frac{\delta L(t, T)}{1 + \delta L(t, T)} \sigma_L^i(t, T),$$

and the properties

$$\sigma_B^i(t, t) = \sigma_L^i(t, t) = 0,$$

one obtains

$$\sigma_B^i(t, t + \delta) = 0.$$

Suppose we impose the condition

$$\sigma_B^i(t, T) = 0 \quad \text{for} \quad T \in (t, t + \delta).$$

Show that

$$\sigma_B^i(t, T) = \sum_{i=1}^{k} \frac{\delta L(t, T - k\delta)}{1 + \delta L(t, T - k\delta)} \sigma_L^i(t, T - k\delta),$$

where k is the largest integer less than or equal to $(T - t)/\delta$.

8

Interest Rate Derivatives: Bond Options, LIBOR and Swap Products

This chapter provides an exposition on the pricing models of some commonly traded interest rate derivatives, like bond options, range notes, interest rate caps, swaps and swaptions, etc. In the past few decades, many innovative and exotic interest rate derivatives have been developed to meet the particular needs of institutional and retail investors. For traders on these derivatives, they always quest for more efficient and robust procedures for pricing and hedging these exotic products. Compared to equity and foreign exchange derivatives, the pricing and hedging of interest rate derivatives pose greater challenges. This is because the payoff functions of most interest rate derivatives depend on interest rates at multiple time points. It is then necessary to develop dynamic models that describe the stochastic evolution of the whole yield curve. Also, the volatilities of these interest rates may differ quite substantially from the short-term rates to the long-term rates. The development of efficient procedures in the calibration of the parameter functions in the dynamic models of interest rates using traded market prices of interest rate derivatives is still under active research.

When we price equity derivative products under stochastic interest rates, the modeling of the joint dynamics of the underlying asset price and interest rates is required in the pricing procedure. In those pricing models where interest rates are used only in discounting the cash flows, the use of the forward measure technique allows one to isolate the discounting effect from the joint evolution of the asset price and interest rates. While the money market account is used as the numeraire in the risk neutral measure, the forward measure is an equivalent martingale measure where the bond price is used as the numeraire. In Sect. 8.1, we illustrate the use of the forward measure in pricing equity options under stochastic interest rates. We also examine the expectation of the short rate and the LIBOR processes under the forward measure and the financial interpretation behind the results.

In Sect. 8.2, we consider the pricing of the two most popular classes of interest rate sensitive derivatives, namely, bond options and range notes. The underlying asset in a T-maturity bond option is a T'-maturity bond, where $T' > T$. For a range note, the buyer receives interest payments that are proportional to the amount of time in which a reference interest rate lies inside a corridor (range). The pricing models of both products exhibit nice analytic tractability when the underlying interest rate dynamics are governed by either a Gaussian HJM model or an affine term structure model. We illustrate how to apply various techniques to the change of numeraire that lead to the effective valuation of these exotic interest rate products.

The caplet is a call option on the LIBOR, which is one of the most popular LIBOR derivative products traded in over the counter. An interest rate cap provides protection for the buyer against the LIBOR rising above a preset level, called the cap rate, at a series of reset times. Thus, a cap can be visualized as a portfolio of caplets. In Sect. 8.3, we show how to derive the price formulas of caps under the Gaussian HJM model, where a single risk neutral measure is applied to all forward rates at the reset dates. Unfortunately, the analytic representation of the cap price formula is quite daunting. This causes the calibration of the volatility functions in the Gaussian HJM model to the market implied cap volatilities too cumbersome. The market convention for pricing a caplet is to assume a lognormal distribution for the forward LIBOR process. Under this assumption, a caplet can be priced using the Black formula in terms of the forward LIBOR. This paves the direct linkage between the market implied cap volatility and the volatility function used in the Black caplet price formula. While the HJM approach is based on the instantaneous forward interest rates (which are not directly observable), the market LIBOR models introduced in Sect. 8.3 are based on the market interest rates (LIBORs). To each forward LIBOR process, the Lognormal LIBOR model assigns a forward measure defined with respect to the relevant settlement date of the associated forward rate. The Black caplet price formula can be shown to be well justified under the framework of the Lognormal LIBOR model.

The last section deals with swaps, swaptions and cross-currency swap products. Since swap payments can be visualized as a portfolio of discount bonds, the swap rates and discount rates are closely related. The value of a forward swap can be shown to be proportional to the difference of the prevailing swap rate and the fixed swap rate. The proportional factor is the value of a portfolio of discount bonds, commonly called the annuity numeraire. A swap measure is defined using the annuity numeraire as the numeraire asset. We propose the Lognormal Swap Rate model, under which the forward swap rate process is a lognormal martingale. A swaption can be priced using the Black formula under the Lognormal Swap Rate model. One can relate the volatilities of the forward swap rates to the bond price volatilities based on the frozen weights approximation approach. We also show how to find an analytic approximation price formula for a swaption priced under the Lognormal LIBOR model. Finally, we discuss the pricing and hedging issues of cross-currency swap products. We consider the extension of the Lognormal Market models to the two-currency setting. The imposition of the properties of lognormal martingales for both the domestic and foreign interest rate markets leaves only the possibility of specifying lognormality for a single forward exchange rate at one specified maturity. Alternatively, one may specify the lognormal martingales for the market rates for one currency world and assume lognormality for the exchange rates at all maturities under the respective forward measures. We illustrate how to price a (cross-currency) differential swap using the two-currency market models. The differential swap is faced with the risks of the joint dynamics of the exchange rates and their correlation. We also discuss the construction of the associated dynamic hedging strategy for this cross-currency swap product.

8.1 Forward Measure and Dynamics of Forward Prices

We examine the pricing of European style equity derivatives under the assumption of stochastic interest rates, where the underlying asset price process is correlated with the stochastic interest rate process. The analytic procedure of deriving the price formulas of T-maturity derivatives under stochastic interest rates can be performed effectively if an appropriate martingale pricing measure is chosen. This new pricing measure, commonly called the T-forward measure, is an equivalent martingale measure where the bond price $B(t, T)$ is chosen as the numeraire. Under the T-forward measure Q_T, the forward price of a forward contract is the expectation of the value of the forward price at maturity T, implying that the forward prices are Q_T-martingales. Also, the stochastic differential equations that describe the dynamics of the instantaneous and discrete forward rates are shown to have nice analytic representation under Q_T. We derive the Radon–Nikodym derivative that effects the change of measure from the risk neutral measure Q to this T-forward measure Q_T. We manage to obtain closed form price formulas of European equity options under the Gaussian HJM model of stochastic interest rates. Interestingly, noting that futures contracts traded in exchange are endowed with the marking to market mechanism, the futures prices can be shown to be Q-martingales. We consider the difference between the price processes of the forward and futures on the same underlying asset under stochastic interest rates, and from which we can deduce the conditions under which the forward and futures price processes are identical. Finally, an appropriate forward measure is found under which the forward LIBOR process is a martingale under this forward measure.

8.1.1 Forward Measure

First of all, we assume the existence of a risk neutral measure Q under which discounted price processes are Q-martingales, thus implying the absence of arbitrage opportunities. Let $f(X_T)$ denote the time-T payoff of a derivative, where the price process of the underlying asset is modeled by the stochastic state variable X_t and f is the payoff function with dependence on X_T. According to the risk neutral valuation principle, the value of the European style derivative at time $t, t < T$, is given by

$$V_t = E_Q^t[e^{-\int_t^T r_u\,du} f(X_T)], \tag{8.1.1}$$

where E_Q^t denotes the expectation under Q conditional on the filtration $\mathcal{F}_t$. Recall that Q is the measure that uses the money market account $M(t)$ as the numeraire. To perform the above expectation calculation, it is necessary to find the joint distribution of the two stochastic state variables, r_t and X_t, under the measure Q. Not only is it highly cumbersome to find the joint distribution, we also have to evaluate a double integral in the subsequent expectation calculation.

It was explained in Sect. 3.2 that choosing the money market account as the numeraire is not unique under the risk neutral valuation framework. For the above pricing problem, it would be more convenient to use the bond price $B(t, T)$ as the numeraire asset. Let Q_T denote the equivalent measure under which all security prices normalized with respect to $B(t, T)$ are Q_T-martingales. By invoking the risk neutral valuation principle and observing $B(T, T) = 1$, the time-t price X_t and time-T price X_T of an asset are related by

$$\frac{X_t}{B(t,T)} = E^t_{Q_T}[X_T], \quad t < T, \tag{8.1.2a}$$

where $E^t_{Q_T}$ is the expectation under Q_T conditional on the filtration $\mathcal{F}_t$. The joint dynamics of the discount process and asset price process is not required since the dependence on the discount process is eliminated by taking advantage of the property: $B(T, T) = 1$. Recall that $X_t/B(t, T)$ is the time-t forward price of forward delivery of X_T at time T, so Q_T is termed the *T-forward measure*. The term is consistent with the observation that the forward price is the expectation of X_T under this forward measure. Let F_t denote the time-t forward price and observe that $F_T = X_T$. We then have

$$F_t = E^t_{Q_T}[F_T], \tag{8.1.2b}$$

that is, the forward price is a martingale under Q_T.

Expectation of Future Short Rate under Forward Measure

The expectation hypothesis postulates that the instantaneous forward rate $F(t, T)$ is an unbiased estimator of the future short rate r_T. Actually, the hypothesis can be shown to be valid under the forward measure (however, not so under the actual probability measure). To prove the result, recall that

$$\begin{aligned} B(t,T)E^t_{Q_T}[r_T] &= E^t_Q\left[e^{-\int_t^T r_u\,du}r_T\right] \\ &= E^t_Q\left[-\frac{\partial}{\partial T}e^{-\int_t^T r_u\,du}\right] \\ &= -\frac{\partial}{\partial T}\left\{E^t_Q\left[e^{-\int_t^T r_u\,du}\right]\right\} \\ &= -\frac{\partial B}{\partial T}(t,T), \end{aligned}$$

so that

$$E^t_{Q_T}[r_T] = -\frac{1}{B(t,T)}\frac{\partial B}{\partial T}(t,T) = F(t,T). \tag{8.1.3}$$

Change of Measure from Q to Q_T

We would like to illustrate how to effect the change of measure from the risk neutral measure Q to the T-forward measure Q_T. Let the dynamics of the T-maturity discount bond price under Q be governed by

$$\frac{dB(t,T)}{B(t,T)} = r(t)\,dt - \sigma_B(t,T)\,dZ(t), \tag{8.1.4}$$

where $Z(t)$ is Q-Brownian. By integrating the above equation and observing $\frac{M(t)}{M(0)} = \int_0^t r(u)\,du$, we obtain

$$\frac{B(t,T)}{M(t)} = \frac{B(0,T)}{M(0)} \exp\left(-\int_0^t \sigma_B(u,T)\,dZ(u) - \frac{1}{2}\int_0^t \sigma_B(u,T)^2\,du\right).$$

The Radon–Nikodym derivative $\frac{dQ_T}{dQ}$ conditional on $\mathcal{F}_T$ is found to be [see (3.2.4)]

$$\begin{aligned}\frac{dQ_T}{dQ} &= \frac{B(T,T)}{B(0,T)} \Big/ \frac{M(T)}{M(0)} \\ &= \exp\left(-\int_0^T \sigma_B(u,T)\,dZ(u) - \frac{1}{2}\int_0^T \sigma_B(u,T)^2\,du\right).\end{aligned} \tag{8.1.5}$$

For a fixed T, we define the process

$$\xi_t^T = E_Q\left[\frac{dQ_T}{dQ}\bigg|\mathcal{F}_t\right] \tag{8.1.6}$$

and since $M(0) = 1$ and $B(0,T)$ is known at time t, we obtain

$$\begin{aligned}\xi_t^T &= \frac{1}{B(0,T)} E_Q\left[\frac{B(T,T)}{M(T)}\bigg|\mathcal{F}_t\right] = \frac{B(t,T)}{B(0,T)M(t)} \\ &= \exp\left(-\int_0^t \sigma_B(u,T)\,dZ(u) - \frac{1}{2}\int_0^t \sigma_B(u,T)^2\,du\right).\end{aligned} \tag{8.1.7}$$

By virtue of the Girsanov Theorem and observing the result in (8.1.7), we deduce that the process

$$Z^T(t) = Z(t) + \int_0^t \sigma_B(u,T)\,du \tag{8.1.8}$$

is Q_T-Brownian.

As an example, consider the Vasicek model where the short rate is modeled by

$$dr(t) = \alpha[\gamma - r(t)]\,dt + \rho\,dZ(t),$$

where $Z(t)$ is Q-Brownian. The corresponding volatility function $\sigma_B(t,T)$ of the discount bond price process is known to be

$$\sigma_B(t,T) = \frac{\rho}{\alpha}[1 - e^{-\alpha(T-t)}].$$

Under the T-forward measure Q^T, the dynamics of $r(t)$ are given by

$$dr(t) = \alpha\left\{\gamma - \frac{\rho^2}{\alpha^2}[1 - e^{-\alpha(T-t)}] - r(t)\right\}dt + \rho\, dZ^T(t), \tag{8.1.9}$$

where $Z^T(t)$ as defined by (8.1.8) is Q_T-Brownian. We integrate the above equation to obtain

$$\begin{aligned} r(t) &= r(s)e^{-\alpha(t-s)} + \left(\gamma - \frac{\rho^2}{\alpha^2}\right)[1 - e^{-\alpha(t-s)}] \\ &+ \frac{\rho^2}{2\alpha^2}\left[e^{-\alpha(T-t)} - e^{-\alpha(T+t-2s)}\right] + \rho\int_s^t e^{-\alpha(t-u)}\, dZ^T(u). \end{aligned} \tag{8.1.10}$$

Under Q_T, the distribution of $r(t)$ conditional on $\mathcal{F}_s$ is normal with the mean and variance

$$\begin{aligned} E_{Q_T}\left[r(t)\middle|\mathcal{F}_s\right] &= r(s)e^{-\alpha(t-s)} + \left(\gamma - \frac{\rho^2}{\alpha^2}\right)[1 - e^{-\alpha(t-s)}] \\ &+ \frac{\rho^2}{2\alpha^2}[e^{-\alpha(T-t)} - e^{-\alpha(T+t-2s)}] \end{aligned} \tag{8.1.11a}$$

$$\mathrm{var}_{Q_T}\left(r(t)\middle|\mathcal{F}_s\right) = \frac{\rho^2}{2\alpha}[1 - e^{2\alpha(t-s)}], \quad s \le t \le T. \tag{8.1.11b}$$

8.1.2 Pricing of Equity Options under Stochastic Interest Rates

We will now find the price formula of a European call option with maturity date T under stochastic interest rate economy. Let S_t be the price process of an asset and $F_t = S_t/B(t,T)$ be its forward price. Let the dynamics of S_t and $B(t,T)$ under the risk neutral measure Q be governed by

$$\frac{dS_t}{S_t} = r_t\, dt + \sum_{i=1}^m \sigma_S^i(t)\, dZ_i(t) \tag{8.1.12a}$$

$$\frac{dB(t,T)}{B(t,T)} = r_t\, dt - \sum_{i=1}^m \sigma_B^i(t,T)\, dZ_i(t), \tag{8.1.12b}$$

where $\mathbf{Z}(t) = (Z_1(t)\cdots Z_m(t))^{\mathrm{T}}$ is an m-dimensional standard Q-Brownian process. The volatility functions $\sigma_S^i(t)$ and $\sigma_B^i(t,T)$, $i = 1, \cdots, m$, are assumed to be deterministic. Nice analytic tractability is exhibited when the dynamics of the asset price and bond price processes follow the above Gaussian framework. Using Ito's lemma, the dynamics of F_t under Q is computed as follows:

$$\begin{aligned} dF_t &= \frac{dS_t}{B(t,T)} - \frac{S_t}{B(t,T)^2}dB(t,T) - \frac{dS_t\, dB(t,T)}{B(t,T)^2} + \frac{S_t}{B(t,T)^3}[dB(t,T)]^2 \\ &= \frac{S_t}{B(t,T)}\left\{\left[r_t\, dt + \sum_{i=1}^m \sigma_S^i(t)\, dZ_i(t)\right]\left[r_t\, dt - \sum_{i=1}^m \sigma_B^i(t,T)\, dZ_i(t)\right]\right. \end{aligned}$$

$$+\sum_{i=1}^{m}\sigma_S^i(t)\sigma_B^i(t,T)\,dt+\sum_{i=1}^{m}\sigma_B^i(t,T)^2\,dt\Bigg\}$$

$$=F_t\left\{\sum_{i=1}^{m}[\sigma_S^i(t)+\sigma_B^i(t,T)][dZ_i(t)+\sigma_B^i(t,T)\,dt]\right\}.$$

Next, we define

$$Z_i^T(t)=Z_i(t)+\int_0^t\sigma_B^i(u,T)\,du,\quad i=1,\cdots,m,$$

and $\mathbf{Z}(t)=(Z_1^T(t)\cdots Z_m^T(t))^{\mathrm{T}}$ is known to be an m-dimensional Brownian process under the T-forward measure Q_T. The dynamics of F_t can then be expressed as

$$\frac{dF_t}{F_t}=\sum_{i=1}^{m}[\sigma_S^i(t)+\sigma_B^i(t,T)]\,dZ_i^T(t),\tag{8.1.13}$$

which assures that F_t is a Q_T-martingale.

In summary, the probability density of the forward price F_T conditional on F_t is lognormally distributed under Q with mean

$$\overline{\mu}_F=\frac{1}{T-t}\int_t^T\sum_{i=1}^{m}\sigma_B^i(u,T)[\sigma_S^i(u)+\sigma_B^i(u,T)]\,du\tag{8.1.14a}$$

and term volatility $\overline{\sigma}_F$, where

$$\overline{\sigma}_F^2=\frac{1}{T-t}\int_t^T\sum_{i=1}^{m}[\sigma_S^i(u)+\sigma_B^i(u,T)]^2\,du.\tag{8.1.14b}$$

When the probability measure is changed from Q to Q_T, the lognormal distribution for F_T has zero mean and the same term volatility as above.

Under the forward measure Q_T and observing $F_T=S_T$, the value of the European call option at time t is given by

$$c(S,t)=B(t,T)E_{Q_T}[\max(F_T-X,0)|S_t=S],\tag{8.1.15}$$

where X is the strike price. Conditional on $F_t=F$ where $F=S/B(t,T)$, F_T is lognormal distributed with term volatility $\overline{\sigma}_F$ over the time period $(t,T]$, we then obtain

$$c(S,t)=B(t,T)[FN(d_1)-XN(d_2)],\tag{8.1.16}$$

where

$$d_1=\frac{\ln\frac{F}{X}+\frac{\overline{\sigma}_F^2}{2}(T-t)}{\overline{\sigma}_F\sqrt{T-t}},\quad d_2=d_1-\overline{\sigma}_F\sqrt{T-t}.$$

The above form of representation of the call value has dependence on the forward price, and this analytic form is commonly called the *Black formula*. The analytic pricing procedure can be done more effectively when the forward price is used as the underlying state variable. In the call price formula, the dependence on the stochastic interest rate dynamics appears both in the discount factor $B(t, T)$ and the term volatility $\overline{\sigma}_F$ of the forward price.

8.1.3 Futures Process and Futures-Forward Price Spread

Consider a futures contract with maturity T on an underlying asset whose price process is S_t. Write f_t as the futures price process and let the dates of settlement over the period be $t_i, i = 1, 2, \cdots, n$, where the current time t is taken to be t_0 and the maturity date T is taken to be t_n. The sum of discounted cash flows occurring on the settlement dates is given by

$$\sum_{i=1}^{n} e^{-\int_{t_0}^{t_i} r(u)\,du}(f_{t_i} - f_{t_{i-1}}).$$

The value of the futures contract at time t is given by the expectation of the above cash flows under the risk neutral measure Q. Note that the futures value at time t is zero, so we have

$$E_Q^t\left[\sum_{i=1}^{n} e^{-\int_{t_0}^{t_i} r(u)\,du}(f_{t_i} - f_{t_{i-1}})\right] = 0, \tag{8.1.17}$$

where E_Q^t denotes the expectation under Q conditional on filtration $\mathcal{F}_t$. We consider the continuous limit by taking $\max_i(t_i - t_{i-1})$ to go to zero. Also, we define the discount factor D_s by

$$D_s = e^{-\int_0^s r(u)\,du}.$$

The continuous limit of (8.1.17) becomes

$$E_Q^t\left[\int_t^T D_s\,df_s\right] = 0.$$

Provided that $r(t)$ is positive and bounded, we have

$$\lim_{\delta t \to 0} E_Q^t\left[\int_t^{t+\delta t} D_s\,df_s\right] = D_t E_Q^t[df_t] = 0$$

so that the stochastic differential df_t has zero drift. We deduce that f_t is a Q-martingale. Since $f_T = S_T$, the futures price is given by

$$f_t = E_Q^t[f_T] = E_Q^t[S_T]. \tag{8.1.18}$$

In Sect. 1.4, we showed that the futures price equals the forward price when the interest rate is constant. However, under a stochastic interest rate economy, the futures-forward price spread can be expressed as

$$\begin{aligned} f_t - F_t &= E_Q^t[S_T] - \frac{S_t}{B(t,T)} \\ &= \frac{E_Q^t[S_T]E_Q^t\left[e^{-\int_t^T r(u)\,du}\right] - E_Q^t\left[e^{-\int_t^T r(u)\,du} S_T\right]}{B(t,T)} \\ &= -\frac{\mathrm{cov}_Q^t\left[e^{-\int_t^T r(u)\,du}, S_T\right]}{B(t,T)}. \end{aligned} \tag{8.1.19}$$

Hence, the spread becomes zero when the discount process and the price process of the underlying asset are uncorrelated.

Suppose the dynamics of S_t and $B(t,T)$ under Q are governed by (8.1.12a,b), and since futures price and forward price are equal at maturity T, one can use (8.1.18) to obtain

$$f_t = E_Q^t[F_T] = F_t e^{\overline{\mu}_F(T-t)}, \tag{8.1.20}$$

where $\overline{\mu}_F$ is given by (8.1.14a).

Dynamics of the Forward LIBOR Process

We define the forward LIBOR process $L_t[T, T+\delta]$ at time t applied over the future period $(T, T+\delta]$ by

$$L_t[T, T+\delta] = \frac{1}{\alpha}\frac{B(t,T) - B(t,T+\delta)}{B(t,T+\delta)}, \quad t \le T, \tag{8.1.21}$$

where α is the accrual factor for the period $(T, T+\delta]$. Consider the quantity $\frac{1}{\alpha}[B(t,T) - B(t,T+\delta)]$ which can be seen as a multiple of the difference of two bond prices, hence it is seen to be a tradeable asset. Suppose the $(T+\delta)$-maturity bond is used as the numeraire, we normalize this tradeable asset by $B(t, T+\delta)$ accordingly. The corresponding normalized tradeable asset is seen to be $L(t,T)$, and it is a martingale under the $(T+\delta)$-forward measure $Q_{T+\delta}$.

Consider the standard LIBOR payment, the LIBOR is observed at time T and the payment at the end of the accrual period at $T+\delta$ is αL_T. Using the property of $Q_{T+\delta}$-martingale, the time-t value of the LIBOR payment is seen to be

$$V_L(t) = B(t, T+\delta)E_{Q_{T+\delta}}^t[\alpha L_T] = B(t,T) - B(t,T+\delta). \tag{8.1.22}$$

Let $\mathbf{Z}^{T+\delta}(t) = (Z_1^{T+\delta}(t) \cdots Z_m^{T+\delta}(t))^{\mathrm{T}}$ be an m-dimensional Brownian process under $Q_{T+\delta}$. Since $L_t[T, T+\delta]$ is a martingale under $Q_{T+\delta}$, the dynamics of L_t under $Q_{T+\delta}$ are governed by

$$dL_t = L_t \sum_{i=1}^m \sigma_L^i(t)\,dZ_i^{T+\delta}(t), \tag{8.1.23}$$

where $\sigma_L^i(t)$ is the ith component of the volatility function of the forward LIBOR process. Using the change of measure from $Q_{T+\delta}$ to Q, it is relatively straightforward to deduce the dynamics of L_t under Q. Suppose we define $Z_i(t)$ by

$$Z_i^{T+\delta}(t) = Z_i(t) + \int_0^t \sigma_B^i(u, T+\delta)\, du, \quad i = 1, 2, \cdots, m,$$

then $(Z_1(t) \cdots Z_m(t))^{\mathrm{T}}$ is an m-dimensional Brownian process under the risk neutral measure Q [see (8.1.13)]. Under Q, the dynamics of L_t are then given by

$$dL_t = L_t \left[\sum_{i=1}^m \sigma_B^i(u, T+\delta)\sigma_L^i(t) \right] dt + L_t \sum_{i=1}^m \sigma_L^i(t)\, dZ_i(t), \tag{8.1.24}$$

which agrees with the result obtained earlier in Sect. 7.4.3 [see (7.4.26)].

8.2 Bond Options and Range Notes

In this section, we illustrate the versatility of the "change of numeraire" technique in pricing bond options and range notes. In Sect. 8.2.1, we consider the pricing of a European option maturity at T whose underlying asset is a T'-maturity bond, with $T' > T$. The underlying bond may be zero-coupon or coupon bearing. The earliest version of the bond option pricing models uses the extension of the Black–Scholes pricing framework by assuming that the bond price follows a lognormal distribution. However, the aging of the bond price implies that the instantaneous rate of return on the bond is distributed with a variance that decreases when the maturity of the bond is approached, a feature that distinguishes it from the price process of an equity. Since the bond price is dependent on the evolution of interest rates, the more reasonable pricing approach should relate the bond price process to the term structure of interest rate evolution.

Range notes are structured products convenient for investors who hold the view that interest rates will fall within a certain range (corridor). These notes pay at the end of a defined period an interest payment that equals a prespecified reference interest rate (commonly LIBOR) times the number of days where the reference rate lies inside a corridor. For example, a floating range note pays coupon rates that are linked to the three-month spot LIBOR plus 200 basis points spread. These corridor products offer investors the opportunity to sell volatility for an enhanced yield if rates fall within the specified range. They are structured to reflect an investor's view that differs from the forward rate curves, thus providing opportunities for investors to exploit the arbitrage if they believe that the actual realized rates would not match with the rates as predicted by the forward curves. The pricing of the range notes under the multi-factor Gaussian HJM model is considered in Sect. 8.2.2.

8.2.1 Options on Discount Bonds and Coupon-Bearing Bonds

When the underlying bond pays no coupons, it is quite straightforward to obtain a closed form price formula of the European bond option, provided that the interest rate dynamics are governed by either the Gaussian Heath–Jarrow–Morton (HJM) model or an affine term structure model. When the bond pays discrete coupons between T

and T', we illustrate an effective decomposition technique to derive closed form price formulas when the short rate process is assumed to be Markovian. Under the general multifactor term structure models, the decomposition technique fails. However, we manage to derive the analytic approximation formulas for pricing European options on coupon bearing bonds.

Options on Discount Bonds under Gaussian HJM Models

Due to the nice analytic tractability associated with the Gaussian HJM model, it is quite straightforward to derive the price formula of a European call option maturing at T on a T'-maturity discount bond, $T' > T$. According to the Gaussian HJM formulation, the dynamics of the forward rate $F(t, T)$ under the risk neutral measure Q is governed by

$$dF(t,T') = \left(\sum_{i=1}^{m} \sigma_F^i(t,T') \int_t^{T'} \sigma_F^i(t,u)\,du\right) dt + \sum_{i=1}^{m} \sigma_F^i(t,T')\,dZ_i(t), \quad (8.2.1)$$

where the volatility vector function $\boldsymbol{\sigma}_F(t,T') = (\sigma_F^1(t,T') \cdots \sigma_F^m(t,T'))^{\mathrm{T}}$ is deterministic and $\mathbf{Z}(t) = (Z_1(t) \cdots Z_m(t))^{\mathrm{T}}$ is an m-dimensional Q-Brownian process. The bond price process $B(t, T')$ under Q then follows (see Sect. 7.4.1)

$$\frac{dB(t,T')}{B(t,T')} = r(t)\,dt - \sum_{i=1}^{m}\left(\int_t^{T'} \sigma_B^i(t,u)\,du\right) dZ_i(t), \quad (8.2.2)$$

where $\sigma_F^i(t,T) = \frac{\partial \sigma_B^i}{\partial T}(t,T)$. Define $F^B(t;T,T') = B(t,T')/B(t,T)$, which is the price of the T-maturity forward on the T'-maturity bond. The T-forward measure is defined by using $B(t,T)$ as the numeraire, and the dynamics of $B(t,T)$ follows a similar process as defined in (8.1.12b). According to (8.1.13), the dynamics of $F^B(t)$ under the T-forward measure Q_T can be expressed as

$$\begin{aligned}\frac{dF^B(t)}{F^B(t)} &= \sum_{i=1}^{n}\left(\int_t^{T} \sigma_B^i(t,u)\,du - \int_t^{T'} \sigma_B^i(t,u)\,du\right) dZ_i^T(t)\\ &= \sum_{i=1}^{n}\left(-\int_T^{T'} \sigma_B^i(t,u)\,du\right) dZ_i^T(t),\end{aligned} \quad (8.2.3)$$

where $\mathbf{Z}^T(t) = (Z_1^T(t) \cdots Z_m^T(t))^{\mathrm{T}}$ is an m-dimensional Q_T-Brownian process and

$$dZ_i^T(t) = dZ_i(t) + \int_t^T \sigma_B^i(t,u)\,du, \quad i = 1, 2, \cdots, n.$$

Under Q_T, F_t^B is seen to be lognormally distributed with zero drift and

$$\begin{aligned}B(T,T') &= F^B(T;T,T')\\ &= F^B(t;\ T,T')\exp\left(-\int_t^T \int_T^{T'} \sum_{i=1}^{m} \sigma_B^i(s,u)\,du\,dZ_i^T(s)\right.\\ &\qquad \left.-\frac{1}{2}\int_t^T \int_T^{T'} \sum_{i=1}^{m} \sigma_B^i(s,u)^2\,du\,ds\right).\end{aligned} \quad (8.2.4)$$

The term variance over (t, T) is given by

$$\overline{\sigma}^2(t, T)(T - t) = \text{var}_T(\ln B(T, T')|\mathcal{F}_t) = \int_t^T \int_T^{T'} \sum_{i=1}^m \sigma_B^i(s, u)^2 \, du \, ds. \tag{8.2.5}$$

The value of the European bond call option is found to be

$$\begin{aligned} c(t; T, T') &= B(t, T) E_{Q_T}^t[\max(F^B(T; T, T') - X, 0)] \\ &= B(t, T)[F^B(t, T, T')N(d_1) - XN(d_2)] \\ &= B(t, T')N(d_1) - B(t, T)XN(d_2), \end{aligned} \tag{8.2.6}$$

where X is the strike price and

$$d_1 = \frac{\ln \frac{B(t,T)}{B(t,T')X} + \frac{1}{2}\overline{\sigma}^2(t, T)(T - t)}{\overline{\sigma}(t, T)\sqrt{T - t}}, \quad d_2 = d_1 - \overline{\sigma}(t, T)\sqrt{T - t}.$$

Options on Discount Bonds under Affine Term Structure Models

Similarly, closed form price formulas of European options on a discount bond can be derived under the assumption of an affine term structure interest rate model. The bond option price can be expressed in terms of conditional distribution of the bond price at option's maturity T under the T-forward measure Q_T and the T'-forward measure $Q_{T'}$. Let the price of the T'-maturity bond under the affine term structure model be governed by

$$B(t, T') = \exp(a(\tau') + \mathbf{b}^{\mathrm{T}}(\tau')\mathbf{Y}(t)), \quad \tau' = T' - t, \tag{8.2.7}$$

where $\mathbf{Y}(t)$ is the m-dimensional stochastic state vector, $a(\tau')$ and $\mathbf{b}(\tau')$ are parameter functions [see (7.3.19)]. Let X be the strike price of the bond call option. We write $Q_T[B(T, T') > X]$ as the conditional probability of the event $\{B(T, T') > X\}$ under the T-forward measure, and similar definition for $Q_{T'}[B(T, T') > X]$ under the T'-forward measure. The value of the European bond call option can be expressed as

$$\begin{aligned} & c(t; T, T') \\ &= E_Q^t\left[e^{-\int_t^T r(u)\,du} \max(B(T, T') - X, 0)\right] \\ &= B(t, T)\left\{E_{Q_T}^t\left[B(T, T')\mathbf{1}_{\{B(T,T')>X\}}\right] - XE_{Q_T}\left[\mathbf{1}_{\{B(T,T')>X\}}\right]\right\} \\ &= B(t, T')Q_{T'}[B(T, T') > X] - XB(t, T)Q_T[B(T, T') > X]. \end{aligned} \tag{8.2.8}$$

Since $\{B(T, T') > X\}$ is equivalent to

$$\{a(\tau') + \mathbf{b}^{\mathrm{T}}(\tau')\mathbf{Y}(t) > \ln X\},$$

the above forward probabilities can be computed provided that the conditional distribution of the affine diffusion process under the two forward measures are known. For example, the above formulation can be used to derive the closed form price formula of a European bond call option under the extended CIR model (see Maghsoodi, 1996).

Options on Coupon-Bearing Bonds

We consider a T-maturity bond option whose underlying bond is coupon bearing that pays off cash payment A_i at time $T_i, i = 1, 2, \cdots, n$, where $T < T_1 < \cdots < T_n = T'$. The payment stream consists of coupon payments at $T_1, \cdots, T_{n-1}$ while the last payment at T' is the sum of the last coupon and par. Under the assumption that the discount bond prices are functions of the short rate r that follows a Markovian process, Jamshidian (1989) showed that an option on the coupon bearing bond can be decomposed into a portfolio of bond options.

Conditional on $r(t) = r$, the price of the coupon bearing bond B_c at time t is given by

$$B_c(r, t; T') = \sum_{i=1}^{n} A_i B(r, t; T_i), \tag{8.2.9}$$

where $B(r, t; T_i)$ is the price of the unit par T_i-maturity discount bond. Consider a European call option on the bond with strike price X, since all discount bond prices are decreasing functions of r, so the call option will be exercised when the short rate at option's maturity T is less than some threshold value r^*, where r^* satisfies

$$\sum_{i=1}^{n} A_i B(r^*, T; T_i) = X. \tag{8.2.10}$$

For notational convenience, we set

$$X_i = B(r^*, T; T_i), \quad i = 1, 2, \cdots, n.$$

It is seen that when $r < r^*$, then

$$\sum_{i=1}^{n} A_i B(r, T; T_i) > X \quad \text{and} \quad B(r, T; T_i) > X_i, \quad i = 1, 2, \cdots, n,$$

and the inequalities are all reversed when $r > r^*$, so

$$\max\left(\sum_{i=1}^{n} A_i B(r, T; T_i) - X, 0\right) = \sum_{i=1}^{n} A_i \max(B(r, T; T_i) - X_i, 0).$$

The value of the bond call option is given by

$$\begin{aligned} c(t; T, T') &= E_Q^t\left[e^{-\int_t^T r(u)\,du} \max\left(\sum_{i=1}^{n} A_i B(r, t; T_i) - X, 0\right)\right] \\ &= A_i \sum_{i=1}^{n} E_Q^t\left[e^{-\int_t^T r(u)\,du} \max(B(r, t; T_i) - X_i, 0)\right]. \end{aligned} \tag{8.2.11}$$

Since the ith term in the above expression can be interpreted as the value of the call option on the T_i-maturity discount bond with strike price X_i, the call option on the coupon bearing bond can be decomposed into a portfolio of call options on discount bonds. All these constituent options have the same maturity date T and each individual strike price is obtained by properly distributing the original strike price X according to

$$X = \sum_{i=1}^{n} A_i X_i.$$

The above decomposition works when all discount bond prices depend on a single stochastic state variable: a short rate process that is Markovian. Under this scenario, all discounted bond prices are instantaneously perfectly correlated. The successful solution of r^* becomes impossible if the bond price depends not only on $r(t)$ but also on the path of the interest rate. Barber (2005) proposed an approximation approach to deal with coupon bond option valuation for non-Markovian short rate processes.

When the number of state variables in the interest rate term structure model is more than one, the Jamshidian decomposition technique cannot be applied. We consider two analytic approximation approaches to deal with coupon bond option pricing in multifactor models, namely, the stochastic duration approach (Wei, 1997; Munk, 1999) and the affine approximation method (Singleton and Umantsev, 2002).

Stochastic Duration Approach

The most common definition of duration that measures the sensitivity of percentage change in bond price to yield change is given by $-\frac{1}{B}\frac{\partial B}{\partial R}$, where R is the yield to maturity of the bond. This duration measure makes good sense only when the yield curve is flat and moves in a parallel manner. When the interest rate is stochastic, a better definition of risk measure is $-\frac{1}{B}\frac{\partial B}{\partial r}$. Recall that a coupon bearing bond can be treated as a portfolio of bonds whose value is $B_c(r,t) = \sum_{i=1}^{n} A_i B(r,t;T_i)$. We define the portfolio weight w_i, $i = 1, 2, \cdots, n$, by

$$w_i(r,t) = \frac{A_i B(r,t;T_i)}{\sum_{i=1}^{n} A_i B(r,t;T_i)},$$

with dependence on r and t. We then have

$$\begin{aligned} -\frac{1}{B_c}\frac{\partial B_c}{\partial r} &= -\frac{\sum_{i=1}^{n} A_i \frac{\partial B}{\partial r}(r,t;T_i)}{\sum_{i=1}^{n} A_i B(r,t;T_i)} \\ &= -\sum_{i=1}^{n} w_i(r,t)\left[\frac{1}{B(r,t;T_i)}\frac{\partial B}{\partial r}(r,t;T_i)\right]. \end{aligned} \tag{8.2.12}$$

The stochastic duration D is defined with respect to a proxy discount bond whose time to maturity is equal to D. The risk measure of this proxy bond is given by $-\frac{1}{B(r,t;t+D)}\frac{\partial B}{\partial r}(r,t;t+D)$, which is then matched to the risk measure of the coupon bearing bond as defined in (8.2.12).

As an illustrative example, we use the class of one-factor affine term structure models to illustrate the calculation of the stochastic duration D. The discount bond price of an affine model admits solution of the form

$$B(r, t; T) = \exp(a(\tau) + b(\tau)r), \quad \tau = T - t. \tag{8.2.13}$$

By substituting into the equation that matches the risk measure, we obtain the following algebraic equation for the determination of D:

$$b(D) = \sum_{i=1}^{n} w_i(r, t) b(T_i - t). \tag{8.2.14}$$

For example, consider the Vasicek model whose $b(\tau)$ is known to be

$$b(\tau) = -\frac{1 - e^{-\alpha\tau}}{\tau},$$

one can solve for D in closed form

$$D = -\frac{1}{\alpha} \ln\left(\sum_{i=1}^{n} w_i(r, t; T_i) e^{-\alpha(T_i - t)}\right). \tag{8.2.15}$$

The stochastic duration approximation approach approximates the call option on a coupon bearing bond by another call option on the proxy discount bond with the same risk measure. In other words, the proxy bond's time to maturity equals the stochastic duration of the coupon bearing bond. As the proxy discount bond and the coupon bearing bond should have the same value at current time t, the par value of the proxy discount bond equals

$$P = \frac{\sum_{i=1}^{n} A_i B(r, t; T_i)}{B(r, t; t + D)}.$$

Let $c(t; T, T', X)$ and $\widehat{c}(t; T, T', X)$ denote the value of the European call on the coupon bearing bond and the proxy discount bond, respectively. Here, T is the option maturity date, T' is the bond maturity date, $T' > T$ and X is the strike price. The stochastic duration approximation approach can be stated as

$$c(t; T, T', X) \approx P\widehat{c}(t; T, t + D, X/P). \tag{8.2.16}$$

The analytic formula for $\widehat{c}(t; T, t + D, X/P)$ can be deduced either from (8.2.6) (under the Gaussian HJM models) or from (8.2.8) (under the affine term structure models).

What would be the pricing error in the above approximation? Write $T^* = t + D$, and let $B_c(t)$ be the price of the coupon bearing bond and $E^t_{Q_{T^*}}$ denote the expectation under the T^*-forward measure conditional on information $\mathcal{F}_t$. The error in the stochastic duration approximation is seen to be (Munk, 1999)

$$\begin{aligned}
& c(t; T, T', X) - \frac{B_c(t)}{B(t, T^*)}\widehat{c}\left(t; T, T^*, \frac{B(t, T^*)}{B_c(t)}X\right) \\
&= B(t, T^*)E^t_{Q_{T^*}}\left[\frac{\max(B_c(T) - X, 0)}{B(T, T^*)}\right] \\
&\quad - \frac{B_c(t)}{B(t, T^*)}B(t, T^*)E^t_{Q_{T^*}}\left[\frac{\max\left(B(T, T^*) - \frac{B(t,T^*)}{B_c(t)}X, 0\right)}{B(T, T^*)}\right] \\
&= B(t, T^*)E^t_{Q_{T^*}}\left[\max\left(\frac{B_c(T)}{B(T, T^*)} - \frac{X}{B(T, T^*)}, 0\right)\right. \\
&\qquad \left. - \max\left(\frac{B_c(t)}{B(t, T^*)} - \frac{X}{B(T, T^*)}, 0\right)\right].
\end{aligned}$$

When the bond call option is deep-in-the-money with $B_c(t) \gg X$, the probability of exercise of the option is close to one. Under this scenario, the above approximation error is close to zero since

$$\begin{aligned}
& c(t; T, T', X) - \frac{B_c(t)}{B(t, T^*)}\widehat{c}\left(t; T, T^*, \frac{B(t, T^*)}{B_c(t)}X\right) \\
&\approx B(t, T^*)\left\{E^t_{Q_{T^*}}\left[\frac{B_c(T)}{B(T, T^*)}\right] - \frac{B_c(t)}{B(t, T^*)}\right\} = 0. \qquad (8.2.17)
\end{aligned}$$

As shown by the numerical calculations performed by Munk (1999), the pricing approximation errors are most significant when the bond option is currently at-the-money.

Affine Approximation Method

When the underlying bond is coupon bearing, which is visualized as a portfolio of discount bonds, the value of the European bond call option is given by

$$\begin{aligned}
c(t; T) &= \sum_{i=1}^{n} A_i E^t_Q[B(t, T_i)\mathbf{1}_{\{B_c(T)>X\}}] - XE^t_Q[\mathbf{1}_{\{B_c(T)>X\}}] \\
&= \sum_{i=1}^{n}\{A_i B(t, T_i)Q_{T_i}[B_c(T) > X] \\
&\qquad - XB(t, T)Q_T[B_c(T) > X]\}, \qquad (8.2.18)
\end{aligned}$$

where $Q_{T_i}[A]$ denotes the probability of occurrence of event A under the forward measure Q_{T_i}. Note that

$$\begin{aligned}
& Q_{T_i}[B_c(T) > X] \\
&= Q_{T_i}\left[\sum_{i=1}^{n} A_i B(T; T_i) > X\right] \\
&= Q_{T_i}\left[\sum_{i=1}^{n} A_i e^{a(T_i-T)}e^{\mathbf{b}^{\mathrm{T}}(T_i-T)\mathbf{Y}(T_i-T)} > X\right].
\end{aligned}$$

The boundary of the exercise set $\{B_c(T) > X\}$ in the m-dimensional $(Y_1, \cdots, Y_m)$-plane is in general concave. In the affine approximation method proposed by Singleton and Umantsev (2002), the concave exercise boundary is approximated by a hyperplane, $\beta_1 Y_1 + \cdots + \beta_n Y_n = \alpha$. In this case, the computation of the forward probabilities as defined in (8.2.18) reduces to the evaluation of $Q_{T_i}[\beta_1 Y_1(T_i - T) + \cdots + \beta_n Y_n(T_i - T) > \alpha], i = 1, 2, \cdots, n$. Under this affine approximation, the evaluation procedure of the call option on a coupon bearing bond becomes identical to that of the option on a discount bond.

8.2.2 Range Notes

We start with the description of the product nature of a range note and also fix the mathematical notation. Let t denote the valuation date of the range note. Let T_0 be the previous coupon payment date ($T_0 \leq t$) and the N future coupon dates are $T_j, j = 1, \cdots, n$. Based on certain day-count conventions, we denote the number of days and length of time period (in year) between the successive coupon dates T_j and T_{j+1} by n_j and δ_j, respectively. We set $(T_j, T_{j+1}]$ be the $(j+1)$th compounding period, $j = 0, 1, \cdots, n-1$. In particular, for the current period, we write n_0^- and n_0^+ as the number of days between T_0 and t and between t and T_1, respectively (see Fig. 8.1).

The *spot* LIBOR $L(t, t+\delta)$ at time t for a given compounding period δ is defined as

$$L(t, t+\delta) = \frac{1}{\delta}\left[\frac{1}{B(t, t+\delta)} - 1\right], \tag{8.2.19}$$

where $B(t, t+\delta)$ is the time-t price of a discount bond with maturity date at $t+\delta$.

Recall that interests are accrued on a daily basis. We let $T_{j,i}$ denote the date that corresponds to i days after T_j and $\delta_{j,i}$ denote the length (in year) of the compounding period starting at $T_{j,i}$. Further, we write $[R_\ell(T_{j,i}), R_u(T_{j,i})]$ as the prespecified range for the ith day of the $(j+1)$th compounding period so that interest is accrued on date $T_{j,i}$ provided that

$$R_\ell(T_{j,i}) \leq L(T_{j,i}, T_{j,i} + \delta_{j,i}) \leq R_u(T_{j,i}).$$

Let s_j denote the spread over the reference LIBOR paid by the floating range note during the $(j+1)$th compounding period and D_j be the number of days in a year for the $(j+1)$th compounding period. We write $\mathcal{D}(T_j, T_{j+1})$ as the number of days in the

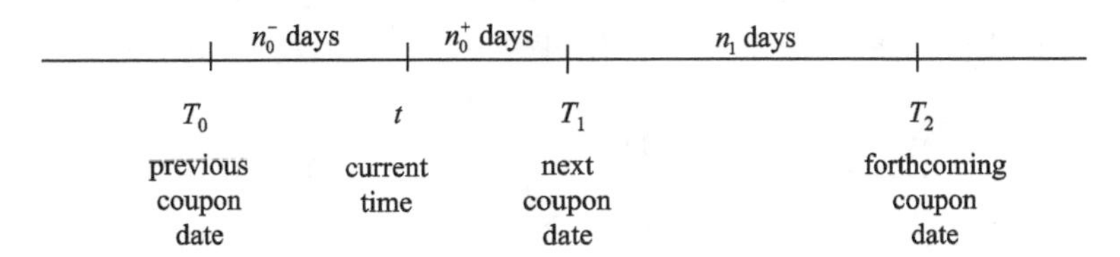

Fig. 8.1. The coupon dates and day counting between coupon dates.

$(j+1)$th compounding period that the reference LIBOR lies inside the prespecified range, that is,

$$\mathcal{D}(T_j, T_{j+1}) = \sum_{i=1}^{n_j} \mathbf{1}_{\{R_\ell(T_{j,i}) \le L(T_{j,i}, T_{j,i}+\delta_{j,i}) \le R_u(T_{j,i})\}}. \tag{8.2.20}$$

Now, the value of the $(j+1)$th coupon at time T_{j+1} is then given by

$$c_{j+1}(T_{j+1}) = \frac{L(T_j, T_j+\delta_j) + s_j}{D_j} \mathcal{D}(T_j, T_{j+1}). \tag{8.2.21}$$

Let $c_{j+1}(t)$ denote the time-t value of the $(j+1)$th coupon. The value of the floating range note is then given by

$$V_f(t) = B(t, T_N) + \sum_{j=0}^{n-1} c_{j+1}(t), \tag{8.2.22}$$

where $B(t, T_n)$ is the time-t value of the unit par payment at the final maturity date T_n. We would like to price the floating range note by assuming the pure discount bond price process to be governed by the multifactor Gaussian HJM model as defined in (8.1.12b).

Delayed Range Digital Options

It can be seen from (8.2.20)–(8.2.21) that the coupons can be decomposed into a portfolio of delayed range digital options. A delayed range digital option provides a terminal payoff equal to 1 paid at payment time T_p when the underlying spot LIBOR $L(T, T+\delta)$ lies inside a prespecified range (R_ℓ, R_u) at observation time T, $T \le T_p$. Let $V_r(t; T, T_p, \delta)$ denote the time-t value of the delayed range digital option, then its terminal payoff at time T_p is defined by

$$V_r(T_p) = \mathbf{1}_{\{R_\ell \le L(T, T+\delta) \le R_u\}}. \tag{8.2.23}$$

Accordingly, $V_r(t)$ is then given by

$$\begin{aligned} &V_r(t; T, T_p, \delta) \\ &= B(t, T_p) E^t_{Q_{T_p}} \left[\mathbf{1}_{\{R_\ell \le L(T, T+\delta) \le R_u\}} \right] \\ &= B(t, T_p) Q_{T_p} \left[\ln \frac{1}{1+\delta R_u} \le \ln B(T, T+\delta) \le \ln \frac{1}{1+\delta R_\ell} \right], \end{aligned}$$

where $Q_{T_p}[A]$ denotes the probability that event A occurs under the T_p-forward measure. Suppose the bond price process $B(t, T)$ is governed by (8.1.12b), it can be shown that under the measure Q_{T_p}, we obtain the following analytic representation of $B(T, T+\delta)$:

$$\begin{aligned} \ln B(T, T+\delta) = \ln &\left(\frac{B(t, T+\delta)}{B(t, T)} \right) - \frac{1}{2} g(t, T, T+\delta) + \ell(t, T, T+\delta, T_p) \\ &+ \sum_{i=1}^{m} \int_t^T \left[\sigma_B^i(u, T) - \sigma_B^i(u, T+\delta) \right] dZ_i^{T_p}(u)\, du, \end{aligned} \tag{8.2.24}$$

where $\mathbf{Z}^{T_p}(t) = (Z_1^{T_p}(t) \cdots Z_m^{T_p}(t))^{\mathrm{T}}$ is an m-dimensional Brownian process under Q_{T_p} and

$$g(t, T, T+\delta) = \sum_{i=1}^{m} \int_t^T \left[\sigma_B^i(u, T) - \sigma_B^i(u, T+\delta)\right]^2 du$$

$$\ell(t, T, T+\delta, T_p) = \sum_{i=1}^{m} \int_t^T \left[\sigma_B^i(u, T) - \sigma_B^i(u, T+\delta)\right] \left[\sigma_B^i(u, T) - \sigma_B^i(u, T_p)\right] du.$$

Thus, $\ln B(T, T+\delta)$ is a univariate normal distribution with

$$\text{mean} = \ln\left(\frac{B(t, T+\delta)}{B(t, T)}\right) - \frac{1}{2} g(t, T, T+\delta) + \ell(t, T, T+\delta, T_p),$$

and

$$\text{standard deviation} = \sqrt{g(t, T, T+\delta)}.$$

We then obtain (Nunes, 2004)

$$V_r(t; T, T_p, \delta) = B(t, T_p)\left[N(h(R_\ell)) - N(h(R_u))\right], \tag{8.2.25}$$

where

$$h(r) = \frac{\ln \frac{B(t,T)}{B(t,T+\delta)(1+\delta r)} + \frac{1}{2} g(t, T, T+\delta) - \ell(t, T, T+\delta, T_p)}{\sqrt{g(t, T, T+\delta)}}.$$

Valuation of the Time-t Value of the Coupons

The time-t value of the first coupon $c_1(t)$ can be easily evaluated since $L(T_0, T_0+\delta)$ is already known at time t. Since $L(T_0, T_0+\delta)$ and $L(T_{0,i}, T_{0,i}+\delta_{0,i})$, $i = 1, \cdots, n_0^-$, are measurable with respect to $\mathcal{F}_t$, we have

$$\begin{aligned}
c_1(t) &= \frac{L(T_0, T_0+\delta_0) + s_0}{D_0} B(t, T_1) E_{Q_{T_1}}^t \left[\mathcal{D}(T_0, T_1)\right] \\
&= \frac{L(T_0, T_0+\delta_0) + s_0}{D_0} \Big\{ B(t, T_1) E_{Q_{T_1}}^t \left[\mathcal{D}(T_0, t)\right] \\
&\quad + \sum_{i=n_0^+ + 1}^{n_0} B(t, T_1) E_{Q_{T_1}}^t \left[\mathbf{1}_{\{R_\ell(T_{0,i}) \le L(T_{0,i}, T_{0,i}+\delta_{0,i}) \le R_u(T_{0,i})\}}\right] \Big\} \\
&= \frac{L(T_0, T_0+\delta_0) + s_0}{D_0} \Bigg[B(t, T_1) \mathcal{D}(T_0, t) \\
&\quad + \sum_{i=n_0^+ + 1}^{n_0} V_r(t; T_{0,i}, T_1, \delta_{0,i}) \Bigg].
\end{aligned} \tag{8.2.26}$$

The valuation of the time-t value of the subsequent coupons is more complicated. Consider

$$
\begin{aligned}
& c_{j+1}(t) \\
&= B(t, T_{j+1}) E^t_{Q_{j+1}}\left[\frac{L(T_j, T_j+\delta_j)+s_j}{D_j}\mathcal{D}(T_j, T_{j+1})\right] \\
&= \left(\frac{s_j}{D_j} - \frac{1}{\delta_j D_j}\right) B(t, T_{j+1}) \sum_{i=1}^{n_j} E^t_{Q_{T_{j+1}}}\left[\mathbf{1}_{\{R_\ell(T_{j,i}) \le L(T_{j,i}, T_{j,i}+\delta_{j,i}) \le R_u(T_{j,i})\}}\right] \\
&\quad + \frac{B(t, T_{j+1})}{\delta_j D_j} \\
&\quad \sum_{i=1}^{n_j} E^t_{Q_{T_{j+1}}}\left[\frac{1}{B(T_j, T_{j+1})}\mathbf{1}_{\{R_\ell(T_{j,i}) \le L(T_{j,i}, T_{j,i}+\delta_{j,i}) \le R_u(T_{j,i})\}}\right].
\end{aligned} \tag{8.2.27}
$$

The first term can be expressed in terms of a portfolio of delayed range digital options, similar to that in (8.2.26). However, the evaluation of the second term requires the knowledge of the joint distribution of $\frac{1}{B(T_j, T_{j+1})}$ and $\mathbf{1}_{\{R_\ell(T_{j,i}) \le L(T_{j,i}, T_{j,i}+\delta_{j,i}) \le R_u(T_{j,i})\}}$. The details of the tedious calculation of the expectation of the joint distribution can be found in Nunes (2004). Eberlein and Kluge (2006) proposed using the adjusted forward measure $Q_{T_j, T_{j+1}}$ that avoids dealing with the joint distribution. The hint of this approach is outlined in Problem 8.23.

8.3 Caps and LIBOR Market Models

The two most simple interest rate derivatives traded in the financial market are the caplet and floorlet. A caplet guarantees that the interest rate charged on a floating rate loan at any given time will be the minimum of the prevailing floating rate (say, LIBOR) and a preset cap rate. If the rate rises above the cap rate, the holder receives cash flow from the issuer which exactly compensates the additional interest expense incurred beyond the cap rate; if otherwise, then no cash flow results. The caplet can be considered a call option on the floating LIBOR with the cap rate as the strike price.

On the other hand, a floorlet guarantees the holder to receive the maximum of the prevailing floating rate and the preset floor rate on a floating rate deposit. The holder is guaranteed to have a minimum rate level for his or her floating rate deposit. A floorlet can be seen as a put option on the floating LIBOR with the floor rate as the strike price.

A collar agreement specifies both the upper ceiling and lower floor for the floating interest rate that will be charged on a floating rate loan. When the interest rate lies between the bounds, the prevailing floating rate is charged; otherwise, the ceiling (floor) rate is charged when the floating rate is above (below) the upper (lower) bound. Therefore, a collar can be considered as a combination of long position in a caplet and short position in a floorlet. Usually the collar is structured in such a way that the price of the caplet offsets exactly the price of the floorlet so that the initial value of the collar is zero.

A caplet over the period $(T_{i-1}, T_i]$ possesses an optionality feature similar to the T_{i-1}-maturity European put option on a T_i-maturity discount bond since the exercise decision date is T_{i-1} while the payment is delivered at T_i. A cap can be considered a portfolio of European put bond options. Using the pricing technique developed in Sect. 8.2.1, we are able to derive the analytic price formula of a cap under the Gaussian HJM framework.

8.3.1 Pricing of Caps under Gaussian HJM Framework

Suppose a company is indebted in a floating LIBOR loan and has to pay interest payments at a series of coupon dates $T_1, T_2, \cdots, T_n$. Let $L_{i-1}(t)$ denote the forward LIBOR contracted at time t, where $t < T_{i-1}$, for the period $(T_{i-1}, T_i]$, $i = 1, 2, \cdots, n$. These LIBOR rates $L_0, L_1, \cdots, L_{n-1}$ are reset at $T_0, T_1, \cdots, T_{n-1}$, respectively, and let the accrual factor (year fraction) associated with the time interval $(T_{i-1}, T_i]$ be denoted by α_i, $i = 1, 2, \cdots, n$. To protect itself from rising interest rates, the company would like to lock the LIBOR at a preset ceiling rate L_c. The protection can be achieved by entering into an interest rate cap contract. The cap is a series of caplets, where the holder of the cap receives $\mathcal{N}\alpha_i \max(L_{i-1} - L_c, 0)$ at each payment date T_i, $i = 1, 2, \cdots, n$. Here, $\mathcal{N}$ is the notional principal of the cap and L_c is the ceiling rate. Likewise, a floor is a series of floorlets, where the floor holder is entitled to receive $\mathcal{N}\alpha_i \max(L_f - L_{i-1}, 0)$ at T_i, $i = 1, 2, \cdots, n$, where L_f is the preset floor rate. In our subsequent discussion, we take the notional value $\mathcal{N}$ to be unity.

Cap as a Portfolio of Bond Options

Consider the pricing model of a caplet whose ceiling rate is L_c. Recall that the cap holder is entitled to receive at time T_i an amount equal to $\alpha_i \max(L_{i-1}(T_{i-1}) - L_c, 0)$. The present value of this payment at time T_{i-1} is

$$\begin{aligned}
&\frac{\alpha_i}{1 + \alpha_i L_{i-1}(T_{i-1})} \max(L_{i-1}(T_{i-1}) - L_c, 0) \\
&= \max\left(1 - \frac{1 + \alpha_i L_c}{1 + \alpha_i L_{i-1}(T_{i-1})}, 0\right). \qquad (8.3.1)
\end{aligned}$$

Recall that the time-t price of the discount T_i-maturity bond $B(t; T_i)$ and the forward LIBOR $L_{i-1}(t)$ observe the following relation at time T_{i-1}:

$$\frac{1}{B(T_{i-1}, T_i)} = 1 + \alpha_i L_{i-1}(T_{i-1}).$$

The quantity $\frac{1+\alpha_i L_c}{1+\alpha_i L_{i-1}(T_{i-1})}$ can be considered as the value at time T_{i-1} of a discount bond that pays off $1 + \alpha_i L_c$ at time T_i. Hence, the payoff function in (8.3.1) is the same as that from a put option with expiration date T_{i-1} on a discount bond with maturity date T_i. The par value of the bond is $1 + \alpha_i L_c$ and the strike price of the put option is unity. As a consequence, an interest rate cap can be considered as a portfolio of European put options on discount bonds.

The time-t value of the caplet that pays $\alpha_i \max(L_{i-1}(T_{i-1}) - L_c, 0)$ at time T_i can be expressed as

$$C_i(t; T_{i-1}, T_i) = B(t, T_i) E^t_{Q_{T_i}}[\alpha_i \max(L_{i-1}(T_{i-1}) - L_c, 0)].$$

Since $L_{i-1}(T_{i-1})$ is $\mathcal{F}_{T_{i-1}}$-measurable, we may write

$$C_i(t; T_{i-1}, T_i) = B(t, T_{i-1}) E^t_{Q_{T_{i-1}}}[B(T_{i-1}, T_i)\alpha_i \max(L_{i-1}(T_{i-1}) - L_c, 0)],$$

and by virtue of (8.3.1) it can be simplified as

$$C_i(t; T_{i-1}, T_i) = B(t, T_{i-1}) E^t_{Q_{T_{i-1}}}[\max(1 - (1 + \alpha_i L_c) B(T_{i-1}, T_i), 0].$$

Assuming that the bond price process $B(t, T)$ under the risk neutral measure Q follows the Gaussian HJM model as defined in (8.1.12b), the time-t value of the caplet is given by

$$\begin{aligned} &C_i(t; T_{i-1}, T_i) \\ &= B(t, T_{i-1}) N(-d_2^{(i)}) - (1 + \alpha_i L_c) B(t, T_i) N(-d_1^{(i)}), \quad t < T_{i-1}, \end{aligned} \tag{8.3.2}$$

where

$$d_2^{(i)} = \frac{\ln \frac{(1+\alpha_i L_c) B(t,T_i)}{B(t,T_{i-1})} - \frac{1}{2}\overline{\sigma}_i^2(t)(T_{i-1} - t)}{\overline{\sigma}_i(t)\sqrt{T_{i-1} - t}}, \quad d_1^{(i)} = d_2^{(i)} + \overline{\sigma}_i(t)\sqrt{T_{i-1} - t},$$

$$\overline{\sigma}_i^2(t) = \frac{1}{T_{i-1} - t} \int_t^{T_{i-1}} \int_{T_{i-1}}^{T_i} \sum_{j=1}^m \sigma_F^j(s, u)\, du\, ds.$$

Since the interest rate cap is a series of caplets, the time-t value of the cap is then equal to

$$C(t; T_0, T_1, \cdots, T_n) = \sum_{i=1}^n C_i(t; T_{i-1}, T_i), \quad t < T_0. \tag{8.3.3}$$

8.3.2 Black Formulas and LIBOR Market Models

The analytic expression of the above cap price formula (8.3.3) appears to be quite daunting. This is not surprising since we have used the model that is based on (unobservable) instantaneous forward rates to price a derivative on discrete LIBORs. The complexity in the analytic formula leads to implementation difficulties in calibrating volatility functions to market cap data. This motivates the development of the so-called market interest rate models whose underlying state variables are observable market interest rates. We illustrate the convenience of the Black formula and implied Black volatilities in pricing caplets and these help construct the Lognormal LIBOR Market models.

Black Formula

The fixed income market has been adopting the Black formula [see (8.1.16)] for pricing a caplet/cap by assuming that the underlying forward LIBOR process is lognormally distributed with zero drift under some "market probability" Q_m. In its simplest form, let σ_{i-1}^L denote the constant Black volatility of the forward LIBOR process $L_{i-1}(t)$. Suppose $L_{i-1}(t)$ is governed by

$$dL_{i-1}(t) = L_{i-1}(t)\sigma_{i-1}^L \, dZ_t^m, \tag{8.3.4}$$

where Z_t^m is a Brownian process under Q_m. The Black formula for the time-t value of the caplet that pays $\alpha_i \max(L_{i-1}(T_{i-1}) - L_c, 0)$ at time T_i is given by

$$\begin{aligned} c_i^B(t; T_{i-1}, T_i) &= \alpha_i B(t, T_i) E_{Q_m}^t [\max(L_{i-1}(T_{i-1}) - L_c, 0)] \\ &= \alpha_i B(t, T_i)[L_{i-1}(t)N(d_1^{i-1}) - L_c N(d_2^{i-1})], \end{aligned} \tag{8.3.5}$$

where

$$d_1^{i-1} = \frac{\ln \frac{L_{i-1}(t)}{L_c} + \frac{\left(\sigma_{i-1}^L\right)^2}{2}(T_{i-1} - t)}{\sigma_{i-1}^L \sqrt{T_{i-1} - t}}, \quad d_2^{i-1} = \frac{\ln \frac{L_{i-1}(t)}{L_c} - \frac{\left(\sigma_{i-1}^L\right)^2}{2}(T_{i-1} - t)}{\sigma_{i-1}^L \sqrt{T_{i-1} - t}}.$$

For the cap, which is a series of caplets, its time-t value is given by

$$c^B(t; T_0, T_1, \cdots, T_n) = \sum_{i=1}^{n} c_i^B(t; T_{i-1}, T_i), \quad t < T_0. \tag{8.3.6}$$

The "market practice" assumes that the forward LIBOR for any maturity has a lognormal probability distribution.

Implied Black Volatilities

In the financial markets, cap prices are quoted in terms of implied cap volatilities, similar to the implied equity volatility for equity options. Note that caps instead of caplets are traded. There are two types of implied Black volatilities that can be defined, one with reference to the caps directly while the other is related to the deduced prices of caplets obtained by "boot-strapping" of the cap prices.

Let the tenor structure consist of a fixed set of dates $T_0, T_1, \cdots, T_n$, and $t \leq T_0$. For each i, $i = 1, 2, \cdots, n$, we have the market price $c^{mkt}(t; T_0, \cdots, T_i)$ of the traded cap with tenor $\{T_0, T_1, \cdots, T_i\}$. According to (8.3.3), the "derived" market price of the caplet over $(T_{i-1}, T_i]$ can be defined by

$$c_i^{mkt}(t; T_{i-1}, T_i) = c^{mkt}(t; T_0, \cdots, T_i) - c^{mkt}(t; T_0, \cdots, T_{i-1}). \tag{8.3.7}$$

The implied flat Black cap volatility is the volatility implied by the market cap prices. Consider the cap with tenor $\{T_0, \cdots, T_i\}$, we define the flat volatility σ_i^f, $i = 1, 2, \cdots, n$, by solving

$$\mathcal{C}_i^{mkt}(t;T_0,\cdots,T_i) = \sum_{k=1}^{i} \mathcal{C}_i^B(t;T_{k-1},T_k,\sigma_i^f). \tag{8.3.8}$$

Here, the volatility values in the Black caplet formula [see (8.3.5)] are set to be σ_i^f. On the other hand, the implied spot volatilities $\sigma_1^s,\cdots,\sigma_n^s$ are defined in terms of "derived" caplet prices. They are obtained by solving

$$\mathcal{C}_i^{mkt}(t;T_{i-1},T_i) = \mathcal{C}_i^B(t;T_{i-1},T_i,\sigma_i^s), \quad i=1,2,\cdots,n. \tag{8.3.9}$$

Lognormal LIBOR Market Models

We wish to construct an arbitrage free model of the LIBOR process such that the Black formula of the caplet is produced in a logically consistent manner. The various formulations of these models were discussed by Miltersen, Sandmann and Sondermann (1997), Brace, Gatarek and Musiela (1997) and Jamshidian (1997).

We have seen that the forward LIBOR $L_i(t)$ is a martingale under the forward measure $Q_{T_{i+1}}$, where the discount bond prices $B(t,T_{i+1})$ is a numeraire. Assuming that we are given an initial term structure of interest rates, specified by a family of discount bond price $B(0,T_i)$, $i=0,1,\cdots,n$, with the property that $B(0,T_i) > B(0,T_{i+1})$ (implying positive interest rates). Let $\boldsymbol{\sigma}_i^L(t) = (\sigma_{i,1}^L(t)\cdots\sigma_{i,m}^L(t))$ be a set of bounded adapted m-dimensional processes, $i=1,2,\cdots,m$, representing the volatilities of the forward LIBOR $L_i(t)$. Let $\mathbf{Z}^i(t) = (Z_1^i(t)\cdots Z_m^i(t))$ be a family of the m-dimensional Brownian processes under the respective forward measure Q_{T_i}, $i=1,2,\cdots,n$. The Lognormal LIBOR Market (LLM) model assumes that $L_i(t)$ satisfies the following stochastic differential equation under the T_{i+1}-forward measure $Q_{T_{i+1}}$ [see also (8.1.23)]

$$dL_i(t) = L_i(t)\sum_{j=1}^{m} \sigma_{i,j}^L(t)\,dZ_j^{i+1}(t) \tag{8.3.10}$$

with the initial condition:

$$L_i(0) = \frac{B(0,T_i)-B(0,T_{i+1})}{\alpha_{i+1}B(0,T_{i+1})}, \quad i=0,1,\cdots,n-1.$$

Caplet Pricing Revisited

Under the LLM model, we reconsider the pricing of the ith caplet that pays $\alpha_i\max(L_{i-1}(T_{i-1})-L_c,0)$ at time T_i, $i=1,2,\cdots,n$. The price of the ith caplet is given by

$$\mathcal{C}_i^{LLM}(t;T_{i-1},T_i) = \alpha_i B(t,T_i) E_{Q_{T_i}}^t[\max(L_{i-1}(T_{i-1})-L_c,0)]. \tag{8.3.11}$$

Suppose $L_{i-1}(t)$ is governed by (8.3.10), $L_{i-1}(t)$ is lognormally distributed under Q_{T_i}, where

$$L_{i-1}(T_{i-1}) = L_{i-1}(t)\ \exp\left(\int_t^{T_{i-1}} \sum_{j=1}^m \sigma_{i-1,j}^L(u)\, dZ_j^i(u) - \frac{1}{2}\int_t^{T_{i-1}} \sum_{j=1}^m [\sigma_{i-1,j}^L(u)]^2\, du\right). \quad (8.3.12)$$

We then obtain

$$c_i^{LLM}(t; T_{i-1}, T_i) = \alpha_i B(t, T_i)[L_{i-1}(t)N(d_1^i) - L_c N(d_2^i)], \quad (8.3.13)$$

where

$$d_1^i = \frac{\ln \frac{L_{i-1}(t)}{L_c} + \frac{v_{i-1}^2(t)}{2}}{v_{i-1}(t)}, \quad d_2^i = d_1^i - v_{i-1}(t),$$

$$v_{i-1}^2(t) = \int_t^{T_{i-1}} \left[\sum_{j=1}^m \sigma_{i-1,j}^L(u)\right]^2 du.$$

In order to obtain the Black-type formula for caplet pricing, it is necessary to have the lognormality property under the appropriate forward measure for each LIBOR. If we take the volatility of $L_{i-1}(t)$ to be a constant scalar, then we recover the Black caplet price formula [see (8.3.5)].

LIBOR in Arrears

In standard LIBOR derivatives, the claim on the LIBOR L_{i-1} for period $(T_{i-1}, T_i]$ is paid at the end of the interval, namely, at time T_i. However, in a LIBOR-in-arrears payment, the LIBOR L_i over period $(T_i, T_{i+1}]$ is paid at time T_i, the starting time of the interval. We would like to find the price $\widetilde{V}_L(t)$ of this LIBOR-in-arrears payment at the current time t, $t < T_i$, using the LLM model. Since the payment is made at T_i, it is natural to use $B(t, T_i)$ as the numeraire so that

$$\widetilde{V}_L(t) = B(t, T_i) E_{Q_{T_i}}^t [\alpha_{i+1} L_i(T_i)]. \quad (8.3.14a)$$

However, since $L_i(T_i)$ is a $Q_{T_{i+1}}$-martingale, it is preferable to use the forward measure $Q_{T_{i+1}}$. Applying the appropriate change of measure, we obtain

$$\widetilde{V}_L(t) = \alpha_{i+1} B(t, T_i) E_{Q_{T_{i+1}}}^t \left[L_i(T_i) \frac{dQ_{T_i}}{dQ_{T_{i+1}}}\right], \quad (8.3.14b)$$

where $\frac{dQ_{T_i}}{dQ_{T_{i+1}}}$ is the Radon–Nikodym derivative that effects the change of measure from Q_{T_i} to $Q_{T_{i+1}}$. Note that $\frac{dQ_{T_i}}{dQ_{T_{i+1}}}$ is given by the ratio of the numeraires:

$$\frac{dQ_{T_i}}{dQ_{T_{i+1}}} = \frac{B(T_i, T_i)}{B(T_i, T_{i+1})} \frac{B(t, T_{i+1})}{B(t, T_i)} = \frac{1 + \alpha_{i+1} L_i(T_i)}{1 + \alpha_{i+1} L_i(t)}.$$

Combining the above results, we obtain

$$\begin{aligned}\widetilde{V}_L(t) &= \alpha_{i+1} B(t,T_i) E^t_{Q_{T_{i+1}}}\left[\frac{L_i(T_i)+\alpha_{i+1}L_i^2(T_i)}{1+\alpha_{i+1}L_i(t)}\right]\\ &= \alpha_{i+1}B(t,T_i)L_i(t)\left\{1+\frac{\alpha_{i+1}}{L_i(t)}\frac{\mathrm{var}_{Q_{T_{i+1}}}[L_i(T_i)]}{1+\alpha_{i+1}L_i(t)}\right\}. \qquad (8.3.15)\end{aligned}$$

The last factor $1+\frac{\alpha_{i+1}}{L_i(t)}\frac{\mathrm{var}_{Q_{T_{i+1}}}[L_i(T_i)]}{1+\alpha_{i+1}L_i(t)}$ is recognized as the *convexity correction factor*. Convexity adjustment arises due to the slight mismatch between the date of payment and the "appropriate" forward LIBOR. Based on the lognormal distribution assumption of LIBOR process under the LLM model [see (8.3.12)], we obtain

$$\widetilde{V}_L(t) = \alpha_{i+1}B(t,T_i)L_i(t)\left\{1+\frac{\alpha_{i+1}L_i(T_i)}{1+\alpha_{i+1}L_i(t)}\left[e^{v_i^2(t)}-1\right]\right\}, \qquad (8.3.16)$$

where

$$v_i^2(t) = \int_t^{T_i}\left[\sum_{j=1}^m \sigma_{i,j}^L(u)\right]^2 du.$$

Terminal Measure

When we price a LIBOR derivative whose payoff depends on different LIBOR values, it is more convenient to work with the LIBOR model that puts different LIBORs under a single measure instead of having each LIBOR under its respective forward measure. A natural choice would be the terminal measure Q_{T_n} with the longest maturity bond as the numeraire. Let $\mathbf{Z}^n(t) = (Z_1^n(t)\cdots Z_m^n(t))^{\mathrm{T}}$ be an m-dimensional Q_{T_n}-Brownian process and $\boldsymbol{\sigma}_{n-1}^L(t) = (\sigma_{n-1,1}^L(t)\cdots\sigma_{n-1,m}^L(t))^{\mathrm{T}}$ be an m-dimensional volatility vector function. Under the terminal measure, the terminal LIBOR $L_{n-1}(t)$ is a martingale. The Lognormal LIBOR Market model postulates

$$\frac{dL_{n-1}(t)}{L_{n-1}(t)} = \sum_{j=1}^m \sigma_{n-1,j}^L(t)\,dZ_j^n(t). \qquad (8.3.17)$$

Starting from the vector Q_{T_n}-Brownian process $\mathbf{Z}^n(t)$, we show how to apply the Girsanov Theorem to find the successive vector Brownian processes under the respective measures $Q_{T_{n-1}}, Q_{T_{n-2}}, \cdots, Q_{T_1}$.

As a general step, we consider the Radon–Nikodym derivative $\zeta_i^{i+1}(t)$ that effects the change from the measure Q_{T_i} to the measure $Q_{T_{i+1}}$. Recall that

$$\begin{aligned}\zeta_i^{i+1}(t) &= \left.\frac{dQ_{T_i}}{dQ_{T_{i+1}}}\right|_{\mathcal{F}_t} = \frac{B(t,T_i)/B(0,T_i)}{B(t,T_{i+1})/B(0,T_{i+1})}\\ &= \frac{B(0,T_{i+1})}{B(0,T_i)}[1+\alpha_{i+1}L_i(t)],\end{aligned}$$

we then deduce that

$$d\zeta_i^{i+1}(t) = \frac{B(0, T_{i+1})}{B(0, T_i)} \alpha_{i+1}\, dL_i(t). \tag{8.3.18}$$

Assuming that the dynamic of $L_i(t)$ is governed by

$$\frac{dL_i(t)}{L_i(t)} = \sum_{j=1}^{m} \sigma_{i,j}^L(t)\, dZ_j^{i+1}(t),$$

where $\mathbf{Z}^{i+1}(t) = (Z_1^{i+1}(t) \cdots Z_m^{i+1}(t))^{\mathrm{T}}$ is an m-dimensional $Q_{T_{i+1}}$-Brownian process, we then have

$$d\zeta_i^{i+1}(t) = \zeta_i^{i+1}(t) \sum_{j=1}^{m} \gamma_{i,j}(t)\, dZ_j^{i+1}(t), \tag{8.3.19}$$

where

$$\gamma_{i,j}(t) = \frac{\alpha_{i+1} L_i(t)}{1 + \alpha_{i+1} L_i(t)} \sigma_{i,j}^L(t), \quad j = 1, 2, \cdots, m.$$

The solution to (8.3.19) is found to be

$$\zeta_i^{i+1}(t) = \exp\left(\int_0^t \sum_{j=1}^{m} \gamma_{i,j}(u)\, dZ_j^{i+1}(u) - \frac{1}{2} \int_0^t \sum_{j=1}^{m} \gamma_{i,j}(u)^2\, du \right),$$
$$0 < t < T_i. \tag{8.3.20}$$

By virtue of the Girsanov Theorem, the above solution allows us to define the Q_{T_i}-Brownian process $\mathbf{Z}^i(t) = (Z_1^i(t) \cdots Z_m^i(t))^{\mathrm{T}}$ which is derived from the m-dimensional $Q_{T_{i+1}}$-Brownian process $\mathbf{Z}^{i+1}(t)$ by the following relation

$$dZ_j^i(t) = dZ_j^{i+1}(t) - \gamma_{i,j}(t)\, dt, \quad t \in [0, T_i], \quad j = 1, 2, \cdots, m. \tag{8.3.21}$$

Given the vector process $\mathbf{Z}^n(t)$ which is Brownian under the terminal measure Q_{T_n}, then $L_{n-2}(t)$ follows the process

$$\frac{dL_{n-2}(t)}{L_{n-2}(t)} = \sum_{j=1}^{m} \sigma_{n-2,j}^L(t) \left(dZ_j^n(t) - \frac{\alpha_{n-1} L_{n-2}(t) \sigma_{n-2,j}^L}{1 + \alpha_{n-1} L_{n-2}(t)}\, dt \right)$$

under the terminal measure Q_{T_n}. We see that while $L_{n-2}(t)$ is a martingale under the forward measure $Q_{T_{n-1}}$, it would not be a martingale under Q_{T_n}. Deductively, we obtain that $L_i(t)$, $i = n-2, n-3, \cdots, 0$, follows the process

$$\frac{dL_i(t)}{L_i(t)} = \sum_{j=1}^{m} \left[-\sum_{k=i+1}^{n} \frac{\alpha_{k+1} L_k(t)}{1 + \alpha_{k+1} L_k(t)} \sigma_{k,j}^L(t)\, dt + \sigma_{k,j}^L(t)\, dZ_j^n(t) \right]. \tag{8.3.22}$$

Generalization
Instead of using the terminal forward measure Q_{T_n}, the LIBOR model formulation of $L_i(t)$ can be generalized to any forward measure $Q_{T_k}, k \neq i$. For simplicity of discussion, we take $m = 1$ so that the driving stochastic process of $L_i(t)$ is modeled by a scalar Brownian process. In a similar manner, it can be shown that the dynamics of $L_i(t)$ under Q_{T_k} can be expressed as

$$\frac{dL_i(t)}{L_i(t)} = \sigma_i^L(t)\,dZ^k(t) + \sigma_i^L(t)\gamma_{ik}(t)\,dt, \tag{8.3.23}$$

where $Z^k(t)$ is Q_{T_k}-Brownian and

$$\gamma_{ik}(t) = \begin{cases} \displaystyle\sum_{j=k}^{i} \frac{\alpha_{j+1}L_j(t)}{1+\alpha_{j+1}L_j(t)}\sigma_j^L(t) & k < i \\ 0 & k = i \\ \displaystyle -\sum_{j=k+1}^{n} \frac{\alpha_{j+1}L_j(t)}{1+\alpha_{j+1}L_j(t)}\sigma_j^L(t) & k > i \end{cases}.$$

The calculation steps used to establish the above result are outlined in Problem 8.36.

In general, it is impossible to solve for $L_i(t)$ in closed form since the drift terms in the stochastic differential equation (8.3.23) contains nonlinear function of $L_i(t)$. By adopting some ingenious analytic approximations, various analytic lognormal approximations of the forward LIBORs have been obtained in the literature [say, Daniluk and Gatarek (2005); see Problem 8.33].

8.4 Swap Products and Swaptions

Swap contracts involve the exchange of cash flows between the two parties according to some specified rules of calculating the cash flows. For example, in the cross-currency swaps, the payments of the two legs are dependent on the floating interest rates in two different currency worlds. Many new swap products with various optionality features have been structured to meet specific needs of the end users. Forward swaps and swaptions are forward contracts and options, respectively, whose underlying asset is a swap. A swaption gives the holder the right to enter into a specified interest rate swap at some future time while the holder of a forward swap faces the obligation to enter into the underlying swap. We first show how the forward swap rate is related to a portfolio of traded discount bond prices. Similar to the Lognormal LIBOR Market model, we construct the Lognormal Swap Rate model to price a swaption. Using the swap measure where the annuity (a portfolio of discount bond prices) is used as the numeraire, the forward swap rate is a martingale under the swap measure. As a consequence, the price of a swaption admits the Black type formula. The Swap Rate model and the LIBOR model are incompatible, so there will be no simple swaption price formula under the Lognormal LIBOR Market model.

We illustrate how to obtain an analytic approximation formula of a swaption under the Lognormal LIBOR Market model. Finally, we consider a special type of cross-currency swap, called the differential swap, where at least one of the floating legs is referenced to the floating interest rate of a foreign currency world while all swap payments among the two counterparties are denominated in the domestic currency world. We present detailed discussion on the pricing and hedging of the differential swaps.

8.4.1 Forward Swap Rates and Swap Measure

We present the derivation of the forward swap rate using the notion of forward measures and the time-t value of the forward swap, and subsequently the forward swap measure is defined using a portfolio of bonds as the numeraire. The Black formula for a swaption using the swap measure can be obtained under the so-called Lognormal Swap Rate model.

Forward Payer Swap

We would like to find the value of a forward start fixed-for-floating interest rate swap with unit notional. We consider a tenor structure $T_0 < T_1 < \cdots < T_n$, where $\{T_0, \cdots, T_{n-1}\}$ are the reset dates at which the relevant LIBOR $\{L_0, \cdots, L_{n-1}\}$ are determined. The payment dates are $\{T_1, \cdots, T_n\}$ at which the floating leg payer pays $\alpha_i[L_{i-1}(T_{i-1}) - K]$ at T_i, where α_i is the accrual factor of the time period $(T_{i-1}, T_i]$, $i = 1, 2, \cdots, n$, and K is the preset fixed rate (see Fig. 7.4 for the cash flows of this vanilla interest rate swap). The time-t value of this forward payer swap, $t < T_0$, is given by

$$\begin{aligned} s_{0,n}(t) &= E_Q^t\left[\sum_{i=1}^{n} \alpha_i e^{-\int_t^{T_i} r_u\,du}[L_{i-1}(T_{i-1}) - K]\right] \\ &= \sum_{i=1}^{n} \alpha_i B(t, T_i) E_{Q_{T_i}}^t\left[L_{i-1}(T_{i-1}) - K\right], \end{aligned} \tag{8.4.1}$$

where Q is the risk neutral measure and Q_{T_i} is the T_i-forward measure with $B(t, T_i)$ as the numeraire. Since

$$E_{Q_{T_i}}^t[L_{i-1}(T_{i-1})] = L_{i-1}(t) = \frac{B(t, T_{i-1}) - B(t, T_i)}{\alpha_i B(t, T_i)},$$

we obtain

$$s_{0,n}(t) = \sum_{i=1}^{n}[B(t, T_{i-1}) - B(t, T_i)] - K\sum_{i=1}^{n} \alpha_i B(t, T_i). \tag{8.4.2}$$

Likewise, we define the time-t value of the forward swap rate $K_t[T_0, T_n]$ to be the fixed swap rate K chosen in the manner that $s_{0,n}(t)$ becomes zero. We then obtain

$$K_t[T_0, T_n] = \frac{B(t, T_0) - B(t, T_n)}{\sum_{i=1}^n \alpha_i B(t, T_i)}. \tag{8.4.3}$$

This gives the same formula that is derived based on the notion that an interest rate swap is a portfolio of Forward Rate Agreements [see (7.1.4)]. Substituting $K_t[T_0, T_1]$ into (8.4.2), we obtain

$$\begin{aligned} s_{0,n}(t) &= \{K_t[T_0, T_n] - K\} \sum_{i=1}^n \alpha_i B(t, T_i) \\ &= \{K_t[T_0, T_n] - K\} \widehat{B}(t; T_0, T_n). \end{aligned} \tag{8.4.4}$$

The sum $\widehat{B}(t; T_0, T_n) = \sum_{i=1}^n \alpha_i B(t, T_i)$ is commonly called the present value of the basis point (PVBP or PV01 for short) in the financial market. Note that the value of the forward swap is proportional to the difference of the prevailing swap rate and the fixed swap rate, and the proportional factor is PVBP.

The forward swap rates are expressed in terms of discount bond prices. Since interest rate swaps become more frequently traded than the discount bonds in the financial markets, market quotes on the forward swap rates are more commonly available. In the reverse sense, we may use the market quotes on the swap rates to extract the market view on the discount curve. This reverse transformation can be accomplished using the *bootstrap method*. Given the discount bond price $B(t, T_0)$ and all forward swap rates $K_t[T_0, T_i]$, $i = 1, 2, \cdots, n$, we start with the relation

$$K_t[T_0, T_1] = \frac{B(t, T_0) - B(t, T_1)}{\alpha_1 B(t, T_1)}$$

to obtain

$$B(t, T_1) = \frac{B(t, T_0)}{1 + \alpha_1 K_t[T_0, T_1]}.$$

Once $B(t, T_1)$ is obtained, we can deductively derive the other discount bond prices as follows:

$$B(t, T_i) = \frac{B(t, T_0) - K_t[T_0, T_i]\sum_{j=1}^{i-1} \alpha_j B(t, T_j)}{1 + \alpha_i K_t[T_0, T_i]}, \quad i = 2, 3, \cdots, n. \tag{8.4.5}$$

Swap Measure and Black Formula for Swaptions

One can develop the Black formula for a swaption, similar to that for a caplet, by suitably defining a swap measure. Consider a T_0-maturity payer swaption with strike swap rate X, the holder of this swaption has the right but not the obligation to enter into a swap with fixed rate X at time T_0. That is, the net payment at time T_i is $L_{i-1}(T_{i-1}) - X$, $i = 1, 2, \cdots, n$. The swaption is exercised only when $K_{T_0}[T_0, T_n] > X$ and the payoff of the swaption to the holder is

$$\begin{aligned} &\max(K_{T_0}[T_0, T_n] - X, 0) \sum_{i=1}^n \alpha_i B(T_0, T_i) \\ &= \widehat{B}(T_0; T_0, T_n) \max(K_{T_0}[T_0, T_n] - X, 0). \end{aligned} \tag{8.4.6}$$

To define the so-called swap measure, we take $\widehat{B}(t; T_0, T_n)$ as the numeraire. Now, under the swap measure $Q_{s0,n}$ associated with the *annuity numeraire* $\widehat{B}(t; T_0, T_n)$, all derivative prices normalized by $\widehat{B}(t; T_0, T_n)$ are martingales under $Q_{s0,n}$. Let $V(t; T_0, T_n, X)$ be the value of the above payer swaption with strike swap rate X, we then have

$$\frac{V(t; T_0, T_n, X)}{\widehat{B}(t; T_0, T_n)} = E_{Q_{s0,n}}[\max(K_{T_0}[T_0, T_n] - X, 0)]. \tag{8.4.7}$$

The swap rate model assumes that the forward swap rate $K_t[T_0, T_n]$ is a lognormal martingale under $Q_{s0,n}$. That is,

$$\frac{dK_t[T_0, T_n]}{K_t[T_0, T_n]} = \sigma^K_{0,n}(t)\, dZ_t^{0,n}, \tag{8.4.8}$$

where $Z_t^{0,n}$ is $Q_{s0,n}$-Brownian. By evaluating the expectation in (8.4.7), we obtain the Black formula for the payer swaption as follows:

$$V(t; T_0, T_n, X) = \widehat{B}(t; T_0, T_n)\{K_t[T_0, T_n]N(d_1) - XN(d_2)\}, \tag{8.4.9}$$

where

$$d_1 = \frac{\ln \frac{K_t[T_0,T_n]}{X} + \frac{\widehat{\sigma}^K_{0,n}(t,T_0)^2}{2}(T_0 - t)}{\widehat{\sigma}^K_{0,n}(t, T_0)\sqrt{T_0 - t}}, \quad d_2 = d_1 - \widehat{\sigma}^K_{0,n}(t, T_0)\sqrt{T_0 - t},$$

$$\widehat{\sigma}^K_{0,n}(t, T_0)^2 = \frac{1}{T_0 - t}\int_t^{T_0} \sigma^K_{0,n}(u)^2\, du.$$

Volatilities of Forward Swap Rates

One may relate the volatility $\sigma^K_{0,n}$ of the forward swap rate $K_t[T_0, T_n]$ to the bond price volatilities. Let $\sigma^i_B(t, T_i)$ denote the volatility of the discount bond price process $B(t, T_i)$, $i = 1, 2, \cdots, n$, and define the weight factor w_j by

$$w_j(t) = \frac{\alpha_j B(t, T_j)}{\sum_{k=1}^n \alpha_k B(t, T_k)}, \quad j = 1, 2, \cdots, n. \tag{8.4.10}$$

Since $B(t, T_i)/\widehat{B}(t; T_0, T_n)$ is a martingale under $Q_{s0,n}$ and noting $\sum_{j=1}^n w_j = 1$, we then have

$$\begin{aligned}\frac{d\left(\frac{B(t,T_i)}{\widehat{B}(t;T_0,T_n)}\right)}{\frac{B(t,T_i)}{\widehat{B}(t;T_0,T_n)}} &= \left[\sigma^i_B(t, T_i) - \sum_{j=1}^n w_j(t)\sigma^j_B(t, T_j)\right] dZ_t^{0,n}\\ &= \sum_{j=1}^n w_j(t)\left[\sigma^i_B(t, T_i) - \sigma^j_B(t, T_j)\right] dZ_t^{0,n}.\end{aligned}$$

Taking $i = 0$ and $i = n$ in the above equation and subtracting the resulting expressions, we obtain

$$
\begin{aligned}
dK_t[T_0, T_n] &= \left\{ \frac{B(t, T_0)}{\widehat{B}(t; T_0, T_n)} \sum_{j=1}^{n} w_j(t)[\sigma_B^0(t, T_0) - \sigma_B^j(t, T_j)] \right. \\
&\quad \left. - \frac{B(t, T_n)}{\widehat{B}(t; T_0, T_n)} \sum_{j=1}^{n} w_j(t)[\sigma_B^n(t, T_0) - \sigma_B^j(t, T_j)] \right\} dZ_t^{0,n} \\
&= K_t[T_0, T_n] \left\{ \sum_{j=1}^{n} w_j(t)[\sigma_B^0(t, T_0) - \sigma_B^j(t, T_j)] \right. \\
&\quad \left. + \frac{B(t, T_n)}{B(t, T_0) - B(t, T_n)} [\sigma_B^0(t, T_0) - \sigma_B^n(t, T_n)] \right\} dZ_t^{0,n}
\end{aligned}
$$

so that

$$
\begin{aligned}
\sigma_{0,n}^K(t) &= \sum_{j=1}^{n} w_j(t)[\sigma_B^0(t, T_0) - \sigma_B^j(t, T_j)] \\
&\quad + \frac{B(t, T_n)}{B(t, T_0) - B(t, T_n)} [\sigma_B^0(t, T_0) - \sigma_B^n(t, T_n)]. \qquad (8.4.11)
\end{aligned}
$$

The above formula relates the forward swap rate volatility $\sigma_{0,n}^K(t)$ to the volatilities of the discount bond prices under the Gaussian HJM framework. Alternatively, one may be interested in relating $\sigma_{0,n}^K(t)$ to the volatilities of the LIBOR under the Lognormal LIBOR Market model.

First, let us express $K_t[T_0, T_n]$ in terms of a weighted sum of LIBORs. By rewriting the forward swap rate as

$$
K_t[T_0, T_n] = \frac{\displaystyle\sum_{i=1}^{n} \alpha_i B(t, T_i) \left[\frac{B(t, T_{i-1}) - B(t, T_i)}{\alpha_i B(t, T_i)} \right]}{\displaystyle\sum_{i=1}^{n} \alpha_i B(t, T_i)},
$$

so we obtain

$$
K_t[T_0, T_n] = \sum_{i=1}^{n} w_i(t) L_{i-1}(t), \qquad (8.4.12)
$$

where $w_i(t)$ is defined as in (8.4.10). We make the approximation of "freezing" the weights at the values at the current time t_0 so that

$$
K_t[T_0, T_n] \approx \sum_{i=1}^{n} w_i(t_0) L_{i-1}(t), \quad t_0 < t < T_0. \qquad (8.4.13a)
$$

This leads to a simple analytic representation of the differential of $K_t[T_0, T_n]$, where

$$dK_t[T_0, T_n] \approx \sum_{i=1}^{n} w_i(t_0)\, dL_{i-1}(t), \quad t_0 < t < T_0. \tag{8.4.13b}$$

The forward LIBOR process $L_{i-1}(t)$ is assumed to follow the lognormal process:

$$\frac{dL_{i-1}(t)}{L_{i-1}(t)} = \sigma_i^L(t)\, dZ^i(t),$$

where $Z^i(t)$ is Q_{T_i}-Brownian, $i = 1, 2, \cdots, n$. Further, we assume $dZ^i(t)\, dZ^j(t) = \rho_{ij}\, dt$, where ρ_{ij} is the correlation coefficient between the Brownian processes $Z^i(t)$ and $Z^j(t)$. The approximate percentage quadratic variation of $K_t[T_0, T_n]$ is then given by

$$\left(\frac{dK_t[T_0, T_n]}{K_t[T_0, T_n)}\right)\left(\frac{dK_t[T_0, T_n]}{K_t[T_0, T_n]}\right)$$
$$\approx \frac{\sum_{i=1}^{n}\sum_{j=1}^{n} w_i(t_0)w_j(t_0)L_{i-1}(t)L_{j-1}(t)\rho_{ij}\sigma_i^L(t)\sigma_j^L(t)\, dt}{K_t[T_0, T_n]^2}. \tag{8.4.14}$$

The above approximate variance can be substituted in the swaption price formula (8.4.9). This gives an approximate value of the swaption as priced under the LIBORs and their volatilities.

The accuracy of this "freezing" weights approximation procedure has been examined theoretically and through numerical tests (Brigo and Liinev, 2005). Hull and White (2000) proposed another slightly more elaborate procedure of deriving a similar approximation by *not* freezing the weights in the first step when the differential of $K_t[T_0, T_n]$ is calculated. Both approximations are shown to exhibit sufficiently high accuracy when compared to numerical results obtained from direct Monte Carlo simulation procedures. This is not surprising since the Lognormal LIBOR Market model and the Lognormal Swap Market model are distributionally "close". More recent developments in the theory and calibration of the Swap Rate Market models can be found in Galluccio and Ly (2007).

8.4.2 Approximate Pricing of Swaption under Lognormal LIBOR Market Model

The price formula of a swaption is seen to admit the Black type formula when priced under the Lognormal Swap Market model when the annuity $\widehat{B}(t; T_0, T_n)$ is used as the numeraire. As the Swap Market model and LIBOR Market model are incompatible, the forward swap rate does not yield a lognormal volatility structure under the LIBOR Market model. These inconsistencies lead to difficulties in the calibration of the volatility functions that fits for both the cap and swaption markets. Though a forward swap rate and a forward LIBOR could not have a lognormal volatility structure simultaneously under a common measure, the inconsistency is small. Here, we present the lognormal approximation of the forward swap rate under the Lognormal LIBOR Market model and apply the approximation to price a swaption under the LIBOR model.

First, we let A denote the event that the swaption is exercised at T_0 and $\mathbf{1}_A$ is the corresponding indicator random variable. Note that event A occurs if and only if

$$K_{T_0}[T_0, T_n] = \frac{B(T_0, T_0) - B(T_0, T_n)}{\sum_{i=1}^{n} \alpha_i B(T_0, T_i)} > X$$

$$\Leftrightarrow 1 > (1 + \alpha_n X) B(T_0, T_n) + \sum_{i=1}^{n-1} \alpha_i X B(T_0, T_i)$$

$$= B_{fix}(T_0; T_0, T_n, X), \tag{8.4.15}$$

where $B_{fix}(t; T_0, T_n, X)$ is the time-t value of the unit par fixed-coupon bond paying fixed rate X on the payment dates: $T_1, \cdots, T_n$ and par at T_n. More precisely, the cash payment c_i paid at T_i by the fixed-coupon bond is given by

$$c_i = \begin{cases} \alpha_i X & i = 1, 2, \cdots, n-1 \\ 1 + \alpha_n X & i = n \end{cases}.$$

The payoff of the swaption at expiry T_0 can be written as [see (8.4.6)]

$$\max(K_{T_0}[T_0, T_n] - X, 0) \sum_{i=1}^{n} \alpha_i B(T_0, T_i)$$

$$= \left[1 - \sum_{i=1}^{n} c_i B(t, T_i) \right] \mathbf{1}_A$$

$$= \sum_{i=1}^{n} \alpha_i [L_{i-1}(T_0) - X] B(T_0, T_i) \mathbf{1}_A. \tag{8.4.16}$$

At time $t < T_0$, the time-t value of the swaption is given by

$$V(t; T_0, T_n, X) = \sum_{i=1}^{n} \alpha_i B(t, T_i) E_{Q_{T_i}}[(L_{i-1}(T_0) - X) \mathbf{1}_A]. \tag{8.4.17}$$

The remaining procedures amount to the evaluation of

$$E_{Q_{T_i}}[(L_{i-1}(T_0) - X) \mathbf{1}_A], \quad i = 1, 2, \cdots, n,$$

under the Lognormal LIBOR Market model. Recall that

$$\mathbf{1}_A = 1 \quad \Leftrightarrow \quad B_{fix}(T_0; T_0, T_n, X) < 1,$$

and the difficulties in the expectation calculations stem from the dependence of $B_{fix}(T_0; T_0, T_n, X)$ on all LIBORs since

$$B_{fix}(T_0; T_0, T_n, X) = \sum_{i=1}^{n} c_i B(T_0, T_i)$$
$$= \sum_{i=1}^{n} c_i \prod_{j=1}^{i} \frac{1}{1+\alpha_j L_{j-1}(T_0)}. \qquad (8.4.18)$$

The Lognormal LIBOR Market model assumes that the dynamics of $L_{i-1}(t)$ under Q_{T_i} is governed by

$$\frac{dL_{i-1}(t)}{L_{i-1}(t)} = \sigma_i^L(t)\, dZ^i(t), \quad t < T_0,$$

where $Z^i(t)$ is Q_{T_i}-Brownian. Upon integration, we obtain

$$L_{i-1}(T_0) = L_{i-1}(t) \exp\left(-\frac{1}{2}\int_t^{T_0} \sigma_i^L(u)^2\, du + \int_t^{T_0} \sigma_i^L(u)\, dZ^i(u)\right). \qquad (8.4.19)$$

Writing

$$\eta_i = \int_t^{T_0} \sigma_i^L(u)\, dZ^i(u)$$

and using the change of measure formula

$$dZ^i(u) = dZ^j(u) + \sum_{k=1}^{i} \frac{\alpha_k L_{k-1}(u)}{1+\alpha_k L_{k-1}(u)} \sigma_k^L(u)\sigma_i^L(u)\, du$$
$$- \sum_{k=1}^{j} \frac{\alpha_k L_{k-1}(u)}{1+\alpha_k L_{k-1}(u)} \sigma_k^L(u)\sigma_j^L(u)\, du,$$

we wish to derive the appropriate distribution of η_i under Q_{T_j}, $j \neq i$. Brace, Gatarek and Musiela (1997) proposed the "frozen coefficients" approximation by taking

$$L_{k-1}(u) \approx L_{k-1}(t) \quad \text{for} \quad t \le u \le T_0.$$

This gives

$$\int_t^{T_0} \frac{\alpha_k L_{k-1}(u)}{1+\alpha_k L_{k-1}(u)} \sigma_k^L(u)\sigma_i^L(u)\, du \approx \frac{\alpha_k L_{k-1}(t)}{1+\alpha_k L_{k-1}(t)} \Delta_{ki}, \qquad (8.4.20)$$

where

$$\Delta_{ki} = \int_t^{T_0} \sigma_k^L(u)\sigma_i^L(u)\, du.$$

We define ξ_i^j, $i, j = 1, 2, \cdots, n$, $j \neq i$, by

$$\xi_i^j = \int_t^{T_0} \sigma_i^L(u)\, dZ^j(u) + \mu_i^j, \qquad (8.4.21)$$

where

$$\mu_i^j = \sum_{k=1}^{i} \frac{\alpha_k L_{k-1}(t)}{1+\alpha_k L_{k-1}(t)} \Delta_{ki} - \sum_{k=1}^{j} \frac{\alpha_k L_{k-1}(t)}{1+\alpha_k L_{k-1}(t)} \Delta_{kj}.$$

The distribution of $\eta_1, \cdots, \eta_n$ conditional on $\mathcal{F}_t$ under the forward measure Q_{T_j}, $j = 1, 2, \cdots, n$, is now approximated by the distribution of $\xi_1^j, \cdots, \xi_n^j$, which is seen to be a correlated vector of Gaussian random variable $N(\boldsymbol{\mu}^j, \Delta)$. Here, the mean vector is $\boldsymbol{\mu}^j = (\mu_1^j, \cdots, \mu_n^j)$ and the correlation matrix is Δ, whose (k, i)th-component is $\Delta_{ki}, k, i = 1, 2, \cdots, n$.

Brace, Gatarek and Musiela (1997) made a further approximation on the matrix Δ by arguing that the first eigenvalue of Δ is significantly larger than the second one. Thus Δ can be approximated as a rank-one matrix and its components can be expressed as

$$\Delta_{ki} = \gamma_k \gamma_i, \quad k, i = 1, 2, \cdots, n, \tag{8.4.22}$$

for some positive constants $\gamma_1, \cdots, \gamma_n$. Define

$$\delta_i = \sum_{k=1}^{i} \frac{\alpha_k L_{k-1}(t)}{1+\alpha_k L_{k-1}(t)} \gamma_k, \quad i = 1, 2, \cdots, n,$$

then μ_i^j can be approximated by

$$\mu_i^j = \gamma_i(\delta_i - \delta_j), \quad i, j = 1, 2, \cdots, n. \tag{8.4.23}$$

Under the forward measure Q_{T_j}, we observe that $B_{fix}(T_0; T_0, T_n, X)$ can be approximated as

$$\begin{aligned} & B_{fix}(T_0; T_0, T_n, X) \\ & \approx \sum_{i=1}^{n} c_i \prod_{k=1}^{i} \frac{1}{1+\alpha_k L_{k-1}(t) \exp(\gamma_k(\widetilde{x}_k^j + \delta_k - \delta_j) - \gamma_k^2/2)}, \end{aligned} \tag{8.4.24}$$

where $\widetilde{x}_k^j, k = 1, \cdots, n$, are identically distributed standard normal variables. The property of zero correlation among $\widetilde{x}_k^j$ stems from the approximate rank-one property of the correlation matrix Δ.

Define the set of functions

$$f_j(x) = \sum_{i=1}^{n} c_i \prod_{k=1}^{j} \frac{1}{1+\alpha_k L_{k-1}(t) \exp(\gamma_k(x + \delta_k - \delta_j) - \gamma_k^2/2)}, \quad j = 1, 2, \cdots, n.$$

Taking $\widetilde{x}$ to be a univariate standard normal random variable, we make the approximation that

$$f_j(\widetilde{x}) \approx B_{fix}(T_0; T_0, T_n, X) \tag{8.4.25}$$

under Q_{T_j}, $j = 1, 2, \cdots, n$. Accordingly, the event $\{\mathbf{1}_A = 1\}$ can be approximately equivalent to $f_j(\widetilde{x}) < 1$ under Q_{T_j}. The function $f_j(x)$ can be shown to be monotonically decreasing with respect to x and $f(-\infty) < 1 < f(\infty)$. Hence, there is a unique solution x_j^* such that $f_j(x_j^*) = 1$. It now becomes straightforward to obtain

$$\begin{aligned} &E_{Q_{T_j}}[(L_{j-1}(T_0) - X)\mathbf{1}_A] \\ &\approx L_{j-1}(t)N(\gamma_j - x_j^*) - XN(-x_j^*), \quad j = 1, 2, \cdots, n. \end{aligned} \tag{8.4.26}$$

Finally, the time-t value of the swaption priced under the Lognormal LIBOR Market model is approximately given by

$$\begin{aligned} &V(t; T_0, T_n, X) \\ &\approx \sum_{i=1}^{n} \alpha_i B(t, T_i)[L_{i-1}(t)N(\gamma_i - x_i^*) - XN(-x_i^*)], \quad t < T_0. \end{aligned} \tag{8.4.27}$$

8.4.3 Cross-Currency Swaps

Cross-currency swaps are characterized by cash flows in the exchange payments that are dependent on floating interest rates in multiple currency worlds. The swap parties have exposure to both interest rate risk and exchange rate risk. In this section, we consider the product nature, uses, pricing and hedging of the differential swaps. As part of the pricing procedure, we also present the various formulations of the cross-currency Lognormal LIBOR Market models.

Differential Swaps

A differential swap is an exchange of floating rate cash flows denominated in domestic currency based on the domestic and foreign floating LIBORs. The unique feature is that the foreign LIBORs are applied to the domestic principal to determine the payments on one leg of the differential swap. For example, a five-year differential swap entitles one party to pay interest payments based on the six-month LIBOR over each six-month period while receiving interest payments based on the six-month Euro LIBOR minus margin over the same period. Here, both interest rates are applied on the same domestic principal and interest payments are denominated in the domestic currency. Noting that all payments are in domestic currency, are the two parties immune to the exchange rate risk?

Consider an asset manager who is seeking to enhance the returns on a low-yielding dollar-denominated investment. The manager enters into a differential swap which entitles him to receive six-month Euro LIBOR minus margin and pay six-month dollar LIBOR (see Fig. 8.2). The dollar LIBOR cash flows from the investment are used to match the dollar LIBOR payments under the differential swap. In essence, the investor receives the returns at the rate of Euro LIBOR minus margin. The investor would benefit provided that the six-month Euro LIBOR minus margin exceeds the six-month dollar LIBOR. The differential swap allows the investor to speculate on interest rate differentials across the two currencies without incurring exchange rate risk.

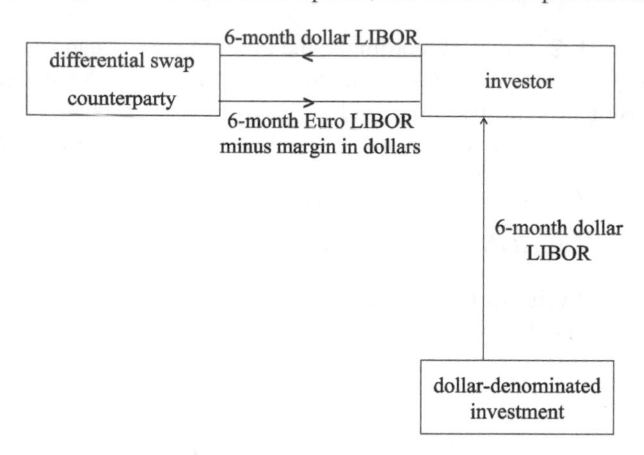

Fig. 8.2. An asset manager uses a differential swap to enhance the returns on a low-yielding dollar-denominated investment.

Dynamic Hedging Strategy

Let the tenor structure be $\{T_0, T_1, \cdots, T_n\}$, where the differential swap is entered at the trade date T_0, payment dates are at $T_i, i = 1, 2, \cdots, n$, and the swap terminates after the final payment date T_n. The domestic and foreign LIBORs over the interval $(T_{i-1}, T_i]$ are reset at $T_{i-1}, i = 1, 2, \cdots, n$. At time T_i, the buyer pays an interest amount that equals the domestic notional times the domestic LIBOR L_{i-1} times the accrual factor α_i of the interval $(T_{i-1}, T_i]$, and in return the buyer receives an interest amount that equals domestic notional times the foreign LIBOR $\widetilde{L}_{i-1}$ times α_i. All cash flows are denominated in domestic currency. Let $B(t, T)$ and $\widetilde{B}(t, T)$ denote the time-t prices of the domestic and foreign unit par T-maturity bonds. Assuming unit notional for convenience, and recalling

$$L_{i-1}(T_{i-1}) = \frac{1}{\alpha_i}\left[\frac{1}{B(T_{i-1}, T_i)} - 1\right] \quad \text{and} \quad \widetilde{L}_{i-1}(T_{i-1}) = \frac{1}{\alpha_i}\left[\frac{1}{\widetilde{B}(T_{i-1}, T_i)} - 1\right],$$

the net cash flow in units of domestic currency to the buyer at time T_i is

$$\frac{1}{\widetilde{B}(T_{i-1}, T_i)} - \frac{1}{B(T_{i-1}, T_i)}.$$

To initiate the differential swap, a nonzero "up front cost" (may be positive or negative) is determined at time T_0. Alternatively, the cost is spread over the term of the swap as the margin of the differential swap.

Jamshidian (1994) advocated the following dynamic hedging strategy using a portfolio of domestic and quanto roll bonds. Define a domestic roll bond over $(T_{i-1}, T_i]$ which promises to pay at maturity T_i an amount of $\frac{1}{B(T_{i-1},T_i)}$ domestic dollars. Correspondingly, a quanto roll bond promises to pay at maturity T_i an amount of $\frac{1}{\widetilde{B}(T_{i-1},T_i)}$ domestic dollars. It is seen that the buyer of the differential swap is equivalent to short holding of a portfolio of $(T_{i-1}, T_i]$ domestic roll bond and long holding of a portfolio of $(T_{i-1}, T_i]$ quanto roll bond, summing over $i = 1, 2, \cdots, n$. After the payment date T_i, the ith domestic and quanto roll bonds are dropped off from the portfolio.

Let the current time t fall between the interval $[T_{K-1}, T_K)$. Since $T_{K-1} \leq t < T_K$, both the values $B(T_{K-1}, T_K)$ and $\widetilde{B}(T_{K-1}, T_K)$ are known so that the time-t value (in domestic currency) of the domestic and foreign roll bonds are

$$\frac{B(t, T_K)}{B(T_{K-1}, T_K)} \quad \text{and} \quad \frac{B(t, T_K)}{\widetilde{B}(T_{K-1}, T_K)},$$

respectively. Hence, the time-t value of the net interest payment at time T_K is given by

$$B(t, T_K)\left[\frac{1}{\widetilde{B}(T_{K-1}, T_K)} - \frac{1}{B(T_{K-1}, T_K)}\right], \quad T_{K-1} \leq t < T_K.$$

The hedging strategy for the payment at time T_K is longing one unit of domestic roll bond and shorting one unit of foreign roll bond. Next, we face the harder job of hedging the payments at $T_{K+1}, T_{K+2}, \cdots, T_n$. The net payout at time $T_i, i = K+1, \cdots, n$, consists of the domestic roll bond payout $\frac{1}{B(T_{i-1}, T_i)}$ minus the quanto roll bond payout $\frac{1}{\widetilde{B}(T_{i-1}, T_i)}$. To replicate the domestic roll bond payout at T_i, we simply hold $\frac{1}{B(T_{i-1}, T_i)}$ units of T_i-maturity unit par bond. Its value at time T_{i-1} is exactly unity so the present value at time t is $B(t, T_{i-1}), t < T_{i-1}$. The quanto roll bond payout at time T_i is exposed to the foreign interest rate risk, so it is necessary to hold foreign discount bonds of different maturities in the replicating portfolio. These foreign bonds are dynamically adjusted until time T_{i-1} such that the net present value of them is maintained to be zero throughout all times earlier than T_{i-1}. In addition, we hold T_i-maturity domestic bond which is used to replicate the quanto roll bond payout in domestic currency at time T_i. At the reset time T_{i-1}, the positions of the foreign bonds are liquidated. The T_i-maturity domestic bond is kept until time T_i. The resulting portfolio has hedged against the exchange rate risk dynamically since the net present value of the foreign bonds are balanced off.

Let $V_i(t)$ denote the time-t value of the portfolio that is constructed to replicate the quanto roll bond payoff $\frac{1}{\widetilde{B}(T_{i-1}, T_i)}$ at payment date $T_i, t \leq T_{i-1}$. It is then necessary to enforce the terminal condition:

$$V_i(T_{i-1}) = \frac{B(T_{i-1}, T_i)}{\widetilde{B}(T_{i-1}, T_i)}. \tag{8.4.28}$$

Let $n_i^d(t)$ denote the number of units held at time t in the portfolio of long position of T_i-maturity domestic discount bond. Similarly, let $n_i^{f,e}(t)$ and $n_i^{f,\ell}(t)$ denote the number of units of long position of the earlier maturity T_{i-1}-maturity foreign discount bond and short position of the later maturity T_i-maturity foreign discount bond, respectively. Let $X(t)$ denote the exchange rate at time t. The dynamic replicating strategy is to choose

$$n_i^d(t) = \frac{V_i(t)}{B(t, T_i)}$$

$$n_i^{f,e}(t) = \frac{V_i(t)}{\widetilde{B}(t, T_{i-1})X(t)}$$

$$n_i^{f,\ell}(t) = \frac{V_i(t)}{\widetilde{B}(t, T_i)X(t)} \tag{8.4.29}$$

so that

$$V_i(t) = n_i^d(t)B(t, T_i) + X(t)\big[n_i^{f,e}(t)\widetilde{B}(t, T_{i-1}) - n_i^{f,\ell}(t)\widetilde{B}(t, T_i)\big]. \tag{8.4.30}$$

It remains to find $V_i(t)$ in order to determine the composition of the replicating portfolio. The replicating portfolio is taken to be self-financing so that the differential change in portfolio value $dV_i(t)$ over differential time interval dt is attributed to the differential change in the bond values and exchange rate. Assuming the portfolio weights of the three bonds to be "frozen" over dt and using (8.4.30), we obtain

$$dV_i(t) = n_i^d(t)\, dB(t, T_i) + [X(t) + dX(t)][n_i^{f,e}(t)\, d\widetilde{B}(t, T_{i-1}) - n_i^{f,\ell}(t)\, d\widetilde{B}(t, T_i)].$$

Substituting the relations in (8.4.29) into the above equation, we have

$$\begin{aligned}
\frac{dV_i(t)}{V_i(t)} &= \frac{dB(t, T_i)}{B(t, T_i)} + \frac{d\widetilde{B}(t, T_{i-1})}{\widetilde{B}(t, T_{i-1})} - \frac{d\widetilde{B}(t, T_i)}{\widetilde{B}(t, T_i)} \\
&\quad + \frac{d\widetilde{B}(t, T_{i-1})}{\widetilde{B}(t, T_{i-1})}\frac{dX(t)}{X(t)} - \frac{d\widetilde{B}(t, T_i)}{\widetilde{B}(t, T_i)}\frac{dX(t)}{X(t)} \\
&= \frac{dB(t, T_i)}{B(t, T_i)} + \frac{d\widetilde{B}(t, T_{i-1})}{\widetilde{B}(t, T_{i-1})} - \frac{d\widetilde{B}(t, T_i)}{\widetilde{B}(t, T_i)} \\
&\quad + \mathrm{cov}\left(\frac{d\widetilde{B}(t, T_{i-1})}{\widetilde{B}(t, T_{i-1})} - \frac{d\widetilde{B}(t, T_i)}{\widetilde{B}(t, T_i)}, \frac{dX(t)}{X(t)}\right).
\end{aligned}$$

Integrating the above stochastic differential equation and applying the terminal condition (8.4.28), we obtain

$$\begin{aligned}
V_i(t) = {} & \frac{B(t, T_i)\widetilde{B}(t, T_{i-1})}{\widetilde{B}(t, T_i)} \\
& \exp\left(\int_t^{T_{i-1}} \mathrm{cov}\left(\frac{d\widetilde{B}(u, T_i)}{\widetilde{B}(u, T_i)} - \frac{d\widetilde{B}(u, T_{i-1})}{\widetilde{B}(u, T_{i-1})},\right.\right. \\
& \left.\left.\frac{dX(u)}{X(u)} + \frac{d\widetilde{B}(u, T_i)}{\widetilde{B}(u, T_i)} - \frac{dB(u, T_i)}{B(u, T_i)}\right)\right).
\end{aligned} \tag{8.4.31}$$

The time-t value $S_{diff}(t)$ of the differential swap is the sum of the time-t value of the net payments at $T_K, T_{K+1}, \cdots, T_n$, so we obtain

$$S_{diff}(t) = B(t, T_K)\left[\frac{1}{\widetilde{B}(T_{K-1}, T_K)} - \frac{1}{B(T_{K-1}, T_K)}\right] + \sum_{i=K+1}^{n} \left[V_i(t) - B(t, T_{i-1})\right], \quad T_{K-1} \leq t < T_K. \qquad (8.4.32)$$

We have derived the explicit trading strategies of constructing the dynamically rebalanced replicating portfolio. As a consequence, we also obtain the above model-independent pricing formula.

It is relatively straightforward to derive the pricing formula of a differential swap under the Gaussian type models (see Problem 8.41). How about the pricing of the differential swap under the Lognormal LIBOR Market models? It is shown below that a closed form price formula can also be derived, though in the pricing procedure it is necessary to construct the cross-currency version of the Lognormal LIBOR Market models.

Cross-Currency LIBOR Models

Recall that under the deterministic volatility assumption, the Lognormal LIBOR Market model takes the forward LIBORs as lognormal martingales under the forward measure with respect to the end of the respective accrual periods. When the market model is extended to the cross-currency setting, there are no arbitrage requirements to be imposed on the simultaneous lognormality assumption on the forward LIBORs in the two currencies and forward exchange rate at all maturities. More precisely, suppose lognormal dynamics are assumed for the forward LIBORs in both currency worlds at all maturities, the forward exchange rate linking the two currencies can only be chosen to be lognormal for one maturity, while the dynamics of the exchange rates at other maturities are given by no arbitrage conditions. If one chooses to have lognormality at all maturities for the forward exchange rates and forward LIBOR in one currency, then the dynamics of the forward LIBORs in the other currency are determined by no arbitrage conditions (Schlögl, 2002).

First of all, we would like to establish the measure link between the domestic and foreign LIBOR markets. Considering the time-T_i domestic and foreign forward probability measures, denoted by Q_{T_i} and $\widetilde{Q}_{T_i}$ respectively, the dynamics of the domestic forward LIBOR process $L_{i-1}(t)$ and the foreign forward LIBOR process $\widetilde{L}_{i-1}(t)$ are governed by

$$dL_{i-1}(t) = L_{i-1}(t)\sigma_{i-1}^{L}(T)\, dZ^{T_i}(t) \qquad (8.4.33a)$$

$$d\widetilde{L}_{i-1}(t) = \widetilde{L}_{i-1}(t)\widetilde{\sigma}_{i-1}^{L}(t)\, d\widetilde{Z}^{T_i}(t), \qquad (8.4.33b)$$

where $Z^{T_i}(t)$ and $\widetilde{Z}^{T_i}(t)$ are scalar Brownian processes under Q_{T_i} and $\widetilde{Q}_{T_i}$, respectively, $\sigma_{i-1}^{L}(t)$ and $\widetilde{\sigma}_{i-1}^{L}(T)$ are deterministic volatility functions. $T_0 < T_1 < \cdots < T_n$. Here, $\{T_0, T_1, \cdots, T_n\}$ is the tenor structure. Both $L_{i-1}(t)$ and $\widetilde{L}_{i-1}(t)$ are lognormal martingales under their respective measures Q_{T_i} and $\widetilde{Q}_{T_i}$, $i = 1, 2, \cdots, n$. How to establish the measure link between $dZ^{T_i}(t)$ and $d\widetilde{Z}^{T_i}(t)$?

Note that the spot exchange rate process $X(t)$ is not a tradeable security in either the domestic or foreign market. On the other hand, the T_i-maturity foreign bond $\widetilde{B}(t, T_i)$ converted into domestic currency at the spot exchange rate $X(t)$ can be considered as a domestic asset. We define the time-T_i forward exchange rate process by

$$X(t, T_i) = \frac{X(t)\widetilde{B}(t, T_i)}{B(t, T_i)}, \tag{8.4.34}$$

which is the domestic currency price of the T_i-maturity foreign bond normalized by the T_i-maturity domestic bond price. Here, $X(t, T_i)$ is seen to be a martingale under Q_{T_i}. In a similar manner, $\frac{1}{X(t,T_i)}$ is a martingale under $\widetilde{Q}_{T_i}$. The dynamics of $X(t, T_i)$ and $\frac{1}{X(t,T_i)}$ are expressed as

$$dX(t, T_i) = X(t, T_i)\sigma_i^X(t)\, dZ^{T_i}(t) \tag{8.4.35a}$$

$$d\left(\frac{1}{X(t, T_i)}\right) = \frac{1}{X(t, T_i)}\sigma_i^{1/X}(t)\, d\widetilde{Z}^{T_i}(t), \tag{8.4.35b}$$

where $\sigma_i^X(t)$ and $\sigma_i^{1/X}(t)$ are the corresponding deterministic volatility functions. The Radon–Nikodym derivative that links $\widetilde{Q}_{T_i}$ and Q_{T_i} is given by

$$\frac{d\widetilde{Q}_{T_i}}{dQ_{T_i}} = \frac{X(T_i)\widetilde{B}(T_i, T_i)B(0, T_i)}{X(0)\widetilde{B}(0, T_i)B(T_i, T_i)} = \frac{X(T_i, T_i)}{X(0, T_i)},$$

and restricting to the information $\mathcal{F}_t$, we have

$$\left.\frac{d\widetilde{Q}_{T_i}}{dQ_{T_i}}\right|_{\mathcal{F}_t} = \frac{X(t, T_i)}{X(0, T_i)}.$$

Based on the dynamics of $X(t, T_i)$ defined in (8.4.35a), we obtain

$$\frac{d\widetilde{Q}_{T_i}}{dQ_{T_i}} = \exp\left(\int_0^{T_i} \sigma_i^X(u)\, dZ^{T_i}(u) - \frac{1}{2}\int_0^{T_i} \sigma_i^X(u)^2\, du\right).$$

By the Girsanov Theorem, we deduce that

$$d\widetilde{Z}^{T_i}(t) = dZ^{T_i}(t) - \sigma_i^X(t)\, dt. \tag{8.4.36}$$

Supposc the volatility $\sigma_i^X(t)$ of the forward exchange rate at the specific maturity T_i has been specified, then the volatility $\sigma_j^X(t)$ at another maturity T_j, $T_j \neq T_i$, is fixed since the links between the forward measures of different maturities have been fixed through the forward LIBOR volatilities via no arbitrage conditions. Let $\sigma_j^L(t)$ and $\widetilde{\sigma}_j^L(t)$ be the deterministic volatility functions of the domestic forward LIBOR $L_j(t)$ and foreign forward LIBOR $\widetilde{L}_j(t)$, respectively. In Sect. 8.3.2, we established

$$dZ^{T_j}(t) = dZ^{T_{j+1}}(t) - \frac{\alpha_{j+1}L_j(t)}{1+\alpha_{j+1}L_j(t)}\sigma_j^L(t)\,dt \tag{8.4.37a}$$

$$d\widetilde{Z}^{T_j}(t) = d\widetilde{Z}^{T_{j+1}}(t) - \frac{\alpha_{j+1}\widetilde{L}_j(t)}{1+\alpha_{j+1}\widetilde{L}_j(t)}\widetilde{\sigma}_j^L(t)\,dt. \tag{8.4.37b}$$

Combining (8.4.36) and (8.4.37a,b), we can derive the appropriate relation between $\sigma_i^X(t)$ and $\sigma_j^X(t)$, $j \neq i$. In particular, $\sigma_i^X(t)$ and $\sigma_{i+1}^X(t)$ are related by

$$\sigma_i^X(t) = \frac{\alpha_{i+1}\widetilde{L}_i(t)}{1+\alpha_{i+1}\widetilde{L}_i(t)}\widetilde{\sigma}_i^L(t) - \frac{\alpha_{i+1}L_i(t)}{1+\alpha_{i+1}L_i(t)}\sigma_i^L(t) + \sigma_{i+1}^X(t). \tag{8.4.38}$$

The links between the forward measures in the domestic and foreign currency worlds are summarized in Fig. 8.3 (Schlögl, 2002).

Alternatively, we may specify the forward LIBORs in one of the currency worlds to be lognormal, together with lognormality assumptions for all forward exchange rates. According to (8.4.36), the dynamics of the forward LIBORs will be fixed via no arbitrage conditions. It will be shown below that such specification of the domestic

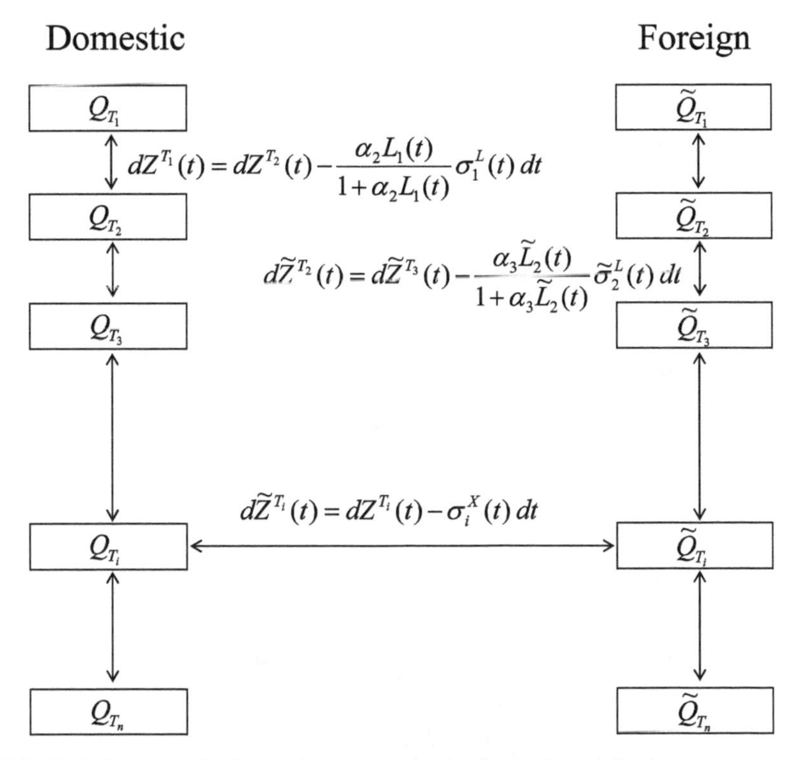

Fig. 8.3. Links between the forward measures in the domestic and foreign currency worlds. Lognormal dynamics are specified for all LIBORs in the domestic and foreign currency worlds while lognormal dynamics is specified for the forward exchange rate at one specified maturity.

and foreign forward LIBORs leads to an effective valuation approach for pricing the differential swaps.

Pricing of Differential Swaps
Assuming the notional to be unity, we would like to find the time-0 value of the differential swap to the buyer who receives interest payments based on the foreign LIBORs but pays out interest payments based on the domestic LIBORs, all cash flows are denominated in domestic currency. By expressing the values of the interest payments as the discounted expectation under the relevant forward measures, we obtain

$$S_{diff} = \sum_{i=0}^{n-1} \alpha_{i+1} B(0, T_{i+1}) E_{Q_{T_{i+1}}}\big[\widetilde{L}_i(T_i) - L_i(T_i)\big]. \tag{8.4.39}$$

Since $L_i(t)$ is a lognormal martingale under $Q_{T_{i+1}}$, we have

$$E_{Q_{T_{i+1}}}\big[L_i(T_i)\big] = L_i(0).$$

It suffices to evaluate $E_{Q_{T_{i+1}}}[\widetilde{L}_i(T_i)]$. We consider two different specifications of the cross-currency LIBOR model.

(i) The *foreign* forward LIBORs and forward exchange rates at all maturities are specified.
In this case, we have

$$\begin{aligned} d\widetilde{L}_i(t) &= \widetilde{L}_i(t)\widetilde{\sigma}_i^L(t)\, d\widetilde{Z}^{T_{i+1}}(t) \\ &= \widetilde{L}_i(t)\widetilde{\sigma}_i^L(t)\big[dZ^{T_{i+1}}(t) - \sigma_i^X(t)\, dt\big], \end{aligned} \tag{8.4.40}$$

so that

$$\begin{aligned} \widetilde{L}_i(t) = \widetilde{L}_i(0) &\exp\left(-\int_0^t \widetilde{\sigma}_i^L(u)\left[\sigma_i^X(u) + \frac{\widetilde{\sigma}_i^L(u)}{2}\right] du\right) \\ &\exp\left(\int_0^t \widetilde{\sigma}_i^L(u)\, dZ^{T_{i+1}}(u)\right), \end{aligned}$$

thus giving

$$E_{Q_{T_{i+1}}}\big[\widetilde{L}_i(t)\big] = \widetilde{L}_i(0)\exp\left(-\int_0^{T_i} \widetilde{\sigma}_i^L(u)\sigma_i^X(u)\, du\right).$$

The value of the differential swap can be expressed as

$$\begin{aligned} S_{diff} &= \sum_{i=0}^{n-1} \alpha_{i+1} B(0, T_{i+1}) \\ &\left[\widetilde{L}_i(0)\exp\left(-\int_0^{T_i} \widetilde{\sigma}_i^L(u)\sigma_i^X(u)\, du\right) - L_i(0)\right]. \end{aligned} \tag{8.4.41}$$

(ii) The *domestic* forward LIBORs and forward exchange rates at all maturities are specified.
Using the relation in (8.4.38), the dynamics of the foreign forward LIBOR $\widetilde{L}_i(t)$ are governed by

$$\begin{aligned}\frac{d\widetilde{L}_i(t)}{\widetilde{L}_i(t)} &= \widetilde{\sigma}_i^L(t)\,d\widetilde{Z}^{T_{i+1}}(t)\\ &= \frac{1+\alpha_{i+i}\widetilde{L}_i(t)}{\alpha_{i+1}\widetilde{L}_i(t)}\bigg[\sigma_i^X(t)-\sigma_{i+1}^X(t)\\ &\qquad + \frac{\alpha_{i+1}L_i(t)}{1+\alpha_{i+1}L_i(t)}\sigma_i^L(t)\bigg]\,d\widetilde{Z}^{T_{i+1}}(t), \qquad (8.4.42)\end{aligned}$$

where the deterministic volatility functions $\sigma_i^X(t), \sigma_{i+1}^X(t)$ and $\sigma_i^L(t)$ are specified. No analytic solution exists for the above stochastic differential equation. However, if one argues that $\frac{\alpha_{i+1}L_i(t)}{1+\alpha_{i+1}L_i(t)}$ has a low variance under $Q_{T_{i+1}}$ so that

$$\frac{\alpha_{i+1}L_i(t)}{1+\alpha_{i+1}L_i(t)} \approx \frac{\alpha_{i+1}L_i(0)}{1+\alpha_{i+1}L_i(0)},$$

then

$$d\widetilde{L}_i(t) \approx \left[\frac{1}{\alpha_{i+1}} + \widetilde{L}_i(t)\right]\widehat{\sigma}_i^L(t)\,d\widetilde{Z}^{T_i+1}(t), \qquad (8.4.43)$$

where the deterministic function $\widehat{\sigma}_i^L(t)$ is given by

$$\widehat{\sigma}_i^L(t) = \sigma_i^X(t) - \sigma_{i+1}^X(t) + \frac{\alpha_{i+1}L_i(0)}{1+\alpha_{i+1}L_i(0)}\sigma_i^L(t).$$

Based on the above approximation, the approximate solution to $\widetilde{L}_i(t)$ can be found to be

$$\begin{aligned}\widetilde{L}_{i+1}(t) \approx & -\frac{1}{\alpha_{i+1}} + \left[\frac{1}{\alpha_{i+1}} + \widetilde{L}_i(0)\right]\\ & \exp\left(\int_0^t \widehat{\sigma}_i^L(u)\,d\widetilde{Z}^{T_{i+1}}(u) - \int_0^t \widehat{\sigma}_i^L(u)\left[\sigma_{i+1}^X(u) + \frac{\widehat{\sigma}_i^L(u)}{2}\right]du\right),\end{aligned}$$

thus giving

$$\begin{aligned}&E_{Q_{T_{i+1}}}[\widetilde{L}_i(t)]\\ &\approx \left[\frac{1}{\alpha_{i+1}} + \widetilde{L}_i(0)\right]\exp\left(-\int_0^t \widehat{\sigma}_i^L(u)\sigma_{i+1}^X(u)\,du\right) - \frac{1}{\alpha_{i+1}}. \qquad (8.4.44)\end{aligned}$$

8.5 Problems

8.1 Show that the risk neutral measure Q and the T-forward measure Q_T are identical if and only if the short rate r_t is deterministic. Also, show that the instantaneous forward rate is given by

$$F(t, T) = E_Q\left[r_T \frac{e^{-\int_t^T r_u\, du}}{B(t, T)}\right],$$

where $B(t, T)$ is the time-t price of the T-maturity discount bond.

8.2 Consider a contingent claim whose payoff $X_{T'}$ is known at time T' but it is payable at a later time T, $T > T'$. Show that the time-t value of the contingent claim is given by

$$\begin{aligned} X_t &= E_Q^t\left[e^{-\int_t^T r_u\, du} X_{T'}\right] \\ &= E_Q^t\left[e^{-\int_t^{T'} r_u\, du} X_{T'} B(T', T)\right], \end{aligned}$$

where $B(T', T)$ is the time-T' price of the T-maturity discount bond. Give a financial interpretation of the result.
Hint: Use the tower property of conditional expectations.

8.3 Show that the process ξ_t^T, $t \geq 0$, defined in (8.1.7), is a positive Q-martingale and the expectation of ξ_t^T under Q is one. Here, Q is the risk neutral measure.

8.4 The dynamics of the instantaneous forward rate $F(t, T)$ under the risk neutral measure Q is governed by

$$dF(t, T) = \sum_{i=1}^m \frac{\partial \sigma_B^i}{\partial T}(t, T)\sigma_B^i(t, T)\, dt + \sum_{i=1}^m \frac{\partial \sigma_B^i}{\partial T}(t, T)\, dZ_i(t),$$

where $(Z_1(t) \cdots Z_m(t))^{\mathrm{T}}$ is an m-dimensional Brownian process under Q. Let $(W_1(t) \cdots W_m(t))^{\mathrm{T}}$ be an m-dimensional Brownian process under the T'-forward measure $Q_{T'}$ and write

$$\dot{\sigma}_B(t, T) = \frac{\partial \sigma_B^i}{\partial T}(t, T).$$

Show that the dynamics of $F(t, T)$ under $Q_{T'}$ is given by

$$dF(t, T) = \sum_{i=1}^m \dot{\sigma}_B^i(t, T)\, dW_i(t) + \sum_{i=1}^m \dot{\sigma}_B^i(t, T)\left[\sigma_B^i(t, T') - \sigma_B^i(t, T)\right] dt.$$

In particular, when $T' = T$, we obtain

$$dF(t, T) = \sum_{i=1}^n \dot{\sigma}_B^i(t, T)\, dW_i(t).$$

8.5 Let $R(t, t+\delta)$ denote the yield to maturity over the period $(t, t+\delta]$ of a discount bond maturing at $t + \delta$, and $f(0, t, t + \delta)$ be the forward rate observed at time zero over the period $(t, t+\delta]$. Let $\sigma_B(t, T)$ denote the volatility function of the discount bond price process $B(t, T)$. Show that $R(t, t+\delta)$ can be expressed as (El-Karoui and Geman, 1994)

$$R(t, t+\delta) = f(0, t, t+\delta) + \frac{1}{\delta} \int_0^t [\sigma_B(u, t+\delta) - \sigma_B(u, t)] \, dZ(u) + \frac{1}{2\delta} \int_0^t [\sigma_B(u, t+\delta) - \sigma_B(u, t)]^2 \, du,$$

where $Z(u)$ is a Brownian process under the t-forward measure Q_t. Furthermore, when the bond price volatilities are deterministic, show that

$$E_{Q_t}[R(t, t+\delta)] = f(0, t, t+\delta) + \frac{\delta}{2} \text{var}(R(t, t+\delta)).$$

8.6 The forward rate over the future period $(T_1, T_2]$ as observed at the earlier time t can be computed from bond prices by the formula

$$\frac{1}{T_2 - T_1} \ln \frac{B(t, T_2)}{B(t, T_1)}.$$

On the other hand, the futures rate is given by the expectation of the T_2-maturity short rate observed at T_1 under the risk neutral measure conditional on the information at t. Find the spread between the forward rate and the futures rate, assuming that the interest rate dynamics follows the Hull–White model.

8.7 Under the risk neutral measure Q, the dynamics of the price process of an asset $S(t)$ and the discount bond price process are governed by

$$dS(t) = S(t)[r(t) \, dt + \sigma_1(t) \, dZ_1(t) + \sigma_2(t) \, dZ_2(t)]$$
$$dB(t, T') = B(t, T')[r(t) \, dt + \sigma_B(t, T') \, dZ_1(t)],$$

where $Z_1(t)$ and $Z_2(t)$ are uncorrelated standard Q-Brownian processes, $\sigma_B(t, t) = 0$ and the volatilities are time dependent functions.

(a) Find $\text{cov}(S(t), B(t, T'))$.

(b) Suppose the bond price $B(t, T)$ is used as the numeraire in the T-forward measure Q_T. Show that the solutions of the above stochastic differential equations are given by (Nielsen and Sandmann, 1996)

$$S(t) = S(t_0) \frac{B(t, T)}{B(t_0, T)} \exp\left(-\frac{1}{2} \int_{t_0}^t [\sigma_1(u) - \sigma_B(u, T)]^2 + \sigma_2^2(u) \right) du + \int_{t_0}^t [\sigma_1(u) - \sigma_B(u, t)] \, dZ_1^T(u) + \int_{t_0}^t \sigma_2(u) \, dZ_2^T(u)$$

$$\frac{B(t, T')}{B(t_0, T')} = \frac{B(t, T)}{B(t_0, T)} \exp\left(-\frac{1}{2} \int_{t_0}^t [\sigma_B(u, T') - \sigma_B(u, T)]^2 \, du \right) + \int_{t_0}^t [\sigma_B(u, T') - \sigma_B(u, T)] \, dZ_1^T(u),$$

where $Z_1^T(t)$ and $Z_2^T(t)$ are uncorrelated standard Wiener processes under the T-forward measure Q^T.

8.8 The price function of a European call option under stochastic interest rates can also be solved using the partial differential equation approach. Let the asset value process S_t and the short rate process r_t be governed by

$$\begin{aligned}\frac{dS_t}{S_t} &= r_t\,dt + \sigma_S(t)\,dZ_S\\ dr_t &= \mu_r(t)\,dt + \sigma_r(t)\,dZ_r,\end{aligned}$$

where $dZ_S\,dZ_r = \rho\,dt$. Let λ_r denote the market price of risk of the short rate. Let $c(S,r,t)$ denote the time-t value of a European call option with strike price X and maturity date T.

(a) Show that the governing equation $c(S,r,t)$ is given by

$$\begin{aligned}&\frac{\partial c}{\partial t} + \frac{\sigma_S^2(t)}{2}S^2\frac{\partial^2 c}{\partial S^2} + \rho\sigma_S(t)\sigma_r(t)S\frac{\partial^2 c}{\partial S\partial r} + \frac{\sigma_r^2(t)}{2}\frac{\partial^2 c}{\partial r^2}\\ &+ rS\frac{\partial c}{\partial S} + \left[\mu_r(t) - \lambda_r\sigma_r(t)\right]\frac{\partial c}{\partial r} - rc = 0.\end{aligned}$$

(b) Let $B(r,t;T)$ denote the discount bond price, which satisfies

$$\frac{\partial B}{\partial t} + \frac{\sigma_r^2(t)}{2}\frac{\partial^2 B}{\partial r^2} + \left[\mu_r(t) - \lambda_r\sigma_r(t)\right]\frac{\partial B}{\partial r} - rB = 0.$$

The bond price function admits a solution of the form

$$B(r,t) = a(t)r + b(t).$$

Defining $\xi = S/B$ and $u = c/B$, show that $u(\xi,t)$ satisfies

$$\frac{\partial u}{\partial t} + \left[\frac{\sigma_S^2(t)}{2} - \rho\sigma_S(t)\sigma_r(t)a(t) + \frac{\sigma_r^2(t)a^2(t)}{2}\right]\xi^2\frac{\partial^2 u}{\partial \xi^2} = 0.$$

(c) Show that the value of the European equity call option under stochastic interest rates is given by

$$c(S,r,t) = SN(d_+) - XB(r,t)N(d_-),$$

where

$$\begin{aligned}d_\pm &= \frac{\ln\frac{S}{B(r,t)X} \pm \frac{\overline{\sigma}^2}{2}(T-t)}{\overline{\sigma}\sqrt{T-t}}\\ \overline{\sigma}^2 &= \frac{1}{T-t}\int_t^T \left[\sigma_S^2(u) - 2\rho\sigma_S(u)\sigma_r(u)a(u) + \sigma_r^2(u)a^2(u)\right]du.\end{aligned}$$

8.9 Let $X(t)$ denote the exchange rate process in units of the domestic currency per one unit of the foreign currency, and let $r_d(t)$ and $r_f(t)$ denote the domestic and foreign riskless interest rate, respectively. Also, let $B_d(t, T)$ and $B_f(t, T)$ denote the T-maturity domestic and foreign bond price processes, respectively, $t \leq T$. Under the domestic risk neutral measure Q_d, the dynamics of $X(t)$ and $B_d(t, T')$ are governed by

$$dX(t) = X(t)\left\{[r_d(t) - r_f(t)]\,dt + \sum_{i=1}^{n} \sigma_X^i(t)\,dZ_d^i(t)\right\}$$

$$dB_d(t, T) = B_d(t, T)\left[r_d(t)\,dt + \sum_{i=1}^{n} \sigma_d^i(t, T)\,dZ_d^i(t)\right],$$

where $(Z_d^1(t) \cdots Z_d^n(t))^{\mathrm{T}}$ is an n-dimensional Brownian process under Q_d. Let $(\sigma_f^1(t, T) \cdots \sigma_f^n(t, T))^{\mathrm{T}}$ be the vector volatility function of $B_f(t, T)$. Show that the dynamics of $B_f(t, T)$ under Q_d is given by

$$dB_f(t, T) = B_f(t, T)\left\{\left[r_f(t) - \sum_{i=1}^{n} \sigma_f^i(t, T)\sigma_X^i(t)\right] dt + \sum_{i=1}^{n} \sigma_f^i(t, T)\,dZ_d^i(t)\right\}.$$

By using the T-maturity domestic bond price $B_d(t, T)$ as the numeraire, find the Radon–Nikodym derivative that effects the change of measure from Q_d to the T-maturity domestic forward measure Q_d^T. Solve for the solution of the exchange rate process (Nielsen and Sandmann, 2002).

8.10 Under the risk neutral measure, S_t follows the Geometric Brownian process where

$$\frac{dS_t}{S_t} = r_t\,dt + \sigma\,dZ_t.$$

Let $B(t, T)$ be the time-t value of the T-maturity discount bond. Under the forward measure Q_T where $B(t, T)$ is used as the numeraire, show that

$$E_{Q_T}^t\left[\frac{S_T}{S_t^*}\right] = \begin{cases} \frac{B(t,t^*)}{B(t,T)}, & t \leq t^* \\ \frac{S_t}{S_{t^*}\,B(t,T)}, & t > t^*. \end{cases}$$

Give a financial interpretation of the above result.

8.11 We would like to examine the credit yield spreads of the floating rate debt and fixed rate debt under the Merton risky bond model with stochastic interest rate (Ikeda, 1995). Let A_t and r_t denote the firm asset value and short rate, respectively, whose dynamics are governed by

$$\frac{dA_t}{A_t} = \mu\, dt + \sigma_A\, dZ_A(t), \quad dr_t = \alpha(\gamma - r_t)\, dt + \sigma_r\, dZ_r(t),$$

and $dZ_A\, dZ_r = \rho\, dt$. For the defaultable fixed rate debt, the payoff at debt's maturity T is given by $\min(A_T, F)$, where F is the fixed par value. However, the par at maturity of a floating rate debt is stochastic, whose value is given by

$$F_T = De^{\int_0^T r_u\, du},$$

for some constant D. The stochastic par is equivalent to the value of the money market account invested for T years with initial deposit D. Let $B(r, t; T)$ denote the price of the corresponding default free discount bond and let $\sigma_B(r, t; T)$ denote the corresponding bond volatility.

(a) Show that the value of the defaultable fixed rate debt at the current time t conditional on $A_t = A$ and $r_t = r$ is given by

$$V_X(A, r, t) = AN(-h_X) + FB(r, t; T)N(h_X - \overline{\sigma}_X),$$

where

$$h_X = -\frac{\ln k_X}{\overline{\sigma}_X} + \frac{\overline{\sigma}_X}{2}, \quad k_X = \frac{B(r, t; T)F}{A},$$

$$\overline{\sigma}_X^2 = \int_t^T [\sigma_A^2 + 2\rho\sigma_A\sigma_B(r, u; T) + \sigma_B^2(r, u; T)]\, du.$$

(b) Write $D_t = De^{\int_0^t r(u)\, du}$, which is known at time t from the actual realization of the interest rate process from 0 to t. Show that the value of the defaultable floating rate debt is given by

$$V_L(A, r, t) = AN(-h_L) + D_t N(h_L - \sigma_A\sqrt{\tau}),$$

where

$$h_L = -\frac{\ln k_L}{\sigma_A\sqrt{\tau}} + \frac{\sigma_A\sqrt{\tau}}{2}, \quad k_L = \frac{D_t}{A} \text{ and } \tau = T - t.$$

Interestingly, the bond volatility does not enter into $V_L(A, r, t)$. Why?

Hint: Let $Y(r, t; T)$ denote the value of the money market account with initial deposit D. The evolution dynamics of $Y(t)$ are given by $dY = r_t Y\, dt$. Use $B(r, t; T)$ and $Y(r, t; T)$ as the corresponding numeraire in the fixed rate debt and floating rate debt valuation, respectively.

8.12 Consider a European call option with strike price X maturing at T on a futures whose underlying asset is a T'-maturity discount bond. Derive the value of this option under the Gaussian HJM term structure model.

Hint: Use the result in (8.1.20) for the relation between the forward price and futures price, and compute $\overline{\mu}_F$ as defined in (8.1.14a).

8.13 Let the dynamics of the short rate $r(t)$ be governed by the extended Vasicek model

$$dr(t) = [\phi(t) - \alpha r(t)]\,dt + \sigma_r\,dZ(t).$$

Show that the value of the European call option with strike price X maturity at T on a T'-maturity discount bond is given by

$$c(t; T, T') = B(t, T')N(d_1) - XB(t, T)N(d_2),$$

where $B(t, T)$ is the discount bond price,

$$d_1 = \frac{1}{\widehat{\sigma}} \ln \frac{B(t, T')}{XB(t, T)} + \frac{\widehat{\sigma}}{2}, \quad d_2 = d_1 - \widehat{\sigma},$$

$$\widehat{\sigma} = \frac{\sigma_r}{\alpha}\left[1 - e^{-\alpha(T'-T)}\right]\sqrt{\frac{1 - e^{-2\alpha(T-t)}}{2\alpha}}.$$

Also, show that the put-call parity relation between the prices of the European put and call options on the same underlying discount bond is given by

$$c(t; T, T') + XB(t, T) = p(t; T, T') + B(t, T').$$

8.14 Under the risk neutral measure Q, assume that the bond price $B(t, T)$ is related to the short rate r_t by

$$B(t, T) = e^{-a(\tau)-b(\tau)r_t}, \quad \tau = T - t.$$

Define the following probability measures Q_T and Q_{T^*}, where

$$\frac{dQ_T}{dQ} = \frac{e^{-\int_t^T r_u\,du}}{B(t, T)}$$

$$\frac{dQ_{T^*}}{dQ} = \frac{e^{-\int_t^T r_u\,du} B(T, T^*)}{B(t, T^*)}.$$

Consider a European call option on the T^*-maturity discount bond $B(t, T^*)$ maturing at time T, $T < T^*$. Let X be the strike price of the bond option and define

$$\overline{r} = \frac{\ln X - a(T^* - T)}{b(T^* - T)}.$$

Conditional on $r_t = r$, show that the price of the bond option is given by

$$c(r, t) = B(t, T^*)Q_{T^*}[r_T < \overline{r}] - XB(t, T)Q_T[r_T < \overline{r}].$$

8.15 Suppose the forward rate $F(t, T)$ under the risk neutral measure is governed by

$$dF(t, T) = \left(\sum_{j=1}^{J} \sigma_F^j(t, T) \int_t^T \sigma_F^j(t, T)\,du\right) dt + \sum_{j=1}^{J} \sigma_F^j(t, T)\,dZ_j(t),$$

and consider the coupon bond with n coupon payments whose weights $w_i(r,t)$ are defined by (8.2.12), $i = 1, 2, \cdots, n$, show that the stochastic duration $D(t)$ of the coupon bond is the solution to the following equation (Munk, 1999)

$$\sum_{j=1}^{J}\left[\int_0^{D(t)} \sigma_F^j(t, t+u)\,du\right]^2 = \sum_{j=1}^{J}\left(\sum_{i=1}^{n} w_i(r,t)\int_0^{T_i-t} \sigma_F^j(t, t+u)\,du\right)^2.$$

Here, T_i is the payment date of the ith coupon.

8.16 Suppose the forward rate volatility under the one-factor HJM model takes the form

$$\sigma_F(t,T) = \sigma_1 + \sigma_2[1 - e^{-\alpha(T-t)}],$$

show that the Jamshidian decomposition technique (Jamshidian, 1989) cannot be used to price an option on a coupon bearing bond with coupon payment date $T_i, i = 1, 2, \cdots, n$. Find an approximation price formula to the European call option on a coupon bearing bond whose weight $w_i(r,t)$ is defined by (8.2.12).

Hint: The short rate process under the given volatility structure is non-Markovian. The stochastic duration is the solution $D(t)$ to the following equation

$$\begin{aligned}
&(\sigma_1+\sigma_2)^2 D(t)^2 + \frac{\sigma_2^2}{\alpha^2}[1 - e^{-\alpha D(t)}]^2 \\
&= (\sigma_1+\sigma_2)^2\left[\sum_{i=1}^{n} w_i(r,t)(T_i - t)\right]^2 \\
&+ \frac{\sigma_2^2}{\alpha^2}\left\{\sum_{i=1}^{n} w_i(r,t)[1 - e^{-\alpha(T_i-t)}]\right\}^2.
\end{aligned}$$

8.17 Let $Y(t;\tau)$ denote the yield at time t for a discount bond with a fixed time to maturity τ. The average of the constant maturity yield $Y(t;\tau)$ over a prespecified time period $(0, T]$ is given by

$$A(T) = \frac{1}{T}\int_0^T Y(t;\tau)\,dt.$$

We would like to price a European call option on the average yield $A(T)$ whose terminal payoff at time T is given by

$$c(T) = \max(A(T) - X, 0),$$

where X is the strike price. We specify the risk neutral dynamics of the short rate process to be the continuous Ho–Lee model, where

$$r(t) = f(0,t) + \frac{\sigma^2 t^2}{2} + \sigma Z(t),$$

where $f(0, t)$ is the initial term structure of the forward rates and $Z(t)$ is a standard Brownian process under the risk neutral measure Q. Show that the price of the average-yield call option at time t, $t < T$, can be expressed as (He and Takahashi, 2000)

$$c(t) = \frac{1}{T\tau}\frac{B(0,T)}{B(0,t)}e^{-\sigma^2(T^3-t^3)/6}E_Q^t\left[e^{-\sigma\int_t^T Z(u)\,du}c(T)\right],$$

where

$$c(T) = \max\left(\int_0^T \frac{B(0,u)}{B(0,u+\tau)}\exp\left(\frac{\sigma^2}{2}(u+\tau)u\tau + \sigma\tau Z(u)\right)du \right.$$
$$\left. - (1+X\tau)T, 0\right).$$

8.18 Consider a European call bond option maturity on T_0 whose underlying bond pays $A_i \geq 0$ at time T_i, $1 \leq i \leq n$, where $0 < T_0 < T_1 < \cdots < T_n$. Assume that the zero-coupon bond price $B(t, T)$ follows the one-factor HJM model, where

$$\frac{dB(t,T)}{B(t,T)} = r(t)\,dt + \sigma_B(t,T)\,dZ_t,$$

and the deterministic volatility function satisfies

$$\sigma_B(t,t_2) - \sigma_B(t,t_1) = f(t_1,t_2)g(t), \quad g(t) > 0.$$

Show that the time-0 price of the European call option on the coupon bearing bond with strike price X is given by (Henrard, 2003)

$$\sum_{i=1}^n A_i B(0,T_i)N(x+\alpha_i) - XB(0,T_0)N(x+\alpha_0),$$

where x is the (unique) solution of

$$\sum_{i=1}^n A_i B(0,T_i)\exp\left(-\frac{\alpha_i^2}{2} - \alpha_i x\right) = XB(0,T_0),$$

and $\alpha_i > 0$, $i = 0, 1, \cdots, n$, is given by

$$\alpha_i^2 = \int_0^T [\sigma_B(u,T_i) - \sigma_B(u,T)]^2\,du.$$

Here, T is the expiration date of the underlying bond.

8.19 The no arbitrage bond price process is assumed to follow the m-dimensional Gaussian HJM model. Let $t < T_0 < T_1 < \cdots < T_n$, and α_i be the accrual factor over the interval $(T_{i-1}, T_i]$, $i = 1, 2, \cdots, n$.

(a) A floor with preset floor rate L_f is a series of floorlets, where the payoff at T_i is $\alpha_i \max(L_f - L_{i-1}(T_{i-1}), 0)$, $i = 1, 2, \cdots, n$. Find the time-t value of the floor.

(b) A caplet in arrears pays $\alpha_i \max(L_{i-1}(T_{i-1}) - L_c, 0)$ at T_{i-1}, that is, the payment date coincides with the LIBOR reset date. Note that the usual caplet pays the same amount at T_i. Find the time-t value of this caplet.

(c) A cap with deferred caplets has all caplet payments made at the terminal date T_n. Find the time-t value of this deferred cap.

8.20 Let the exchange rate process $X(t)$, T-maturity domestic and foreign bond processes $B_d(t, T)$ and $B_f(t, T)$ be the same as those defined in Problem 8.9. Find the time-t value of the LIBOR spread option which pays at time $T + \delta$

$$\alpha[L_d(T; \delta) - L_f(T; \delta)],$$

where $t < T$, $L_d(T; \delta)$ and $L_f(T; \delta)$ are the domestic and foreign LIBORs over the time interval $(T, T + \delta]$, respectively, observed at time T.

8.21 Assume that the T-maturity discounted bond price process $B(t, T)$ follows the one-factor Gaussian HJM under the risk neutral measure Q:

$$\frac{dB(t, T)}{B(t, T)} = r_t\, dt - \sigma_B(t, T)\, dZ_t.$$

A caption is a call option on a cap, whose terminal payoff at time T is given by

$$\max\left(\sum_{i=1}^{n} C_i(T; T_{i-1}, T_i) - X, 0\right), \quad T < T_0 < T_1 < \cdots < T_n.$$

Here, $C_i(T; T_{i-1}, T_i)$ is the time-T value of a caplet with payment on the LIBOR L_{i-1} at time T_i and X is the strike price, $i = 1, 2, \cdots, n$. Since a cap can be visualized as a series of put options on the zero-coupon bonds, a caption is seen as a compound call on a put. By applying Jamshidian's decomposition technique (Jamshidian, 1989) for an option on a coupon-bearing bond, find the time-t value of the caption, $t < T$.

8.22 Consider a floor on composition, where the composition is defined as

$$\prod_{i=1}^{n-1}(1 + \alpha_{i+1} L_i).$$

Here, L_i is reset at time T_i and α_{i+1} is the accrual factor over the time interval $(T_i, T_{i+1}]$. The payment of this floor at maturity T_n is

$$\max\left(\prod_{i=1}^{n-1}(1 + \alpha_{i+1} L_i), K\right).$$

Assuming that the discounted bond price process follows the same process as in Problem 8.21, show that the time-0 value of this floor on composition is given by (Henrard, 2005)

$$\mathcal{F}(0; T_1, \cdots, T_n) = B(0, T_1)N(k+\sigma) + KB(0, T_n)N(-k), \quad 0 \le T_1,$$

where

$$\sigma^2 = \sum_{i=1}^{n-1}\sum_{j=1}^{n-1}\int_0^{\min(T_i,T_j)} \left[\sigma_B(u, T_{i+1}) - \sigma_B(u, T_i)\right] \left[\sigma_B(u, T_{j+1}) - \sigma_B(u, T_j)\right] du,$$

and

$$k = \frac{1}{\sigma}\left[\ln\frac{B(0, T_1)}{KB(0, T_n)} - \frac{\sigma^2}{2}\right].$$

Hint: Use the identity

$$\sigma_B(u, T_n) - \sigma_B(u, T_i) = \sum_{j=1}^{n-1}[\sigma_B(u, T_{j+1}) - \sigma_B(u, T_j)]$$

to show that

$$\sigma^2 = \sum_{i=1}^{n-1}\int_0^{T_i} [\sigma_B(u, T_{i+1}) - \sigma_B(u, T_i)] [2\sigma_B(u, T_n) - \sigma_B(u, T_{i+1}) - \sigma_B(u, T_i)]\, du.$$

8.23 For $T' < T$, define the adjusted forward measure $Q_{T',T}$ by (Eberlein and Kluge, 2006)

$$\frac{dQ_{T,T'}}{dQ_T} = \frac{F(T', T', T)}{F(0, T', T)} = \frac{B(0, T)}{B(0, T')B(T', T)},$$

where $F(t, T', T) = \frac{B(t,T')}{B(t,T)}$ denotes the bond forward price process. Restricting this density to $\mathcal{F}_t$ for $t < T'$, show that

$$\left.\frac{dQ_{T,T'}}{dQ}\right|_{\mathcal{F}_t} = \frac{B(t, T')}{M(t)B(0, T')} = \left.\frac{dQ_T}{dQ}\right|_{\mathcal{F}_t}$$

[see (8.1.6)]. However, for $T' < t < T$, show that

$$\left.\frac{dQ_{T,T'}}{dQ}\right|_{\mathcal{F}_t} = \frac{B(t, T)}{B(T', T)M(t)B(0, T')}.$$

Using the adjusted forward measure, show that [see (8.2.27)]

$$E^t_{Q_{T_{j+1}}}\left[\frac{1}{B(T_j,T_{j+1})}\mathbf{1}_{\{R_\ell(T_{j,i})\le L(T_{j,i},T_{j,i}+\delta_{j,i})\le R_u(T_{j,i})\}}\right]$$
$$= E^t_{Q_{T_j,T_{j+1}}}\left[\mathbf{1}_{\{R_\ell(T_{j,i})\le L(T_{j,i},T_{j,i}+\delta_{j,i})\le R_u(T_{j,i})\}}\right].$$

8.24 Suppose the forward LIBOR $L(t, T)$ satisfies the following stochastic differential equation under the risk neutral Q-measure

$$dL(t,T) = \mu(t,T)L(t,T)\,dt + L(t,T)\sum_{i=1}^{n}\sigma_i(t,T)\,dZ_i(t),$$

where $\sigma_i(t,T), i = 1, 2, \cdots, n$ are deterministic volatility functions and $\mathbf{Z}(t) = (Z_1(t)\cdots Z_n(t))^{\mathrm{T}}$ is Q-Brownian. Define

$$F_B(t,T,T+\delta) = \frac{B(t,T+\delta)}{B(t,T)},$$

which is the time-t price of the T-forward on the $T+\delta$-maturity discount bond. Under the Q_T-measure, show that $F_B(t)$ satisfies

$$dF_B(t) = -F_B(t)[1 - F_B(t)]\sum_{i=1}^{n}\sigma_i(t,T)\,dZ_i^T(t),$$

where $\mathbf{Z}^T(t) = (Z_1^T(t)\cdots Z_n^T(t))^{\mathrm{T}}$ is Q_T-Brownian. Let $V(x,t)$ denote the forward price of the T-maturity put option on the $(T+\delta)$-maturity bond, where x is the forward bond price. Show that V satisfies

$$\frac{\partial V}{\partial t} + \frac{1}{2}\left[x^2(1-x^2)\sum_{i=1}^{n}\sigma_i^2(t,T)\right]\frac{\partial^2 V}{\partial x^2} = 0,$$

with the terminal condition: $V(x,T) = \max(X - x, 0)$. Here, X is the strike price of the put. Solve for $V(x,t)$ (Miltersen, Sandmann and Sondermann, 1997).

8.25 Suppose the cap rate of a cap and the floor rate of a floor are both set equal to L. Let $\mathcal{C}(t; T_0, \cdots, T_n)$ and $\mathcal{F}(t; T_0, \cdots, T_n)$ denote the time-t value of the cap and floor, respectively. Prove the following cap-floor parity relation:

$$\mathcal{C} - \mathcal{F} = \sum_{i=1}^{n}[B(t;T_{i-1}) - (1+\alpha_i L)B(t;T_i)].$$

8.26 The holder of a reverse floater is entitled to receive the LIBOR $L_i(T_i)$ while pay $\max(K - L_i(T_i), K')$ at time $T_{i+1}, i = 1, 2, \cdots, n-1$, with respect to a unit principal, $K > 0$ and $K' > 0$. Using the Lognormal LIBOR Market model as defined by (8.3.10), find the fair value of the reverse floater at time $t, t < T_0$.

8.27 Suppose we define the modified forward LIBOR $L_i^m(t)$ and futures LIBOR $L_i^f(t)$ by

$$1 + \alpha_{i+1} L_i^m(t) = E_{Q_{T_i}}^t \left[\frac{1}{B(T_i, T_{i+1})} \right]$$
$$1 + \alpha_{i+1} L_i^f(t) = E_Q^t \left[\frac{1}{B(T_i, T_{i+1})} \right],$$

respectively. Here, Q_{T_i} and Q are the T_i-forward measure and risk neutral measure, respectively. Assuming that the discount bond price $B(t, T)$ follows the Gaussian HJM process, show that $L^m(t)$ and $L^f(t)$ satisfy the following stochastic differential equations:

$$dL_i^m(t) = \frac{1 + \alpha_{i+1} L_i^m(t)}{\alpha_{i+1}} \sum_{j=1}^{m} \left[\sigma_B^j(t, T_j) - \sigma_B^j(t, T_{j+1}) \right] dZ_j^{T_i}(t),$$
$$dL_i^f(t) = \frac{1 + \alpha_{i+1} L_i^f(t)}{\alpha_{i+1}} \sum_{j=1}^{m} \left[\sigma_B^j(t, T_j) - \sigma_B^j(t, T_{j+1}) \right] dZ_j(t),$$

where $\mathbf{Z}^{T_i}(t) = (Z_1^{T_i}(t) \cdots Z_m^{T_i}(t))^{\mathrm{T}}$ is an m-dimensional Q_{T_j}-Brownian process and $\mathbf{Z}(t) = (Z_1(t) \cdots Z_m(t))^{\mathrm{T}}$ is an m-dimensional Q-Brownian process.

8.28 Consider the European put option with expiration date T_{i-1} and strike price X written on a T_i-maturity discount bond, $0 < X < 1$. Recall that the ith caplet is equivalent to the put bond option, deduce the price function of this put bond option using the caplet price formula under the Lognormal LIBOR Market model [see (8.3.10)].

8.29 We would like to price the floor on the composition defined in Problem 8.22 using the LIBOR Market model. Now, we assume that the LIBOR $L_i(t)$ follows the *arithmetic* Brownian process:

$$dL_i(t) = \sigma_i^L(t)\, dZ_i^{T_{i+1}}(t) \quad i = 1, 2, \cdots, n-1.$$

Here, $\sigma_i^L(t)$ is the deterministic volatility function and $Z_i^{T_{i+1}}(t)$ is $Q_{T_{i+1}}$-Brownian. By making the "frozen coefficient" assumption in the drift term of the stochastic differential equation of $L_i(t)$ under the terminal measure Q_{T_n}, show that the time-0 value of the floor on composition is given by (Henrard, 2005)

$$\mathcal{F}(0; T_1, \cdots, T_n) = B(0, T_1) N(k + \sigma) + K B(0, T_n) N(-k), \quad 0 \le T_1,$$

where

$$\sigma^2 = \boldsymbol{\lambda}^T S \boldsymbol{\lambda} \quad \text{and} \quad k = \frac{1}{\sigma}\left[\ln \frac{B(0,T_1)}{KB(0,T_n)} - \frac{\sigma^2}{2}\right],$$

$$\boldsymbol{\lambda} = \left(\frac{1}{L(0,T_2) + \frac{1}{\alpha_2}} \cdots \frac{1}{L(0,T_n) + \frac{1}{\alpha_n}}\right)^{\mathrm{T}},$$

α_{i+1} is the accrual factor of $(T_i, T_{i-1}), i = 1, 2, \cdots, n-1$, and the (i, j)th entry of the matrix S is given by

$$(S)_{ij} = \int_0^{\min(T_i,T_j)} \sigma_i^L(u)\sigma_j^L(u)\,du, \quad i, j = 1, 2, \cdots, n-1.$$

8.30 Consider the Lognormal LIBOR Market (LLM) model for the LIBOR $L_i(t)$, $i = 0, 1, \cdots, n-1$, defined on the tenor structure $\{T_0, T_1, \cdots, T_n\}$ where $0 < T_0 < T_1 < \cdots < T_n$. Let $v_i(t)$ denote the scalar volatility function of $L_i(t)$, and write $L_i(0) = L_{i,0}, i = 0, 1, \cdots, n-1$. Now, $L_i(t)$ satisfies

$$\frac{dL_i(t)}{L_i(t)} = v_i(t)\,dZ_i(t), \quad 0 < t < T_i, \quad i = 0, 1, \cdots, n-1,$$

where $Z_i(t)$ is Q_{T_i}-Brownian. We write $\rho_{ij}(t)$ as the correlation coefficient between $Z_i(t)$ and $Z_j(t)$ such that $\rho_{ij}(t)\,dt = dZ_i(t)\,dZ_j(t)$. Define

$$X_i(t) = \ln \frac{L_i(t)}{L_{i,0}} + \int_0^t \frac{v_i^2(u)}{2}\,du, \quad 0 < t < T_i, \quad i = 0, 1, \cdots, n-1.$$

Under the terminal measure Q_{T_n}, show that $X_i(t)$ satisfies the following stochastic differential equation

$$dX_i(t) = -\sum_{j=i}^{n-1} \frac{\alpha_j v_j(t) v_i(t) \rho_{ij}(t) L_{i,0} e^{X_j(t) - \frac{1}{2}\int_0^t v_j^2(u)\,du}}{1 + \alpha_{j+1} L_{j,0} e^{X_j(t) - \frac{1}{2}\int_0^t v_j^2(u)\,du}}\,dt$$
$$+ v_i(t)\,d\widetilde{Z}_i^n(t), \quad 0 < t < T_i, \quad i = 0, 1, \cdots, n-1,$$

with initial condition: $X_i(0) = 0$, and $\widetilde{Z}_i^{n+1}(t)$ is Q_{T_n}-Brownian under the Girsanov transformation: $Q_{T_i} \to Q_{T_n}$.

Hint: Note that $\widetilde{Z}_1^n, \cdots, \widetilde{Z}_n^n$ are Q_{T_n}-Brownian and

$$d\widetilde{Z}_i^n\,d\widetilde{Z}_j^n = \rho_{ij}(t)\,dt$$

since an equivalent change of a probability measure preserves the correlation between the Brownian processes. We write formally

$$\frac{dL_i(t)}{L_i(t)} = \widehat{\mu}_i(t)\,dt + \sigma_i(t)\,d\widetilde{Z}_i^n,$$

and subsequently $\widehat{\mu}_i(t)$ is determined. Recall

$$B(t, T_i) = B(t, T_{i+1})[1 + \alpha_{i+1} L_i(t)]$$

and observe that

$$\begin{aligned}
&-B(t, T_{i+1})\widehat{\mu}_i(t) L_i(t)\, dt \\
&= dB(t, T_{i+1})\, dL_i(t) \\
&= B(t, T_{i+1}) L_i(t) \sum_{j=i}^{n-1} \frac{\alpha_{j+1} L_j(t)}{1 + \alpha_{j+1} L_j(t)} \sigma_i(t) \sigma_j(t) \rho_{ij}(t)\, dt.
\end{aligned}$$

8.31 Suppose $L_i(t)$ satisfies the LLM model as in Problem 8.30. Here, we would like to find the stochastic differential equation of $L_i(t)$ under the spot LIBOR measure $Q_{\widetilde{M}}$ whose numeraire is the discrete money market account process $\widetilde{M}$. The discrete dynamics of $\widetilde{M}$ is defined by

$$\widetilde{M}(T_j) = \prod_{k=0}^{j-1} \frac{1}{B(T_k, T_{k+1})} \quad \text{with } \widetilde{M}(T_0) = 1, \quad j = 1, 2, \cdots, n.$$

Show that the dynamics of $L_i(t)$ under the spot LIBOR measure $Q_{\widetilde{M}}$ is given by

$$\frac{dL_i(t)}{L_i(t)} = \sum_{j=0}^{i-1} \frac{\alpha_{j+1} v_j(t) v_i(t) \rho_{ij}(t) L_j(t)}{1 + \alpha_{j+1} L_j(t)}\, dt + v_i(t)\, d\widetilde{Z}_i(t),$$

where $\widetilde{Z}_i(t)$ is $Q_{\widetilde{M}}$-Brownian.

8.32 Consider the time-t value of the LIBOR-in-arrears payment [see (8.3.14a,b)], show that

$$\begin{aligned}
\widetilde{V}_L(t) &= B(t, T_j) E^t_{Q_j}[\alpha_{j+1} L_j(T_j)] \\
&= \alpha_j L_j(t) + \int_0^\infty 2\alpha_j \alpha_{j+1} C_j(t; T_j, T_{j+1}, X)\, dX,
\end{aligned}$$

where $C_j(t; T_j, T_{j+1}, X)$ is the time-t value of the caplet with strike X, resetting at T_j and paying $\max(L_j(T_j) - X, 0)$ at T_{j+1}.
Hint: For any twice-differentiable function $V(y)$ defined for $y \geq 0$, we have

$$V(y) = V(0) + yV'(0) + \int_0^\infty V''(X) \max(y - X, 0)\, dX.$$

8.33 Suppose the dynamics of $L_i(t)$ under the forward measure Q_{T_k} is governed by (8.3.23), show that the distribution of the LIBOR $L_i(T)$ under Q_{T_k} admits the following lognormal approximation (Daniluk and Gataret, 2005):

$$L_i(T) \approx L_i(0)\exp\bigg(\int_0^T \sigma_{ik}(t,T)\,dZ^k(t) - \frac{1}{2}\int_0^T \sigma_i^L(t)^2\,dt + \int_0^T \mu_{ik}(t,T)\,dt\bigg),$$

where $T \le \min(T_i, T_k)$, $C_{ik}(t,T) = \int_t^T \sigma_i^L(u)\sigma_k^L(u)\,du$ and

(i) $i < k$

$$\sigma_{ik}(t,T) = \sigma_i^L(t) - \sum_{j=i+1}^{n} C_{ij}(t,T)\frac{K_j(0)}{1+\alpha_{j+1}K_j(0)}\sigma_j^L(t)$$

$$\mu_{ik}(t,T) = \sum_{j=i+1}^{k} K_j(0)\sigma_j^L(t)\left[-\sigma_i^L(t) + \frac{C_{ij}(t,T)}{1+\alpha_{j+1}L_j(0)}\sum_{\ell=j}^{k} K_\ell(0)\sigma_\ell^L(t)\right].$$

(ii) $i > k$

$$\sigma_{ik}(t,T) = \sigma_i^L(t) + \sum_{j=k+1}^{i} C_{ij}(t,T)\frac{K_j(0)}{1+\alpha_{j+1}K_j(0)}\sigma_j^L(t)$$

$$\mu_{ik}(t,T) = \sum_{j=k+1}^{i} K_j(0)\sigma_j^L(t)\left[\sigma_i^L(t) + \frac{C_{ij}(t,T)}{1+\alpha_{j+1}L_j(0)}\sum_{\ell=k+1}^{j-1} K_\ell(0)\sigma_\ell^L(t)\right].$$

Hint: Use integration by parts to show that

$$\int_0^T K_j(t)\sigma_j^L(t)\sigma_i^L(t)\,dt = \int_0^T K_j(0)\sigma_j^L(t)\sigma_i^L(t)\,dt + \int_0^T C_{ij}(t,T)\,dK_j(t).$$

Subsequently, show that

$$dK_j(t) = \frac{K_j(t)\sigma_j^L(t)}{1+\alpha_{j+1}L_j(t)}\big\{\big[\gamma_{kj}(t) - K_j(t)\sigma_j^L(t)\big]\,dt + dZ^k(t)\big\}.$$

Applying the "frozen" LIBOR technique, show that

$$dK_j(t) \approx \frac{K_j(0)\sigma_j^L(t)}{1+\alpha_{j+1}L_j(0)}\left[dZ^k(t) - \sum_{\ell=j}^{k} K_\ell(0)\sigma_\ell^L(t)\,dt\right] \text{ for } j < k$$

$$dK_j(t) \approx \frac{K_j(0)\sigma_j^L(t)}{1+\alpha_{j+1}L_j(0)}\left[dZ^k(t) + \sum_{\ell=j}^{j-1} K_\ell(0)\sigma_\ell^L(t)\,dt\right] \text{ for } j > k.$$

8.34 Consider the forward payer swap settled in advance, that is, each reset date is also a settlement date. The LIBOR $L_i(t)$ reset at T_i is used to determine the cash flow at T_i. Suppose the payments made at T_i are discounted by the factor $\frac{1}{1+\alpha_{i+1}L_i(T_i)}$ so that the floating cash flow is defined to be

$$\frac{\alpha_{i+1}L_i(T_i)}{1+\alpha_{i+1}L_i(T_i)}$$

while the fixed cash flow is

$$-\frac{K\alpha_{i+1}}{1+\alpha_{i+1}L_i(T_i)}.$$

Here, α_{i+1} is the accrual factor over $(T_i, T_{i+1}]$. Show that the time-t value of this forward payer swap is equal to that of the vanilla swap with the same fixed swap rate K.

8.35 Consider a swap with reset dates $T_0, T_1, \cdots, T_{n-1}$ and payment dates $T_1, T_2, \cdots, T_n$. A trigger swap is a contract where the holder has to enter into a swap with fixed swap rate K over the remaining period $[T_i, T_n]$ when $L_i(T_i) > K_i$, where K_i is the trigger level set for date T_i. Define i^* to be

$$i^* = \min_{1\leq m\leq n} \{m : L_m(t_m) > K_m\}.$$

Here, T_{i^*} is a stopping time. For $t < T_1$, show that the time-t value of the trigger swap can be expressed as

$$\begin{aligned} V_{trig} &= \sum_{m=1}^{n-1} B(t, T_m) E^t_{Q^T_m}[\mathbf{1}_{\{i^*=m\}}] \\ &\quad - \sum_{m=1}^{n-1} B(t, T_n) E^t_{Q^T_m}[\mathbf{1}_{\{i^*=m\}}] \\ &\quad - \sum_{m=1}^{n-1}\sum_{j=m}^{n-1} K\alpha_j B(t, T_{j+1}) E^t_{Q^T_{j+1}}[\mathbf{1}_{\{i^*=m\}}]. \end{aligned}$$

8.36 Use the following relations,

$$dZ(t) = dZ^i(t) + \sigma_B(t, T_i)\,dt = dZ^k(t) + \sigma_B(t, T_k)\,dt$$

and

$$\sigma_B(t, T_i) - \sigma_B(t, T_{i+1}) = \frac{\alpha_{i+1}L_i(t)}{1+\alpha_{i+1}L_i(t)}\sigma_i^L(t, T),$$

to show the result in (8.3.23).

8.37 Suppose we write the price function of the swaption as

$$V(t; T_0, T_n, X) = N(d_1)[B(t, T_0) - B(t, T_n)] - \sum_{i=1}^{n} XN(d_2)\alpha_i B(t, T_i),$$

[see (8.4.9)], the resulting expression reveals a hedging strategy of the swaption using discount bonds with varying maturities. The replicating portfolio $\Pi_s(t)$ consists of a long position in $N(d_1)$ units of T_0-maturity bond, a short position in $N(d_1)$ units of T_n-maturity bond and $\alpha_i N(d_2)X$ units of T_i-maturity bond, $i = 1, 2, \cdots, n$. Under the assumption of deterministic volatility function, show that the replicating portfolio $\Pi_s(t)$ is self-financing.
Hint: Check whether

$$d\Pi_s(t) = N(d_1)\, dB(t, T_0) - N(d_1)\, dB(t, T_n) - \sum_{i=1}^{n} XN(d_2)\alpha_i\, dB(t, T_i).$$

8.38 Consider the family of forward swap rates $K_t[T_i, T_n], i = 0, 1, \cdots, n-1$, with the common terminal payment date T_n. We would like to express the dynamics of $K_t[T_i, T_n]$ under the terminal forward measure Q_{T_n}. Since $K_t[T_{n-1}, T_n]$ is simply the LIBOR $L_{n-1}(t)$, we have

$$\frac{dK_t[T_{n-1}, T_n]}{K_t[T_{n-1}, T_n]} = \frac{dL_{n-1}(t)}{L_{n-1}(t)} = \sigma_{n-1,n}^K(t)\, dZ_t^{T_n},$$

where $Z_t^{T_n}$ is Q_{T_n}-Brownian. Derive the relation

$$\begin{aligned}\frac{\widehat{B}(t; T_{n-2}, T_n)}{B(t, T_n)} &= \alpha_{n-2} + \alpha_{n-1}[1 + \alpha_{n-2}K_t[T_{n-2}, T_n]) \\ &\quad + \alpha_n(1 + \alpha_{n-1}K_t[T_{n-1}, T_n])\{1 + \alpha_{n-2}K_t(T_{n-2}, T_n)\},\end{aligned}$$

and show that

$$dK_t[T_{n-2}, T_n] = \mu_{n-2}(t)\, dt + \sigma_{n-2,n}^K K_t[T_{n-2}, T_n]\, dZ_t^{T_n},$$

where

$$\widehat{B}(t; T_j, T_n) = \sum_{i=j+1}^{n} \alpha_i B(t, T_i), \quad j < n,$$

$$\mu_{n-2}(t) = -\frac{\alpha_n \sigma_{n-1,n}^K(t) K_t[T_{n-1}, T_n] \alpha_{n-1} \sigma_{n-2,n}^K(t) K_t[T_{n-2}, T_n]}{\alpha_{n-1} + \alpha_n\{1 + \alpha_{n-1}K_t[T_{n-1}, T_n]\}}.$$

In general, deduce the relation

$$\frac{\widehat{B}(t; T_i, T_n)}{B(t, T_n)} = \alpha_i + \sum_{k=i+1}^{n} \alpha_k \prod_{j=i+1}^{k} (1 + \alpha_j K_t[T_j, T_n]), \quad i = n-2, n-3, \cdots, 0,$$

and express the dynamics of $K_t[T_i, T_n]$ under the terminal measure Q_{T_n}.

8.39 Suppose the short rate r_t is governed by the Vasicek model

$$dr_t = \alpha(\gamma - r_t)\,dt + \rho\,dZ_t,$$

where Z_t is a Brownian process under the risk neutral measure Q. Show that the stochastic differential equation of the swap rate $K_t[T_0, T_n]$ under the swap measure $Q_{S_{0,n}}$ with the annuity numeraire $\widehat{B}(t; T_0, T_n)$ is given by (Schrager and Pelsser, 2006)

$$\frac{dK_t[T_0, T_n]}{K_t[T_0, T_n]} = \rho \frac{\partial K_t[T_0, T_n]}{\partial r_t}\, dZ_t^{0,n},$$

where

$$\frac{\partial K_t[T_0, T_n]}{\partial r_t} = -b(t, T_0)D(t, T_0) + b(t, T_n)D(t, T_n) + K_t[T_0, T_n]\sum_{i=1}^{n} \alpha_i b(t, T_i)D(t, T_n),$$

$$b(t, T) = \frac{1}{\alpha}[1 - e^{-\alpha(T-t)}],$$

$$D(t, T_i) = \frac{B(t, T_i)}{\widehat{B}(t; T_0, T_n)}, \quad i = 1, 2, \cdots, n.$$

8.40 Consider a Constant Maturing Swap (CMS) caplet whose payoff at payment date T_p takes the form

$$\max(K_{T_0}[T_0, T_n] - s_{cap}, 0),$$

where the par swap rate over the tensor $\{T_0, \cdots, T_n\}$ is set at T_0, $T_p > T_0$ and s_{cap} is some pre-set constant cap value. As usual, $\{T_0, \cdots, T_{n-1}\}$ are the reset dates at which the relevant LIBOR $\{L_0, \cdots, L_{n-1}\}$ are determined. Using the annuity $\widehat{B}(t; T_0, T_n)$ as the numeraire, show that the time-t value of the CMS caplet is given by

$$V_{CMS}(t; T_0, T_n) = \widehat{B}(t; T_0, T_n) E^t_{Q_{S_{0,n}}}\left[\frac{\max(K_{T_0}[T_0, T_n] - s_{cap}, 0)B(T_0, T_p)}{\widehat{B}(T_0; T_0, T_n)}\right],$$

where $Q_{S_{0,n}}$ is the swap measure with the annuity numeraire. Using the relation

$$E^t_{Q_{S_{0,n}}}\left[\frac{B(T_0, T_p)}{\widehat{B}(T_0; T_0, T_n)}\right] = \frac{B(t, T_p)}{\widehat{B}(t; T_0, T_n)},$$

express $V_{CMS}(t; T_0, T_n)$ in terms of the price function of a European swaption plus a convexity adjustment term due to payment of the caplet payoff at a later date T_p. Determine the form of this convexity adjustment.

8.41 Under the risk neutral measure, suppose the dynamics of the domestic interest rate r_d and foreign interest rate r_f follow the mean-reversion processes:

$$\begin{aligned} dr_d &= k_d(\mu_d - r_d)\,dt + \sigma_d\,dZ_d \\ dr_f &= k_f(\mu_f - r_f)\,dt + \sigma_f\,dZ_f, \end{aligned}$$

and the exchange rate X follows

$$\frac{dX}{X} = (r_d - r_f)\,dt + \sigma_X\,dZ_X.$$

Let ρ_{df} and ρ_{fX} denote the correlation coefficient between the domestic and foreign interest rates, and between the foreign interest rate and exchange rate, respectively. Taking the notional to be unity and assuming that the interest payment to be paid at T_1 is known at time t, show that the sum of the present values at time t of all future interest payments based on the foreign LIBORs in a differential swap is given by (Wei, 1994; Chang, Chung and Yu, 2002)

$$\begin{aligned} V_f &= r_f(t)B_d(t,T_1) \\ &\quad + \sum_{i=1}^{n-1}\left[\frac{B_d(t,T_{i+1})B_f(t,T_i)}{B_f(t,T_{i+1})}\exp\left(-b_i^{(1)} + b_i^{(2)} - b_i^{(3)}\right)\right. \\ &\qquad\qquad \left. - B_d(t,T_{i+1})\right], \qquad t < T_1, \end{aligned}$$

where $B_d(t,T_i)$ and $B_f(t,T_i)$ are the T_i-maturity domestic and foreign bond discount prices, respectively, and

$$\begin{aligned} b_i^{(1)} &= \frac{\sigma_d\sigma_f\rho_{df}}{k_dk_f}\left\{\frac{1+e^{-k_f(T_{i+1}-t)} - e^{-k_f(T_i-t)} - e^{-k_f(T_{i+1}-T_i)}}{k_f}\right. \\ &\quad + \frac{1}{k_d+k_f}\left[e^{-k_d(T_{i+1}-t)-k_f(T_i-t)} + e^{-(k_d+k_f)(T_{i+1}-T_i)}\right. \\ &\qquad\qquad \left.\left. - e^{-(k_d+k_f)(T_{i+1}-t)} - e^{-k_d(T_{i+1}-T_i)}\right]\right\}, \\ b_i^{(2)} &= \frac{\sigma_f^2}{k_f^3}\left[1 - \frac{3}{2}e^{-k_f(T_{i+1}-T_i)} + e^{-k_f(T_{i+1}-t)} - e^{-k_f(T_i-t)}\right. \\ &\qquad \left. - \frac{1}{2}e^{-2k_f(T_{i+1}-t)} + \frac{1}{2}e^{-2k_f(T_{i+1}-T_i)} + \frac{1}{2}e^{-k_f(T_{i+1}+T_i-2t)}\right], \\ b_i^{(3)} &= \frac{\sigma_f\sigma_X\rho_{fX}}{k_f^2}\left[1 - e^{-k_f(T_{i+1}-T_i)} + e^{-k_f(T_{i+1}-t)} - e^{-k_f(T_i-t)}\right]. \end{aligned}$$

Using financial arguments, explain why ρ_{df} and ρ_{fX} enter into the pricing formula, and the differential swap would be worth something even if the domestic and foreign forward rates are the same and evolve with perfect correlation.

8.42 A differential swap may involve three currency worlds: interest payments are calculated based on the floating LIBOR of the first two currencies but the actual payments are denominated in a third currency. Show that this type of three-currency differential swap can be decomposed into a portfolio of standard type differential swaps which involve only two currency worlds.

References

Ahn, D.M., Dittmar, R.F., Gallant, A.R., "Quadratic term structure models: Theory and evidence," *Review of Financial Studies*, vol. 15 (2002) pp. 243–288.

Andersen, L., Broadie, M., "Primal-dual simulation algorithm for pricing multi-dimensional American options," *Management Science*, vol. 50(9) (2004) pp. 1222–1234.

Andreasen, J., "The pricing of discretely sampled Asian and lookback options: A change of numeraire approach," *Journal of Computational Finance*, vol. 2 (1998) pp. 5–30.

Babbs, S.H., Nowman, K.B., "Kalman filtering of generalized Vasicek term structure models," *Journal of Financial and Quantitative Analysis*, vol. 34 (1999) pp. 115–130.

Balduzzi, P., Das, S.R., Foresi, S., Sundaram, R.K., "A simple approach to three-factor affine term structure models," *Journal of Fixed Income*, vol. 6 (1996) pp. 43–53.

Barber, J.R., "Bond option valuation for non-Markovian interest rate process," *Financial Review*, vol. 40 (2005) pp. 519–532.

Barone-Adesi, G., Whaley, R.E., "Efficient analytic approximation of American option values," *Journal of Finance*, vol. 42(2) (1987) pp. 301–320.

Barraquand, J., Pudet, T., "Pricing of American path-dependent contingent claims," *Mathematical Finance*, vol. 6(1) (1996) pp. 17–51.

Beaglehole, D.R., Tenney, M.S., "General solutions of some interest rate-contingent claim pricing equation," *Journal of Fixed Income* (1991) pp. 69–83.

Bensaid, B., Lesne, J.P., Pages, H., Scheinkman, J., "Derivative asset pricing with transaction costs," *Mathematical Finance*, vol. 2 (April 1992) pp. 63–86.

Bingham, N.H., Kiesel, R., *Risk-neutral valuation: Pricing and hedging of financial derivatives*, Springer, Heidelberg, second edition (2004)

Black, F., "Fact and fantasy in the use of options," *Financial Analysts Journal*, vol. 31 (July 1975) pp. 36–72.

Black, F., "The pricing of commodity contracts," *Journal of Financial Economics*, vol. 3 (1976) pp. 167–179.

Black, F., "How we came up with the option formula," *Journal of Portfolio Management*, vol. 7 (Winter 1989) pp. 4–8.

Black, F., Cox, J.C., "Valuing corporate securities: Some effects of bond indenture provisions," *Journal of Finance*, vol. 31 (May 1976) pp. 351–367.

Black, F., Karasinski, P., "Bond and option pricing when short rates are lognormal," *Financial Analysts Journal* (July 1991) pp. 52–59.

Black, F., Scholes, M., "The pricing of option and corporate liabilities," *Journal of Political Economy*, vol. 81 (1973) pp. 637–659.

Black, F., Derman, E., Toy, W., "A one-factor model of interest rates and its application to Treasury bond options," *Financial Analysts Journal* (Jan. 1990) pp. 33–39.

Bouaziz, L., Briys, E., Crouhy, M., "The pricing of forward-starting Asian options," *Journal of Banking and Finance*, vol. 18 (1994) pp. 823–839.

Boyle, P.P., "A lattice framework for option pricing with two state variables," *Journal of Financial and Quantitative Analysis*, vol. 23 (March 1988) pp. 1–12.

Boyle, P.P., "New life forms on the option landscape," *Journal of Financial Engineering*, vol. 2(3) (1993) pp. 217–252.

Boyle, P., Broadie, M., Glasserman, P., "Monte Carlo methods for security pricing," *Journal of Economic Dynamics and Control*, vol. 21 (June 1997) pp. 1267–1321.

Boyle, P.P., Tian, Y., "An implicit finite difference approximation to the pricing of barrier options," *Applied Mathematical Finance*, vol. 5 (1998) pp. 17–43.

Brace, A., Musiela, M., "A multifactor Gauss Markov implementation of Heath, Jarrow, and Morton," *Mathematical Finance*, vol. 4 (1994) pp. 259–283.

Brace, A., Gatarek, D., Musiela, M., "The market model of interest rate dynamics," *Mathematical Finance*, vol. 7 (1997) pp. 127–147.

Breeden, D., Litzenberger, R., "Prices of state-contingent claims implicit in option prices," *Journal of Business*, vol. 51 (1979) pp. 621–651.

Brennan, M.J., Schwartz, E.S., "Finite difference method and jump processes arising in the pricing of contingent claims," *Journal of Financial and Quantitative Analysis*, vol. 13 (Sept. 1978) pp. 461–474.

Brennan, M.J., Schwartz, E.S., "A continuous time approach to the pricing of bonds," *Journal of Banking and Finance*, vol. 3 (1979) pp. 133–155.

Brennan, M.J., Schwartz, E.S., "Analyzing convertible bonds," *Journal of Financial and Quantitative Analysis*, vol. 15 (Nov. 1980) pp. 907–929.

Brenner, M., Courtadon, G., Subrahmanyan, M., "Options on the spot and options on futures," *Journal of Finance*, vol. 40 (Dec. 1985) pp. 1303–1318.

Brigo, D., Liinev, J., "On the distributional distance between the lognormal LIBOR and swap market models," *Quantitative Finance*, vol. 5(5) (2005), pp. 433–442.

Broadie, M., Detemple, J., "American capped call options on dividend-paying assets," *Review of Financial Studies*, vol. 8 (1995) pp. 161–191.

Broadie, M., Detemple, J., "American option valuation: New bounds, approximations, and a comparison of existing methods," *Review of Financial Studies*, vol. 9 (1996) pp. 1211–1250.

Broadie, M., Glasserman, P., "Pricing American-style securities using simulation," *Journal of Economic Dynamics and Control*, vol. 21 (1997) pp. 1323–1352.

Broadie, M., Glasserman, P., Kou, S., "A continuity correction for discrete barrier option," *Mathematical Finance*, vol. 7 (1997), pp. 325–349.

Broadie, M., Glasserman, P., Kou, S.G., "Connecting discrete and continuous path dependent options," *Finance and Stochastics*, vol. 3 (1999) pp. 55–82.

Bunch, D.S., Johnson, H., "A simple and numerically efficient valuation method for American puts using a modified Geske–Johnson approach," *Journal of Finance*, vol. 47 (June 1992) pp. 809–816.

Bunch, D.S., Johnson, H., "The American put option and its critical stock price," *Journal of Finance*, vol. 55(5) (2000) pp. 2333–2356.

Cairns, A.J.G., *Interest rate models*, Princeton University Press, Princeton (2004).

Carr, P., Bandyopadhyay, A., "How to derive Black–Scholes equation correctly?" *Working paper of Banc of America Securities and University of Illinois* (2000).

Carr, P., Chesney, M., "American put call symmetry," *Working paper of Morgan Stanley and Groupe HEC* (1996).

Carr, P., Jarrow, R., Myneni, R., "Alternative characterizations of American put options," *Mathematical Finance*, vol. 2 (1992) pp. 87–106.

Chan, K.C., Karolyi, G.A., Longstaff, F.A., Sanders, A.B., "An empirical comparison of alternative models of short-term interest rate," *Journal of Finance*, vol. 47 (July 1992) pp. 1209–1227.

Chang, C.C., Chung, S.L., Yu, M.T., "Valuation and hedging of differential swaps," *Journal of Futures Markets*, vol. 22 (2002) pp. 73–94.

Chen, R.R., "Exact solutions for futures and European futures options on pure discount bonds," *Journal of Financial and Quantitative Analysis*, vol. 27 (Mar. 1992) pp. 97–107.

Chesney, M., Jeanblanc-Picqué, M., Yor, M., "Brownian excursions and Parisian barrier options," *Advances in Applied Probabilities*, vol. 29 (1997) pp. 165–184.

Cheuk, T.H.F., Vorst, T.C.F., "Currency lookback options and observation frequency: A binomial approach," *Journal of International Money and Finance*, vol. 16(2) (1997) pp. 173–187.

Chiarella, C., Kwon, O.K., "Finite dimensional affine realizations of HJM models in terms of forward rates and yields," *Review of Derivatives Research*, vol. 6 (2003) pp. 129–155.

Chu, C.C., Kwok, Y.K., "Reset and withdrawal rights in dynamic fund protection," *Insurance: Mathematics and Economics*, vol. 34(2) (2004) pp. 273–295.

Chung, S.L., Shackleton, M., Wojakowski, R., "Efficient quadratic approximation of floating strike Asian option values," *Finance*, vol. 24 (2003) pp. 49–62.

Clément, E., Lamberton, D., Protter, P., "An analysis of a least squares regression algorithm for American option pricing," *Finance and Stochastic*, vol. 6 (2002) pp. 449–471.

Conze, A., Viswanathan, "Path dependent options: The case of lookback options," *Journal of Finance*, vol. 46(5) (1991) pp. 1893–1907.

Cox, D.R., Miller, H.D., *The theory of stochastic processes*, Chapman and Hall, London (1995)

Cox, J.C., Ross, S.A., "The valuation of options for alternative stochastic processes," *Journal of Financial Economics*, vol. 3 (1976) pp. 145–166.

Cox, J.C., Ross, S., Rubinstein, M., "Option pricing: A simplified approach," *Journal of Financial Economics*, vol. 7 (Oct. 1979) pp. 229–264.

Cox, J.C., Ingersoll, J.E. Jr., Ross, S.A., "An analysis of variable rate loan contracts," *Journal of Finance*, vol. 35 (1980) pp. 389–403.

Cox, J.C., Ingersoll, J.E. Jr., Ross, S.A., "A theory of the term structure of interest rates," *Econometrica*, vol. 53 (March 1985) pp. 385–407.

Curran, M., "Valuing Asian and portfolio options by conditioning on the geometric mean price," *Management Science*, vol. 40(12) (1994) pp. 1705–1711.

Dai, M., Kwok, Y.K., "Knock-in American options," *Journal of Futures Markets*, vol. 24(2) (2004) pp. 179–192.

Dai, M., Kwok, Y.K., "American options with lookback payoff," *SIAM Journal of Applied Mathematics*, vol. 66(1) (2005a) pp. 206–227.

Dai, M., Kwok, Y.K., "Optimal policies of call with notice period requirement for American warrants and convertible bonds," *Asia Pacific Financial Markets*, vol. 12(4) (2005b) pp. 353–373.

Dai, M., Kwok, Y.K., "Options with combined reset rights on strike and maturity," *Journal of Economic Dynamics and Control*, vol. 29 (2005c) pp. 1495–1515.

Dai, M., Kwok, Y.K., "Characterization of optimal stopping regions of American Asian and lookback options," *Mathematical Finance*, vol. 16(1) (2006) pp. 63–82.

Dai, M., Kwok, Y.K., "Optimal multiple stopping models of reload options and shout options," to appear in *Journal of Economic Dynamics and Control* (2008).

Dai, M., Kwok, Y.K., Wu, L., "Options with multiple reset rights," *International Journal of Theoretical and Applied Finance*, vol. 6(6) (2003) pp. 637–653.

Dai, M., Kwok, Y.K., Wu, L., "Optimal shouting policies of options with strike reset right," *Mathematical Finance*, vol. 14(3) (2004) pp. 383–401.

Dai, Q., Singleton, K.J., "Specification analysis of affine term structure models," *Journal of Finance*, vol. 55 (2000) pp. 1943–1978.

Dai, M., Wong, H.Y., Kwok, Y.K., "Quanto lookback options," *Mathematical Finance*, vol. 14(3) (2004) pp. 445–467.

Daniluk, A., Gatarek, D., "A fully lognormal LIBOR market model," *Risk* (Sept. 2005) pp. 115–118.

Das, S.R., Foresi, S., "Exact solutions for bond and option prices with systematic jump risk," *Review of Derivatives Research*, vol. 1 (1996) pp. 7–24.

Davis, M.H.A., Panas, V.G., Zariphopoulou, T., "European option pricing with transaction costs," *SIAM Journal of Control*, vol. 31 (1993) pp. 470–493.

Dempster, M.A.H., Hutton, J.P., "Pricing American stock options by linear programming," *Mathematical Finance*, vol. 9 (1999) pp. 229–254.

Derman, E., Kani, I., "Stochastic implied tress: Arbitrage pricing with stochastic term and strike structure of volatility," *International Journal of Theoretical and Applied Finance*, vol. 1 (1998) pp. 61–110.

Dewynne, J.N., Howison, S.D., Rupf, I., Wilmott, P., "Some mathematical results in the pricing of American options," *European Journal of Applied Mathematics*, vol. 4 (1993) pp. 381–398.

Dothan, U.L., "On the term structure of interest rate," *Journal of Financial Economics*, vol. 6 (1978) pp. 59–69.

Duan, J., "The Garch option pricing model," *Mathematical Finance*, vol. 5 (1995) pp. 13–32.

Duffie, D., Huang, C.F., "Implementing Arrow–Debreu equilibria by continuous trading of a few long-lived securities," *Econometrica*, vol. 53(6) (1985) pp. 1337–1356.

Duffie, D., Kan, R., "A yield-factor model of interest rates," *Mathematical Finance*, vol. 6(4) (1996) pp. 379–406.

Dupire, B., "Pricing with a smile," *Risk*, vol. 7(1) (Jan. 1994) pp. 18–20.

Eberlein, E., Kluge, W., "Valuation of floating range notes in Lévy term structure models," *Mathematical Finance*, vol. 16(2) (2006) pp. 237–254.

El-Karoui, N., Geman, H., "A probabilistic approach to the valuation of floating rate notes with an application to interest rate swaps," *Advances in Options and Futures Research*, vol. 1 (1994) pp. 41–64.

Evans, J.D., Kuske, R., Keller, J.B., "American options on assets with dividends near expiry," *Mathematical Finance*, vol. 12(3) (2002) pp. 219–237.

Feller, W., "Two singular diffusion problems," *Annals of Mathematics*, vol. 54 (1951) pp. 173–182.

Figlewski, S., Gao, B., "The adaptive mesh model: A new approach to efficient option pricing," *Journal of Financial Economics*, vol. 53 (1999) pp. 313–351.

Fong, H.G., Vasicek, O.A., "Fixed-income volatility management," *Journal of Portfolio Management* (Summer 1991) pp. 41–46.

Forsyth, P.A., Vetzal, K.R., "Quadratic convergence of a penalty method for valuing American options," *SIAM Journal of Scientific Computing*, vol. 23 (2002) pp. 2096–2123.

Forsyth, P.A., Vetzal, K.R., Zvan, R., "Convergence of lattice and PDE methods for valuing path dependent options using interpolation," *Review of Derivatives Research*, vol. 5 (2002) pp. 273–314.

Fouque, J.P., Papanicolaou, G., Sircar, K.R., *Derivatives in financial markets with stochastic volatility*, Cambridge University Press, Cambridge, U.K. (2000)

Fu, M.C., Laprise, S.C., Madan, D.B., Su, Y., Wu, R., "Pricing American options: A comparison of Monte Carlo simulation approaches," *Journal of Computational Finance*, vol. 4(3) (Spring 2001) pp. 39–88.

Galluccio, S., Ly, J.M., "Theory and calibration of swap market models," *Mathematical Finance*, vol. 17 (Jan. 2007) pp. 111–141.

Gao, B., Huang, J.Z., Subrahmanyam, M., "The valuation of American barrier options using the decomposition technique," *Journal of Economic Dynamics and Control*, vol. 24(11–12) (2000) pp. 1783–1827.

Garman, M., "Recollection in tranquillity", in *From Black–Scholes to Black Holes: New Frontiers in Options*, Risk Magazine, Ltd, London (1992) pp. 171–175.

Garman, M.B., Kohlhagen, S.W., "Foreign currency option values," *Journal of International Money and Finance*, vol. 2 (1983) pp. 231–237.

Geman, H., El Karoui, N., Rochet, J.C., "Changes of numeraire, changes of probability measure and option pricing," *Journal of Applied Probability*, vol. 32 (1995) pp. 443–458.

Gerber, H.U., Shiu, E.S.W., "From perpetual strangles to Russian options," *Insurance: Mathematics and Economics*, vol. 15 (1994) pp. 121–126.

Geske, R., "The valuation of compound options," *Journal of Financial Economics*, vol. 7 (1979) pp. 375–380.

Geske, R., Johnson, H.E., "The American put option valued analytically," *Journal of Finance*, vol. 39 (1984) pp. 1511–1524.

Glasserman, P., *Monte Carlo methods in financial engineering*, Springer, New York (2004).

Glasserman, P., Yu, B., "Number of paths versus number of basis functions in American option pricing," *Annals of Applied Probability*, vol. 14(4) (2004) pp. 2090–2119.

Goldman, M.B., Sosin, H.B., Gatto, M.A., "Path dependent options: Buy at the low, sell at the high," *Journal of Finance*, vol. 34(5) (1979) pp. 1111–1127.

Grannan, E.R., Swindle, G.H., "Minimizing transaction costs of option hedging strategies," *Mathematical Finance*, vol. 6 (Oct. 1996) pp. 341–364.

Grant, D., Vora, G., Weeks, D.E., "Simulation and the early-exercise option problem," *Journal of Financial Engineering*, vol. 5(3) (1996) pp. 211–227.

Harrison, J.M., Kreps, D.M., "Martingales and arbitrage in multiperiod securities markets," *Journal of Economic Theory*, vol. 20 (1979) pp. 381–408.

Harrison, J.M., Pliska, S.R., "A stochastic calculus model of continuous trading: Complete markets," *Stochastic Processes and Their Applications*, vol. 15 (1983) pp. 313–316.

He, H., Takahashi, A., "A variable reduction technique for pricing average-rate options," *International Review of Finance*, vol. 1(2) (2000) pp. 123–142.

Heath, D., Jarrow, R., Morton, A., "Bond pricing and the term structure of interest rates: A new methodology for contingent claims valuation," *Econometrica*, vol. 60 (Jan. 1992) pp. 77–105.

Henderson, V., Wojakowski, R., "On the equivalence of floating and fixed-strike Asian options," *Journal of Applied Probability*, vol. 39 (2002) pp. 391–394.

Henrard, M., "Explicit bond option and swaption formula in Heath–Jarrow–Morton one factor model," *International Journal of Theoretical and Applied Finance*, vol. 6(1) (2003) pp. 57–72.

Henrard, M., "LIBOR market model and Gaussian HJM explicit approaches to option on composition," *Working paper* (2005)

Heston, S.L., "A closed-form solution for options with stochastic volatility with applications to bond and currency options," *Review of Financial Studies*, vol. 6 (1993) pp. 327–343.

Heston, S., "Discrete-time versions of continuous-time interest rate models," *Journal of Fixed Income* (Sept. 1995) pp. 86–88.

Heston, S., Zhou, G., "On the rate of convergence of discrete-time contingent claims," *Mathematical Finance*, vol. 10(1) (2000) pp. 53–75.
Heynen, R., Kat, H., "Partial barrier options," *Journal of Financial Engineering*, vol. 3(3/4) (1994a) pp. 253–274.
Heynen, R., Kat, H., "Selective memory," *Risk*, vol. 7 (Nov. 1994b) pp. 73–76.
Heynen, R.C., Kat, H.M., "Lookback options with discrete and partial monitoring of the underlying price," *Applied Mathematical Finance*, vol. 2 (1995) pp. 273–284.
Heynen, R.C., Kat, H.M., "Discrete partial barrier options with a moving barrier," *Journal of Financial Engineering*, vol. 5 (1996) pp. 199–209.
Ho, T.S.Y., Lee, S.B., "Term structure movements and pricing interest rate contingent claims," *Journal of Finance*, vol. 41 (Dec. 1986) pp. 1011–1029.
Hodges, S.D., Neuberger, A.J., "Optimal replication of contingent claims under transaction costs," *Review of Futures Markets*, vol. 8 (1989) pp. 222–239.
Hoogland, J., Neumann, C.D., "Asians and cash dividends: Exploiting symmetries in pricing theory," *Technical report of CWI* (2000)
Hsu, H., "Surprised parties," *Risk*, vol. 10 (Apr. 1997) pp. 27–29.
Huang, J.Z., Subrahmanyam, M.G., Yu, G.G., "Pricing and hedging American options: A recursive integration method," *Review of Financial Studies*, vol. 9(1) (1996) pp. 277–300.
Hull, J.C., White, A., "The pricing of options on assets with stochastic volatilities," *Journal of Finance*, vol. 42 (June 1987) pp. 281–300.
Hull, J., White, A., "The use of the control variate technique in option pricing," *Journal of Financial and Quantitative Analysis*, vol. 23(3) (1988) pp. 237–251.
Hull, J., White, A., "Pricing interest-rate-derivative securities," *Review of Financial Studies*, vol. 3(4) (1990) pp. 573–592.
Hull, J., White, A., "Bond option pricing based on a model for the evolution of bond prices," *Advances in Futures and Options Research*, vol. 6 (1993a) pp. 1–13.
Hull, J., White, A., "Efficient procedures for valuing European and American path-dependent options," *Journal of Derivatives* (Fall 1993b) pp. 21–31.
Hull, J., White, A., "Numerical procedures for implementing term structure models I: Single-factor models," *Journal of Derivatives* (Fall 1994) pp. 7–16.
Hull, J., White, A., "Forward rate volatilities, swap rate volatilities, and the implementation of the LIBOR market model," *Journal of Fixed Income*, vol. 10(3) (2000) pp. 46–62.
Ikeda, M., "Default premiums and quality spread differentials in a stochastic interest rate economy," *Advances in Futures and Option Research*, vol. 8 (1995) pp. 175–201.
Imai, J., Boyle, P., "Dynamic fund protection," *North American Actuarial Journal*, vol. 5(3) (2001) pp. 31–49.
Inui, K., Kijima, M., "A Markovian framework in multi-factor Heath–Jarrow–Morton models," *Journal of Financial and Quantitative Analysis*, vol. 33(3) (1998) pp. 423–440.
Jacka, S.D., "Optimal stopping and the American put," *Mathematical Finance*, vol. 1 (April 1991) pp. 1–14.
Jamshidian, F., "An exact bond option formula," *Journal of Finance*, vol. 44 (Mar. 1989) pp. 205–209.
Jamshidian, F., "An analysis of American options," *Review of Futures Markets*, vol. 11(1) (1992) pp. 72–82.
Jamshidian, F., "Hedging quantos, differential swaps and ratios," *Applied Mathematical Finance*, vol. 1 (1994) pp. 1–20.
Jamshidian, F., "A simple class of square-root interest-rate models," *Applied Mathematical Finance*, vol. 2 (1995) pp. 61–72.
Jamshidian, F., "Bond, futures and option evaluation in the quadratic interest rate model," *Applied Mathematical Finance*, vol. 3 (1996) pp. 93–115.

Jamshidian, F., "LIBOR and swap market models and measures," *Finance and Stochastics*, vol. 1 (1997) pp. 293–330.

Jarrow, R., Rudd, A., "Approximate option valuation for arbitrary stochastic processes," *Journal of Financial Economics*, vol. 10 (1982) pp. 347–369.

Jarrow, R.A., Rudd, A., *Option Pricing*, Richard D. Irwin, Homewood (1983).

Jiang, L., Dai, M., "Convergence of binomial tree method for European/American path-dependent options," *SIAM Journal of Numerical Analysis*, vol. 42(3) (2004).

Johnson, H., Stulz, R., "The pricing of options with default risk," *Journal of Finance*, vol. 42 (June 1987) pp. 267–280.

Ju, N., "Pricing an American option by approximating its early exercise boundary as a multi-piece exponential function," *Review of Financial Studies*, vol. 11(3) (1998).

Kamrad, B., Ritchken, P., "Multinomial approximating models for options with *k* state variables," *Management Science*, vol. 37 (Dec. 1991) pp. 1640–1652.

Karatzas, I., "On the pricing of American options," *Applied Mathematics and Optimization*, vol. 60 (1988) pp. 37–60.

Karatzas, I., Shreve, S.E., *Brownian motion and stochastic calculus*, second edition, Springer, New York (1991)

Kat, H., Verdonk, L., "Tree surgery," *Risk*, vol. 8 (Feb. 1995) pp. 53–56.

Kemna, A.G.Z., Vorst, T.C.F., "A pricing method for options based on average asset values," *Journal of Banking and Finance*, vol. 14 (1990) pp. 113–129.

Kevorkian, J., *Partial differential equations. Analytical solution techniques*, Brooks/Cole, Pacific Grove, California (1990).

Kijima, M., Nagayama, I., "Efficient numerical procedures for the Hull–White extended Vasicek model," *Journal of Financial Engineering*, vol. 3(3/4) (1994) pp. 275–292.

Kijima, M., Suzuki, T., "A jump-diffusion model for pricing corporate debt securities in complex capital structure," *Quantitative Finance*, vol. 1 (2001) pp. 1–10.

Kim, I.J., "The analytic valuation of American options," *Review of Financial Studies*, vol. 3(4) (1990) pp. 547–572.

Kim, I.J., Byun, S.J., "Optimal exercise boundary in a binomial option pricing model," *Journal of Financial Engineering*, vol. 3 (1994) pp. 137–158.

Kolkiewicz, A.W., "Pricing and hedging more general double-barrier options," *Journal of Computational Finance*, vol. 5(3) (2002) pp. 1–26.

Krylov, N.V., *Controlled diffusion processes*, Springer, New York (1980)

Kunitomo, N., Ikeda, M., "Pricing options with curved boundaries," *Mathematical Finance*, vol. 2 (Oct. 1992) pp. 275–298.

Kwok, Y.K., Lau, K.W., "Pricing algorithms for options with exotic path dependence," *Journal of Derivatives*, (Fall 2001a) pp. 28–38.

Kwok, Y.K., Lau, K.W., "Accuracy and reliability considerations of option pricing algorithms," *Journal of Futures Markets*, vol. 21 (2001b) pp. 875–903.

Kwok, Y.K., Wu, L., "Effects of callable feature on early exercise policy," *Review of Derivatives Research*, vol. 4 (2000) pp. 189–211.

Kwok, Y.K., Wu, L., Yu, H., "Pricing multi-asset options with an external barrier," *International Journal of Theoretical and Applied Finance*, vol. 1(4) (1998) pp. 523–541.

Kwok, Y.K., Wong, H.Y., Lau, K.W., "Pricing algorithms of multivariate path dependent options," *Journal of Complexity*, vol. 17 (2001) pp. 773–794.

Lai, T.L., Wong, S., "Valuation of American options via basis functions," *IEEE Transactions on Automatic Control*, vol. 49(3) (2004) pp. 374–385.

Langetieg, T., "A multivariate model of the term structure," *Journal of Finance*, vol. 35 (March 1980) pp. 71–97.

Lau, K.W., Kwok, Y.K., "Anatomy of option features in convertible bonds," *Journal of Futures Markets*, vol. 24(6) (2004) pp. 513–532.

Leippold, M., Wu, L., "Asset pricing under the quadratic class," *Journal of Financial and Quantitative Analysis*, vol. 37 (2002) pp. 271–295.

Leland, H.E., "Option pricing and replication with transaction costs," *Journal of Finance*, vol. 40 (Dec. 1985) pp. 1283–1301.

LeRoy, S.F., Werner, J., *Principles of financial economics*, Cambridge University Press, Cambridge, U.K. (2001).

Levy, E., "Pricing European average rate currency options," *Journal of International Money and Finance*, vol. 11 (1992) pp. 474–491.

Levy, E., Mantion, F., "Discrete by nature," *Risk*, vol. 10 (Jan. 1997) pp. 74–75.

Li, A., "The pricing of double barrier options and their variations," *Advances in Futures and Options Research*, vol. 10 (1999) pp. 17–41.

Linetsky, V., "Step options," *Mathematical Finance*, vol. 9 (Jan. 1999) pp. 55–96.

Little, T., Pant, V., "A new integral representation of the early exercise boundary for American put options," *Journal of Computational Finance*, vol. 3(3) (2000) pp. 73–96.

Lo, C.F., Yuen, P.H., Hui, C.H., "Pricing barrier options with square root process," *International Journal of Theoretical and Applied Finance*, vol. 4(5) (2002) pp. 805–818.

Longstaff, F.A., Schwartz, E.S., "Interest rate volatility and the term structure: A two-factor general equilibrium model," *Journal of Finance*, vol. 47 (Sept. 1992) pp. 1259–1282.

Longstaff, F.A., Schwartz, E.S., "A simple approach to valuing risky fixed and floating rate debt," *Journal of Finance*, vol. 50 (July 1995) pp. 789–819.

Longstaff, F.A., Schwartz, E.S., "Valuing American options by simulation: A simple least-squares approach," *Review of Financial Studies*, vol. 14 (2001) pp. 113–147.

Luo, L.S.J., "Various types of double-barrier options," *Journal of Computational Finance*, vol. 4(3) (2001) pp. 125–138.

MacMillan, L.W., "Analytic approximation for the American put option," *Advances in Futures and Options Research*, vol. 1 (Part A 1986) pp. 119–139.

Maghsoodi, Y., "Solution of the extended CIR term structure and bond option valuation," *Mathematical Finance*, vol. 6(1) (1996) pp. 89–109.

Manaster, S., Koehler, G., "The calculation of implied variances from the Black–Scholes model: A note," *Journal of Finance*, vol. 37 (March 1982) pp. 227–230.

Margrabe, W., "The value of an option to exchange one asset for another," *Journal of Finance*, vol. 33 (March 1978) pp. 177–186.

Merton, R.C., "Theory of rational option pricing," *Bell Journal of Economics and Management Sciences*, vol. 4 (Spring 1973) p. 141–183.

Merton, R.C., "On the pricing of corporate debt: The risk structure of interest rates," *Journal of Finance*, vol. 29 (1974) pp. 449–470.

Merton, R.C., "Option pricing when the underlying stock returns are discontinuous," *Journal of Financial Economics*, vol. 3 (March 1976) pp. 125–144.

Meyer, G.H., "Numerical investigation of early exercise in American puts with discrete dividends," *Journal of Computational Finance*, vol. 5(2) (2001) pp. 37–53.

Milevsky, M.A., Posner, S.E., "Asian options, the sum of lognormals, and the reciprocal Gamma distribution," *Journal of Financial and Quantitative Analysis*, vol. 33 (1998) pp. 409–422.

Miltersen, K.R., Sandmann, K., Sondermann, D., "Closed form solutions for term structure derivatives with Log-Normal interest rates," *Journal of Finance*, vol. 52(1) (1997) pp. 409–430.

Munk, C., "Stochastic duration and fast coupon bond option pricing in multi-factor models," *Review of Derivatives Research*, vol. 3 (1999) pp. 157–181.

Myneni, R., "The pricing of the American option," *Annals of Applied Probability*, vol. 2(1) (1992) pp. 1–23.
Neuberger, A., "Option replication with transaction costs—an exact solution for the pure jump process," *Advances in Futures and Options Research*, vol. 7 (1994) pp. 1–20.
Nielsen, J.A., Sandmann, K., "The pricing of Asian options under stochastic interest rates," *Applied Mathematical Finance*, vol. 3 (1996) pp. 209–236.
Nielsen, J.A., Sandmann, K., "Pricing of Asian exchange rate options under stochastic interest rates as a sum of options," *Finance and Stochastics*, vol. 6 (2002) pp. 355–370.
Nielsen, J.A., Sandmann, K., "Pricing bounds on Asian options," *Journal of Financial and Quantitative Analysis*, vol. 38(2) (2003) pp. 449–473.
Nunes, J.P.D., "Multifactor valuation of floating range note," *Mathematical Finance*, vol. 14(1) (2004) pp. 79–97.
Paskov, S., Traub, J.F., "Faster valuation of financial derivatives," *Journal of Portfolio Management* (Fall 1995) pp. 113–120.
Pelsser, A., "Pricing double barrier options using Laplace transforms," *Finance and Stochastics*, vol. 4 (2000) pp. 95–104.
Pliska, S.R., *Introduction to mathematical finance*, Blackwell Publishers, Oxford, U.K. (1997)
Pooley, D.M., Vetzal, K.R., Forsyth, P.A., "Convergence remedies for non-smooth payoff in option pricing," *Journal of Computational Finance*, vol. 6 (Summer 2003) pp. 25–40.
Press, W.H., Teukolsky, S.A., Vetterling, W.T., Flannery, B.P., *Numerical recipes in C: The art of scientific computing*, Second Edition, Cambridge University Press, Cambridge (1992).
Ramaswamy, K., Sundaresan, S.M., "The valuation of options on futures contracts," *Journal of Finance*, vol. 40 (Dec. 1985) pp. 1319–1340.
Rich, D.R., "The mathematical foundations of barrier option-pricing theory," *Advances in Futures and Options Research*, vol. 7 (1994) pp. 267–311.
Ritchken, P., Sankarasubramanian, L., "Volatility structures of forward rates and the dynamics of the term structure," *Mathematical Finance*, vol. 5(1) (1995) pp. 55–72.
Roberts, G.O., Shortland, C.F., "Pricing barrier options with time-dependent coefficients," *Mathematical Finance*, vol. 7 (Jan. 1997) pp. 83–93.
Rogers, L., Shi, Z., "The value of an Asian option," *Journal of Applied Probability*, vol. 32(4) (1995) pp. 1077–1088.
Rossi, A., "The Britten–Jones and Neuberger smile-consistent with stochastic volatility option pricing model: A further analysis," *International Journal of Theoretical and Applied Finance*, vol. 5(1) (2002) pp. 1–31.
Rubinstein, M., "Options for the undecided," in *From Black–Scholes to Black–Holes: New Frontiers in Options*, Risk Magazine, Ltd, London (1992) pp. 187–189.
Schaefer, S.M., Schwartz, E.S., "A two-factor model of the term structure: An approximate analytical solution," *Journal of Financial and Quantitative Analysis*, vol. 19 (Dec. 1984) pp. 413–424.
Schlögl, E., "A multicurrency extension of the lognormal interest rate models," *Finance and Stochastics*, vol. 6 (2002) pp. 173–196.
Schöbel, R., Zhu, J., "Stochastic volatility with an Ornstein–Uhlenbeck process: An extension," *European Finance Review*, vol. 3 (1999) pp. 23–46.
Schrager, D.F., Pelsser, A.A., "Pricing swaptions and coupon bond options in affine term structure models," *Mathematical Finance*, vol. 16 (2006) pp. 673–694.
Selby, M.J.P., Strickland, C., "Computing the Fong and Vasicek pure discount bond price formula," *Journal of Fixed Income* (Sept. 1995) pp. 78–84.
Sepp, A., "Analytical pricing of double-barrier options under a double-exponential jump diffusion process: Applications of Laplace transform," *International Journal of Theoretical and Applied Finance*, vol. 7(2) (2004) pp. 151–175.

Shepp, L., Shiryaev, A.N., "The Russian option: Reduced regret," *Annals of Applied Probability*, vol. 3 (1993) pp. 631–640.

Sidenius, J., "Double barrier options: Valuation by path counting," *Journal of Computational Finance*, vol. 1 (1998) pp. 63–79.

Singleton, K.J., Umantsev, L., "Pricing coupon-bond options and swaptions in affine term structure models," *Mathematical Finance*, vol. 12(4) (2002) pp. 427–446.

Smith, C.W. Jr., "Option pricing—a review," *Journal of Financial Economics*, vol. 3 (1976) pp. 3–51.

Tavella, D., Randall, C., *Pricing financial instruments: The finite difference method*, John Wiley & Sons, New York (2000).

Thompson, G.W.P., "Fast narrow bounds on the value of Asian options," *Working paper of University of Cambridge* (1999).

Tilley, J.A., "Valuing American options in a path simulation model," *Transactions of the Society of Actuaries*, vol. 45 (1993) pp. 499–520.

Tsao, C.Y., Chang, C.C., Lin, C.G., "Analytic approximation formulae for pricing forward-starting Asian options," *Journal of Futures Markets*, vol. 23(5) (2003) pp. 487–516.

Tsitsiklis, L., Van Roy, B., "Regression methods for pricing complex American style options," *IEEE Transactions on Neural Networks*, vol. 12 (2001) pp. 694–703.

Vasicek, O., "An equilibrium characterization of the term structure," *Journal of Financial Economics*, vol. 5 (1977) pp. 177–188.

Wei, J.Z., "Valuing differential swaps," *Journal of Derivatives*, (Spring 1994) pp. 64–76.

Wei, J.Z., "A simple approach to bond option pricing," *Journal of Futures Markets*, vol. 17(2) (1997) pp. 131–160.

Welch, R., Chen, D., "On the properties of the valuation formula for an unprotected American call option with known dividends and the computation of its implied standard deviation," *Advances in Futures and Options Research*, vol. 3 (1988) pp. 237–256.

Whaley, R., "On the valuation of American call options on stocks with known dividends," *Journal of Financial Economics*, vol. 9 (1981) pp. 207–211.

Whalley, E., Wilmott, P., "Counting the costs," *Risk*, vol. 6 (Oct. 1993) pp. 59–66.

Wong, H.Y., Kwok, Y.K., "Sub-replication and replenishing premium: Efficient pricing of multi-state lookbacks," *Review of Derivatives Research*, vol. 6 (2003) pp. 83–106.

Wu, L., Kwok, Y.K., "A front-fixing finite difference method for valuation of American options," *Journal of Financial Engineering*, vol. 6 (1997) pp. 83–97.

Wu, L., Kwok, Y.K., Yu, H., "Asian options with the American early exercise feature," *International Journal of Theoretical and Applied Finance*, vol. 2 (1999) pp. 101–111.

Xu, C., Kwok, Y.K., "Integral price formulas for lookback options," *Journal of Applied Mathematics*, vol. 2005(2) (2005) pp. 117–125.

Zhang, P.G., "Flexible Asian options," *Journal of Financial Engineering*, vol. 3(1) (1994) pp. 65–83.

Zhu, Y.L., Sun, Y., "The singularity-separating method for two-factor convertible bonds," *Journal of Computational Finance*, vol. 3 (1999) pp. 91–110.

Zvan, R., Forsyth, P.A., Vetzal, K.R., "Robust numerical methods for PDE model of Asian options," *Journal of Computational Finance*, vol. 1 (Winter 1998) pp. 39–78.

Zvan, R., Vetzal, K.R., Forsyth, P.A., "PDE methods for pricing barrier options," *Journal of Economic Dynamics and Control*, vol. 24 (2000) pp. 1563–1590.

Author Index

Subject Index

CPSIA information can be obtained
at www.ICGtesting.com
Printed in the USA
BVHW041143161118
533301BV00004B/14/P